Design for Human Ecosystems

LANDSCAPE, LAND USE, AND NATURAL RESOURCES

John Tillman Lyle

ISLAND PRESS
Washington, D.C. • Covelo, California

About Island Press

Island Press is the only nonprofit organization in the United States whose principal purpose is the publication of books on environmental issues and natural resource management. We provide solutions-oriented information to professionals, public officials, business and community leaders, and concerned citizens who are shaping responses to environmental problems.

In 1999, Island Press celebrates its fifteenth anniversary as the leading provider of timely and practical books that take a multidisciplinary approach to critical environmental concerns. Our growing list of titles reflects our commitment to bringing the best of an expanding body of literature to the environmental community throughout North America and the world.

Support for Island Press is provided by The Jenifer Altman Foundation, The Bullitt Foundation, The Mary Flagler Cary Charitable Trust, The Nathan Cummings Foundation, The Geraldine R. Dodge Foundation, The Charles Engelhard Foundation, The Ford Foundation, The Vira I. Heinz Endowment, The W. Alton Jones Foundation, The John D. and Catherine T. MacArthur Foundation, The Andrew W. Mellon Foundation, The Charles Stewart Mott Foundation, The Curtis and Edith Munson Foundation, The National Fish and Wildlife Foundation, The National [illegible] Foundation, The New-Land Foundation, The David and [illegible] Packard Foundation, The Pew Charitable Trusts, Th[illegible]undation, The Winslow Foundation, and individu[illegible]

Library of Congress Cataloging-in-Publication Data

Lyle, John Tillman.
Design for human ecosystems : landscape, land use, and natural resources / John Tillman Lyle.
p. cm.
Originally published: New York : Van Nostrand Reinhold, c1985
Includes bibliographical references and index.
ISBN 1-55963-720-X (pbk.)
1. Land use—Planning. 2. Land use—Planning—Case studies. 3. Human ecology. 4. Conservation of natural resources. 5. Landscape architecture. I. Title.
HD108.6.L95 1999 99-10752
333.73'13—dc21 CIP

Printed on recycled, acid-free paper 

Manufactured in the United States of America
10 9 8 7 6 5 4 3 2 1

CONTENTS

CASE STUDIES

FOREWORD

"Can floating seeds make deep forms?" In a 1992 article, John Tillman Lyle asks readers to think of landscape architecture as floating seeds that disperse, lodge, and sometimes take root, giving physical, designed expression to a particular time and place. If designs are inextricable from the pulse of landscape processes, such as the flows and cycles of water, soil, wind, energy, and species, then they may create deep form. Deep form is different from "shallow" form. Shallow form is a purely visual, stylistic creation that hovers on the surface of the land without connecting to nature's ongoing processes. Designs with deep form have the potential of timelessness, creating a landscape structure that reflects less visible but essential processes, accommodates change, and remains profound over time.

Design for Human Ecosystems itself is a floating seed, one which has resulted in deep form and deep transformations of places, people, and the profession since originally published in 1985. In plants, seeds are a sign of maturity. Likewise, Lyle's book is an indicator of maturation in landscape architecture and planning, according to Forster Ndubisi in *Ecological Design and Planning* (1997).

Professions mature when a new paradigm appears to interpret existing knowledge differently and address a growing need. In 1985, the National Environmental Policy Act and other environmental regulations required that land development decisions be accompanied by clear, accountable decision-making processes to communicate to the public possible repercussions of development. Ian McHarg's suitability mapping approach, as prescribed in the 1969 *Design with Nature,* was widely adopted to fill this need. The dynamics and interrelatedness of landscape structure and processes were often overlooked, however, in typical applications of suitability mapping. Environmental assessment procedures required scientists and designers to join forces and the demand for shared concepts and language grew. Concepts from conservation biology, such as protecting genetic diversity, were only just beginning to penetrate agency design ranks. Scientists were aware that designers' well-intentioned, aesthetic-driven actions, such as revegetating remote trailheads and parking lots with imported nursery stock, were endangering habitat and species survival. Designers decried the resulting restrictions on plant choices and defended the human need for beauty. Better ways of interpreting and incorporating ecological principles, especially those that emphasized the probabilistic, dynamic nature of nature, into landscape design and planning were needed. Likewise, better communication of design process and examples by which ecological principles are successfully integrated into design were also needed to inspire designers and scientists alike. Several experts responded to this need, and a blossoming of books appeared, addressing the importance and practice of integrating humans and nature in design and planning.

Design for Human Ecosystems is recognized as standing out among important publications of the time. The book has been lauded for two central messages. First, Lyle presents a lucid argument that designers must understand ecological order operating at a variety of scales and link this understanding to human values if we are to create durable, responsible, beneficial designs. The argument resonates through Lyle's apparent intellectual ease in crossing discipline boundaries, and he reinforces his position through many illustrative case studies from his professional and academic experience. Second, Lyle submits an ally to his venture—a design method based on Alfred North Whitehead's stages of learning that involves both art and science as mutually inclusive elements. This method, too, is supported with careful historical context, case studies, and experience derived from graduate landscape architecture design studios at California State Polytechnic University, Pomona, where Lyle taught for thirty years. Jon Rodiek wrote in a 1986 review that the book "will mark the beginning of a new educational era . . . ," and he concluded that "in order to appreciate Lyle's work, readers must enlarge their views beyond present day perspectives. Those who can do so may find this work to be one of the more significant efforts yet put forth by any design educator."

Lyle's influence on the landscape architecture profession is palpable—at small and large scales, on land and on people. At an immediate scale, his work drew students and faculty to Cal Poly Pomona, where ecological design principles and methods continue to be practiced and refined through studios and projects similar to those described in *Design for Human Ecosystems.* The studio approach, which requires students to locate funded projects, often converts members of water districts, air pollution control boards, coastal conservancies, Indian tribes, and other organizations into new clients for landscape architecture. Between 1987 and 1995, these projects won six American Society of Landscape Architects professional planning awards.

One studio project, The Center for Regenerative Studies, a local milestone on the Cal Poly Pomona campus, grew into a remarkable reality with far-reaching appeal. On this sixteen-acre parcel, students and faculty explore the built consequences of designing for human ecosystems. A rich dialectic of gardens and buildings, people and plants, teaching and learning merge here in the interest of regenerating, not merely sustaining, resources and spirit for future generations. Joan Safford, current faculty director for the Center and one of those involved since its inception, observed: "There were many heated discussions during which we debated various names for the center. John convincingly proposed and defended Robert Rodale's concept of 'regenerative' as the central idea." Lyle wrote in 1994 that "Regenerative design means replacing the present linear system of throughput flows with cyclical flows at sources, consumption centers, and sinks. . . . (It) has to do with rebirth of life itself, thus with hope for the future." Now, Safford remarks, "it's amazing to remember the overgrazed, compacted cow pasture of a decade ago, overshadowed by an enormous landfill, that now yields armsful of scented, exuberant lavender, sage, rosemary, growing from rejuvenated soils."

The center's deep form is evident in the built (and still scheduled-to-be-built) attention to ecological order. Flows of energy are manifested in south-facing roof gardens and tracking solar collectors. Water and nutrients flow from aquaculture ponds through wetlands, which eventually will serve as waste treatment systems before being cycled into other uses. Stairstepped, used-tire agricultural terraces reflect efficient use of water, cycling of materials, and on-site soil rejuvenation processes and recall agricultural patterns from other arid regions. Although Lyle's unified vision of the place is emerging yet, piece by piece, the deep potential of the place still captures the imagination and optimism of visitors and residents.

Building on *Design for Human Ecosystems*'s foundation and the Center's genesis, Lyle published *Regenerative Design for Sustainable Development* in 1994. The book summarizes numerous techniques and examples of regenerative design that Lyle helped shape or investigated while on sabbatical. Tony Hiss stated in the 1994 awards issue of *Landscape Architecture* that this book marked ". . . the coming of age of the sustainability movement." Zev Naveh, author of *Landscape Ecology: Theory and Application,* wrote: "(Lyle's) design of regenerative systems is in my opinion the most promising and most sustainable way to ensure future agricultural production and therefore also of life."

Since 1994, regenerative design continues to move from theory to application, and from a few isolated projects to a constellation of work. Lyle's planned and built examples can be seen at the Claremont School of Theology and his family garden in California, the Shalom Hill Project in Minnesota, the Glen Helen Ecology Institute and Oberlin College projects in Ohio, the Applied Ministries Educational Resource Center in Kentucky, and the Dean residence in Washington. The Regenerative Design Symposium hosted by Cal Poly Pomona in 1996 demonstrated the far-reaching applications of these ideas, from North America, Asia, and Europe, to the Middle East. Cal Poly Pomona professor Mark von Wodtke continues to bring several of Lyle's projects from plan to reality.

John Lyle writes: "If the floating seeds drop into fertile soil and find a fit with the environment, they sprout and grow and eventually evolve a new culture in that place." Examining the evolution of language and written communication within a profession provides an interesting window on that culture. In 1985, few books focused on ecological design. At this time, the term usually was used in reference to ecology and hard sciences, rather than landscape architecture. Landscape ecology was just emerging in the United States, but texts acknowledged the work of only a handful of landscape architects. *Our Common Future,* the World Commission of Environment and Development report on global environmental degradation, was released in 1987 and hailed sustainable development as a priority goal for all nations. The term *sustainable design* infiltrated the design profession by the early 1990s, with numerous publications instructing designers on techniques of utilizing renewable resources. Public agencies, particularly those with regional-scale concerns, applied the term to their work, most notably the National Park Service, which published *Guiding Principles of Sustainable Design* in 1993. Lyle's offshoot term *regenerative design* gained momentum in the early 1990s and developed a small but loyal following. By the mid-to-late 1990s, authors have applied sustainable design concepts to residential and new town developments and explored psychological motivations behind achieving greater coherence between resource supply and demand. Hundreds of books are now available as references.

Ecological design, sustainable design, regenerative design—although distinctions between these terms are debated still, they largely are used interchangeably now as they share the integration of ecological principles with design. Most new references build on the pioneering groundwork laid by McHarg's *Design with Nature* and Lyle's *Design for Human Ecosystems.* It is unimaginable that these germinal texts at one time went out of print, severing the growing literature from its roots and leaving a tangible void for new students and practitioners.

The niche filled by *Design for Human Ecosystems* is unique and deserves a broader audience. It is a memorable, timeless, and potentially life-changing book. Jeffrey Olson, coprincipal with Lyle in Cal Poly Pomona's third-year graduate landscape architecture studio, notes: "I have seen students who, lost in their purposes, uncertain, or unproductive, read *Design for Human Ecosystems* and, sometimes nearly instantly, become effective and perceptive in their work. The light goes on, the elevator suddenly goes to the top! My view is that one of its most significant contributions is the hierarchy of scales of concern, which, when understood by students, greatly helps to focus their efforts." Mike Stevens, lecturer at the University of Western Australia writes: "Students just drink in Lyle's ecological functioning perspective on how to use site information with enthusiasm." Aerin Martin, a current Cal Poly Pomona student, concurs: "Lyle's prose is quite inspirational and almost commands that one take personal action. His writing is as visionary as his presence was charismatic. These qualities persuade others of ecological design's relevance and importance."

Lyle's life and work changed the course of landscape architecture and will have a long-lasting effect. Dean Freudenberger, professor at the Luther Theological Seminary in St. Paul, writes: "(Lyle's) inspiration, wisdom, persuasiveness, and rich fellowship will result in lasting contributions for the entire environmental movement. (His) contribution is of a magnitude that is impossible to measure at this time." David Orr, chair of the Environmental Studies Program at Oberlin College, adds: ". . . His work will be read twenty-five years from now and he will rank among the important figures of this transitional time." And Zev Naveh states: "I regard him as one of the most creative leaders towards a better future for humankind."

Just as Lyle's life changed the world, his sudden death in 1998 did as well. He had recently retired from teaching to devote more time to consulting and practice. Fully involved in writing and design, he persistently championed regenerative design and his influence continued to spread within this country and overseas. Robert Thayer, author of *Gray World, Green Heart,* summarized Lyle's effect: "He is irreplaceable in the role he served. He was a spokesperson for a new way of looking at and acting in the world."

We have lost the opportunity to read about and listen to John Lyle as he might have reflected later on his life's work. Yet, so many people have been influenced by his work that seeds continue to sprout, grow, and create deep form as a result of his life and work. Lyle traveled extensively to share ideas, but *Design for Human Ecosystems* stands still and shines like a lighthouse. We see the shine of its light and find our own course.

Lyle's life and death changed us and the profession. But what has not changed is the relevance of the ideas in *Design for Human Ecosystems;* the ideas run deep. Philip Pregill and Nancy Volkman, in their text *Landscapes in History,* address the many changes sustained by the profession of landscape architecture, and reflect on the one constant: "Regardless of types of projects, scale of projects, or stylistic fashions, the ability to creatively manipulate multiple criteria, issues, and physical solutions will remain the most critical skill which the profession has to offer. If a study of the past has any value it is to demonstrate that solutions change, but that the so elusive creative process remains the object of greatest value."

John Tillman Lyle's *Design for Human Ecosystems* makes this creative process less elusive and acknowledges the increasing significance of design. We are grateful that Island Press is bringing the book back into the hands of future creators of deep form.

Joan Hirschman Woodward
California State Polytechnic University
Pomona, California

PREFACE

This book describes principles, methods, and techniques for shaping landscape, land use, and natural resources in ways that can make human ecosystems function in the sustainable ways of natural ecosystems. It addresses everyone concerned with such matters: involved citizens as well as landscape architects, planners, and other professionals. Its approach has grown out of my teaching experience and the research and experimental projects that I have carried out with my students and colleagues over the last fifteen years. I have tried to present it as a reasonably cohesive foundation of theory supporting a structure of methods and practical applications. In addition to those from my own experience, the basic ideas on which it is built come from a wide variety of sources and disciplines. Since World War II, there has been an explosion of useful thought and research that seems to me to suggest far more effective ways of landscape design than were ever available before. I have drawn freely on everything that I have found useful, with no regard for disciplinary boundaries. To do so was not only convenient, but necessary, since the theory and literature of landscape architecture and planning are so lacking in this respect. I hope my book will contribute in its own modest way to help remedy the situation. Without a broader theoretical framework to provide purpose, methods and techniques have little meaning.

The book is organized to include both text and case studies. The latter illustrate useful applications of the concepts, principles, methods, and techniques discussed in the text. Most of them are projects that I have participated in, primarily with my two experimental design groups at Cal Poly in Pomona: the Laboratory for Experimental Design and the 606 Studio. Since these groups have provided the crucible for the shaping of my ideas, they have played an important role in the writing of this book. Landscape design does not proceed in the abstract, but in the heat of real issues and the questions, debates, paradoxes, and imbroglios that arise from them. The Laboratory for Experimental Design involves teams of faculty members that carry out landscape design and planning projects of a complex and exploratory nature, mostly for public agencies. Graduate students participate as staff assistants. The 606 Studio projects are undertaken by graduate students, with faculty members supervising. These also address real and pressing issues, again in exploratory ways, mostly with funding by public agencies. Both groups choose their projects with great care, doing only work that may be of significant public benefit and that provides a vehicle for the exploration of ideas we are especially interested in. Thus, it is appropriate, almost necessary, that most of the examples given in this book represent the work of these two groups. All the case studies, including those drawn from other sources, were, in my opinion, the best of their kind for illustrating particular applications.

None of the case studies is reproduced in entirety. Only those portions have been selected that give a general idea of the subject matter of the work and illustrate a particular application. Thus, for example, one of them illustrates a water flow diagram, another includes a complete resource inventory, and yet a third shows a full range of alternatives with a comparative evaluation. Altogether, I believe they cover a broad range, even if they fall short of the full gamut of scales, landscape types, methods, and techniques.

People who write books on design often seem to be trying to present either the first or last word on their subjects. I intend this book to be neither. Rather, I hope it finds a place in the stream of development of ways to design. To at least a limited degree, I have tried to trace the lines of historical growth of design ideas involving natural processes. It is important for us to understand our intellectual and professional roots far better than we do. Design activity needs to be rooted in the solid soil from which all the ideas of our civilization grow. There is much more to be said on the subject than I have attempted here, and I look forward to the dialogue.

Richard Neutra wrote some thirty years ago that "mankind precariously floats to its possible survival on a raft, rather makeshift as yet, and often leaky: Planning and Design." Time has proven him only too right, and it is a desperate plight for a species as reluctant to look beyond tomorrow as our own. Still, the raft remains afloat, somewhat better crafted now, but bouncing about more precariously than ever. The seas have grown angrier, more threatening. We have patched a few of the leaks, but a great many remain. Perhaps this book will plug a few more.

Despite the angry seas, I think I have taken here a stance that is noticeably more optimistic than is usual in books involving ecological issues. That is because the approach that, along with Neutra, I see as the only promising one—the approach through design—requires optimism. One can hardly design anything unless one assumes that the design can be realized and that there will be someone present to put it to use. Thus, my reason for optimism is not so much full confidence in a brighter future as it is that I see no alternative other than gloom and despair with certain defeat to follow, a prospect that I consider unacceptable.

A great many people have been associated with me in the development of these approaches, as well as in shaping the case study projects and the book itself. Every student with whom I have worked during the last fifteen or so years has contributed in some way—with thoughts, insights, discussion, sometimes with challenges or demands for better explanations. Those who worked on the case studies are listed in the credits for each work. They represent only a few among many, however, and all have earned my gratitude. The same is true of our clients—administrators and staff members of the public and private agencies that have supported our work, not only with funds but with interest, expertise, and insights. They are too numerous to list here, but I want especially to mention the planning departments of Los Angeles, Del Mar, Huntington Beach, Lake Elsinore, Palm Springs, and San Diego County; the U.S. National Forest Service and the Bureau of Land Management; the California Division of Beaches and Parks; the Santa Monica Mountains Conservancy; and the Irvine Company.

Perhaps most important of all have been my colleagues during this time both in the Laboratory and the 606 Studio, especially Arthur Jokela, Jeffrey K. Olson, and Mark von Wodke. We have exchanged ideas and been engaged in arguments almost continuously. Without them this book would certainly not be what it is, and probably would not exist at all. All three of them have read the manuscript and offered insightful suggestions.

A great many others have contributed too. Sylvia White read parts of the manuscript, as did my wife, Harriett, and a number of others, and all were extremely helpful. Jack Dangermond, Lewis Hopkins, Mark Sorenson, and Carl Steinitz were generous with their time and work in contributing case studies. John Odam's sensitive work on the design of the book shows very clearly. Bill Bensley redrew a number of the illustrations with a fine hand. Ruth Stratton patiently and very capably typed and retyped. And I am particularly grateful to the California State Polytechnic University for giving me the time, in the form of a sabbatical leave, to finally put it all together.

CHAPTER 1

Introduction

Where Mind and Nature Meet

Every piece of land on earth has a history that goes back some four and a half billion years, and each individual history has had its ups and downs. Read the story in the rocks, and you will find periods of relative stability featuring cycles of bloom and decay, of dormancy, birth, death, and rebirth. Some of these periods are long and some short, but none lasts forever. Breaking and dividing these more or less stable periods are occasional upheavals of sudden change. Over the past twelve thousand or so years, ever since the development of agriculture, and especially during the two centuries since the industrial period began, the upheavals have mostly been brought about by human beings themselves. Our restless passion for change has combined with our technical prowess to alter much of the world's landscape.

Until the middle years of the twentieth century, most people seem to have believed this change was all to the good—a perhaps erratic, but on the whole steady, improvement in the human condition. Since then, the notion of inevitable progress has been called into serious question. Ecologist Ramon Margalef's chilling assertion that "... the evolution of man has not been in the direction of passive adjustment to more mature ecosystems but is actually sustained through a regression of the rest of the biosphere" gives one pause.[1]

Nevertheless, there are exceptions. Either by accident or design, human beings have sometimes created landscapes that are at least as rich and as stable, occasionally as beautiful, as those shaped by nature. Consider the rolling farmlands of Northern Europe or the spectacular terraced slopes in the Andes where the Incas maintained a stable agriculture for several hundred years.

Or consider an extraordinary marsh at the northern end of the Salton Sea. The sea itself was created by the accidental flooding of an irrigation canal in 1905 and is now a thoroughly established part of California's Mojave Desert. The marsh, which is kept wet by runoff water from upstream agricultural irrigation, supports a large and strangely diverse community of birds that has become a subject of intensive study. With the varying salinity levels brought about by the very salty Salton Sea and the relatively fresh runoff water, both fresh- and salt-water birds live there, at least for short periods of time. During the spring and fall migrating seasons, the marsh reminds one of a busy international airport, with all sorts of birds landing and taking off almost continuously, and tens of thousands of others resting on the surface.

[1] See Margalef, 1968; p. 97.

This odd, anomalous, incredibly diverse ecosystem would not have existed at all had it not been for the agricultural development of the surrounding lands. If human beings had not settled there and dug their irrigation canals, the whole valley would still be a desert. It is certainly true, as I will argue later on, that desert landscapes have a beauty and a value of their own, and without a doubt, this is a variable, vulnerable, unstable marsh. But if we measure a landscape by its richness and diversity or by the number of creatures it supports, then human beings have certainly made a contribution to nature here, however inadvertent. They have also contributed to their own well-being, of course, since this is among the world's most productive agricultural regions, even if the farmland may be hardly more stable than the marsh.

If we can create such a rich landscape accidentally, then it seems reasonable to believe that we can do at least as well intentionally, and with proper design make one that is far more sustainable. The science of ecology has given us the information and concepts to work with, even if the precise scientific data that we would like to have in specific cases are usually lacking. By following its principles, even without a great deal of exact data, it is entirely possible to create rich, diverse, productive landscapes that will serve the purposes of both people and nature.

We can best illustrate this point and the underlying principles by exploring a single example in some detail. We will dwell at length on the intricacies of this example because they will serve as an introduction to the concepts and techniques explained in later chapters. Both ecological and social processes stand out sharp and clear in this lucid, relatively simple environment on the southern California coast. The setting is another marsh, but this time a marsh created by nature and degraded by man. In such a place, the time has come for design to eliminate accidents.

AN ARCHETYPE: SAN ELIJO LAGOON

San Elijo Lagoon is a narrow tangle of marshes, mudflats, and shallow channels that push out of the Pacific Ocean into the rolling coastal plain of Southern California, some twenty miles north of San Diego. It is a quiet landscape. Even with development closing in—dikes like long worms dividing its waters, a sewage treatment plant rusting away on an island at one end, a scattering of houses along one edge, a freeway vaulting over—San Elijo still has that sense of airy tranquility that one associates with lagoons everywhere. With rolling hills in the background, it is gentle, unassuming, hardly beautiful, certainly not dramatic. One would not expect it to become a center of conflict.

That is exactly what it was, however, for a number of years. Developers and environmentalists battled over it in what has now become the classic fashion. To understand the conflict, and its eventual resolution, one needs to know something about the lagoon's history.

For 10,000 years or so before urban civilization arrived here, the tides rolled in and out twice each day, mixing with the fresh water that flowed down from the foothills during the winter rainy season. A prolific community of marshgrasses, molluscs, fish of various types and sizes, an array of shorebirds and waterfowl, and a few small mammals thrived in and around the moving waters. For several hundred years, a tribe of Indians also lived on the edge of the lagoon and shared in the bounty, especially the easily harvested shellfish, which helped assure them a leisurely existence.

All this began to change in 1884 when the Santa Fe Railroad line was built along the coast over the lagoon inlet. Though the trestle allowed water to flow beneath it, the inlet could no longer migrate freely in response to the movement of the ocean waves as it had under natural conditions. Thus, when the waves pressed for an inlet location other than that under the trestle, they failed. As a result sand built up in the channel, shutting off the flow of ocean water for part of the year and isolating the waters of the lagoon. Salinity levels rose in the trapped water, and the materials washing into the lagoon began to collect there. The problems were compounded when a coastal highway was built alongside the railroad a few years later, further restricting the movement of water.

With these two handy transportation links giving easy access to the cities of San Diego to the south and Los Angeles about a hundred miles to the north, people began settling on the surrounding coastal plain. Within just a few years the population built up to such a level that a sewage treatment plan was needed. What better place for it than at the edge of the useless lagoon with its murky water and smelly mudflats? The plant was built without further ado on a small island next to the railroad, and large rectangular sections of the lagoon were divided off with dikes to serve as oxidation ponds. These became geometric splotches of pale green as algae fed by the nutrients in the sewage grew at far too fast a pace and decayed on their surfaces.

For several years following, the deterioration continued to be steady but without the drama of sudden change. Then, following World War II, rows of neatly spaced small houses appeared on one of the slopes adjoining the lagoon. Runoff from the newly urbanized area, increased by lawn irrigation and laden with eroded soil, fertilizer, pesticides, and various kinds of debris, collected on the still waters. By the early 1960s, San Elijo was an eyesore and a health hazard. Only small patches of the marshgrasses remained. Though there were few, if any, fish in the water, a good number of migrating birds still stopped to rest and stalk and flutter about on the mudflats, searching for worms as they followed the sun along the Pacific flyway.

Then, in the mid-1960s, a land developer proposed to make the lagoon useful according to his own lights, and profitable as well, by transforming the wetlands into a marina housing development. He bought about half of the wetlands, including the areas nearest the ocean, and worked out a scheme that called for dredging a broad main channel to connect with the ocean and several fairly straight narrower channels reaching out from it to either side. The dredged material would be used to create fingers of land for residential development between the channels. Each finger would have a street down its center and rows of 40-foot-wide lots on either side. Each lot would have one edge fronting on a channel, where a boat could be kept.

Shortly after that, another developer proposed a similar plan for the eastern half of the lagoon, and proposals soon followed for subdividing the still undeveloped portions of the sloping lands overlooking it for single-family tracts.

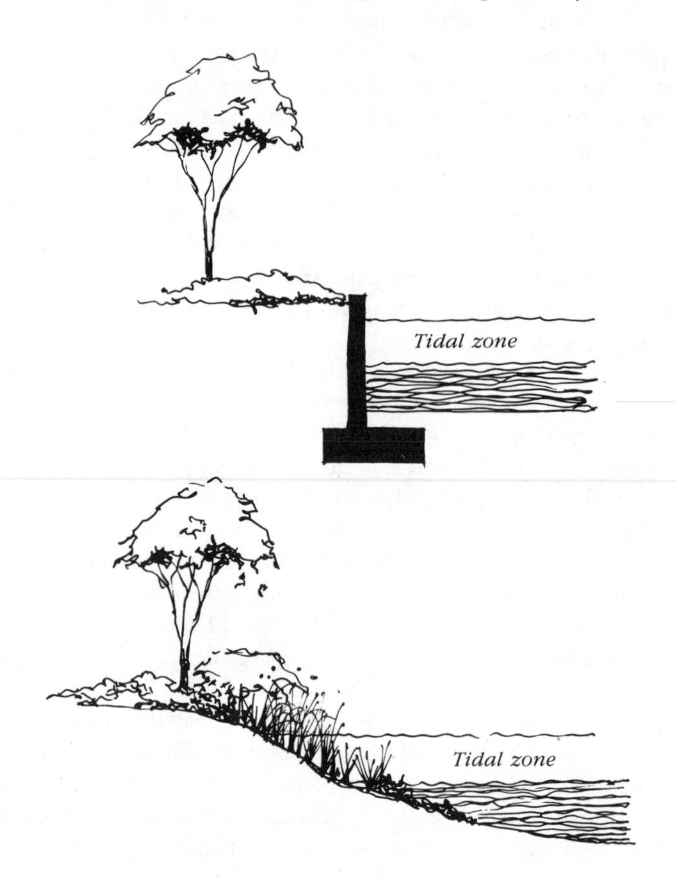

Awareness and Conflict

This was the time when environmental activists were beginning to awaken and make a stir. Several groups banded together to do battle with the developers, and the fate of the lagoon became a hot political issue. Environmentalists, whose main interest was the fate of the bird population, wanted the lagoon left alone. Developers, arguing private property rights and tax base and housing needs, not to mention profits, stood firm. Zoning happened to be on their side, most of the wetlands being zoned R-1. It was a classic developer/environmentalist conflict. With the issue in a stalemate, the Environmental Development Agency of San Diego County asked our Laboratory for Experimental Design to study the problems at hand and recommend land planning policies and guidelines for San Elijo and several other coastal lagoons where similar conflicts were emerging.

After grappling with these issues for some time, it became clear to us that they could not be settled simply by pronouncing in favor of preservation or development. In the light of an analysis of the processes involved, political as well as ecological, the proposals of both sides were seriously flawed. If the lagoon were either developed as a marina subdivision or left entirely alone, it would become an environmental liability, and more and more so as time went on. Either way, its rich potential for contributing to the sustenance and amenity of the human community, as well as its pivotal role in natural systems, would be lost.

If the marina plan were to be carried out, concrete bulkheads would have to be built along the edges of the channels to hold the land in place and provide water deep enough for large boats. The result would be no tidal flats at the water's edge and therefore no marshgrasses. If there are no recurrent marshgrasses to die and decay, there will be no detritus for the shellfish and other small organisms to eat. The energy subsidy that tidal action provides by spreading the detritus around the lagoon to these stationary creatures will thus have gone for naught because there will then be nothing for the small fish and birds to eat, and no small fish for larger fish and birds to eat. Ocean fish will no longer swim into the lagoon to spawn, and it will eventually become a lifeless body of water. The causal chain of unhealthy effects goes on and on.

Somewhat less expected are the consequences of leaving the lagoon alone. Increasing urbanization along the lagoon's upper drainage basin will bring about an inevitable increase in the siltation rate. Silts could be deposited as much as 100 times faster than they would if the lagoon remained in its preurbanization state. It is the common fate of a lagoon, whether people settle around it or not, to be filled with silt and become first a marsh and finally dry land. If human beings had not come on the scene, this process in the San Elijo Lagoon would probably have occurred only over the next 10,000 to 20,000 years. In its present state and at the present rate of urbanization, the process would probably take 20 to 25 years, and then there would be no place left for the fish and birds anyway. The message is clear. The changes brought about by urbanization are so strong and pervasive that no system within the sphere of urban influence can retain the same character that it would have had if human beings were not present at all.

The fatal, all too common flaw in both the marina housing plan and the hands-off approach is that each focuses on a single purpose. Such a narrow focus leads to an overriding concern with just one aspect of a complex system. This means that other aspects are ignored, and the design process becomes overly simplified. The resulting environment, if realized, will also be very simple, eliminating most of the interactions that might otherwise exist, and along with them, a great many benefits.

It is important to recognize that when we make a plan for a piece of land, or even when we alter it in seemingly minor ways—like landscaping a backyard, for example—we are designing an ecosystem. We are laying the ground in which a network of interactions will take place, interactions that will carry on into the future, evolving according to a pattern that is at least partially predictable. A very simple, single-purpose ecosystem will inevitably fail to provide for many diverse, potentially beneficial roles. At the same time, it will lack the stability needed to resist sudden turbulent changes brought about by external events. Houses built on fingers of filled land may be suddenly inundated by a tidal wave or shaken to pieces by surges in the soft fill material beneath them brought about by an earthquake.

Single-purpose systems also tend to deteriorate from the moment of their realization; their attempts to diversify over time as natural systems do conflicts, of course, with their singular purpose. Maintenance becomes a matter of struggling against this trend, and the cost can come very high. Channels will fill with silt and require almost continuous dredging; bulkheads and tidal gates will need frequent repairs.

Public Opinion

Given the number of people interested in environmental issues nowadays and the legal requirements for consulting the public on important issues, public attitudes and values become an essential part of the ecology of a place like San Elijo Lagoon. No plan is likely to get very far without public support, and the most effective and implementable plan is one that considers public attitudes from the beginning so that it can gather support as it develops.

For design in such a complex environment, this approach can run into some serious difficulties. No one can expect everybody to understand the whole tangle of ecological problems. It is therefore necessary to ask how much of an educational effort is necessary and how the designer can best explain the problems without building in his own biases.

In this case, the main vehicle used to measure public attitudes was a questionnaire that reduced the basic information discussed here to a very brief statement and then asked for a ranking of the possible solutions. The results showed a strong preference for returning the lagoon to its natural state and for recreational use. Residential development and boat marinas ranked very low.

Given the state of affairs that existed in the lagoon, the word "natural" was open to a variety of interpretations. Scientists and designers usually define it rather narrowly to mean unaltered by human beings. What the respondents meant in this case, as indicated by the answers to other questions and a series of interviews, was really "natural-appearing."

This sentiment is not unusual. A great many people who value natural resources value them as something to look at. Scenery is an important part of our cultural heritage. Even the Sierra Club states its purpose as "... to explore, enjoy, and protect the nation's *scenic* [my italics] resources." It will probably take generations of environmental education to create a general public understanding of the fact that the importance of nature and its processes goes far beyond scenery. Meanwhile, once the local population's desire for a lagoon they could use for recreation and that looked more or less natural had been accepted, we had an obligation to probe much deeper.

The Benefits of Varied Use

Within the broad mandate to preserve, or more accurately, to create a scenic lagoon, a great many things can be attempted. Considering the lagoon as an ecosystem and examining its flows of materials and energy and its relationships with the surrounding urban complex, we become aware of a variety of contributions that the lagoon might make if carefully managed.

First of all, the estuarine ecosystem is a bounteous protein factory. If the marshgrasses are left to grow around the tidal edges, and the tides are allowed to do their work in spreading the feast of decomposing grasses about the lagoon, then enormous quantities of fish and shellfish will grow there. In fact, a tidal marsh is probably the most efficient food-producing system in nature, about seven times more productive in tons of biomass per acre that the best wheat field. This productivity can be increased severalfold, as has been accomplished in some Japanese estuaries. Pushed to its maximum potential, the lagoon could easily provide over 50,000 pounds of protein per acre per year in the form of oysters or almost five times that amount in the form of mussels. Any urban resources organized to stimulate such productivity in turn will help support the economy of the surrounding concentration of city dwellers. The fact that an improvement can pay its way in dollars and cents is a matter of considerable importance when it comes to deciding what gets done and what does not.

The same food chain that produces the fish and shellfish will also support a large wildlife population. If the food chain is kept healthy, and if the other needs of the birds are provided for, including nesting places and some privacy, then most of them will continue to inhabit the lagoon. When these needs are met, many wild species are quite willing to share their living space with people. In this way, the food chain can be kept intact, and the biological diversity of the coastal community maintained.

As the respondents to the questionnaire perceived, a healthy lagoon has obvious recreational uses as well. Calm waters, if kept reasonably clean by tidal flushing, can be used for small nonpower boating and swimming, while the flatter land along the shore provides a fine setting for fishing, strolling, bicycling, picnicking, and lounging about. Away from these more active areas, bird-watching and nature study can hold sway.

Commercial recreation, even of a fairly intensive sort, is compatible provided that it is well separated from wildlife habitats. A cluster of shops can be built on stilts over the water, or on barges floating in it, to allow estuarine life to go on underneath. Marine businesses will fit nicely here, as will seafood restaurants, which might feature the products of the lagoon on their menus.

Residential development on the overlooking slopes will do no harm if erosion is controlled and nutrient-laden runoff water is prevented from reaching the lagoon. Ideally, houses should be grouped in clusters on sites where runoff can be diverted, slowed, and allowed to percolate only very gradually into the ground water supply, thereby helping form a barrier against the progressive landward movement of underground salt water, which has been a problem in recent years. Natural drainage courses would best be left open and unimpeded so that fresh rainwater can flow into the lagoon unadulterated by contact with urban dregs.

Local people were surprised, and taken aback a bit, to learn that the sewage treatment plant could be made to fit in very comfortably with this whole complex of activities. Several years ago, in response to the Federal Clean Water Act, which forbids introducing water of lesser quality into a natural water body, the local sewer district was ordered to stop dumping effluent into the lagoon. As a result, an ocean outfall was built, and the effluent, after primary treatment, is now pumped through a four-mile-long pipe into the depths of the Pacific Ocean, where it seems to do no harm. On the other hand, neither does it do much good. In the lagoon, this effluent, which consists mainly of water brought over 200 miles by aqueduct from the Colorado River for use in the houses of the district, could be very useful. With careful control, this mixture of fresh water and nutrients can play a beneficial role in the lagoon system, as illustrated by the diagrams. After primary settling, it can be held in isolated biological-treatment ponds and used as a medium for growing aquatic plants. Additional ponds can become nutrient-rich media for growing fish and shellfish. After that, the water will be pure enough to use for irrigating plants in the recreation area. This is a simple, economical, and proven method of sewage treatment that is now working in a number of places.[2] It is not yet widely used, however. The effluent can be purified to a level permitting its periodic controlled release into the lagoon, where it can help stabilize salinity levels and thereby establish optimum conditions for estuarine life. In this way, productivity can be raised far beyond the levels found in nature. We will discuss the technology of biological sewage treatment at some length in Chap. 13.

At the same time, responding to the public's desires, San Elijo can become—in fact, already has become—a key symbolic feature of the urban landscape. Its tranquil vistas can provide a natural focal point for community identity in the midst of an otherwise undifferentiated sprawl. Its visible form can give expression to a special relationship between land and sea, and between these and the human community. It provides, in short, a sense of place.

Joining Human and Natural Systems

Though its sheer intensity and range of activity make the urban environment different in character from the natural environment, they are the same in at least one fundamental respect: Both depend on the same basic processes. In cities, we tend to forget about these vital links because natural sources of energy, food, and water have often become almost totally obscure, having been replaced by artificial single-purpose systems that transport these necessities from distant landscapes. Most of the energy, food, and water consumed by people living in the area of San Elijo Lagoon, for example, come from hundreds of miles away. The cost is high, and the sources are increasingly problematic as growing pop-

[2] See Bastian, 1982; Environmental Protection Agency, 1979 and 1980; Jokela and Jokela, 1978; Reed et al., 1981; and Woodwell, 1977.

CASE STUDY I

San Elijo Lagoon

San Elijo Lagoon includes a network of channels winding through a tranquil, bird-filled stretch of mudflat and salt marsh on the edge of the Pacific Ocean. Although badly degraded by surrounding urbanization, it can be revitalized to work as a sustainable, productive ecosystem once tidal flushing is restored. Realizing its full potential means establishing a complex combination of uses, each located and functioning in such a way as to support the others. These uses include preservation, biotic production, recreation, and urbanization.

Prepared by the Laboratory for Experimental Design, Department of Landscape Architecture, California State Polytechnic University, Pomona. Project Director: John T. Lyle. Consultants: David E. Bess, Andrew Susser, Mark von Wodtke. Graduate Assistants: Bruce Allport, William Cathcart, Robert Cunningham, Patrick Hall, Russell Hunt, Paul Jordan, Craig Neumeyer, Teiichi Jim Oe, Joseph Rodriguez, Dennis Spahr, and David Zapf. Prepared for the County of San Diego as a component of the Integrated Regional Environmental Management Program, funded by the Ford Foundation.

SUITABILITY MODELS

Suitability models are maps—in this case, generated by computer—that define the physical suitability—the fitness of the land—for certain prospective uses. These models use grid-cell mapping, each cell representing a land area 111.1 feet square, and are based on a set of listed concerns. These, in turn, determine the land variables, or characteristics, that will shape the model, including geographically distributed characteristics like variations in slope and vegetation type; those variables are also listed. The computer searches through the data files, rates each grid cell for its combination of desirable attributes, and prints a map showing the aggregated results in terms of levels of suitability. In these maps, the lighter the shade of the cell, the higher the suitability rating.

RESIDENTIAL SUITABILITY MODEL

Major Considerations:
Physical feasibility
Wildlife disruption

Variables Considered (in order of importance):
Water features
Water depth
Vegetation types

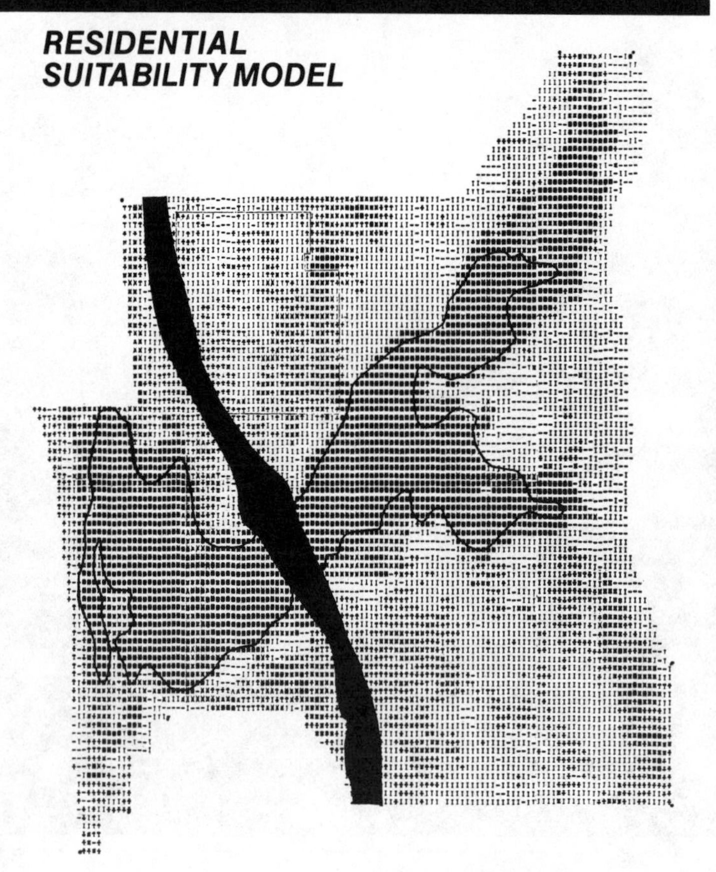

LIGHT RECREATION LAND USE MODEL

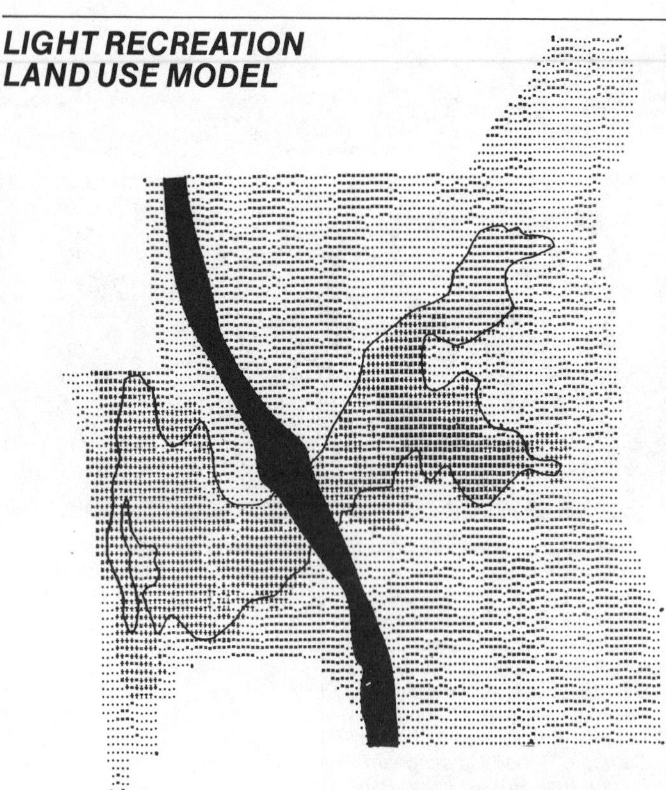

Major Considerations:
Feasibility of construction
Erosion
Landsliding
Wildlife disruption

Variables Considered (in order of importance):
Slope
Soil runoff potential
Soil erodibility
Vegetation types
Geological structure

RECREATION MODEL: SMALL BOATING AND FISHING

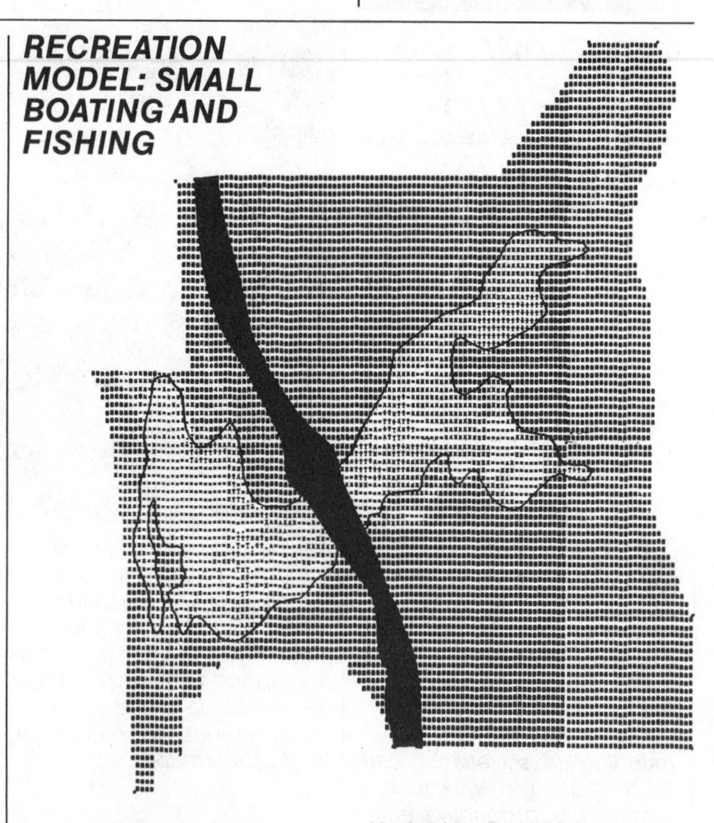

Major Considerations:
Site feasibility
Wildlife disruption
Erosion
Sediment transport

Variables Considered (in order of importance):
Slope
Vegetation types
Water features
Flood plains

PLAN STUDIES

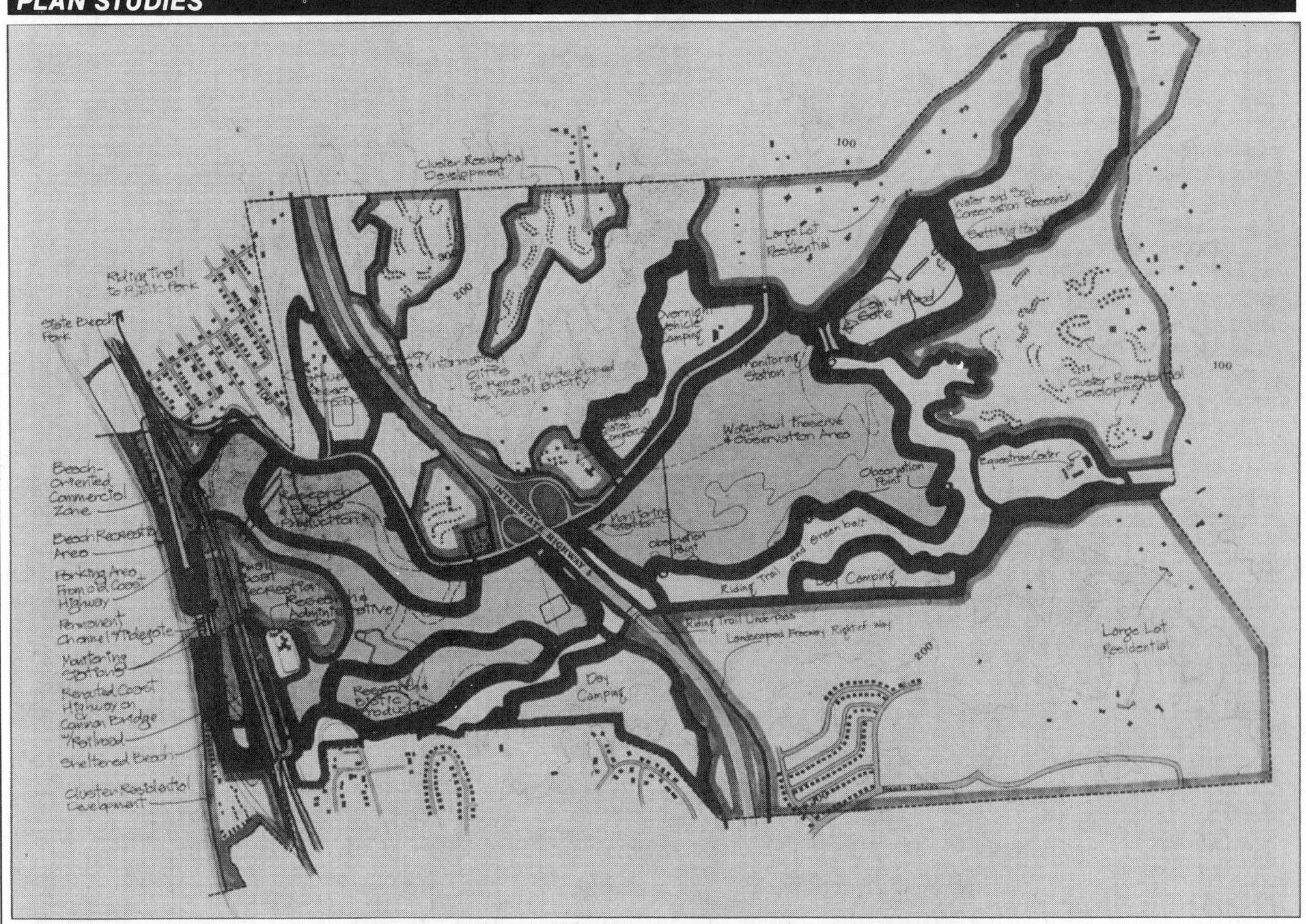

These models were used as guides for design studies that explored various combinations of use and possible locations. Three of the early studies are shown here.

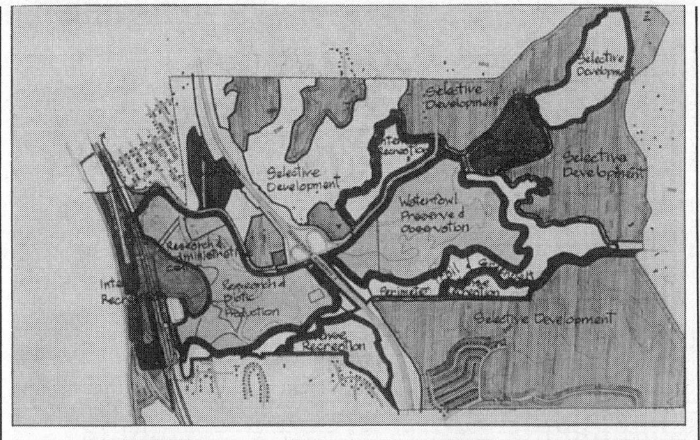

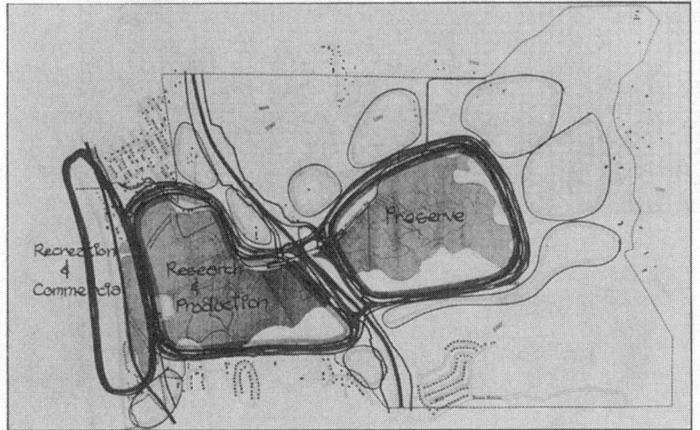

Coastal lagoons are the sensitive and highly productive pivot points between land and sea. These ecologically important environments are seriously threatened by urbanization. Dealing with the complexities of estuarine dynamics required the use of a number of design tools in addition to the suitability models. These included an environmental impact matrix, diagrams analyzing flows of materials and energy, and analytical guides to control by design, as well as some more traditional techniques.

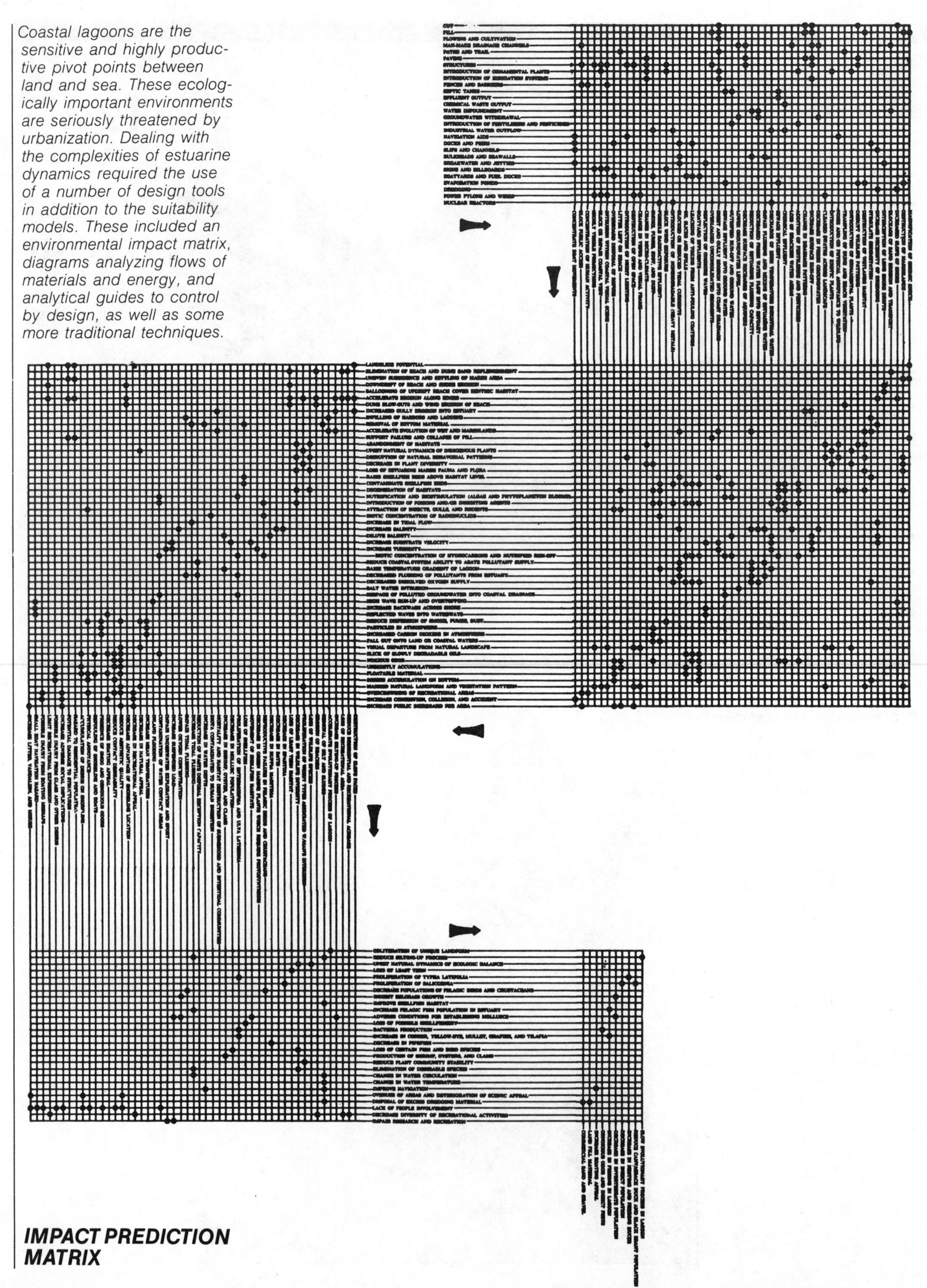

IMPACT PREDICTION MATRIX

THE CHAIN OF EFFECTS

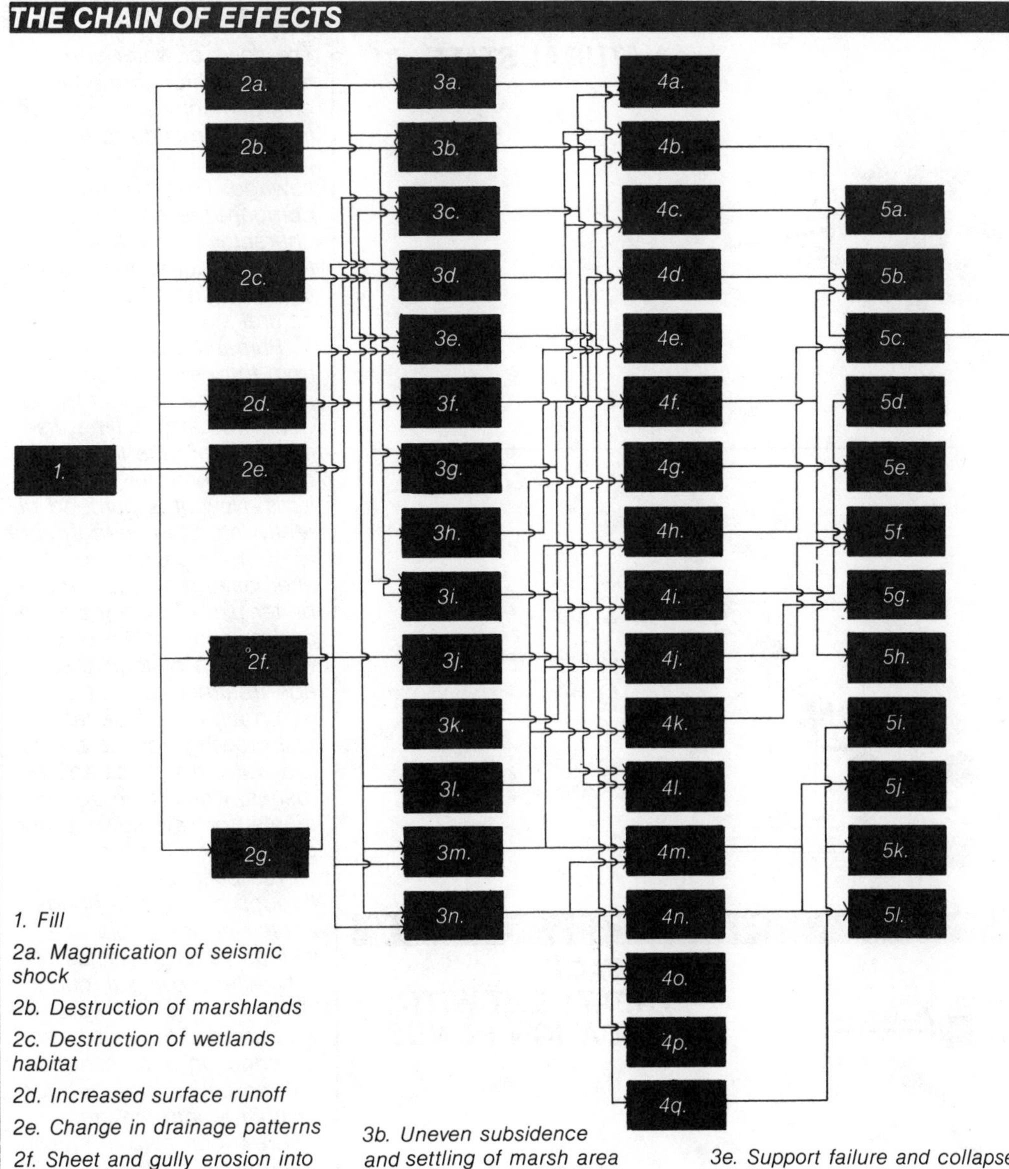

1. Fill

2a. Magnification of seismic shock

2b. Destruction of marshlands

2c. Destruction of wetlands habitat

2d. Increased surface runoff

2e. Change in drainage patterns

2f. Sheet and gully erosion into coast drainage

2g. Overloaded unconsolidated sediments

3a. Landslide potential

3b. Uneven subsidence and settling of marsh area

3c. Increased erosion into estuary

3d. Accelerate evolution of wet marshlands

3e. Support failure and collapse of fill

3f. Abandonment of habitats

3g. Disruption of natural behavioural patterns

3h. Increase in plant diversity

3i. Loss of estuarine marsh flora and fauna

3k. Raise shellfish beds above habitat level

3l. Contaminate shellfish beds

3m. Dilute salinity

3n. Increase substrate velocity

4a. Destruction of slope faces

4b. Increase silting-up process

4c. Accelerate evolutionary process of lagoon

4d. Loss of wildlife species

4e. Proliferation of weedy types associated with man's intrusion

4f. Decrease in wildlife diversity

4g. Decrease in ruppia maritima

4h. Movement of shellfish habitats

4i. Loss of shellfish

4j. Decrease in mollusc population

4k. Increase in shrimp, oyster and clams

4l. Mortality and habitat destruction of submerged and intertidal communities

4m. Increase suspended load

4n. Impair underwater exploration and sport

4o. Hazard to residential population

4p. Potential damage to structures

4q. Limit recreational expression

5a. Obliteration of unique landform

5b. Upset natural dynamics of ecological balance

5c. Adverse conditions for establishing molluscs

5d. Loss of possible shellfishery

5e. Decrease in pipefish

5f. Loss of certain fish and bird species

5g. Production of shrimp, oysters and clams

5h. Elimination of desirable species

5i. Overuse of areas and deterioration of scenic appeal

5j. Lack of people involvement

5k. Decrease diversity of recreational activities

5l. Impair research and recreation

6. Increase in invertebrate population

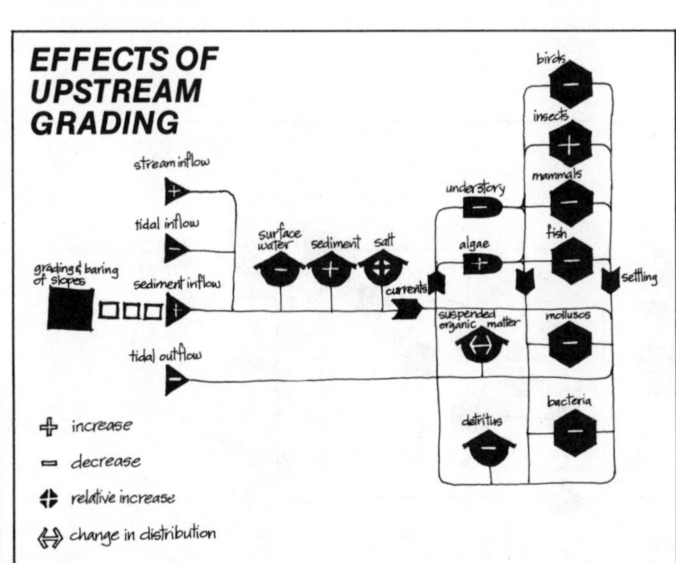

The chains of environmental effects brought about by human actions in the estuarine environment tend to be long and varied. The impact matrix is a tool for tracing them through several levels of the sequence of interactions, beginning with the actions that began the chain. The flow diagram describes, in somewhat different form, the sequence set off by one action, in this case, the filling of wetlands.

FUNCTION: THE FLOW OF WATER AND NUTRIENTS

NATURAL STATE

Escondido Cr.
tidal flushing

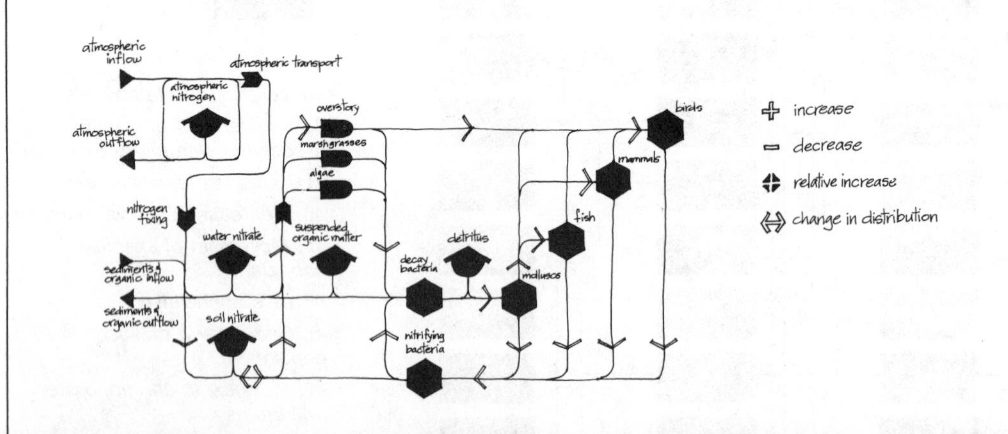

SEWAGE TREATMENT WITH OXIDATION PONDS

domestic & industrial use
from Colorado River
primary treatment
sewage
oxidation ponds
Escondido Cr.
flow reduced upstream
tidal flushing reduced

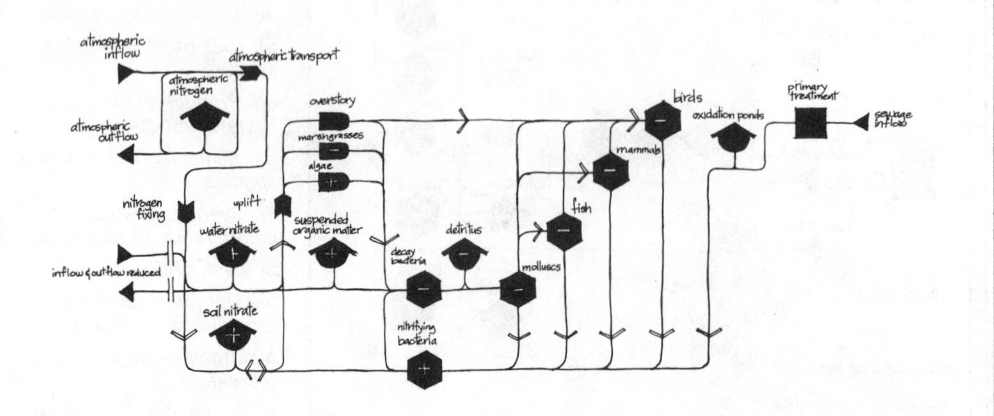

The flows of water and nutrients through the lagoon environment have long been intimately related to the treatment and discharge of sewage. These diagrams compare the ecological character of the flows under four different sets of conditions, beginning with the natural state.

Primarily treated effluent from the sewage treatment plant on the edge of the lagoon was at one time discharged into the lagoon, causing some severe problems. Now, it is pumped directly into the ocean through a four-mile outfall pipe. An alternative that would make better use of both the water and nutrients in the sewage would be a biological sewage treatment process in which algae and certain fast-growing aquatic plants, like water hyacinths and bullrushes, would take up nutrients from the sewage and purify it. The process involves passing sewage through a series of ponds containing these plants as well as various species of fish. After moving through several ponds, the water will be pure enough to be used for irrigation of plantings in the recreation area and for discharge into the lagoon, where it can help to stabilize both the water level and salinity. A more complete description of the biological treatment process is included in Chap. 13.

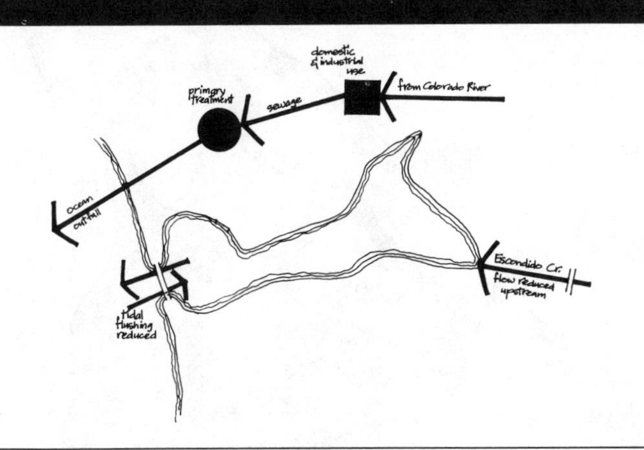

SEWAGE TREATMENT WITH OCEAN OUTFALL

PHOTOSYNTHESIS
Reception and processing of materials and energy through the process of photosynthesis; such as in green plants.

RESPIRATION
Reception and processing of materials and energy through the process of respiration; such as herbivores and carnivores in a grazing food chain.

HUMAN ENTERPRISE
A systematic or purposeful human activity, transforming material and energy; such as agriculture or manufacturing.

INPUT OR OUTPUT
Importation or exportation of material or energy to or from a given system.

STORAGE
Temporary retention of material or energy in a given system; such as the storage of water in a reservoir.

WORKGATE
A material flow acted upon by the energy of some outside force; such as evaporation of water.

(adapted from Lyle and vonWodtke, 1974)

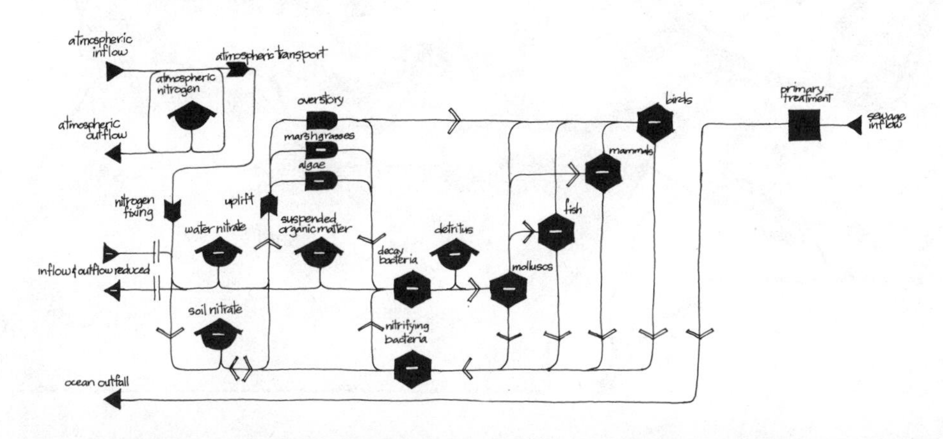

PROPOSED SYSTEM WITH BIOLOGICAL SEWAGE TREATMENT

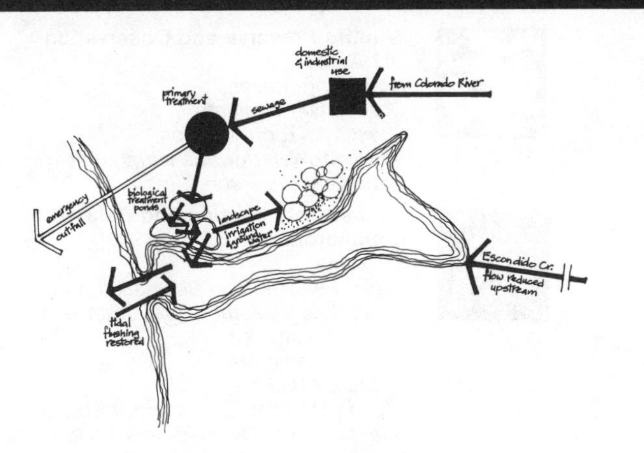

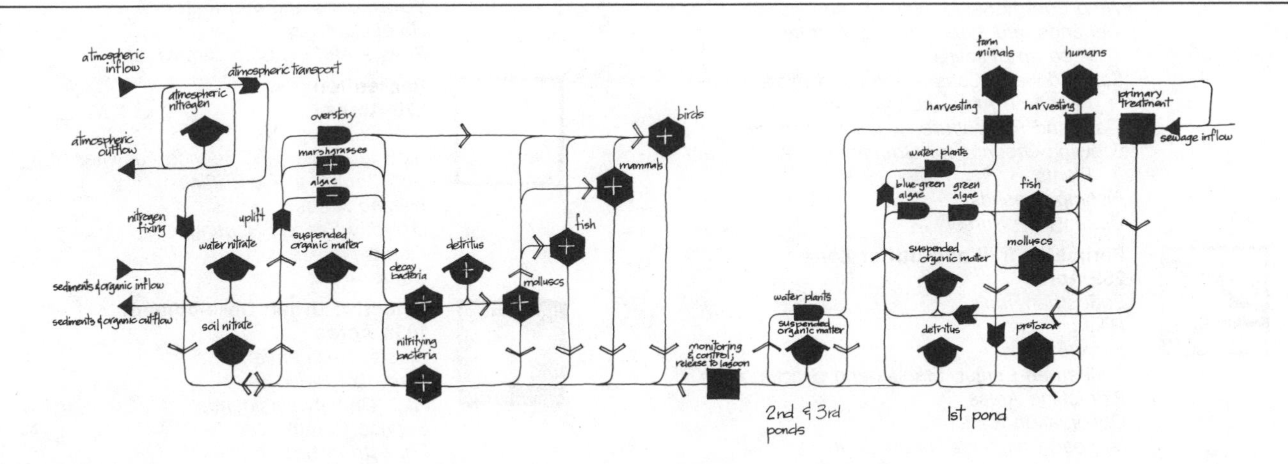

USE FRAME

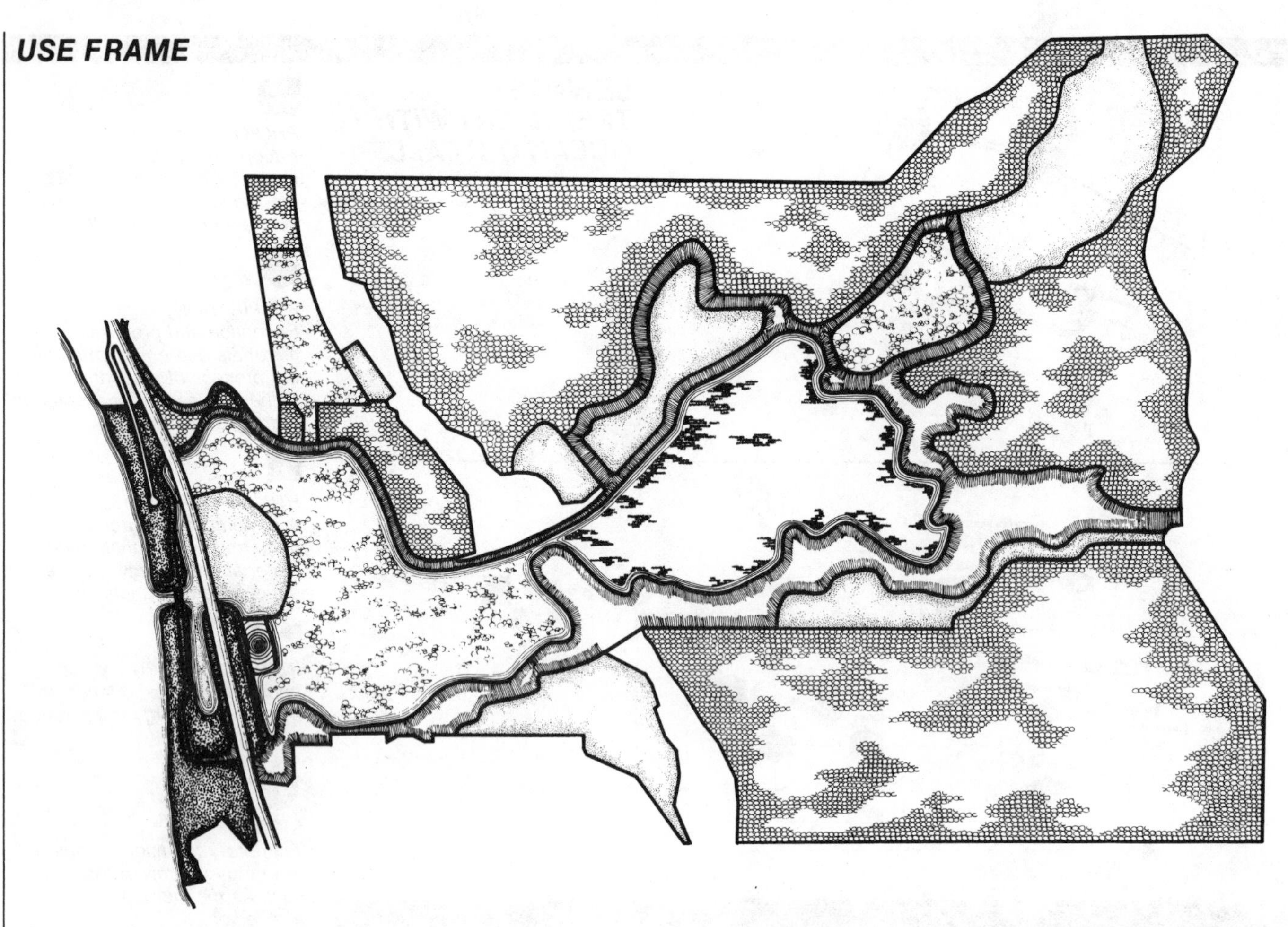

Administration and Research Center
5 Acres
Lagoon Guardianship Center Offices
Lagoon Research Control Center
Lagoon Education Facilities
Tour Center and Information Distribution Facilities
Institutional Research Coordinating Center

Biotic Production, Research and Conservation
250 Acres
Monitoring Stations
Aquaculture Research Facilities
Aquaculture Production Area
Effluent Research Facilities of Sewage Plant
Natural Power Source Research Facilities
Solar Energy Station
Tide and Floodgates
Wind Energy Stations and Windmills
Wetlands and Floodplain Agriculture Research Area and Facilities
Soil and Water Conservation and Production Research Area and Facilities
Plant and Floriculture Research Facilities
Growing Grounds for Ornamentals All Tolerant Plants
Fishing Grounds
Ocean Fish Nursery

Perimeter Trail and Buffer Zone
259 Acres
Equestrian Trails
Hiking Trails
Bicycle Trails
Equestrian Center, Stables and Grazing Area
Picnicking Areas
Observation Stations
Mapped Children's Nature Trail

Wildlife Preserve and Observation
182 Acres
Monitoring Stations
Observation Stations
Coast Life Zoo Operation
Seasonal Waterfowl Sanctuary
Zoological Research

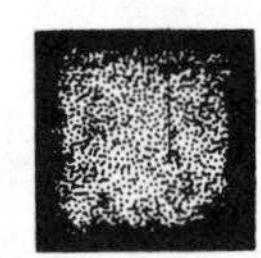

Intensive Recreation and Related Commercial
28 Acres
Public Beaches and Recreation Areas
Small Boat Launching Area (Canoe, Pedalboats, etc.)
Small Sailboat Marina
Sheltered Bathing Beach
Lagoon Oriented Shops and Restaurants
Beach-Tourism Oriented Shops and Restaurants w/outlets for Lagoon Products
Vista Points and Information Areas
Hotel Complex
Private Athletic Club: Tennis, etc.

Recreation
228 Acres
Tent Camping
Bicycle Center and Service Facilities
Golf Course in Flood Plain
Fishing Areas
One-day Season Hunting
Picnicking Areas
Game Areas

Selective Urban Development
1030 Acres
Cluster Residential
Large Lot Residential
High Density Residential
Service Commercial
Equestrian Residential

The Use Frame—the framework for the development and management of the lagoon that grew out of the analyses and studies—describes seven different homogeneous zones and the range of uses permissible in each. The purpose is not to prescribe specific uses, but to define the uses that can be sustained in each zone without seriously affecting the functional integrity of the lagoon.

CONTROL BY DESIGN

Landform Stabilization and Soil Retention

Measure	Description	Decrease	Severe	Moderate	Slight	Slight	Moderate	Severe	Increase
Cultivation	Prevention of erosive forces of water moving downhill by contour plowing (lateral furrows).	Sedimentation Soil storage Water erosion Wind erosion Runoff Resource							Temporary measure requiring constant monitoring and maintenance does not prevent wind erosion. Encourages percolation.
Planting	Use of plant material as soil binder. Method of planting varies: 1. Planting from containers 2. Aerial seeding (large scale) 3. Hydro-seeding (broadcast in liquid mixture by machine)	Sedimentation Soil storage Water erosion Wind erosion Runoff Resource							Effectiveness dependent on species of plants, time of planting. Consideration should be given to use of natives vs. exotics and the need for irrigation.
Jute mesh	Heavy woven jute layer used as surface soil binder, often used in conjunction with planting. Rolled onto slope in strips.	Sedimentation Soil storage Water erosion Wind erosion Runoff Resource							Interim measure while plants become established. Will decompose in a short period of time.
Straw cover	Straw, broadcast over slope, then rolled into surface with sheepsfoot roller forming a compacted, bound surface.	Sedimentation Soil storage Water erosion Wind erosion Runoff Resource							Interim measure while plants become established. Requires monitoring and maintenance.
Rock blanket	Layer of rock applied to surface.	Sedimentation Soil storage Water erosion Wind erosion Runoff Resource							Requires much manual labor for placement. Machinery required may cause incidental compaction.
Sprayed synthetic material	Chemically derived materials applied in liquid or filament form. Applied by machine.	Sedimentation Soil storage Water erosion Wind erosion Runoff Resource							Temporary measure, leaving residue of materials for indefinite periods. Generally hampers plant germination, and can stop percolation when applied heavily.

The Control by Design charts describe the various technical means for carrying out each developmental action and the environmental effects likely to result from each. They provide the means for controlling damage to the lagoon that might result from the development of surrounding areas. This example shows the means and effects of stabilizing landforms.

ulations compete for limited supplies of the essential ingredients of life.

In its natural state, by contrast, the lagoon and the original community it supported never had the option of becoming dependent on distant sources. They had to live on the resources at hand. Over tens of thousands of years, the estuarine ecosystem gained coherence by adapting to the existing conditions of ocean and land. By trial and error, it evolved its own unique trophic structure, every niche filled, for the cycling of materials and the distribution of energy. It learned to share the energy of the tides, which one might expect to be an unsettling influence, to heighten growth and productivity. It found ways of mediating between land and sea, concentrating, controlling, and making the best of the effects of one or the other.

If one is to put all this experience to good use in creating a new natural urban environment—one that will help support the human community on its own local resources—we must search out the secrets of the lagoon's extraordinary productivity. Every ecosystem has certain motivating processes that define its essential character and that provide us with a key for understanding and working with it. Sometimes these processes are obvious from the start, but usually they emerge only after careful analysis. In the estuarine ecosystem, the keys lie in the flow of energy—particularly the tidal subsidy that spreads the food around the lagoon twice a day—and the food web. Together, they largely account for the lagoon's productivity. An essential first step toward reestablishing the essential flow for San Elijo, therefore, will be the restoration of tidal flushing. Given the fact that the existing bridges are there to stay, a built-in mechanical device for preventing the buildup of silt in the channel will probably have to be provided. Several practical techniques for doing so have been proposed. Once such a device is operating, the lagoon should continue flushing and scouring out collecting silts with little human intervention, although some initial dredging might be needed to provide an adequate tidal volume.

The necessary step for reestablishing the food web is the restoration of the marshgrasses. Once tidal flushing is in operation, the grasses will begin to recolonize the tidal areas naturally. Since the natural growth of an extensive population will take a number of years, however, it would be better to hasten the process by a heavy initial planting of the most valuable grass—the Spartina. Once the Spartina covers the tidal areas, the whole food chain, from algae to hawks, will soon reappear.

HOW HUMAN ECOSYSTEMS WORK

The pattern of uses we are envisioning here is far more complex than anyone might have imagined at the beginning when the issue was posed as a simple choice between preservation and development. In this sense, the issues at San Elijo are very much like most other landscape issues that we face in the ever more crowded environment of these last decades of the twentieth century. Though the clarity of an estuarine ecosystem makes the processes and possibilities involved more sharply defined than they are in most places, the issues are typical in that they involve the merging and interacting of human and natural processes.

The poles of preservation and development created by so many conflicts of the 1960s and 1970s are drastic oversimplifications. Without a doubt, there are large areas of the earth that should be preserved in their natural state, and this may very well be the most fundamental environmental issue of all. The wilderness is where our roots are. As Wendell Berry says, "Only if we know how the land was can we know how it is."[3] Granting this, there may very well be other areas, albeit far less extensive, that could be developed in concrete with no great loss, areas that exclude nature entirely.

Eugene Odum has proposed compartmentalization of the total landscape into areas divided according to basic ecological roles. He argues that we need both successionally young ecosystems for their productive qualities and older natural ones for their protective qualities. According to Odum, ". . . the most pleasant and certainly the safest landscape to live in is the one containing a variety of crops, forests, lakes, streams, roadsides, marshes, seashores, and waste-places—in other words, a mixture of communities of different ecological ages."[4] He might well have added houses, gardens, parks, playing fields, offices, and shops. In the interest of achieving or maintaining such a mix, Odum would classify all land in one of four categories:

1. The productive areas, where succession is continually retarded by human controls to maintain high levels of productivity.
2. The protective, or natural, areas, where succession is allowed or encouraged to proceed into the mature, and thus stable if not highly productive, stages.
3. The compromise areas, where some combination of the first two stages exists.
4. The urban industrial, or biologically nonvital, areas.

If we accept this schema a great many of the most pressing, most challenging, and probably even most important landscape issues fall into the third category. "Compromise areas," however, is hardly an adequate term for those places in which human beings and nature might be brought together again after a very long and dangerous period of estrangement. I prefer to call such places "human ecosystems."

Since some might find a hint of humanist arrogance in the whole notion of designing ecosystems, I will hasten to elaborate on my meaning. The truth is that human beings have been designing ecosystems for some twelve thousand years now, ever since they first learned how to cultivate plants. Through all these millenia, they have been habitually, even compulsively, changing the world's landscape. If we are to fulfill our human potentials, we will find it necessary to continue doing so. In the present era, we will have to continue to do so merely to provide the basic necessities for burgeoning populations. The ecosystems shaped by our changes of the landscape will invariably be different in structure and function from the previously existing natural

[3] See Berry, 1977; p. 52.
[4] See Odum, 1969; p. 267.

ones, but they will continue to respond to exactly the same natural forces even though they may be more or less diverse, more or less stable, more or less productive, or have more or less of any number of other qualities. Our creation of new ecosystems has almost always been unintentional—that is, without conscious understanding of how natural processes work and therefore without any way of predicting how the new ecosystem would work, even without any comprehension of the fact that it was actually a system. Not surprisingly then, without conscious control, new systems usually do not work very well. In the San Elijo case, we might call the railroad, the freeway, and the sewage treatment plant all examples of unintentional ecosystem design, and we have seen the results. The developers' and the preservationists' proposals fall into the same general category because, although they do consider some aspects of the lagoon environment, they do not take into account its ecological processes and its interacting, systematic nature.

The point is that if we are going to design ecosystems (and we continually do so whether we want to face all of the implications or not) then it will be best to design them intentionally, making use of all the ecological understanding we can bring to bear. Only then can we shape ecosystems that manage to fulfill all their inherent potentials for contributing to human purposes, that are sustainable, and that support nonhuman communities as well. Not every landscape can fully accomplish all three of these goals, of course, and thus Odum's term, "compromise." There will always be conflicts to be resolved and priorities to be assigned. Intentional design means carrying out conscious choices. What we are trying to do, then, is to gain a measure of control, not in order to dominate nature but to participate creatively in its processes.

Ecosystem design is undoubtedly a difficult undertaking. Nature rarely reveals herself unequivocally, and there is always the risk that we will end up agreeing with Spinoza that "... the attempt to show that nature does nothing in vain ... seems to end in showing that nature, the gods, and man are alike mad." We have to begin by admitting that our tools are still crude, and we do not know enough to do the job with absolute confidence, recognizing at the same time that we will have to do it anyway.

To participate creatively in natural processes and to do so with reasonable hope of success, we need to include as subjects of design not only the visible form of the landscape but its inner workings, the systems that motivate and maintain it. Natural systems are continuously self-organizing (there being nobody available to organize them), and we can draw upon the principles by which they work to make human ecosystems more sustainable. Such an aim requires a knowledge of these systems. Fortunately, the sciences provide a great deal of information, which, while far from complete, is yet enough to get us started.

Generally speaking, we can divide this scientific knowledge into two types. First, there are facts or data concerning the situation at hand. For any given landscape, a great many of these may be available, or very few, depending on how much research has been done. For San Elijo, considerable data were available because of the investigations of marine biologists and earth scientists at several nearby universities. Species populations were fairly well known, for example, as were the compositions of the rock formations around the edges of the lagoon. Even the shellfish species that had been present in the lagoon's natural state were known—the result of analyses of shell middens left by the Indians. More general facts could be extrapolated from research done on similar lagoons along the California coast. Although all this information would obviously be useful in dealing with specific parts of the design, it was far from complete. In practice, of course, it is impossible to have all the facts needed to describe completely even a very small landscape. We have to work with what we can get.

The second type of scientific knowledge might be loosely categorized under the heading of "concepts," a word the dictionary defines as general notions, ideas, or principles conceived in the mind. The science of ecology has developed a number of basic concepts—such as productivity, trophic levels, succession, and energy flow—that help unify and give coherence to the masses of otherwise unrelated facts produced by research. These concepts are large and inclusive and fit the known facts, but since they are conceived in the mind, it is virtually impossible to prove that any of them actually exist in nature. Although research is continuously accumulating new information that furthers our understanding of the interactions that cause succession, the actual existence of succession, for example, remains beyond experimental proof. This discrepancy causes occasional debate among scientists, who sometimes question concepts like succession or believe that science should deal only with theories that can be experimentally proven and thus turned into facts.[5] For purposes of design, however, concepts are indispensable because they can be put to general use. In fact, utility is the criterion of value of a scientific concept[6] but is rarely considered with respect to scientific facts or theories.

For purposes of design, concepts are useful because they provide access to the mechanisms that join all of the facts. They make it possible to work with the forest before the trees. They make it possible to gain a working understanding of an ecosystem even though many of the facts may be unknown. They give us handles with which to grasp the unseeable. They provide us with a basis for developing theories of ecosystem design that allow us to reach into and reshape the inner workings of the landscape.

In this book, we will deal with the roles of both facts and concepts. As in a good design, concepts provide the basis for the larger framework of organization, while facts provide the specifics.

The Ecosystem Concept

The first concept is that of the ecosystem itself. It is a rather new concept, having been first advanced by A. G. Tansley in 1935, but an important one, having become since that time the fundamental principle of all ecological study. Simply defined, an ecosystem is the interacting assemblage of living things and their nonliving environment. Among living things

[5] See Rigler, 1975.
[6] See van Dobben and Lowe-McConnell, 1975.

are human beings themselves, although ecologists usually choose to study ecosystems that exclude man, and human beings usually choose to think of themselves as somehow set apart from ecosystems. This is an important point, and one that is implicit in everything that follows: We human beings are integral, interacting components of ecosystems at every level, and in order to deal adequately with these systems, we have to recognize that simple fact. In most situations, even at the level of the biosphere, we may be an overriding, controlling component, but we are a component nonetheless.

Another important characteristic of the ecosystem is that it can be of any size. That we can consider any landscape of any size is a great convenience for designers, but there are rules to be followed. No ecosystem stands alone. "All ranks of ecosystems are open systems, not closed ones. . . ."[7] This implies that ecosystems are connected by flows of energy and materials. Each system draws in energy and materials from the systems around it and in turn exports to them. In drawing the boundaries of an ecosystem, therefore, we need to consider the flows that link it with its neighbors. Ignoring these connections—these imports and exports of energy and materials—has caused a great many of the disasters of unintentional ecosystem design.

In the shaping of ecosystems, three organizational concepts are of fundamental importance. The first is *scale,* or the relative size of the landscape in question and its connections with larger and smaller systems and ultimately with the whole. It is scale that provides us with an encompassing frame of reference. The second is *design process,* the pattern of thought that we follow in dealing with this frame of reference. The third is the underlying *order* that binds ecosystems together and makes them work. These constitute the three major subjects of this book, which is divided into three parts, each concerned with one of them. For the sake of orientation, I will introduce each concept briefly here before proceeding to treat it in greater detail.

Scale

We need to recognize that every ecosystem is a part—or subsystem—of a larger system and that it in turn includes a number of yet smaller subsystems. It also has necessary linkages to both the larger and the smaller units. San Elijo lagoon, for example, is at the same time a component of a larger watershed unit and a component of an even larger oceanic unit. The water that runs off the land in this eighty-square-mile watershed eventually reaches the lagoon, bringing along everything it has picked up in the interim. This may include silts from eroding slopes, nitrates from fertilized agricultural lands, oil from roads, and any number of other substances that can seriously affect the life of the lagoon. If the lagoon is to operate as a healthy ecosystem, therefore, some control over land use in the watershed will be required. By the same token, all these materials finally flow from the lagoon into the Pacific Ocean, establishing yet another linkage. The lagoon is also linked to the San Diego urban region, even the entire Southern California region, because of all the people who come there for recreation. On a still larger scale, it is tied to Alaska and Central America by the Pacific Flyway. Events in San Elijo can thus seriously affect animal populations thousands of miles away.

Despite all these connections, San Elijo Lagoon is a limited unit of landscape, one of a certain size with definite boundaries, which means that we can deal with it only at a certain scale. The concerns that we can address in detail are likewise limited to those that are appropriate to that scale. Nevertheless, we need to work within the context, or framework, of the larger-scale unit, in this case the watershed, and we need to consider the proposed development projects as smaller-scale units within the framework of the lagoon. Our range of design scales forms a hierarchy that corresponds to the concept of levels of integration in nature or in any other organized system. Certain principles of organization link the levels of this hierarchy and provide guidance for design at any given level. The next four chapters (Part I) will be devoted to this subject—the scales of concern.

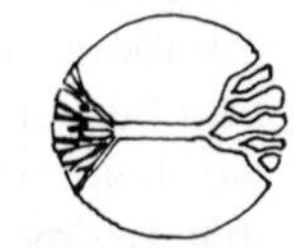

Design Processes

In Part II, we will explore design processes, the vehicles for what we have called creative participation in natural processes. The ways in which we go about design will naturally vary according to the scale of concern and the situation at hand.

At this point I will have to digress briefly in order to clear up a semantic difficulty. The activity of "design," as I am using the term here, means giving form to physical phenomena, and I will use it to represent such activity at every scale. The challenges we face require some broadening and redefinition of the activity of design. According to Erich Jantsch, "Design attempts to find, formalize, and bring optimally into play the innate forms of a process . . . [and] . . . focuses on finding and emphasizing internal factors in evolution, on making them conscious and effective."[8]

This is a departure from the convention of using the term "planning" for landscape shaping at scales larger than that of construction detail. I believe the departure is justified by the very broad, rather indefinite inclusiveness of the term "planning," by the confusion that results from its use, and by the increasing tendency in the environmental design disciplines to associate planning with administrative activity rather than physical form-shaping. This book is about making physical changes in the landscape and not about administrative, legal, or policy-making activity, although, needless to say, it will usually take a great deal of the latter to bring about these changes. Planning and design are thus closely linked and work in tandem, sometimes to the point of being indistinguishable.

In using the term "design" in this sense, I believe I am following, not trying to initiate, a trend. More and more,

[7] See Evans, 1956.

[8] See Jantsch, 1975; p. 44.

we hear of "site design" rather than "site planning." Carl Steinitz refers to "regional landscape design" and justifies the usage of this term by defining design as "intentional change. . . . the landscape and its social patterns are altered by design."[9] And Ian McHarg, of course, entitled his famous work "Design with Nature."[10] In any event, the term "design" carries the connotations of intention, precision, and control that befit the approach I am describing. It also suggests emotional involvement. Jantsch speaks of design as being planning plus love. Consequently, I shall use the term with all these overtones in mind, although with apologies for any confusion it may cause. Likewise I shall use the term "planning" to refer to more strictly administrative and institutional activities, such as the articulation and implementation of policy.

Combining as it does two different modes of thought—analytical use of scientific information and creative exploration (or the left and right sides of the brain, if you will)—ecosystem design can get very complicated. The two modes can work together, but only if the roles of each are clearly established. Especially at the larger scales, design processes are further complicated by the involvement of considerable numbers of people—in some cases, as we shall see, huge numbers. To deal with this complexity in a rational manner, we shall break these processes down into component themes that are more or less common to all of them: formulation, information, models, possibilities, plan evaluation, and management. Then we will examine each of these themes with respect to its content and the analytical or creative orientation associated with it.

The inclusion of management is particularly important in ecosystem design because of the variable future that is a fact of life for any organic entity. The design of ecosystems is probabilistic in that we cannot say what will definitely happen in the future but only what will probably happen. Management deals with this uncertainty in the cybernetic manner, by observing what actually does happen and redesigning as necessary. Thus, being an essential continuation of design by other means, to paraphrase the famous statement on war and politics, management assumes a more creative role than has usually been expected. To repeat, the interlocking relationship between design and management is a *particularly* important feature of any ecosystematic design process.

Order

In the midst of complexity, with its many opportunities for diversion, we need to keep reminding ourselves that the purpose of creating order in human ecosystems is to enable them to fulfill the needs of both their human and other components. But how do we define "order"? There are a great many kinds and degrees of order, although in landscape design, we are most used to thinking in terms of visual order. Ecosystematic order is something else again, although it is usually reflected in what we see.

Here, to return to the concepts of ecology, we can identify three modes of order, each of which provides a key to one aspect of the inner workings of ecosystems. The three modes are structure, function, and location.

Odum defines structure as ". . . the composition of the biological community, including species, their biomass, life histories and spatial distribution, the quantity and distribution of abiotic materials, and the range of conditions like light and climate."[11] Margalef is more succinct: "If we consider the elements and the relations between the elements, we have the structure."[12]

The number and types of the elements present and the ways in which they interact are fundamental to the character of an ecosystem. For analytical purposes, structure can be broken down into substructures. The composition of trophic levels is an example of a substructure, one which, as we have seen, plays an especially important role in San Elijo Lagoon. The relationship of the marshgrasses to the decomposing bacteria and through them to molluscs is unique to coastal inlets and is at the very core of their structure. The birds, though far more visible and more interesting to human beings, are less essential to the ecological structure, and are ultimately dependent on the health of the marshgrasses. In this particular situation, the elements are the species of marshgrass, bacteria, molluscs, shorebirds, and waterfowl, and the relations are their predatory interactions. These we can see as the building blocks and the mortar of the lagoon ecosystem. We will return to the subject of ecosystem structure in Chaps. 11 and 12 to explore its meaning further and examine the ways of shaping the structure of human ecosystems by design.

The second mode of order, function, or the flow of energy and materials, is closely intertwined with structure. According to Odum, the ". . . complex biomass structure is maintained by the total community respiration which continually pumps out disorder."[13] Respiration is fueled by the flow of energy, and keeping this flow going, distributing energy to all the members of the community, is a basic purpose of ecosystem function. At San Elijo, the tides add their force to the "pumping," thus speeding up the flow and increasing the rate of productivity. Every ecosystem has a characteristic pattern of energy flow that corresponds with its structure. The flow begins as solar radiation imparts energy through photosynthesis and goes on from there to reach every living creature in the system. Every individual, and the ecosystem as a whole, has the ability to maintain a state of high internal order, or low entropy, as long as it is continuously supplied with its energy needs. As energy flows through the system, in keeping with the second law of thermodynamics, it is degraded to a more dispersed form with each transformation. Thus as it flows from marshgrass to molluscs to fish and birds, energy is lost. It happens that the marshgrass Spartina and the molluscs are particularly efficient convertors of energy into biomass. The Spartina has a uniquely efficient cellular structure, whereas the mol-

[9] See Steinitz, 1979; p. 3.
[10] See McHarg, 1969.
[11] See Odum, 1971.
[12] See Margalef, 1963; p. 216.
[13] See Odum, 1971; p. 37.

luscs, thanks to the tides, are in the privileged position of not having to forage for food. Most animals spend most of their energy searching for something to eat, but the detritus food of the molluscs is served up to them by waters propelled by tidal energy—hence the enormous productive value of the tidal subsidy. Even so, more energy is lost to unusable heat at each stage than is actually fixed. Consequently, the lagoon system needs continuous energy inputs from sun and tides to maintain itself.

Also essential to ecosystem function are the flows of water and the chemical elements essential to life. In contrast with energy, these are not continuously dissipated but circulate intact along more or less consistent pathways through storage to environment to organisms and back. Thus, the material flows, or biogeochemical cycles, as they are usually called, provide each organism with its needed chemicals and nutrients.

At San Elijo, as in most unintentionally designed manmade ecosystems, the material flows have long been in a state of perpetual dysfunction, not for lack of materials, but because they are directed to the wrong places. During the long period when primarily treated sewage effluent was dumped into the lagoon, the enormous concentration of nutrients from the sewage brought about rapid growth of algae, which used enormous quantities of oxygen from the water, thus denying it to fish and molluscs and depleting their populations. When the algae died at a faster rate than the waters could absorb them or the tides move them out, they decayed on the surface, causing unsightly masses of green scum and unpleasant odors. This is an example of a very common difficulty created by unintentional design. The solution eventually implemented for the "water pollution problem" was the four-mile-long ocean outfall. Now there are only a few occasional algae blooms on the lagoon's surface, mostly caused by fertilizer nitrates in runoff water. But it is not only the nutrients that have been lost for any human purpose; with the freshwater infusions from the sewage cut off as well, the surface level of the lagoon has dropped, leaving dried-up stretches of mudflat around some of its edges. The natural fresh water supply through Escondido Creek was long ago drastically reduced by upstream impoundments.

What we have here, then, is a classic example of the type of once-through flow that has become characteristic of single-purpose, unintentional human ecosystems. Most of the water that flows in over two hundred miles from the Colorado River is used once, mostly for flushing toilets, and then flows out into the ocean. Nutrients flow in, mostly in the form of food from far-flung sources, are used once, and then join the water in household toilets to flow out into the ocean with it. Such a system makes a dramatic contrast with natural systems in which both inputs and outputs are minimal, flows are much slower, and water and nutrients are used and reused over and again to support a diversity of organisms.

The alternative that we propose would redirect the flows to reuse both water and nutrients through biological sewage treatment. Thus, by feeding the primarily treated effluent into a series of ponds in which water hyacinths and other aquatic plants will take up the nutrients, the water will eventually reach a level of purity that will permit its use for irrigating recreational areas and its eventual return to the lagoon. The hyacinths can be harvested for cattle feed and thus eventually be returned to the system as well. Such a pattern of water and nutrient flow is more like that of a natural ecosystem, more efficient and economical. The outfall, incidentally, would still be needed for overflows and emergencies.

One major consideration remains, however, at least for the predictable future. Such a system will not operate itself. Management will have to take the place of the self-regulating mechanisms of a natural estuarine system, which means that a high level of ongoing, creative management of the sort mentioned earlier will be needed.

We will return to the subject of ecosystem function in Chaps. 10 and 13 to delve more deeply into the workings of energy and material flows and the means of shaping them by design.

The third mode of ecosystematic order—locational patterns—usually receives far more attention in design than the other two, although it is less explored in the scientific literature. Usually, the proposed pattern of locations is considered *the* plan. Although this practice follows historical precedent and fits established decision-making patterns, it often results in the less visible aspects of structure and function being ignored. Ideally, the three modes would be considered equally important, so interrelated indeed, that one could not consider one mode without considering the others. Location, nevertheless, remains the most visible of the three, and "the plan" will probably remain the vehicle on which the design of ecosystem structure and function will ride.

The ideal pattern of locations is determined mostly by what is already there. The processes and organisms that we have described are distributed over the landscape in relation to climate and topography. If our purpose is to build on these to develop the sort of human ecosystem we have discussed, then that pattern will have to be respected. At San Elijo, the tidal areas where marshgrasses grow to support the food chain are essential parts of the picture, as are the mudflats where the shorebirds feed and the shallow waters and matted islands where they find cover for nesting.

These, however, are only a few pieces of the puzzle. Other patterns have been superimposed on the natural ones by human beings. The berms that support the railroad, the coastal highway, and the freeway divide the lagoon into three distinct parts. There is also an existing development pattern that includes commercial uses along the highway and residential uses on the upper slopes. There is also a pattern of demand for more open beaches and for more protected, lagoon-side play areas.

Perhaps most difficult to deal with is the pattern of private ownerships, which makes it necessary to show how the land can be used profitably. Whenever public acquisition is recommended, strong justification will be needed.

The most useful tool for sorting out the competing patterns is the *suitability model,* which consists of an analytically derived map showing the relative suitabilities of land increments for given human activities. In the case of San Elijo Lagoon, the complexity of the data made it convenient to use a computer-mapping technique for defining suitabilities. Whether it is produced by hand or by computer, however, there is nothing magical or definitive about a suitability model. The hand or the computer simply combines and aggregates the information that it is given in the ways it is told to and produces a graphic expression of the results.

For example, the first computer-generated map on page 6 shows relative residential suitability. [Since we will discuss computer mapping at length in a later chapter, we will consider here only the models themselves and their role in the design process, not the techniques by which they were derived.] Each of the little symbols, or grid cells, represents an area approximately 111 feet square and gives the estimated level of suitability for the given activity in that area. In each map, the darkest symbol indicates the lowest level of suitability and the lightest, the highest.

These estimates of suitability are based on predictions of future economic costs and environmental impacts. For the first model, that concerned with residential development, the most suitable locations were hypothesized to be those where development costs are likely to be lower, erosion rates less, landsliding improbable, and wildlife populations left undisturbed. Any number of other criteria might have been used, of course, but these were the ones judged most important in this particular case. The dark areas, then, assuming the technical reliability of the models, are those where some combination of high development costs, rapid erosion rates, probable landsliding, and wildlife disruption renders the land unsuitable for residential development.

The series of models that follow shows the relative suitability for various recreational and residential uses. The criteria for each model are different, but in each case the most and least suitable locations are defined.

Suitability models play a pivotal role in ecosystem design, providing a bridge between the consideration of processes and their location on the land. They aggregate complex collections of information concerning natural, social, and economic functions into usable forms. They disclose new

patterns that are extremely difficult, if not impossible, to discern in any other way.

It sometimes happens that a model is used not as a basis for planning, but as a plan in itself. This is a serious mistake. Models are simply expressions of the interactions of clearly stated facts and values. Once they are made, there is still a creative leap to be taken to shape a plan. The models provide a firm footing for this leap, but in the end, the plan will look quite different from the models.

From Models to Plan to Management

Witness the plan for San Elijo that emerged from the long process we have been describing here. It divides the lagoon and its watershed into seven distinct categories of land use. These generally follow the patterns of the suitability models, but the actual configurations are quite different. Moreover, the seven uses bear no resemblance to the traditional zoning categories, because the purpose of the zoning is quite different from that of traditional zoning. Here, we are not trying to promote uniformity of use, but to encourage the greatest diversity that is consistent with the healthy and productive functioning of lagoon processes. Consequently, the definition of uses is as general and as open to ideas as it can reasonably be.

The wetlands themselves are divided into three distinct zones, following the divisions already established by the railroad and the freeway. The inner zone east of the freeway—already the richest, most diverse habitat and the birds' favorite feeding and nesting place—becomes a wildlife preserve. The zone between the freeway and the railroad—where the water is deeper and wildlife far less abundant—will be devoted to biotic production and research. Areas of both natural, or protective, and productive landscape are thus important parts of the plan. The productive area is largely a man-made landscape, but one that is biologically very much alive. Some dredging and shoreline alteration may be needed here to develop the best environment for raising fish and molluscs. The old sewage treatment plant on the lagoon's edge will be refurbished as a biological treatment facility to provide an inflow of fresh water.

The westerly zone—between the railroad and the ocean, already heavily altered—will become an intensive recreation area, with some commercial development. Buildings will rest on piers over the waters of the lagoon so that marine processes can go on undisturbed beneath.

Surrounding the wetlands is a protective buffer zone with a hiking trail, screened from the more sensitive wildlife habitats, and outside that, on the flatter lands, will be several passive recreation areas. Although urban development will be encouraged on the slopes overlooking the lagoon, it will be subject to design controls that limit grading, maintain natural drainage courses and levels of runoff, and require planting for erosion. With such controls, development of these slopes will be entirely compatible with a productive lagoon.

Such a system, however, will continue to work well only if it is managed—man-aged—well. Once the system begins to take shape, ongoing management is to be instituted as one of its essential components. Only management can control the feedback loops needed to augment those that have evolved as internally functioning mechanisms in natural systems. Certain kinds of control are needed to prevent foreign and potentially damaging materials like fertilizers, pesticides, oil residues, or phosphates from entering the lagoon. Human activities can be regulated in such a way as to prevent their interfering with sensitive lagoon processes or populations. Critical indicators of environmental quality, especially water quality, in the lagoon need to be monitored to maintain the stability of the system. When an imbalance appears, or if there is evidence of deterioration or conflict somewhere, some corrective action can be taken. In the absence of such a program, however carefully the initial design may be conceived, the lagoon will eventually return to its present sorry state or worse.

This intentionally designed and managed ecosystem represents a symbiosis of urban and natural processes. Food production, wildlife habitats, recreation, dwelling, resource conservation, water and nutrient recycling, and visual amenity are joined in a network of interdependence. The composition as a whole is very different from the estuarine ecosystem that would still exist at San Elijo had man never arrived on the scene, being more varied in its forms and more intense in its activities. Although it is dependent on human energy and ingenuity for its stability, the reverse is also, to some degree, true. If all goes well, if our models are correct and our design works as it should, and if the management is both imaginative and sound, then human and natural processes will merge indistinguishably into an organic whole, a human ecosystem in the best sense of the term. That, hard as it may be to achieve, is the ideal.

PART I:

Scales of Concern

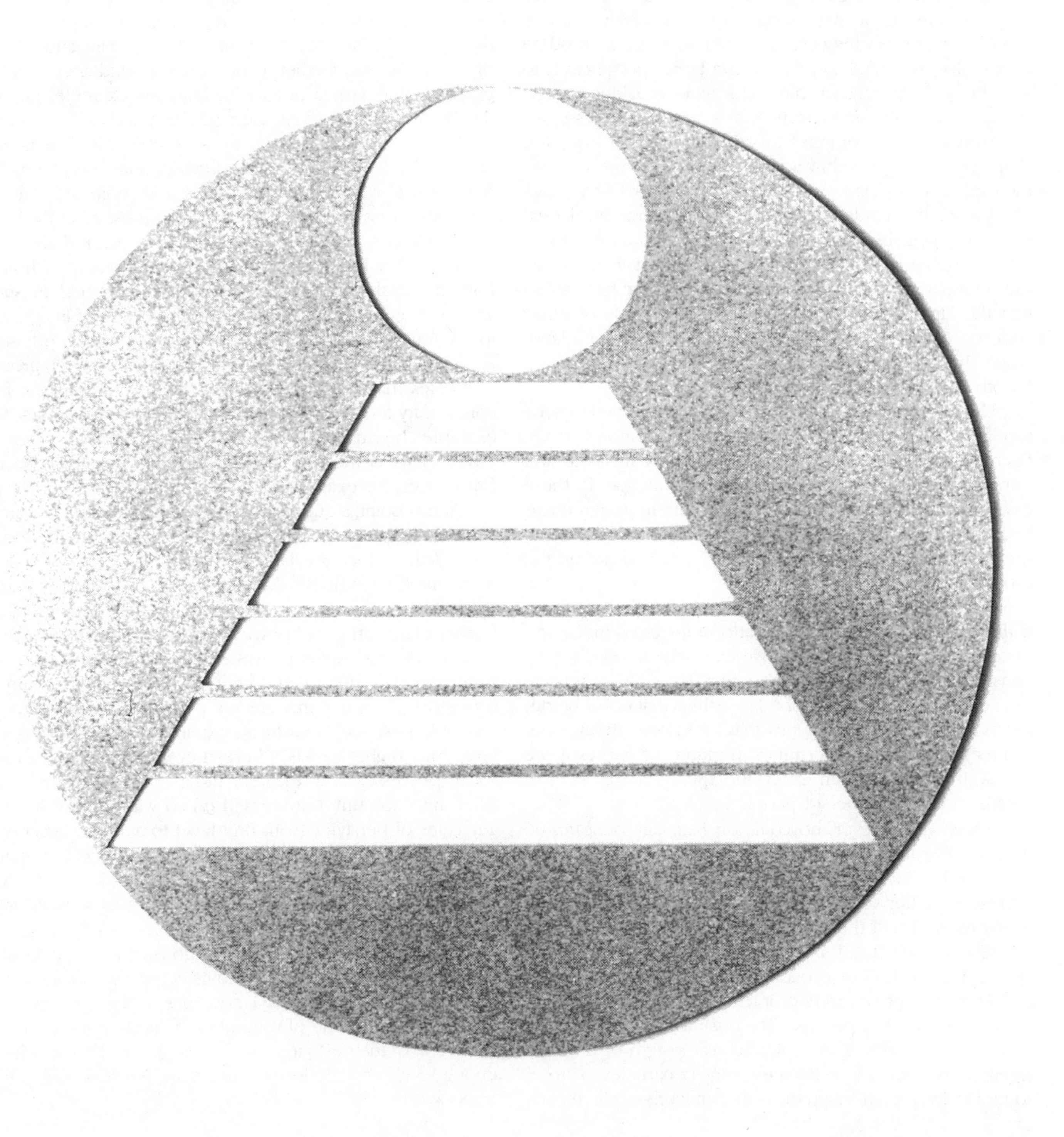

CHAPTER 2

A Hierarchy of Scale

Landscapes, like people, rarely stand alone. Every landscape is joined with all other landscapes in a network of interdependence that extends over the entire earth. Everything, as the saying goes, is indeed related, at some level, to everything else. So when we shape a landscape of any size, we need to place it in a larger perspective, to see the web of relationships and avoid breaking critical strands, and sometimes perhaps to create new ones.

How do we accomplish this? It is a matter of scale. We shape a landscape within a larger frame of reference, and the landscape we shape in turn becomes a frame of reference for the smaller landscapes within it in a hierarchical relationship. Inevitably, at some point we have to draw boundaries, to define our subject, to divide the indivisible. Ideally, the boundaries are determined by topographic features so that the landscape subject will have some sense of inner coherence and unity. A watershed is an example of a landscape that has coherence by at least one criterion, if not by others.

Usually, however, boundaries are determined not by topographic features, but by political and economic ones. Such boundaries are abstractions, ordinarily invisible and unrelated to the physical reality of the landscape. In these cases, the boundaries are hard to deal with in design terms because there is no natural reason why we should be shaping the land on one side of the imaginary line and not on the other.

Whether the boundaries are natural or not, but especially if they are not, a great many relationships cross indiscriminately over them, and these we can only account for by considering the encompassing context. The scale, or relative size, determines the types of relationships that occur within the boundaries and the types that cross over them. Consequently, the scale determines the kinds of concerns we can address in a design effort, the approach, the level of detail, and the number of people involved.

There is, however, no common language of scale, no system of scales to make these determinations easy and describable. We hear a great deal about "the large scale" and "the smaller scale," with no clear notion of what these terms mean. I will thus describe here a system of scales of concern to serve as a frame of reference for this book. This system may well have broader application.

The concept of levels of integration or organization is most helpful for this purpose. The organization of the natural world can be viewed as a hierarchical system of levels of integration that extends from elementary particles to atoms to molecules, then through the organism levels—cells, tissues, organs, organ systems, organisms—and the levels of groupings of organisms—individuals, populations, communities, ecosystems. Each level is composed of several interacting units of the next lower level, representing not the sum of these units but a new, different form of organization. Each of these can be further broken down, and they can be grouped into larger units. Novikoff, for example, groups the levels into four basic tiers: the physical, the chemical, the biological, and the social, in ascending order.[1] He points out that these represent stages of evolutionary development, the physical organization of atoms and molecules having developed first and social organization most recently. Each tier possesses unique properties of behavior that result from different forces of evolutionary development. Although human social organization is generally stratified in such ways and seems to evolve toward higher levels of integration as natural systems do, Novikoff cautions against too hastily applying rules observed in one group to another group. The rules that apply to the organization of animals at the community level may not apply to human communities, for example, because they evolved in response to different forces. This is the logical flaw in such theories as social Darwinism, for example.

At this point, a question arises: Should human ecosystems be considered as belonging to the biological or to the social tier? Clearly, they overlap both and thus point out a flaw in Novikoff's typology. However, since his caution against transferring rules from one level to another remains valid, further clarification is in order. I have already made a case for considering human ecosystems as belonging to the ecosystem level of the biological tier. Since they involve human processes of control that are not present in nature, we can probably best see them as an expansion of this biological level into higher levels where they interact with various social processes. These interactions obviously have to be taken into account, but we will do so without transferring any rules of behavior from one level to another. When we speak of human communities, for example, it is important to remember that we mean a very different kind of social organization from that of communities of wolves or waterfowl.

There are, however, some principles that apply to all levels, and particularly to the relationships among levels. Considering that a landscape and its ecosystem at each scale encompasses a number of landscapes at the next smaller scale and is itself one of a number of landscapes included in the unit of the next larger scale, the matter of relationships among levels of scale looms important. But how, precisely,

[1] See Novikoff, 1945.

do we define these relationships? James Feibleman's "theory of integrative laws" is useful here. The theory includes twelve "laws" and six "rules of explanation" to define the relationships among levels of any hierarchically integrative system or organization. Although some of these laws and rules have no obvious application to design and need not be mentioned here, at least four of them will have a familiar ring to anyone who has dealt with planning hierarchies. The first law, for example, states that "... each level organizes the level below it and one emergent quality."[2] Thus, each level looks to the next higher level for organizational guidance.

Laws 3, 4, and 5 form a set. Law 3 states that the higher level depends upon the lower, and Law 4 that the lower level is directed by the higher. According to Law 5 then, "... for any given organization at any level, its mechanism lies at the level below and its purpose at the level above."[3] In other words, goals are established at each level to be achieved at the level below, but the actual operations or mechanisms by which goal-seeking work is done lie at the

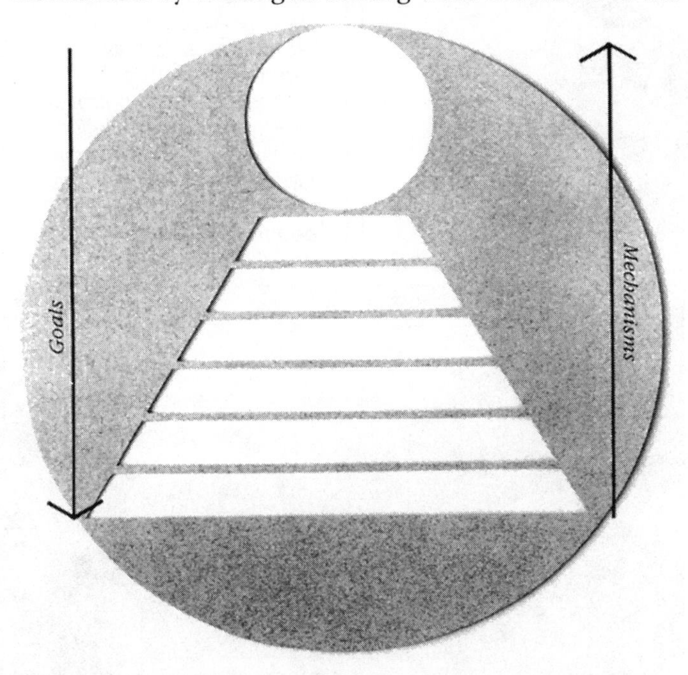

level below that. This is a fundamentally important rule for design, defining as it does the basic working relationship among levels of scale. To examine its meaning, we will explore a case study that involves four scales of concern.

WATER FLOW AS AN INTEGRATING ELEMENT

The most common and visible medium of connections in the landscape is the flow of water. The order of streams, rivers, and bodies of standing water in a landscape tells us a great deal about its structure. And especially in arid climates, the quantities of flow tell us even more about the populations the landscape can support.

In developing a theory of scales of concern, then, it will be useful to take a look at the flow of water through the landscape of a major urban region. We can see how the movement of water ties this region to the larger landscape of which it is a part and to a smaller landscape that is, in turn, an integral part of the region. We can also see how landscape design at these three scales is interlocked. We will also, incidentally, see an example of the rapid growth and ultimate difficulties associated with once-through ecosystems, especially those relying heavily on imported resources.

[2] See Feibleman, 1954; p. 63.
[3] Ibid., p. 64.

Water Flow and Land Use: The Southern California Example

The region that I will discuss here is metropolitan Southern California, a land area of almost 200 square miles that has been saturated by urbanization over the last hundred years, growing from a population of less than 20,000 in 1880 to over 10,000,000 in 1980. The illustrations in the Southern California case study on pages 26–30 show the evolution of land use since 1800, when the population was distributed in three small agricultural colonies, to 1975, when the region was virtually covered by urban development. The parallel column shows the corresponding growth and changes in the water budget.

Bear in mind that these representations were all created long after the fact. Along the path of development, there was little long-range planning and little recognition of the limits that resource supplies might place on growth.

Until 1903, the Southern California population met its water needs within its own watershed. Supplies were generally adequate, although at times the water table was seriously diminished by agricultural irrigation. As the map for the period shows, the amount of land in the region that was in human use before that time was quite small.

In 1903, following a prolonged drought, when the population was just over 100,000 people, the city fathers recognized that the natural watershed would not support a great deal more growth. Water was clearly the limiting factor. After a vigorous campaign for public support and a series of political maneuvers that need not concern us here, construction began on an aqueduct to bring water some 233 miles from the Owens Valley. The source of this water is the snowmelt in the Sierra Nevada. The aqueduct was completed in 1913, just in time for the population boom of the World War I years and the 1920s.

As the first flow diagram shows, by 1917 the aqueduct was furnishing more water than was being taken either from local rainfall or ground supplies. Diagrams of this type give us an easily understandable graphic technique for illustrating such facts. The arrows are vectors that represent quantities of input and output, but great precision is not required in order to get the message across. The data from which these hindsight water budgets were drawn for the early periods are, in fact, rather sketchy. The scale arrow shows the quantities represented by the arrow widths. The circle represents the Los Angeles Metropolitan Area, and the blocks inside show volumes of use for the three basic uses, with the width of the blocks at the same scale as the arrows.

The land-use map for 1925 shows considerable urban expansion since 1900. This period represents an early manifestation of Los Angeles' famous pattern of sprawl. We

THE FLOW OF WATER FROM THE SIERRA NEVADAS TO URBAN SOUTHERN CALIFORNIA

The phenomenal spread of urbanization in the Los Angeles region has been closely related to the periodic expansion of a gargantuan system of water supply that transports water from distant sources. The growth of this system is outlined here, from the time it first drew water from the Owens Valley 233 miles away, then from the Colorado River, which really means from the entire American West, and later from Northern California. The supply of water, which placed a natural limit on the population of the region, was continuously expanded by reaching ever farther into distant landscapes. The result has been uncontrolled growth, and a population that far exceeds the region's resources. The pattern of evolution in land use proceeded from wilderness to agriculture to urbanization.

CASE STUDY II

Land Use and Water Flow in the Southern California Urban Region

Research by the Laboratory for Experimental Design, Department of Landscape Architecture, California State Polytechnic University. Principal Investigator: John T. Lyle. Research Assistants: William Bensley, Peter Hummel, Julianna Riley, Gordon Taoka, and Raymond Walsh. A component of the Focus on L.A. project, funded by the Wells Fargo Foundation.

Southern California's functional watershed, the total area from which it draws water supplies, now covers about one-twelfth of the continental United States.

The historical narrative illustrates the shift that occurred after the first movement away from conservation and careful use of water, toward an attempt to satisfy ever-increasing demands.

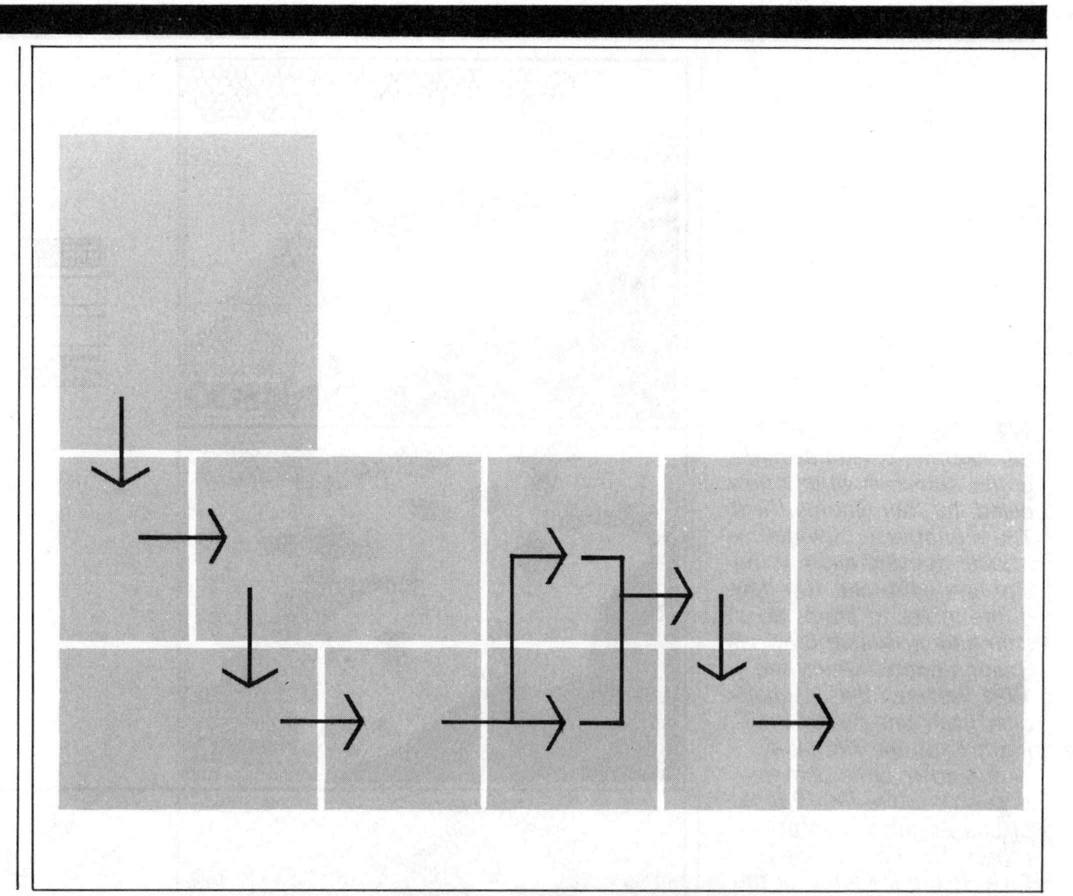

THE CHRONOLOGY OF URBAN GROWTH AND WATER USE IN SOUTHERN CALIFORNIA

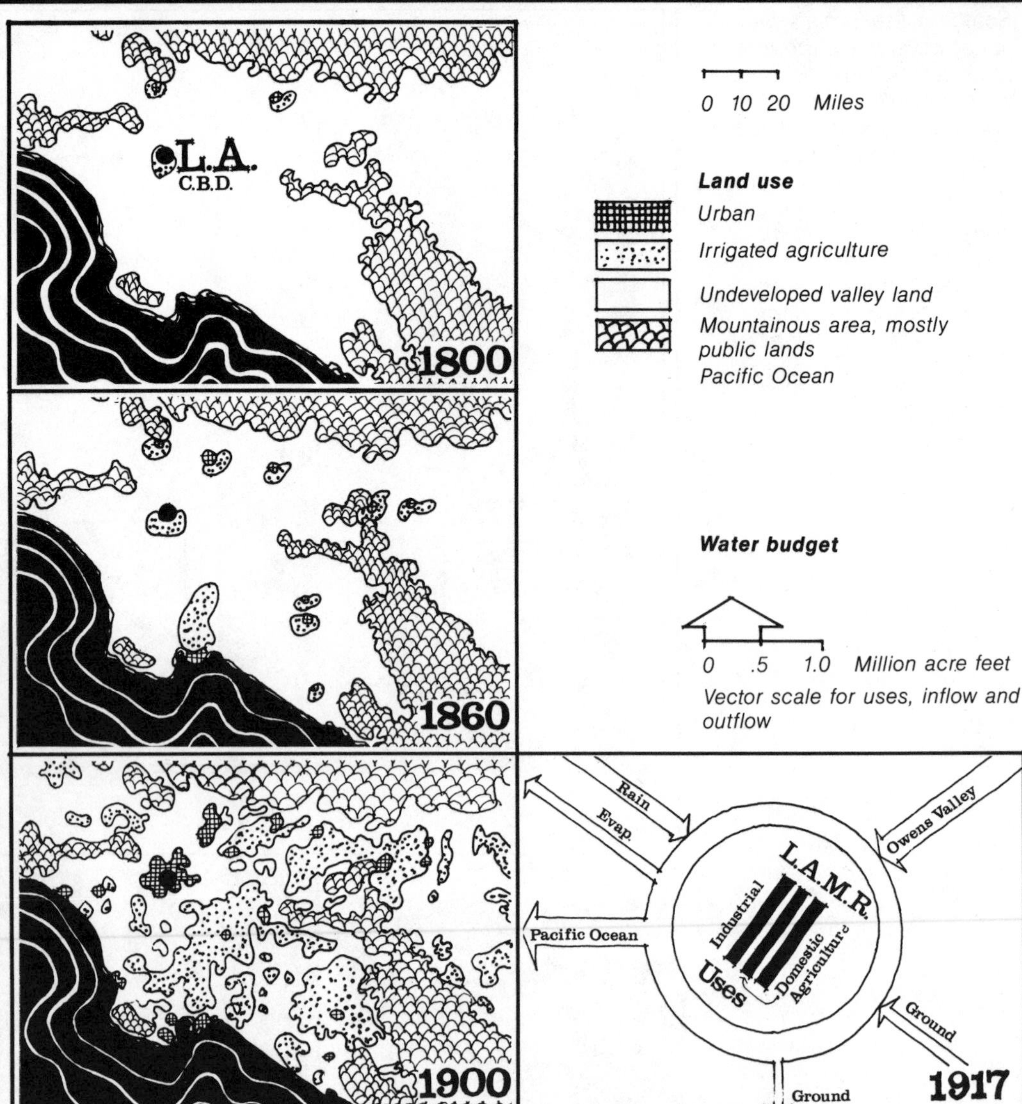

1771 *The Old Mission of San Gabriel is established on the banks of what is now called the San Gabriel River. The sparing use of water is a basic consideration in design and land use. The Law of the Indies, a direct descendant of Roman Law via Spain, controls water use.*
1781 *Between the Porcuincula River and the low rolling hills on the site lately occupied by an Indian village—Yang-na—the Pueblo of Los Angeles is established. The Los Angeles River is the life force of the settlement, and a royal proclamation is signed by King Charles III of Spain, granting the Pueblo ownership of all the river water for all time.*
1800 *Population of Pueblo: 315.*
1810 *San Fernando Mission establishes a system of irrigation. A "Suit at Law" is instituted by the Pueblo against the Padres, who were finally forced to tear down a dam and return to the City of Los Angeles its full rights to the Los Angeles River.*
1830 *Population of town and region: 1200.*
1833 *Secularization Act starts decline of Missions and first signs of a "land boom."*

1850 *The Los Angeles Zanja system is constructed. This is a water main system incorporating a large network of canals, a giant water wheel, a brick storage tank near the Plaza, and a system of hollowed logs that pipe the water to some of the major areas of the city.*

Los Angeles incorporates and passes first anti-water-pollution laws; no clothes to be washed or garbage thrown in the Zanja system.
1852 *Main system can no longer meet water demand; private companies form.*
1861–1862 *A deluge of floods hits the state. The entire valley area from Los Angeles to the ocean toward San Pedro Bay and Ballona becomes a great lake.*

1870 *Population of Los Angeles County: 15,309.*
1874 *A San Fernando ranch of 56,000 acres is subdivided by Senator Charles Maclay, who persuades the Southern Pacific to build a depot. This was a prototype method of development for outlying areas that later became incorporated into Los Angeles.*
1877 *Drought ruins crops. Water resources become a prime concern.*
1885 *Water district expands its irrigation system into higher lands to serve an additional 10,000 acres.*
1900 *Population of Los Angeles: 102,479.*

1895–1905 *Drought years bring about a critical lack of water. William Mulholland, "Father of L.A. municipal water system," is superintendent of the system. "Whoever brings the water, brings the people," he says.*
1905 *L.A. citizens vote 1.5-million-dollar bond issue to purchase Owens Valley land and water rights, setting stage for the Owens Valley Project.*
1913 *On November 5, 233-mile-long Owens Valley Aqueduct is completed. For first time, water from the High Sierra arrives in Southern California.*
1914 *Expansion of agriculture in the San Fernando Valley, with 100,000 acres added as a direct result of Owens Valley water.*

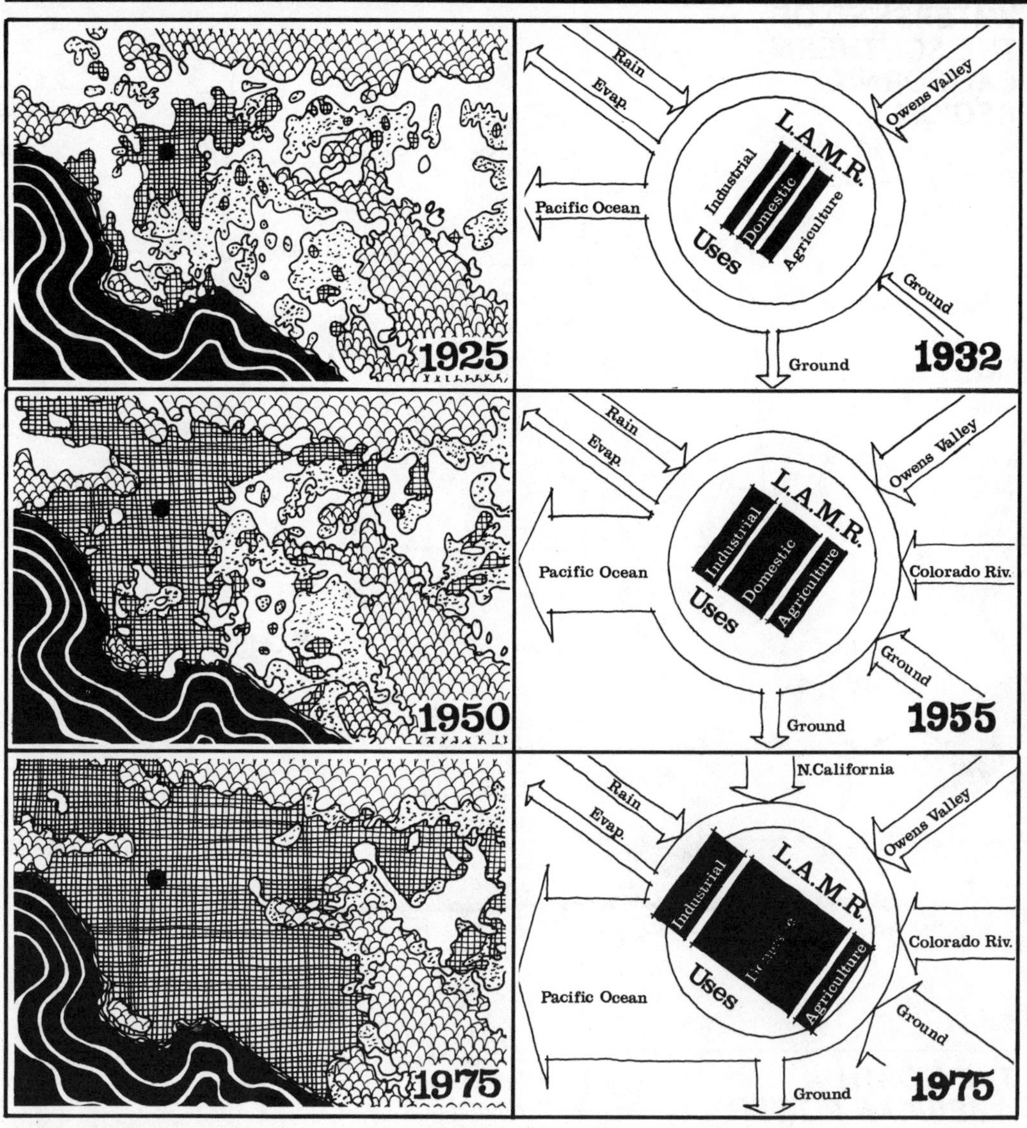

1916–1934 *Drought years, the worst being 1928–1934.*
1924 *Water war with ranchers in Owens Valley.*
1928 *Metropolitan Water District of Southern California formed, a public corporation comprising 13 cities—Burbank, Beverly Hills, Glendale, Pasadena, San Marino, Santa Monica, Los Angeles, Anaheim, Santa Ana, Torrance, Compton, Long Beach, and Fullerton. (The last named four did not join until 1931.) Boulder Canyon Act passed, setting the stage for eventual import of Colorado River water.*
1930 *Population of Los Angeles: 1,238,048.*

1931 *State Engineer Edward Hyatt sends report to legislature on the first "State Water Plan." This cataloged California's water resources and land, its flood control needs, and its irrigation potential. Its aim was to distribute surplus water from the North to areas of need in the central and southern parts of the state.*
1932 *Construction starts on Boulder Dam and the Colorado River Aqueduct.*

1941 *Colorado River Aqueduct is completed: 266 miles long, with 93 miles of tunnels that deliver 750 second-feet of water and supply 14 constituent areas with a population of over 2,000,000.*
1944 *Southern California great drought begins and continues through the sixties.*
1950 *Los Angeles City population: 1,970,358.*
1951 *A. D. Edmonston, State Engineer, presents first complete report on Feather River Project, with power plants, multipurpose dam, and reservoir at Oroville.*

1952 *Arizona vs. California filed before U.S. Supreme Court over water rights.*
1955 *Legislature appropriates funds to begin construction of the State Water Project.*
1957 *Construction begins on Oroville Dam with emergency funds. The first step in state water project is under way.*
1960 *New technology employed in design of pumps to bring water up and over the Tehachapi Mountains at high energy cost.*
1964 *Supreme Court decision (Arizona vs. California) increases Arizona's entitlement to Colorado River water while reducing that of California to less than half its current share.*
1964–1969 *Second barrel of L.A. Aqueduct from Owens Valley constructed; delivery begins in 1969 with 309,000 acre-ft.*
1980 *Energy cost, $.005/kw hr. 3400 kw hrs required to pump 1 acre-ft. Energy cost/acre-ft = $17 + $176 capital expense/acre-ft, for a total of $193/acre-ft. MWD charges, $100/acre ft, wholesale. Average family of five uses 1-acre ft/year.*

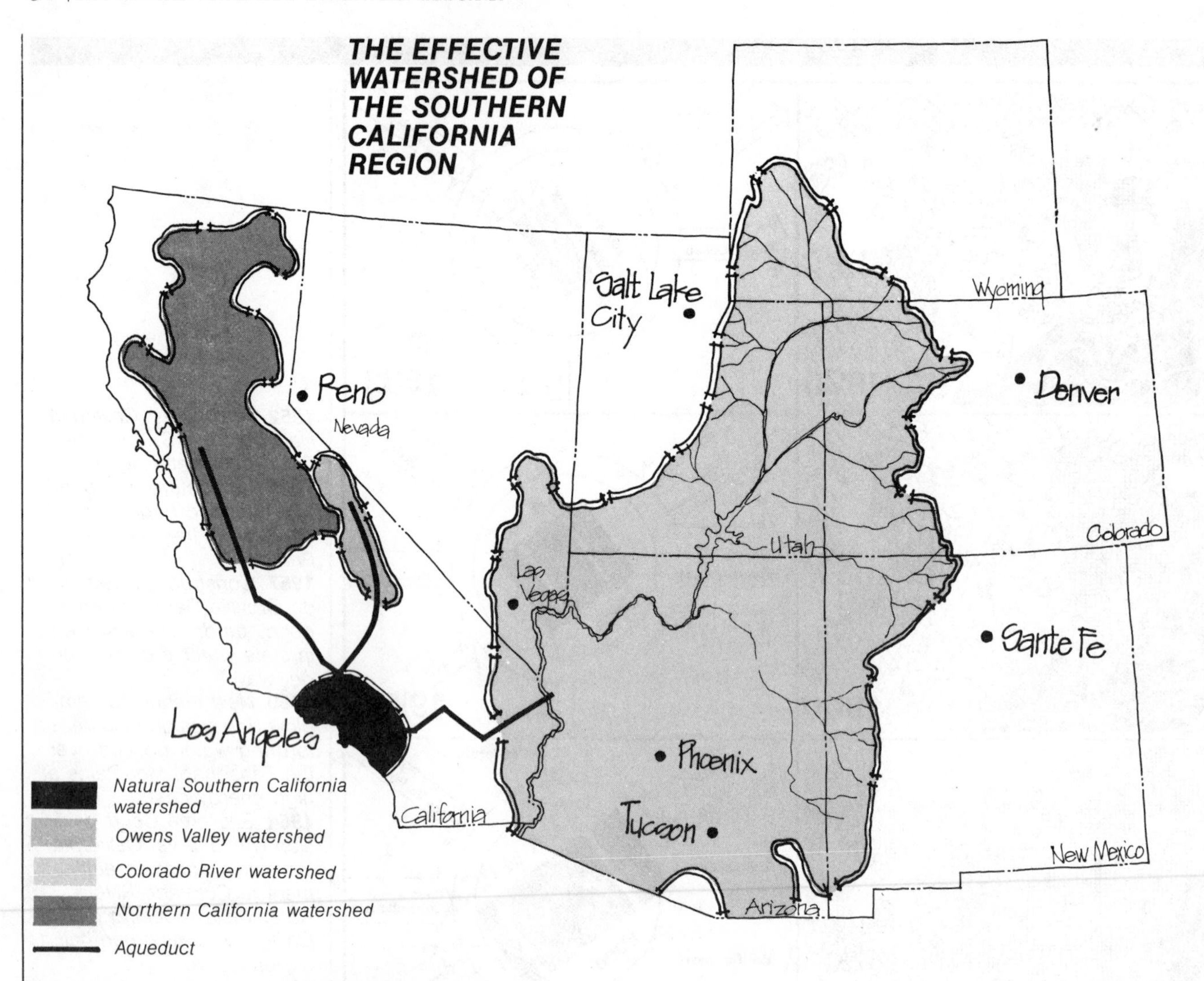

Water is now supplied to Southern California from three distant watersheds through three different aqueducts having a total length of over 800 mi. Competing uses and ecological concerns in all three watersheds are increasingly threatening to limit supplies, perhaps severely. Obviously, there has to be an end to the expansion, but proposals with this aim in mind continue to appear. Probably the most ambitious is the North American Water and Power Alliance, illustrated here. This plan proposes transporting water from the Yukon to help supply various areas of Canada, the United States, and Mexico, including Southern California in particular. The cost has been estimated at over 200 billion dollars. An alternative would be to design for more efficient use and reuse of water.

probably cannot say that the aqueduct caused it, but we certainly can say that without the aqueduct, such growth would not have been possible.

By 1923, another dry spell convinced the city that, if population projections were not to be foiled, another source of water was needed. This time, the desire for more water coincided with the desire for more of another essential resource—energy. The city therefore joined the U.S. Bureau of Reclamation in supporting the construction of a high dam to supply both, at Boulder Canyon on the Colorado River. In 1941, again just in time for a wartime boom, a new 240-mile-long aqueduct began delivering water from the Colorado River to the Los Angeles Metropolitan Region.

The water budget diagrams and the land-use maps tell a tale of enormous increases in water imports during this period, accompanied by rampant urbanization. Agriculture was largely replaced by suburbs in the region during these years.

The next source of water to be tapped was Northern California, which has two-thirds of the state's water but only one-third of its population. The California Water Project, which carries water over four hundred miles from several reservoirs in Northern California to the Los Angeles region was conceived during the boom of the 1950s and completed in 1973. One more small addition to the system was made simultaneously; the Owens Valley aqueduct was extended over 100 miles to the north to draw from the adjacent Mono Lake watershed.

The water budget diagram and the land-use map for 1975 show the picture essentially as it exists today. The repeated expansion of the water supply system has made it possible for urbanization to cover virtually all the buildable land—not to mention a large amount of land that many would consider not buildable—within the Los Angeles Metropolitan region. This growth occurred without any real understanding of the relationship between water and land use, with no consideration of alternatives, and with no thought about the possibilities of conservation or recycling. Had water budgets been developed along with alternative land-use plans before decisions were made in favor of any of these massive water import projects, the Los Angeles region might be a very different place today. The regions from which the water is drawn might be quite different as well.

In all of these regions, major problems have resulted from the export of water. Thousands of acres of once productive farmland in the Owens Valley are now abandoned because they lack water for crops. The water war with Owens Valley residents has gone on intermittently since 1903. A recent salvo was the passing of a referendum to give people of the valley the right to limit pumping of ground water to Los Angeles, a salvo, however, that was later quieted by an adverse court decision. To the north of the Owens Valley, moreover, the water level of Mono Lake has dropped over 20 feet since Los Angeles began taking water from its tributary streams, causing considerable depletion of wildlife populations. Environmental groups are fighting in the courts to limit these exports, and they may succeed in stopping this practice.

In the last 40 years, a great deal of competition has developed for Colorado River water as well. In 1964, after a 12-year legal battle, the U.S. Supreme Court ruled in favor of Arizona's claim to increased entitlement to water from the river, a decision that will reduce California's share by over half in 1985. It is entirely possible that claims by Indian tribes in the upper watershed as well as by agricultural interests may bring further reductions.

Finally, in Northern California, environmental groups are fighting to limit the exports of water. Several dam projects have been halted, and it is very doubtful now that the last increment of the delivery system, a canal to transport water through the agriculturally rich Sacramento Delta, will ever be built. A referendum to support it was rejected by the state's voters.

The most telling issue, however, might be not environmental degradation but energy. The giant pumps that lift the water over the mountains to drop into the Los Angeles Basin use enormous quantities of fuel oil. According to one estimate, the power that drives the pumps for water from the Owens Valley could supply the electrical needs of a city of 100,000 people. Thus energy and water supplies are closely interrelated. And we well know that energy will run short and that its price is rising. Since the cost of fuel oil for these particular pumps is rising fourfold in 1984, the cost of water will rise too, probably to a level at which it can no longer be carelessly wasted.

Although the Los Angeles example is probably extreme, it is not unique. Similar stories could be told about most other cities, which reach farther and farther for the resources to support their growth, whatever the cost to the source landscapes. New York, for example, began importing water over 150 years ago from the Croton watershed, later built reservoirs in the Catskills and Adirondacks, and more recently expanded its supply system to the Delaware River. Almost everywhere, water is a major regional issue, one that usually tends to sprawl over several regions. In fact, the area from which Los Angeles draws water—its functional watershed—covers over one-twelfth of the continental United States. Here regions are meeting, overlapping, and clashing. Similar patterns emerge for other resources, most dramatically, energy. The energyshed of any sizable city straddles the globe.

The map on page 30 shows the natural watershed of the Southern California urban region in relation to the vastly expanded watershed from which it imports water. For almost a century and until just a few years ago, the region imposed its goals on the larger level of integration, which we will call the subcontinental level—a political rather than a natural phenomenon that is clearly in defiance of Feibleman's law. Now, however, the situation seems to be reversing itself. Probably as a result of increasing populations and augmented political influence in the areas from which water is imported, the subcontinental level of integration is beginning to impose its own goals, which include the preservation of wild rivers and wildlife habitats and the development of its own agricultural and industrial potentials. The Southern California region will almost certainly be forced to accept these goals, which dictate a policy of water conservation, and to look to the next smaller level of integration for the operational means of putting it into practice.

Aliso Creek: A Case Study in Descending Scales

This next smaller level is the plan unit, an area of land for which general land uses with definite boundaries can be determined and enforced. We will look now at a particular plan unit and examine some mechanisms that can be used there in working toward regional goals. We will also see how more precise goals are in turn established at this scale for the lower levels. The unit includes 5400 gently rolling, mostly treeless acres in a rapidly urbanizing area of Orange County at the edge of the Southern California Metropolitan Region. It was chosen as a representative portion of the Aliso Creek watershed—which consists of mostly undeveloped land that drains into a minor, intermittent creek that in turn flows to the Pacific Ocean—in order to give some indication of design considerations for other sections of that watershed.

A principal mechanism applied here for working toward a major regional goal—reducing the amount of water imported—was the water budget. This most useful tool for landscape design has been little used or developed until now. It will probably be much more widely used in the future as the importance of water and the need to allocate its use more efficiently becomes clearer. Similar budgeting techniques can be used in allocating other scarce resources.

The five alternative land-use plans that were developed for the Aliso Creek plan unit range from keeping the land in grazing with just one small commercial center to fairly intensive suburbanization with an average of about six people per acre. The land-use comparison chart shows the areas in various uses for each alternative. The corresponding plans for three of the alternatives are shown. Planning decisions are rarely made on the basis of just one factor; in this case, the visual character is an extremely important concern.

All five of the alternative land-use patterns considered for this area can function on less than 10,000 acre-feet of imported water per year, but they distribute its use in very different ways. The first, Option I, would leave most of the area in cattle grazing with 275 acres devoted to a regional park and 100 acres to an office complex. Option II would put 2539 acres into agricultural use, primarily field crops, and leave most of the rest in open space but with 696 acres set aside for low-to-medium-density (one to six units per acre) residential development. This yields a population of about 3427 people, or roughly 0.7 people per acre for the total area. Option III would maintain 3782 acres in open space and develop the rest in medium-density residential use, giving a population of about 14,819. Option IV would permit a much higher overall density and at the same time reserve somewhat less open space, 2549 acres. This yields the largest population of all the options, about 32,000 people. This is essentially the same as the plan originally developed by Orange County planners for the area, and it provides a good example of the common planning practice of establishing densities without considering the resource base. Finally, Option V is also a higher-density scheme but with more open space, 3241 acres, and a consequently smaller population of about 23,000 people. There figures are summarized in the Land Use Comparison Chart, and the distribution of land uses for three of the options (numbers I, IV, and V) is illustrated in the map sketches.

All of the schemes provide for some measure of water reuse for irrigation purposes. Option V is designed for the maximum intensity of water recycling that is now technically and economically feasible using inexpensive, technically simple, and well-proven biological treatment techniques. This is one reason for the larger amount of open space in the plan. Efficient water use requires large land areas both for the collection of water and for its storage, movement, and percolation into the ground supply.

There are any number of possible formats for a water budget, depending on the information being sought. At this scale, in contrast to the rough graphic budgets shown for the region above, more accurate quantities are useful. In this situation, a chart is a more appropriate, if far less graphic, format. Considering the limitations on water imports that are likely to come about in the near future for the reasons previously mentioned, the key number for each option is the quantity of imported water needed. For this reason, the format emphasizes supply and demand. The first three columns, then, give the quantities that can probably be provided by the three immediately available sources: runoff, ground water, and reclaimed water. In this case, ground water was not considered because no information was available concerning quantity or quality of the ground supply. Amounts taken out will equal amounts put in. The quantity and quality of runoff, of course, will vary according to the type and amount of development. The reclaimed water is secondarily treated sewage effluent available from an existing treatment plant in the area. The sewage treated in this plant is actually taken in from another watershed. With present technology and health restrictions, treating this water to drinking-quality standards would be prohibitively expensive. However, with minor advanced treatment, it can be used for landscape irrigation, for filling recreational lakes, for enriching wildlife habitats, and possibly for charging and stabilizing the water level of the intermittent creek within the area. The water budget includes the 1460 acre-feet per year that would be needed to accomplish this for Options III and V.

It may be useful to explain further some of the other numbers in the chart. Perhaps most surprising is the quantity budgeted for irrigation. In the domestic-use column, irrigation and toilet flushing are combined because the available data make it difficult to separate them, but we can generally assume that irrigation, the watering of the residential landscape in this area, typically takes about half of domestic consumption. The amounts for irrigating public parks and golf courses are correspondingly high. Obviously, they can be much reduced by careful plant selection, especially by using more drought-tolerant species and less lawn (lawns are major water consumers). All of the alternatives assume careful selection of plants to balance their contributions to the landscape with their water consumption. Furthermore, if the water-cycling system is developed as recommended, the budget indicates that most of the irrigation need, at

CASE STUDY III

Aliso Creek Development

Urbanization is imminent and inevitable in the Aliso Creek watershed. If development follows the present trends of the area—single-family houses with once-through water systems and surrounded by thirsty lawns and exotic plants—the region's water supply system as described in Case Study II will be further strained. Fortunately, there are alternatives.

Prepared by the 606 Studio; Department of Landscape Architecture, California State Polytechnic University, Pomona. Sandy Vance, Bruce Gay, and George Hanson, with advisers John T. Lyle, Jeffrey K. Olson, and Arthur Jokela.

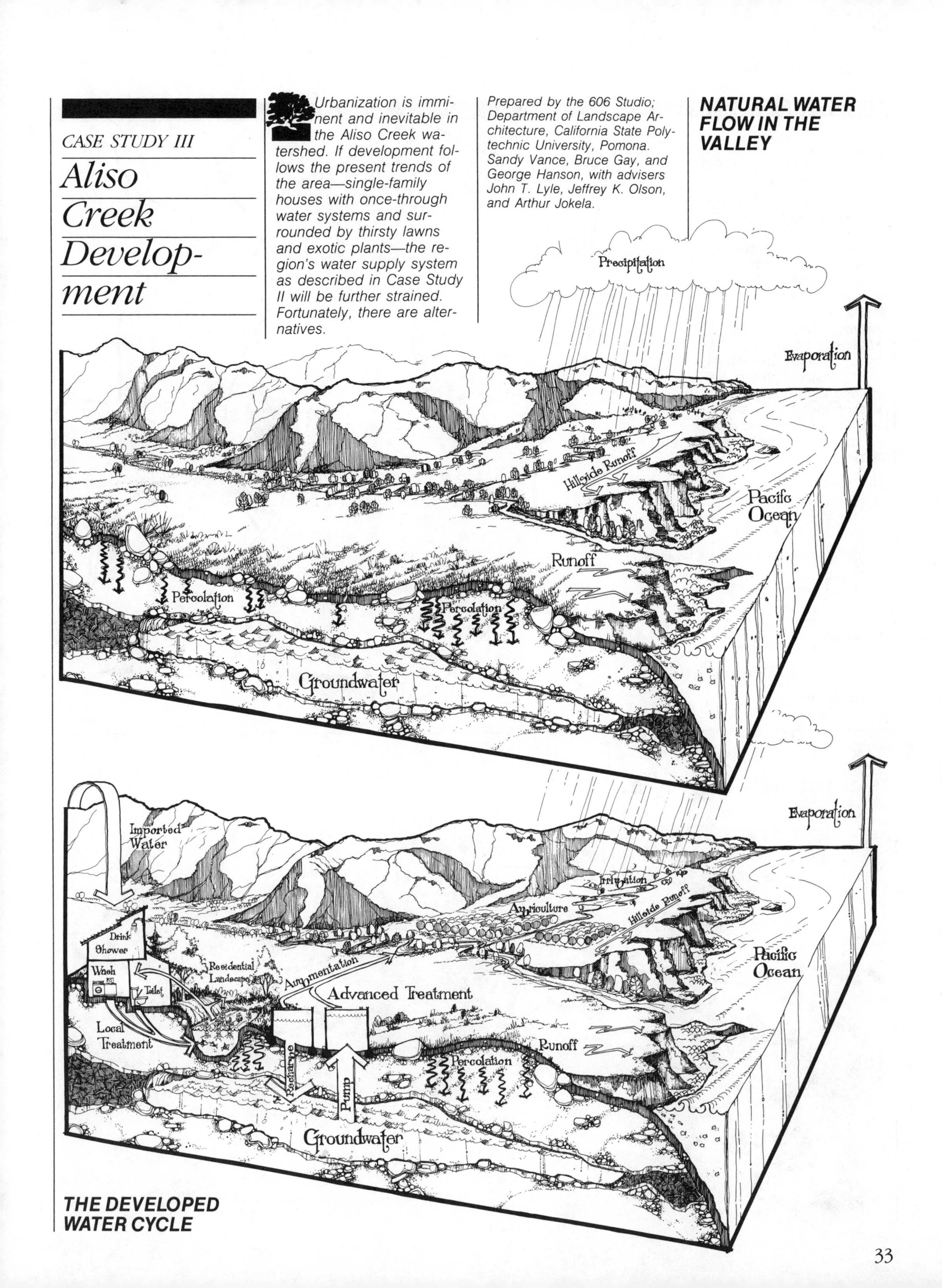

POLICY I: NO GROWTH

Five possible patterns of development and water use are explored here and compared with respect to the numbers of people they would support and the amounts of water they would require. Three of these are shown in more detail.

By designing the system of water flow for the highest economical level of reuse and recycling, using the processing capacities of the landscape itself, per capita water consumption can be reduced to less than half its present level. Doing so will also result in a richer, more diverse, more sustainable landscape.

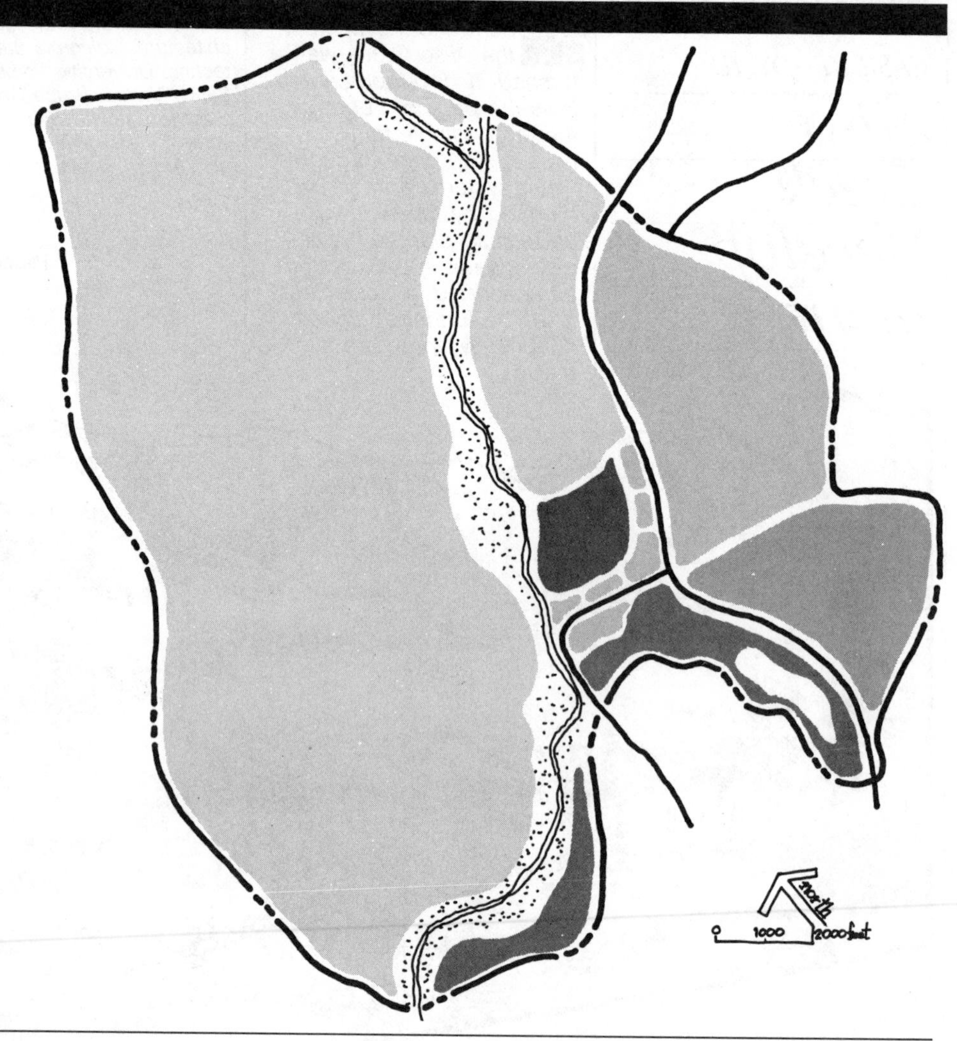

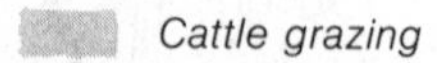
Cattle grazing
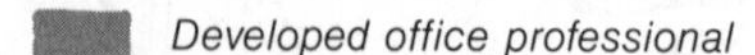
Developed office professional

Creek flood plain
Residential area under construction
Regional park
Undeveloped open space
Secondary highway

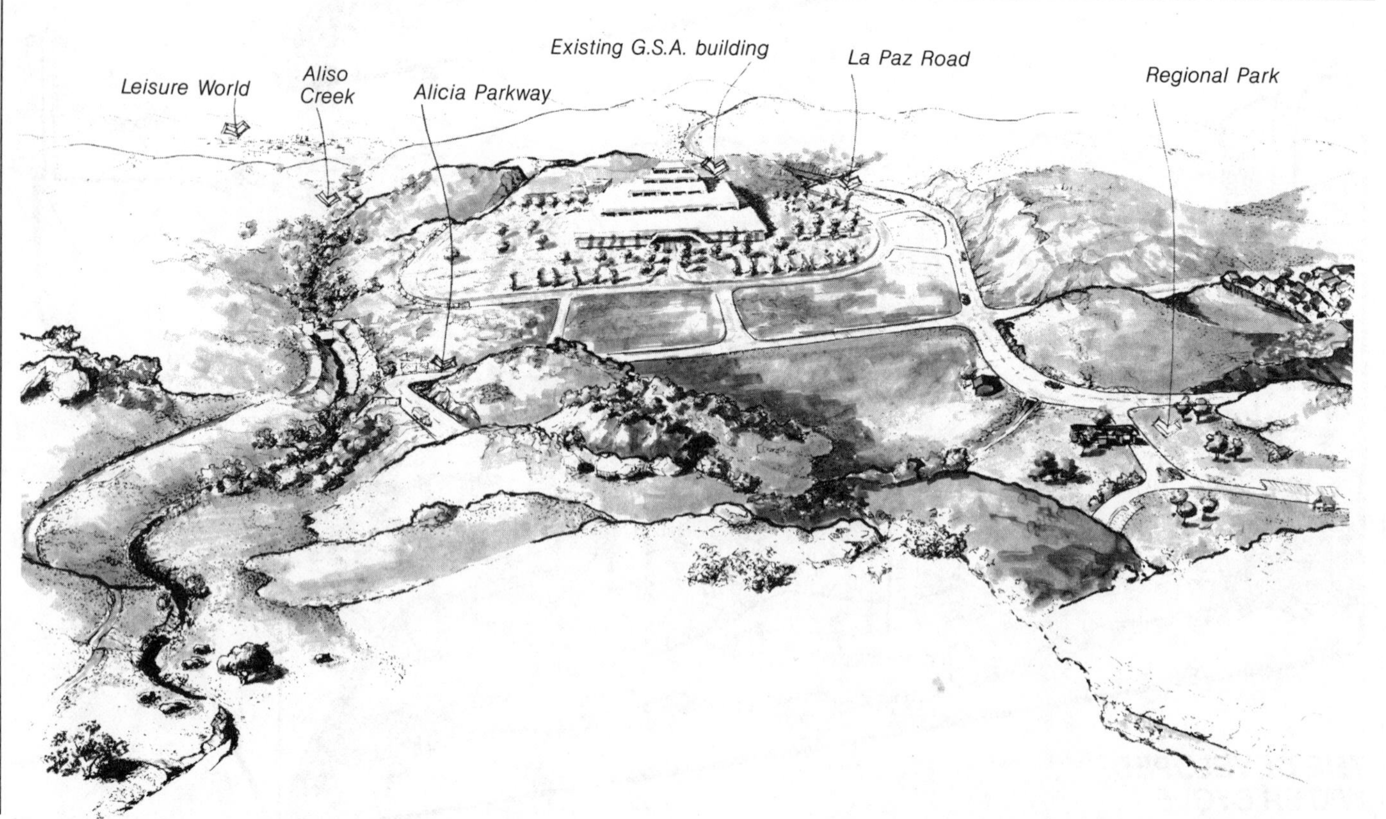

POLICY IV: GROWTH AS PROJECTED

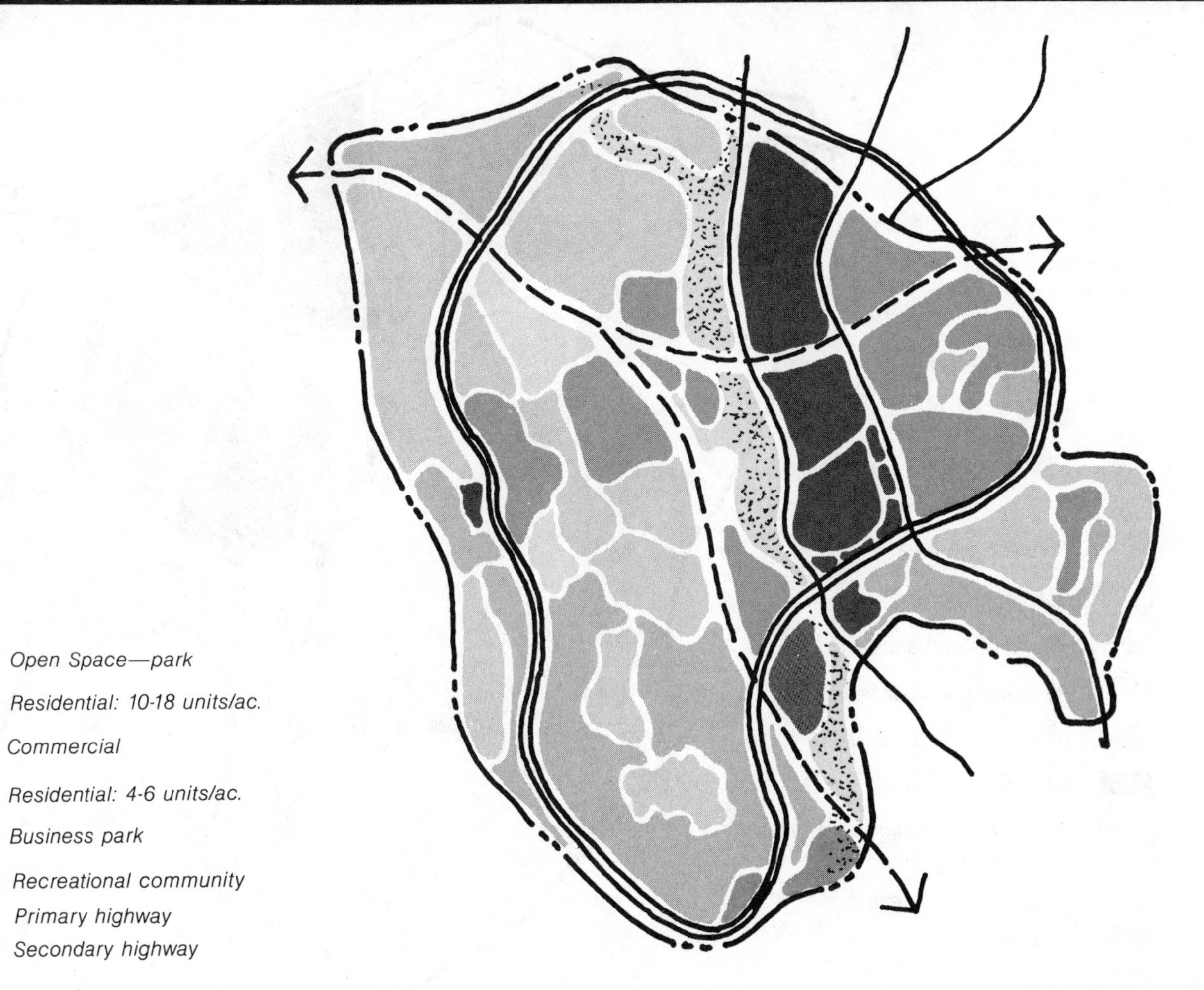

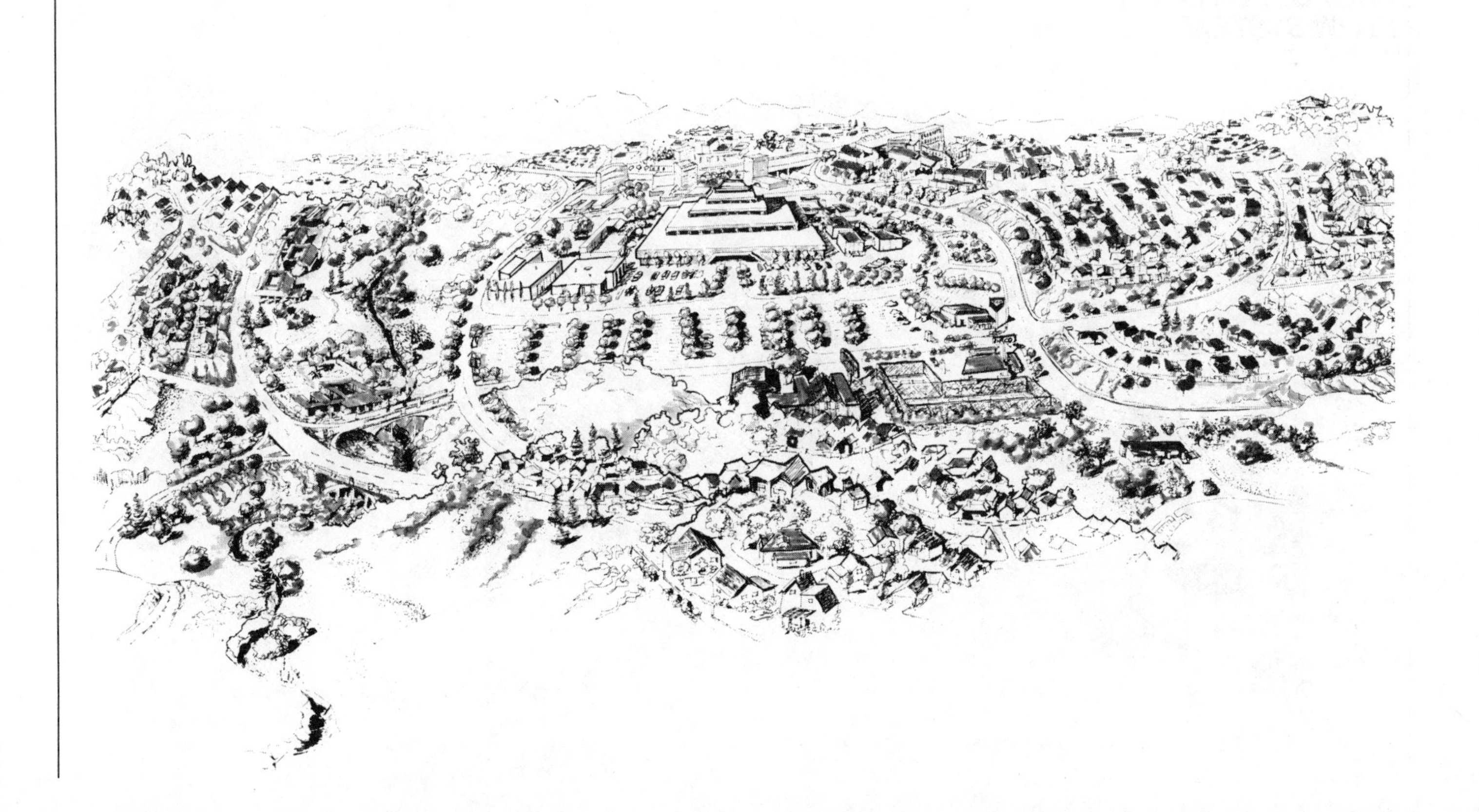

POLICY V: GROWTH BASED ON EFFICIENT WATER USE

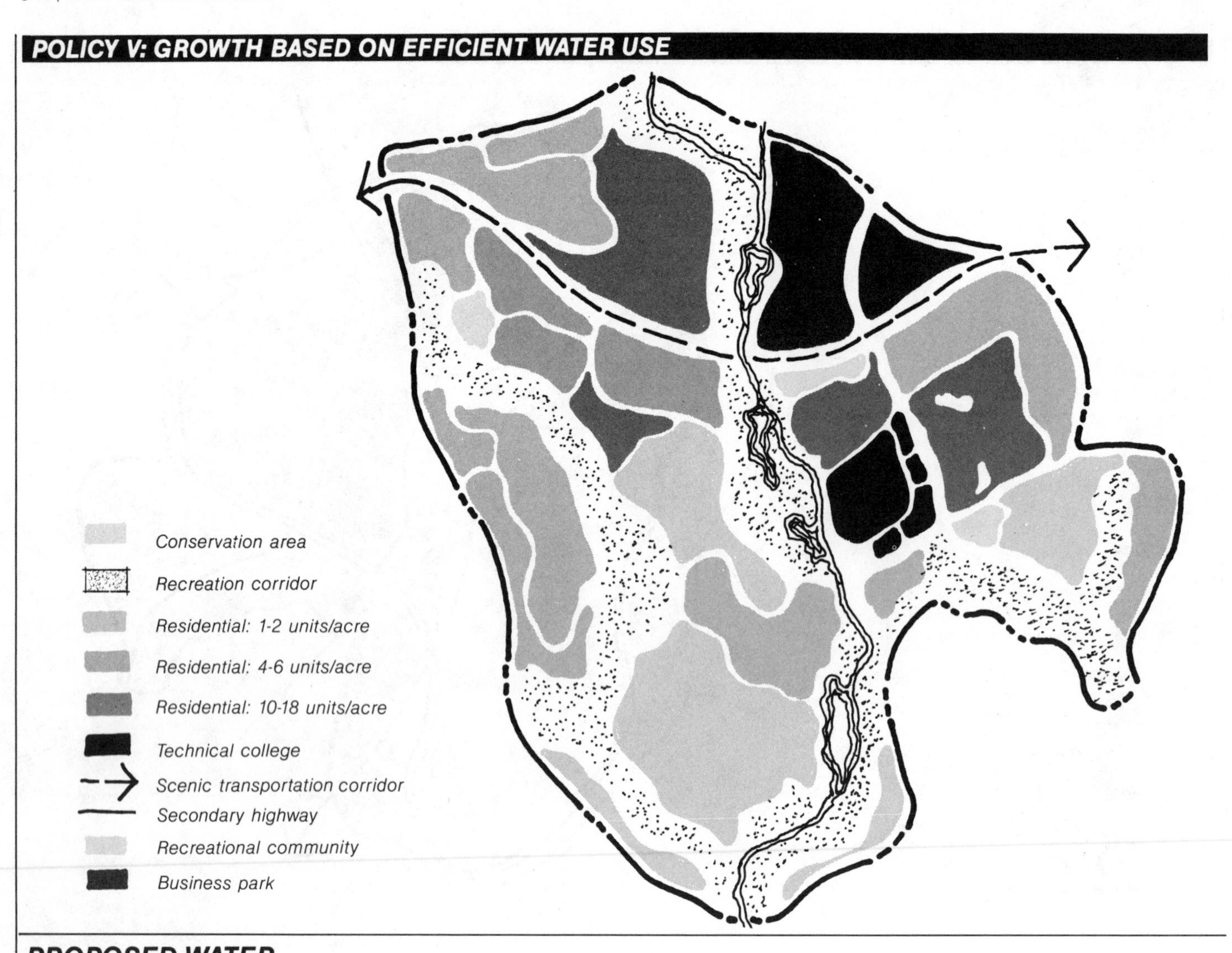

PROPOSED WATER FLOW SYSTEM

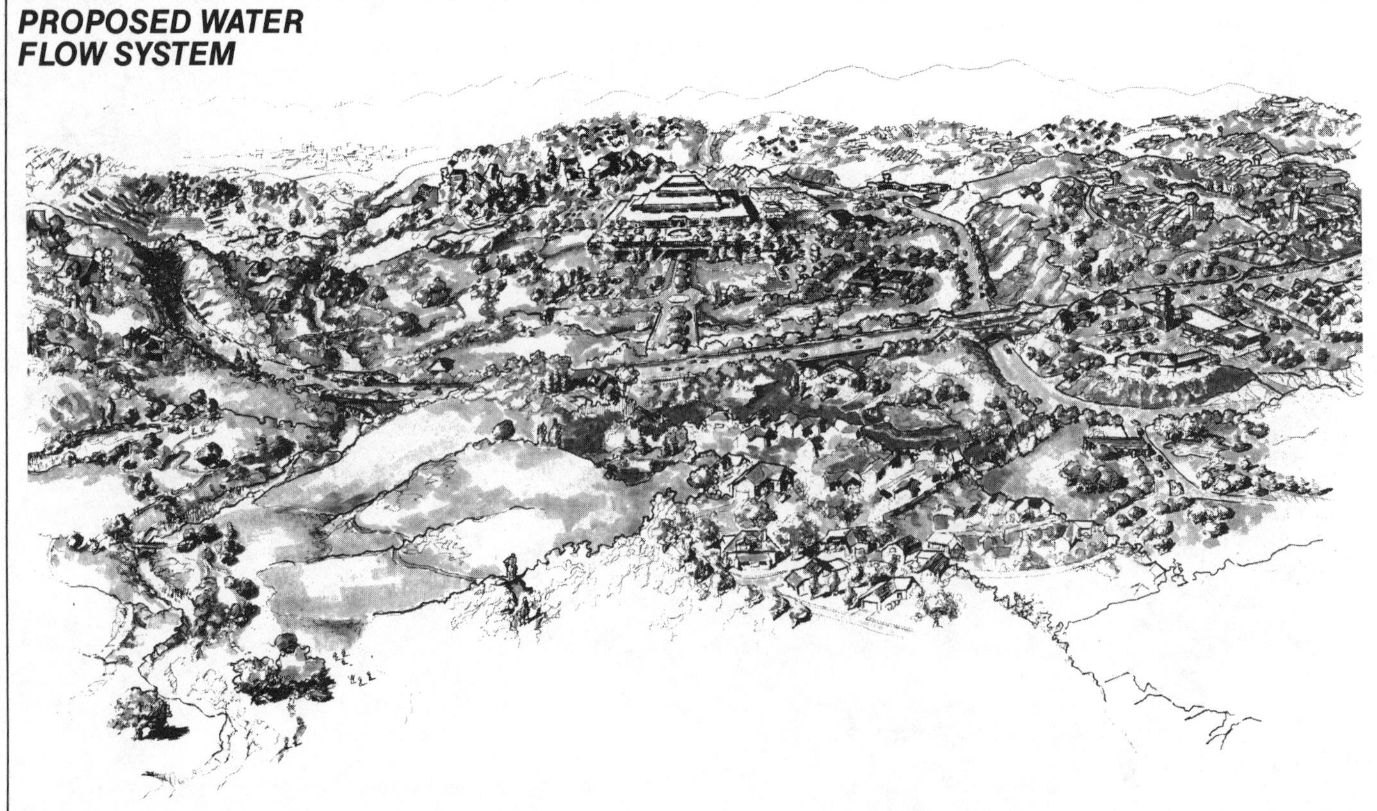

COMPARATIVE LAND USE ANALYSIS

			Policy I no growth	Policy II balanced growth; heavy agriculture	Policy III balanced growth; heavy recreation	Policy IV present county plan	Policy V optimum water management
Gross acreage			5403	5403	5403	5403	5403
Gross density (units/acre)			0	2.4	4.3	5.8	5.4
Net density (units/acre)			0	2.4	5.2	6.8	5.9
Open space acreage			275	2168	3782	2549	3241
Developed acreage			120	696	1621	2854	2162
Dwelling units			0	1071	4631	10,006	7318
Density	Low	1.0 - 2.0	0	96	695	1500	1098
	Med. low	2.1 - 3.5	0	459	926	2004	1464
	Med.	3.6 - 6.5	0	516	1853	4002	2927
	Med. high	6.6 - 10.0	0	0	787	1500	1244
	High	11.0 - 18.0	0	0	370	1000	585
Population			0	3427	14,819	32,020	23,420
Land Use - Acres	Residential		0	440	1081	1728	1351
	Business park		100	56	135	873	481
	Retail/commercial		0	5	22	27	14
	Recreational/comm.		0	11	103	48	90
	Public roads		20	60	194	264	65
	Schools		0	27	86	128	158
Recreations			275	546	1702	334	616
Conservation			0	1622	1324	972	1523
Open space (passive)			0	0	756	1243	1102
Technical college			0	97	97	97	97
Ranching			4231	0	0	0	0
Agricultural			0	2539	0	0	0

COMPARATIVE WATER BUDGET

Supply:

Given condition	Natural run-off	Ground water	Reclaimed water	**Totals**	Differential
Policy I	648		0	**648**	470
Policy II	850		384	**1234**	8185
Policy III	810		1654	**2469**	8781
Policy IV	1659		3857	**5516**	7813
Policy V	983		2623	**3606**	3895

Demand (Residential: Irrigation, toilet – Industrial; Ecosystem: Ground-water recharge – Agricultural; Recreation: Recreational lakes – Miscellaneous):

Totals	Irrigation, toilet	Washing	Drinking, cooking	Commercial	Tree growers	Industrial (including hospitals)	Ground-water recharge	Vegetation	Evapo-transpiration	Creek Recharge	Agricultural	Recreational lakes	Parks, golf courses, etc.	Public services	Miscellaneous	Given condition
1118	0	0	0	0	0	0	0	100	11564	0	0	200	918	0	0	**Policy I**
9419	391	175	12	56	0	415	0	94	3731	0	6217	200	1824	6	29	**Policy II**
11,250	1694	747	50	430	3	826	0	33	4784	1460	0	350	5283	249	125	**Policy III**
13,329	3658	1614	108	252	6	5517	0	51	5095	0	0	200	1116	538	269	**Policy IV**
12,206 **7501**	2677	1181	79	392	4	3234	0	31	6038	1460	0	500	2057	394	197	**Policy V**

A TYPICAL PROJECT SCALE PLAN

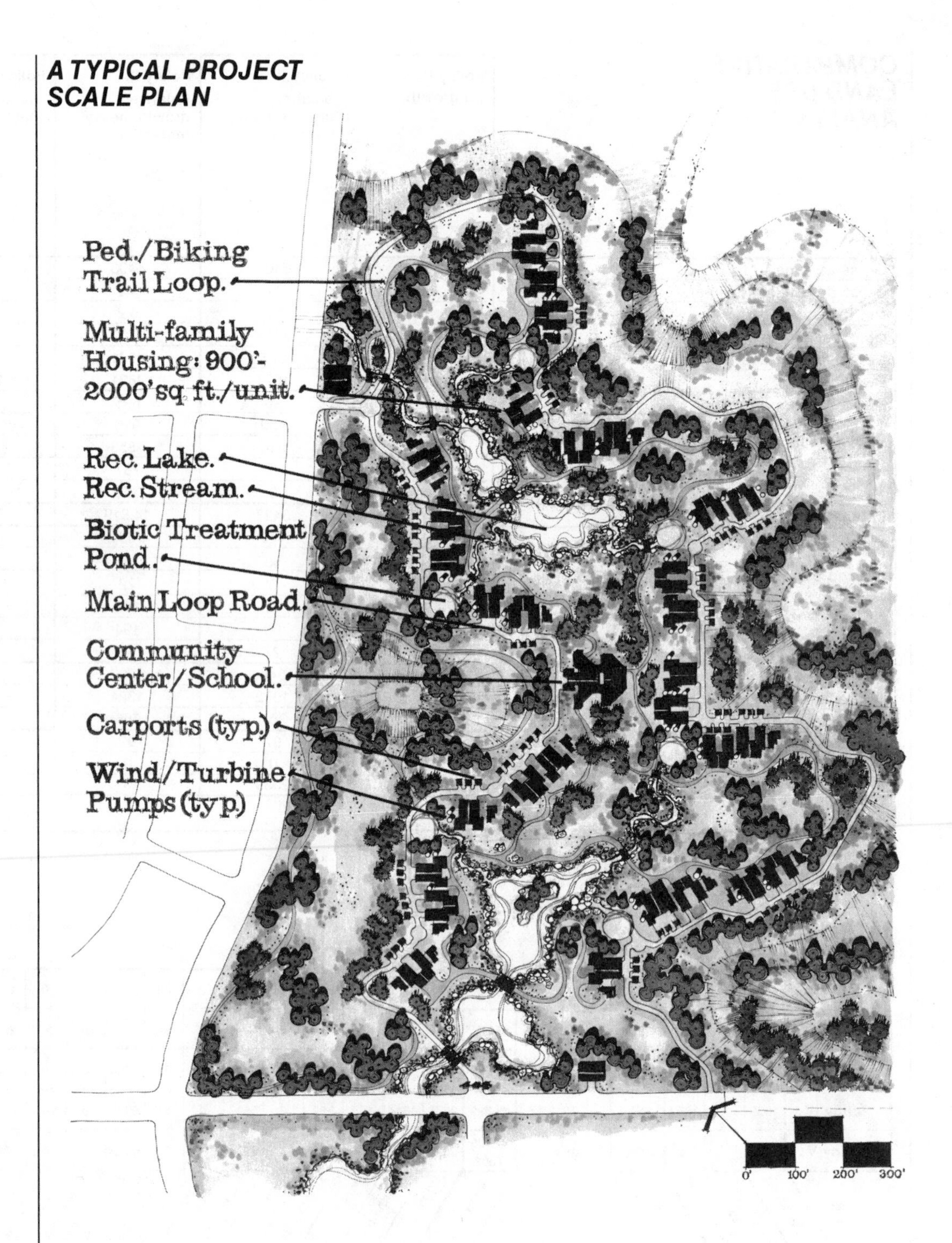

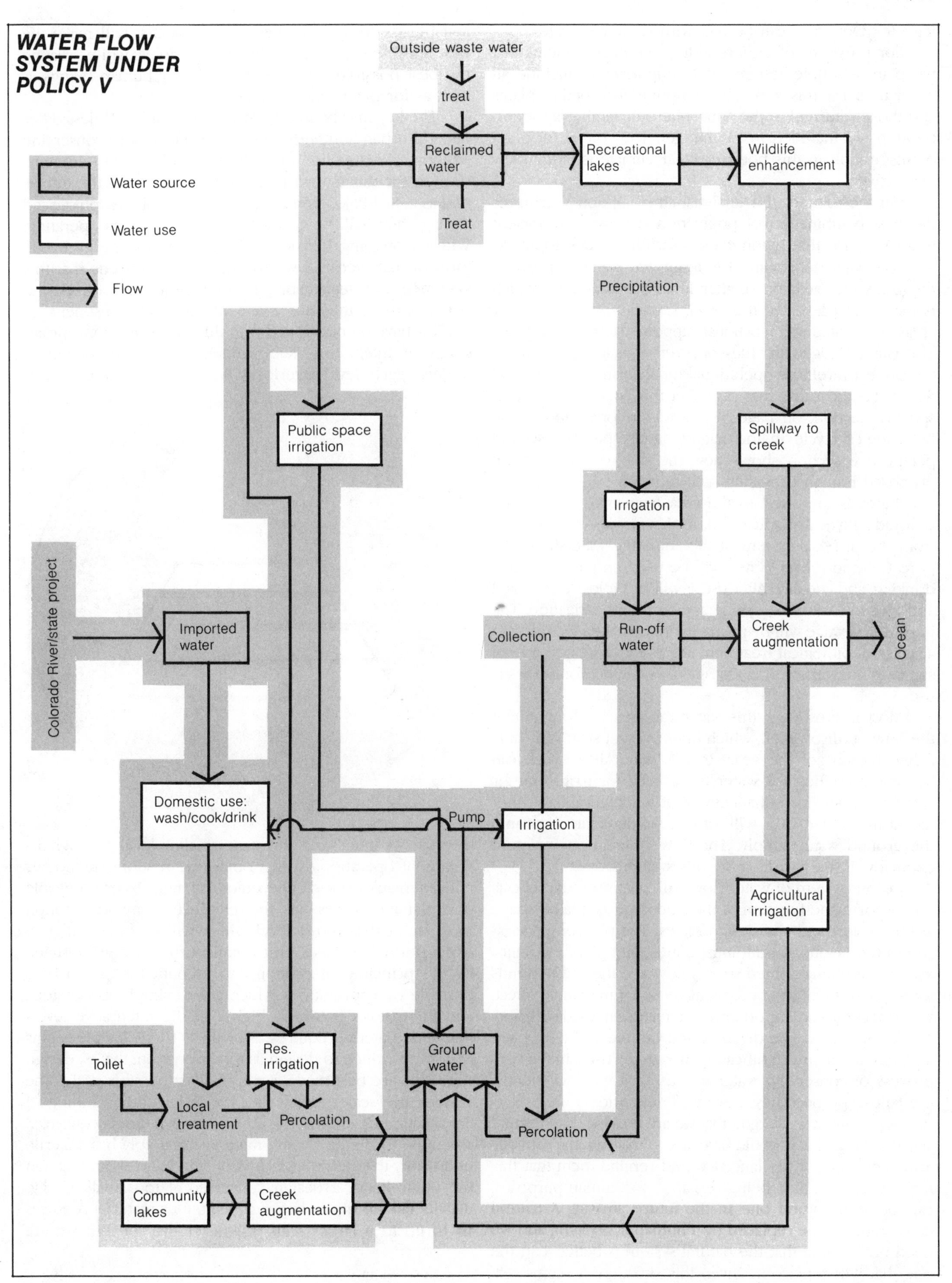
WATER FLOW SYSTEM UNDER POLICY V
Water source
Water use
Flow
Outside waste water
treat
Reclaimed water
Treat
Recreational lakes
Wildlife enhancement
Precipitation
Public space irrigation
Spillway to creek
Irrigation
Colorado River/state project
Imported water
Collection
Run-off water
Creek augmentation
Ocean
Domestic use: wash/cook/drink
Pump
Irrigation
Agricultural irrigation
Toilet
Res. irrigation
Ground water
Local treatment
Percolation
Percolation
Community lakes
Creek augmentation

least for Option V, can be met with reclaimed water.

For purposes of understanding the environments that we plan as whole systems, it is important to include all water uses. For this reason, the amounts absorbed by plants and those returned to the atmosphere through evapotranspiration are included in the budget. The volume of evapotranspiration in this hot, dry, clear climate is particularly instructive.

On the basis of this analysis, since Option V achieves the best combination of population density and efficient water use, it is the option chosen for further development.

Following Feibleman's law again, the plan unit provides the goals for the next smaller level of integration, which is the project level. At this level, specific uses and activity areas are located and functional support systems are defined. The water budget provides one very specific goal, and at the project level, the operational mechanism is a system of water use and reuse that can meet this goal. The first perspective section, on page 33, shows the operation of the hydrological cycle in the natural landscape. The second perspective section shows how the flow of water in the proposed human ecosystem will function.

Water is imported to the area from two sources: the Colorado River Aqueduct and, in the form of raw sewage, from the disposal system of the adjoining watershed. The pure Colorado River water will be used only for washing, drinking, and cooking. After use, it will be biologically treated and reused for toilet flushing and landscape irrigation. The imported sewage will be primarily treated, then fed through a series of biological treatment ponds to purify it to a level suitable for a variety of landscape uses, but not for domestic use.

Most interestingly, this water can be used to augment the flow of Aliso Creek, which in its present state runs only a few months of the year. In this way, Aliso Creek can become a conduit for water to be used for irrigation, for recreation, and for enhancement of wildlife habitats. After use, a major part of it will, of course, percolate back into the ground water supply. The flow diagram shows more explicitly, if less vividly, how this system works.

Such a system of water flow will have profound effects on the form and function of the landscape as it takes shape at the project and site design scales. The plan on page 38 shows the layout of buildings, ponds, and plant elements. As a result of augmented stream flow and the use of ponds for water storage, and by locating these for maximum effect, water becomes an important visual feature in the landscape. At the same time, the diversity and numbers of both plant and animal life are significantly increased. Thus, in the very process of conserving water and using it more efficiently, the landscape becomes more vital, less arid.

Some might well argue that we are creating an unnatural situation here, that it would be better to maintain the naturally arid ecosystem of the land. I would remind them that this landscape, since it is being reshaped for human purposes, cannot be a natural one in the future anyway. A natural ecosystem will be replaced by a human ecosystem, and we need not assume that the natural system is better than the one that will replace it. Given this situation, it seems reasonable to design a human ecosystem that will make the best feasible use of the available resources, while creating the best possible sustainable habitat for human beings as well as for plants and wildlife.

The key may be in the word "sustainable." Unless other parts of urban Southern California shift to more conserving patterns of water use, even the relatively modest amounts needed for this system may not be available in the future. And as at San Elijo, even if the water is available, sophisticated management will be needed to keep the system operating. Without ongoing infusions of water and of energy in the form of management, we would have to concede that this system is not sustainable. I would argue that, given the benefits, these infusions make an excellent investment.

We have demonstrated with this case study appropriate scales of integration: subcontinent—region—plan unit—project. Each level interlocks with the next larger level

through its goals and the next smaller level through its means of operation. At this point, the wisdom of another of Feibleman's principles becomes manifest. His first principle of explanation states that "the reference of any organization must be at the lowest level which will provide sufficient explanation." He goes on to explain that "we are obliged by the principle of economy to account for as much as possible by explanations which are as simple as possible, and this means going no higher in the integrative levels than our material requires us to do."[4] The principle seems to apply as much to design as to explanation; that is, design decisions are best made at the smallest possible scale. The smaller the scale, the more precise an understanding of the situation the designer is likely to have. A design conceived at a scale too large for the material invariably has a sterile uniformity that betrays a lack of detailed understanding. On the other hand, a design conceived at too small a scale usually fails to fit the larger context. Our material requires us to go to a larger scale whenever the concern we are

[4] See Feibleman, 1954; p. 64.

dealing with is too large to be fully comprehended.

Different concerns, having different scope and character, require somewhat different design approaches and different levels of detail. We have still to explore the functional differences between design at one scale and at another.

SEVEN SCALES

In 1969, the Urban Design Committee of the American Institute of Architects published a system of scales that included twelve levels of ascending size:[5]

1. The individual
2. The home
3. The street
4. The neighborhood
5. The area
6. The city
7. The metropolitan area
8. The region
9. The nation
10. The continent
11. The world
12. The solar system

Although this system serves to illustrate the principle of a global (or galactic) organization of scales, it has several problems that inhibit its use for design purposes. Some of the terms, "area" for example, are too vague to be useful. Others, like "city" or "neighborhood" are abstract conceptions that are difficult to delimit in our megalopolitan world. And most of them are social units, determined by groupings of people. Although such units are useful for purposes of social analysis and for some aspects of urban design, they do not necessarily relate to the land itself or to the usual tasks of design.

For landscape design purposes, an operational set of scales has more or less emerged from decades of practice, although it has not, up to now, been articulated as such. Because it has developed from practical applications rather than from theoretical abstractions, this system is flexible, clearly related to political and land tenure processes, and adaptable to project management practices. It is also variable enough in size and scope at each level to encompass other systems based on social units such as that given above or on ecological units. But it is certainly not the only possible system. One might devise any number of sets of hierarchically organized compartments, and the same principles would apply.

We have already introduced the four central levels in this system—the subcontinent, the region, the plan unit, and the project—and shown how they interlock. Above the subcontinental scale, there are few institutional controls and thus few means of implementation. There are numerous examples of national landscape planning, but in most cases these are carried out at the scale we are calling subcontinental. No form of continental planning has really taken shape yet. While recognizing that landscape design at international and continental levels is likely to increase in the future and that levels of integration between the subcontinental and the global are likely to emerge, I will not try to define them here. It should be noted, however, that the global scale, since ecosystems become closed units at that level, is the ultimate frame of reference. Although few decisions are actually made at the global level, its importance as the encompassing source of goals is likely to increase in the future. I will not attempt here to extend the scope of design to the solar system.

At the other end of the scale spectrum, we can descend from the project scale to the site design scale, and from there to the construction level. Here we get into specific form and detail, corresponding roughly to the individual, home, and street levels in the A.I.A. hierarchy.

We might envision these seven integrative levels as a multistory building, pyramid shaped, perhaps, with a broad base. The bottom story—the construction scale—rests firmly on the ground, this being the level at which physical change actually occurs. This is where the man with a shovel is, the one who does the work, and only through him are all the others linked with reality. At the other extreme, the top floor—the global scale—has a vast panoramic view in all directions but is so far removed as to be all but invisible to the man with a shovel. The ground floor is divided into tiny rooms, each of which has only one small window with a restricted view outside, the second floor into larger rooms that lay over and encompass several of the tiny ones below, and the third floor into still larger rooms that lay over several of those on the second level, and so on, up to the seventh level which is undivided, wide open to the panorama. Each room at each of the upper levels is linked by various means of communication with the cluster directly below it

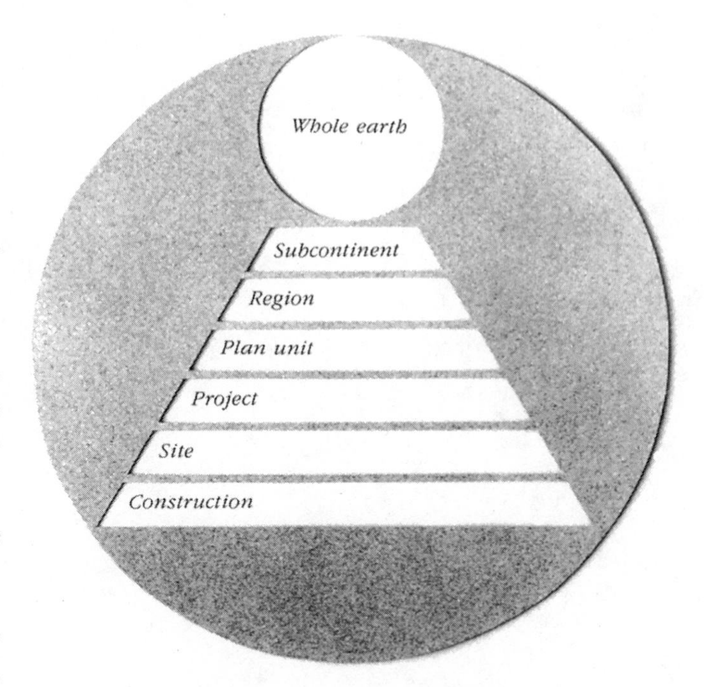

[5] See Brubaker and Sturgis, 1969.

and has doors opening to those adjacent at its own level.

The whole building with its levels and the lines of communication linking them can serve as a basic framework for landscape design, one that encompasses every scale and concern. In keeping with Feibleman's law, we find that at each level of scale a major area of concern emerges to become a principal focus of design at that scale.

I should make it very clear at this point that I am talking about ecosystem organization and thus landscape organization, not political organization. By no means do I mean to describe a system of orders being passed down from above. Participation in the setting of goals is a different matter. The people of a region have the right to participate in setting the goals of their regional ecosystems, and the same is true at every level of integration. Broad participation is assumed. The important point here is that different types of goals and mechanisms are appropriate to each level, and lines of communication are needed to link them together.

It is also important to point out that there are a great many goals other than those derived from natural processes. Functional, esthetic, and social goals emerge at every level. And often, goals of different kinds, or deriving from different levels, come into conflict and create difficult issues.

Thus, the activities that go on at various levels are somewhat different. In the next chapters I will explore each of them at some length, looking at the extent and scope of its role, some of the areas of its concern and responsibility, and some of the tools and approaches appropriate for dealing with it. I will also give some attention to the means of communication between levels.

CHAPTER 3

The Region

We will begin not at the bottom level or at the top, but near the middle, at the regional scale. We do so because the concept of regions has become especially important in planning and design over recent years and because the regional scale is well placed to mediate between the vast general abstractions of the largest scales and the minute specifics of the smaller ones. Beginning at this level will also give us a chance to explore some of the conceptual roots of larger-scale design, most of which grew at the regional level.

Enthusiasm for regional planning has grown steadily through the twentieth century, which is surprising when we consider that there is no way of clearly defining the extent of a region and little general agreement as to what can or should be planned at this scale.

Webster's dictionary defines region with uncharacteristic vagueness as "any more or less extensive, continuous part of a surface or space." Clearly, that is inadequate to help us establish a subject area for study, planning, or design. Nevertheless, the vagueness of the term offers the advantage of flexibility. Anyone wanting to work with a large land area can define it more or less as he chooses. In fact, as Steiner, Brooks, and Struckmeyer have shown, each academic discipline concerned with regions defines the term somewhat differently.[1] Anthropologist John Bennett suggests that a region is "a frame for multidisciplinary research,"[2] whereas economic planner John Friedmann defines regional planning as being "concerned with the ordering of human activities in supra-urban space—that is, in any area which is larger than a single city."[3] The American regional movement of the 1920s defined regions on the basis of historical experience as areas that had evolved a sense of community with common cultural traditions.

Of all the definitions, that of the geographers is perhaps the most useful: "An uninterrupted area possessing some kind of homogeneity in its core, but lacking clearly defined limits."[4] The basis in homogeneity is left then to the researcher or planner or designer; he can define the region according to his concerns. To cite the example given in the last chapter, the Southern California region is defined primarily by the extent of urbanization and secondarily by the watershed. Regions might also be defined by political jurisdictions or combinations thereof, or they might be defined by common cultural characteristics or by physiographic units. The extreme example of regional size is probably the division by Mesarovič and Pestel of the entire world into ten regions for purposes of predicting global patterns of resource depletion.[5] This breakdown was based on "shared tradition, history and style of life, the stage of economic development, socio-political arrangements, and the commonality of major problems. . . ."[6] At their level of analysis and for their purpose, these ten regions can be considered homogeneous, even though they are enormous; one of the ten includes all of Russia and Eastern Europe.

For design purposes, regions are usually much smaller than that, and they are usually pragmatically defined. Richardson describes three different bases for defining a region, depending on one's purpose.[7] The first is homogeneity of some characteristic, which might be visual, ecological, or social, for example. The second is nodality, the nodes being cities. Thus a nodal region is an urban region, like the Southern California region discussed in the last chapter, and the focus is on cities and their interrelations. Third is the planning region, defined according to the geographic extent of the issues at hand. All three of these are commonly used in regional design.

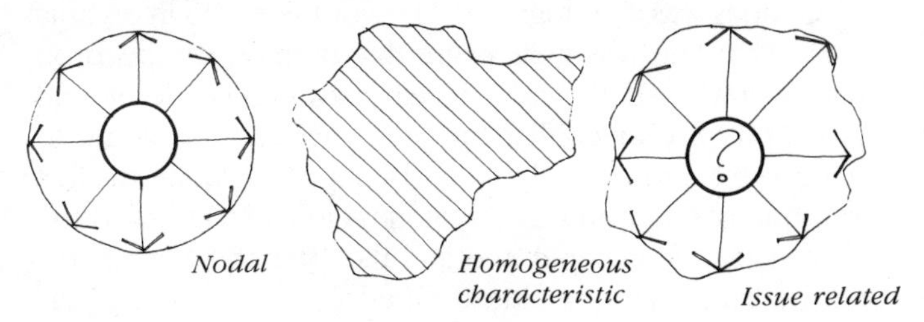

GROWTH AND DIRECTIONS OF REGIONAL PLANNING

At least since the origins of cities, land development has tended to follow regional patterns. Ancient Egypt was, in effect, a linear urban and agricultural region, united by the Nile and defined by the steep bluffs that marked the boundaries between the river's flood plain and the vast desert on both sides. In ancient Greece, the city and the agricultural hinterland that supported it were considered an integral unit, in effect a region, though no such word was used. Usually, an urban/agricultural unit occupied a valley with grazing on the surrounding hillsides, the regional limits defined by the ridges above, a logical pattern that probably

[1] See Steiner, Brooks, and Struckmeyer, 1981.
[2] See Bennett, 1976; p. 309.
[3] See Friedmann, 1964; p. 63.
[4] See Steiner, et al., 1981; p. 1.

[5] See Mesarovič and Pestel, 1974.
[6] Ibid., p. 40.
[7] See Richardson, 1979.

developed incrementally and organically with little conscious forethought. The earliest evidence of design at the regional scale dates from the Renaissance. According to Sigfried Giedion, Leonardo da Vinci was the first to "formulate definite plans based on a comprehension of the physical structure of a region."[8] This structure, as Leonardo understood it, stemmed from the flow of water, which was one of his lifelong obsessions. The beautiful sketch that illustrates his plan to make the River Arno navigable by sweeping a canal in a broad arc around impassable sections, linking the cities of Florence and Pistoria, gives us a clear notion of the breadth of Leonardo's vision. His plans for draining the Pontine Marshes and for digging a system of canals to join Milan with the lakes of Northern Italy have a similar sweeping scope. These schemes, for all their breathtaking vision, were rooted in the Renaissance, with its prevailing assumption of human omnipotence. Leonardo was trying to reorganize natural forces on a regional scale, with no suggestion of the exhaustive analyses of multitudinous factors that accompany design at this scale today.

The world seems not to have been ready for Leonardo's regional design, any more than it was ready for his helicopter, his submarine, or his multilevel cities. There are few indications of regional design after Leonardo until the early part of the twentieth century.

The concept of regional design, as we know it now, seems to have its roots in Patrick Geddes' realization of the fact that cities—in response to the urbanization pressures of the Industrial Revolution—could no longer be viewed as units with distinct boundaries. Rather, they were spreading over the countryside to join other towns in larger accretions that he punished with the appropriately repulsive label of "conurbations."[9]

Geddes was a biologist, and he saw the world in organic terms. Thus, he was well aware that nothing can keep on growing indefinitely, and from that awareness developed his famous prophecy that the industrial era was going to divide itself into an early and a later period, quite distinct from one another, just as the Stone Age had. He called the earlier the Paleotechnic Age and the later the Neotechnic and theorized that the cities of the later would be entirely different from those of the earlier. The crowded, monotonous, noxious cities of the Paleotechnic era would eventually give way to ". . . a synoptic vision of nature . . ." and ". . . constructive conservation of its order and beauty towards the health of cities . . ." The Neotechnic City would involve ". . . the conservation of nature and the increase of our access to her . . .," and it would have ". . . open-air schools for the most part . . ." and ". . . the allotments and the gardens which every city improver must increasingly provide—the whole connected up with tree-planted lanes and blossoming hedgerows, open to birds and lovers."[10]

Basic among the devices Geddes recommended for planning toward the neotechnic city was what he called the "regional survey." Such a survey would include physical factors—soils, geology, vegetation—as well as social factors. He saw it not simply as a practical tool but as serving deeper intellectual purposes. "What if the long-dreamed synthesis of knowledge, which thinkers have commonly sought so much in the abstract—and by help of high and recondite specialisms, logical, mathematical, and the rest, all too apart from this simple world of nature and human life—be really more directly manifest around us, in and along with our surveys of the concrete world?"[11] It is still a question that deserves attention, one that we will return to later.

Perhaps the neotechnic city is happening to some degree, in some ways, in some places. But whatever the success of his prophecies, Geddes has had enormous influence on designers, planners, and theorists, especially those concerned about the relationship between cities and natural landscapes. Members of the Regional Planning Association of America worked to apply some of his ideas on a regional scale, most vividly in their efforts toward a plan for New York State. Their work included an impressive inventory of the state's natural resources. Benton MacKaye, who was a founding member of the Association, and along with Lewis Mumford, one of its most eloquent spokesmen, developed a theory of regional planning that attempted to fit urban growth snugly into the natural scheme of things.

MacKaye saw the region as a battleground in the war between the indigenous and the metropolitan—". . . two 'worlds,' two ideas of life, as distinct perhaps as the Greek and the Roman, and yet interwoven in utmost intimacy."[12] MacKaye defined the indigenous as being composed of the primeval (which today we would call wilderness), the rural (primarily agricultural), and the urban. (Note the similarity to Odum's ecosystem development types discussed earlier.) The metropolitan world, on the other hand, is the artificial face of urbanization, controlled by the machine. Not only is it unnatural, it is overtly antinatural. "Instead of means being adapted to achieve ends, the ends are distorted to fit established means; in lieu of industry being made to achieve culture, culture is made to echo the intonations of industry; oil paints are manufactured not to promote art, art is manufactured to advertise oil paints."[13]

The direction set by the Regional Planning Association was pursued during the 1920s and after by a long series of thinkers and planners. High points were reached in the thirties with the work of the Tennessee Valley Authority and a little later with the National Resources Commission (NRC), which sponsored regional surveys of natural and cultural resources for every part of the United States. The work was carried out by citizens of each region and coordinated and published by the Commission. Unfortunately, the effort ended there. The promising beginnings of TVA and NRC were not carried on. The synthesis of knowledge that Geddes spoke so eloquently about was somewhat lost in the cataclysm of World War II. When the term "regional planning" eventually reemerged in the fifties in the midst of unbounded enthusiasm for growth and development, it was firmly attached to economic development and to urban industrialization as a means of achieving it. Its methods

[8] See Giedion, 1967.
[9] See Geddes, 1915.
[10] See Ibid., pp. 95, 94, and 99.
[11] Ibid., p. 336.
[12] See MacKaye, 1962; p. 56.
[13] Ibid., p. 71.

became highly analytical, growing largely from the abstract theories and mathematical constructs of the neoclassical economists. Von Thunen's general theories of location, developed and expanded by Cristaller and Lösch, provided the spatial components. These theories, founded in the equilibrium theory of economics, had nothing to do with natural resources or the natural character of the landscape. Rather, they were concerned with transportation costs.

We can gain some insight into the assumptions underlying this new field called regional planning, but which might more accurately be called regional *economic* planning, by reading one short quote from Lösch. He begins the development of his theory of regional centers with the assumption of "... a vast plain with an equal distribution of raw materials, and a complete absence of any other inequalities, either political or geographical. We further assume that nothing but self-sufficient farmyards are regularly disposed over that plain."[14] Thus, he eliminates all variables other than economic ones, in the tradition established much earlier by von Thunen, who based his theory of agricultural location on a vast, flat plain, isolated from the rest of the world.

As a result, the emphasis on natural resources so clearly expressed by Geddes, MacKaye, Mumford, and the other early advocates of regional planning was eclipsed for a time by an enthusiasm for economics. In 1964, John Friedmann wrote that "... regional planners are chiefly concerned with economic development or, more precisely, with the spatial incidence of economic growth."[15] Beginning in the late 1960s, regional development theory and practice became increasingly concerned with urban growth and with economic, social, and resource inequalities as bases for regional planning policies. Inequalities in natural resources were seen primarily as problems for economic development.

Through the fifties and sixties, only a handful of designers and planners still considered natural resources as fundamental considerations in shaping the regional landscape.[16] But finally, in 1969, this view reemerged with a flourish in Ian McHarg's *Design with Nature*. The landscape approach has gained steadily since then in stature and sophistication, but it has clearly parted ways with the economic approach. The theoretical foundations are quite different. Landscape designers take the very "inequalities" that Lösch assumes out of existence as their point of departure. The character of the landscape, we say—its natural resources, features, and processes, along with its diversity of human uses—is the most fundamental consideration in shaping its future.

It is not a hopeless division, or even necessarily an unhealthy one. I will return to this question later in this chapter and present a case study that joins the two approaches in a more or less successful union.

PURPOSE AT THE REGIONAL SCALE

What do we actually design at the regional level? The answer is: in reality, very little. The scale is much too large for drawing lines on the ground or determining specific uses. Regional authorities, moreover, rarely have any real power to implement plans. Usually, theirs is a coordinating function, with some policy-making responsibility.

The regional scale is the level of broad patterns that sweep over landscapes large enough to include full ranges of human and natural interactions but still small enough to allow us to perceive the details that give emotional meaning to the whole. At higher levels than the region, details are lost and the landscape becomes an abstraction, whereas at lower levels, detail can obscure the larger picture.

By the definition given earlier, a region is homogeneous in at least one respect, which may be cultural or physical. For design purposes, the homogeneous characteristic is generally the most compelling environmental concern. Often, this is urbanization. We have seen that the early development of the regional concept grew out of the recognition that cities were no longer complete units but were spreading over much larger areas. Now we routinely think of the smallest complete urban unit not as the city but as the urban region that includes the full extent of urban growth. As these regions spread farther out and coalesce, the lines around them become harder to draw. The East Coast megalopolis that extends from Washington to Boston, although clearly a single region as Gottmann convincingly argued a long time ago,[17] is difficult to grasp as a whole.

The Regional Planning Association was originally concerned with urban regions as regions including urbanized areas and their supporting hinterland. Together, these form an ecological unit. As we saw in the last chapter, however, the hinterland becomes harder to define as it spreads, ultimately to encompass the whole earth. Thus, the ecological unit is ultimately the globe.

In practice, smaller areas of more manageable size are usually deducted from larger regions like megalopolis for design study. Examples are the Metland studies carried out by the University of Massachusetts research team[18] and the work of Carl Steinitz and his group in the southeastern sector of the Boston metropolitan area,[19] parts of which are presented as a case study on pages 47–52. Both of these efforts, along with a great many others around the world, focus on the issues of suburbanization, or metropolitanization, that Geddes and MacKaye wrote so eloquently about.

Phillip Lewis has suggested that conurbations (he uses the more benign term of "constellations") of the megalopolitan sort should be a major focus of regional design and has concentrated his own efforts on what he calls the "Circle City constellation," which includes Chicago, Milwaukee, the Fox River Valley, and the Twin Cities.[20] Lewis's work encompasses this entire urban region by using a holistic approach of general "visualization." As compared to the highly analytical approaches of the Massachusetts studies, Lewis uses more intuitive processes and more graphic means of communication.

Economic development planning has generally focused almost entirely on urban regions. The profound importance

[14] See Lösch, 1964; p. 107.
[15] See Friedmann, 1964; p. 83.
[16] See Steinitz, et al., 1969.

[17] See Gottmann, 1961.
[18] See Fabos, 1977, 1979, 1980.
[19] See Steinitz, et al., 1976.
[20] See Lewis, 1982.

of the rural and agricultural landscape has only recently received much attention. In less urbanized landscapes, the bases of homogeneity are more often physical or cultural. The most common basis for physical homogeneity is the watershed unit, also commonly called the "drainage basin," which includes all the land draining into a common stream, river, or lake. Given the great importance of water in shaping the landscape, the watershed is a particularly useful unit for design. During the NRC era, it became clear that watersheds, since they define hydroelectric generating potentials, are also closely related to energy development. The watershed unit became the unit for development planning, not only for the Tennessee Valley Authority but for the entire National Resources Commission development program. The river basin has become the physical unit for most regional development plans in the developing world, usually on a massive scale.

In the case of the Tennessee Valley Authority, the watershed unit was combined with culturally homogeneous units. As is usually the case in rural areas, cultural homogeneity was an important consideration. Economic development, especially when it includes energy infusions on the scale accomplished by the rural electrification projects of TVA, can bring massive social changes and upset long-established cultural traditions. This was a matter of serious concern in the American South, and today it is an even greater concern in most of the less developed countries, where regions often have strong cultural definition.

The Huetar Atlantica region of Costa Rica, which is presented in the case study on pages 54–64, and which we will soon discuss at some length, is such a region. Culturally, Huetar Atlantica is quite different from other parts of Costa Rica. A great many of its residents are black, having descended from immigrants from the West Indies. They have maintained this tradition in much of their music, dress, food, and architecture. Their language remains a rhythmic sort of English rather than Spanish. The region is slow-moving, colorful, and less economically developed than other parts of Costa Rica and is regarded by Costa Ricans of other regions as a wholly different place to which they seldom venture. This cultural distinctiveness demands that for design purposes Huetar Atlantica be treated as a unique whole.

Whether defined by urbanization, landform, or culture, the "emergent quality" at the regional level is usually regional policy. This is the scale at which coordinating principles can be established to guide the development and management of the larger landscape. As a result, most design effort at this scale is directed toward policy making. Among the ways in which regional design supports regional policy are the following:

1. By providing an information base for policy decisions and for design at lower levels
2. By identifying regional landscape patterns and trends
3. By establishing a general framework for development of a region
4. By specific design of linear networks that flow through and unify a region, such as waterways and transportation routes.

THE REGIONAL INFORMATION BASE

The first of these ways, a regional information base, has become an important tool for design and management at this scale. Especially in the complex political and economic environments of the more industrialized nations, such an information base is almost necessarily automated. This automated information base, describing locational variables and their attributes, is commonly called a "Geographic Information System." This is essentially a set of maps stored for manipulation by computer. The maps usually show the natural characteristics of the landscape, such as vegetation and soil types; the various human uses of different areas and the transportation networks connecting them; political and administrative boundaries; and any other characteristics spread across the landscape that might influence decisions.

Geographic Information Systems are used for two different basic purposes. The first and simplest is to retrieve stored data. The data are extensive, complex, and detailed, so much so that retrieving any one fact without the computer's help would be very difficult (the fact retrieved, however, will be the same fact originally stored).

The second basic purpose of a Geographic Information System is to develop interpretations and models from the raw data, that is, to use it to derive new information. In most cases, the new information is a locational model that combines stored data describing different variables to show the distribution of some characteristic of the landscape that is not directly observable.

Case Study: The Metropolitan Boston Information System

The Metropolitan Boston System covers the southeastern sector of the Boston metropolitan area—a landscape that

CASE STUDY IV

Metropolitan Boston: A Modeling Approach to Managing Suburban Growth

The area under consideration here is the southeastern sector of the Boston metropolitan area, which includes eight towns and 756 square kilometers. The rapid suburbanization now occurring is typical of growth patterns around the edges of cities all over the world. The geographic information system developed for this area is designed to project land-use patterns that will result from alternative planning strategies and then to predict the social, economic, and ecological consequences of those strategies. The system is intended as a tool for a wide range of people interested in land use and development, including planners, public agencies, and private developers.

Prepared by the Landscape Architecture Research Office, Harvard University. Principal investigators: Carl Steinitz, Peter Rogers, H. James Brown, William Geizentanner, David Sinton, Frederick Smith, and Douglas Way. Funded by the National Science Foundation under the Research Applied to National Needs (RANN) program.

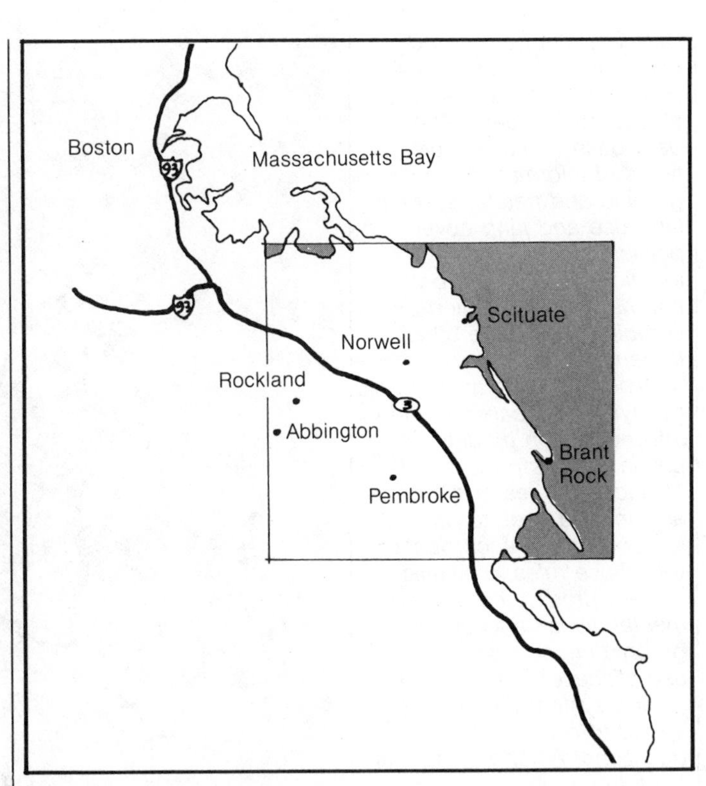

GENERATION OF THE COMPARATIVE PLAN

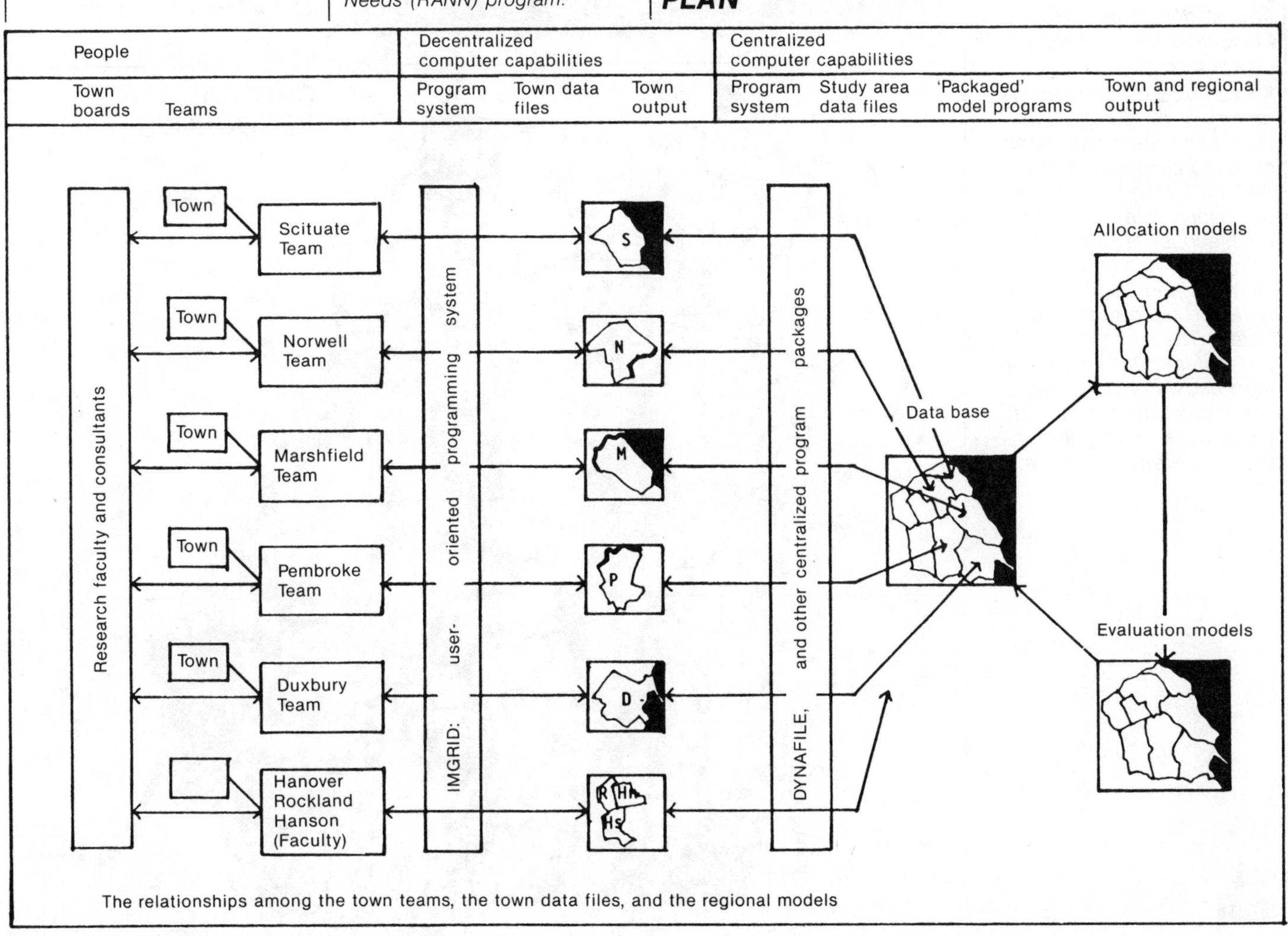

The relationships among the town teams, the town data files, and the regional models

Data are mapped in grid-cell units of 1 hectare (2.47 acres). The study area includes 75,600 cells. The data base contains four types of information: physiographic and natural systems, land use and land cover, political jurisdictions, and functional zones and data that have been derived by combinations of the other three types.

Modeling programs are of two types: allocation models and evaluation models. Allocation models provide sets of decision rules and then assign land uses to particular locations in accordance with those rules. Evaluation models predict the environmental, fiscal, and demographic impacts of the land-use allocations so derived.

The system can thus respond to a wide variety of questions. Among those that have been asked and answered: What should acquisition priorities be for open space? What are the impacts if the industry-zoned area is developed? How can Marshfield retain its visual character?

Examples of information stored in the geographic data base are these maps, showing percentage of surface water, critical natural resources as defined by Massachusetts law, and existing land use. The base year for land use is 1975; data are periodically updated. The system identifies 267 different land types and land uses. For purposes of the existing land use map, these were aggregated into the ten categories shown.

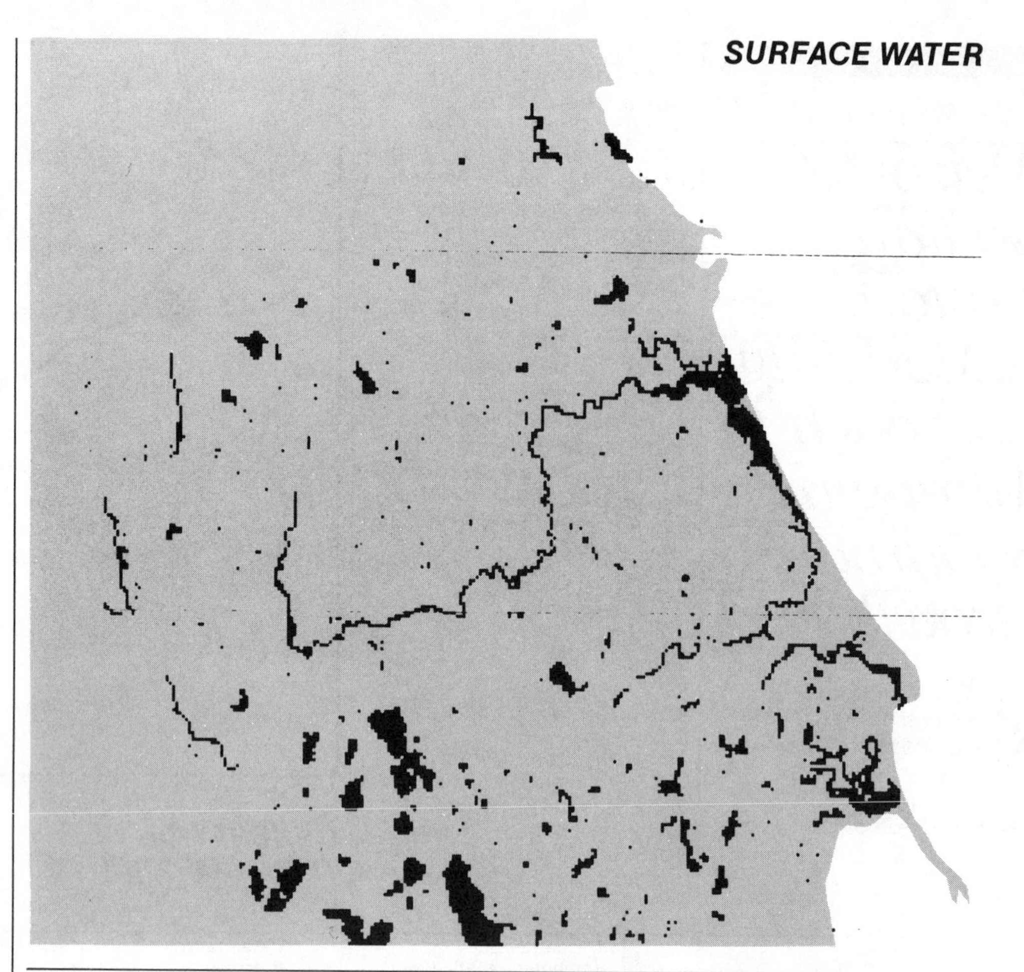

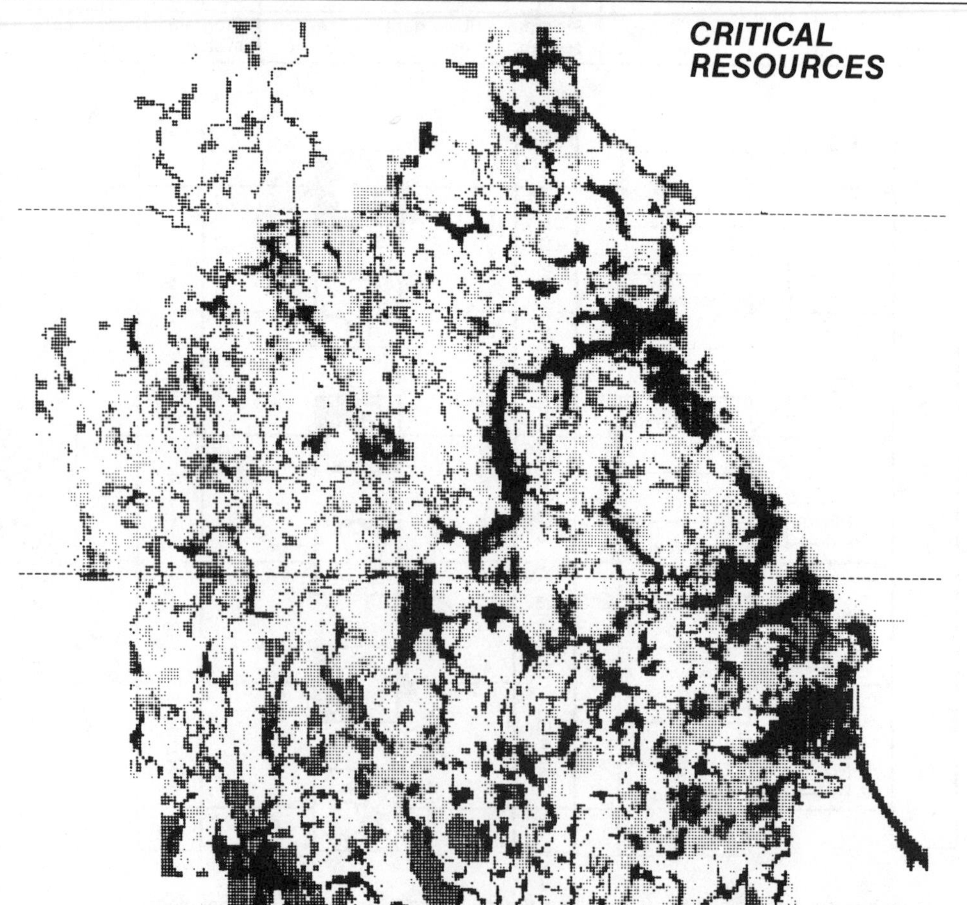

EXISTING LAND USE

The second map shows the 1975 land-use patterns projected to 1985; it was based on allocation models that assume present growth policies and trends will continue. The trend projection provides a baseline condition against which alternative patterns are measured.

The Combined Land Uses for 1985 is one alternative pattern; it represents the results of a series of workshops in which planners, public officials, and citizens from five towns in the study area developed, with the help of researchers, alternative growth management policies. These were combined into a regional development strategy, which was then used as the basis for the 1985 combined land-use pattern. This process is described in the diagram on page 47.

The evaluation models illustrate predictions of the air-quality impacts of the projected land-use trends and of sedimentation patterns for the Combined Plan.

Level	0	1	2	3	4	5	6	7	8	9
Frequency	2409	16623	1936	4361	1009	3001	13367	807	934	836

Blank: ocean, open freshwater, and outside study area

Level 0: other water and wetlands
1: forests
2: agriculture
3: conservation and recreation
4: utilities
5: transportation
6: housing
7: institutions
8: industry
9: commerce

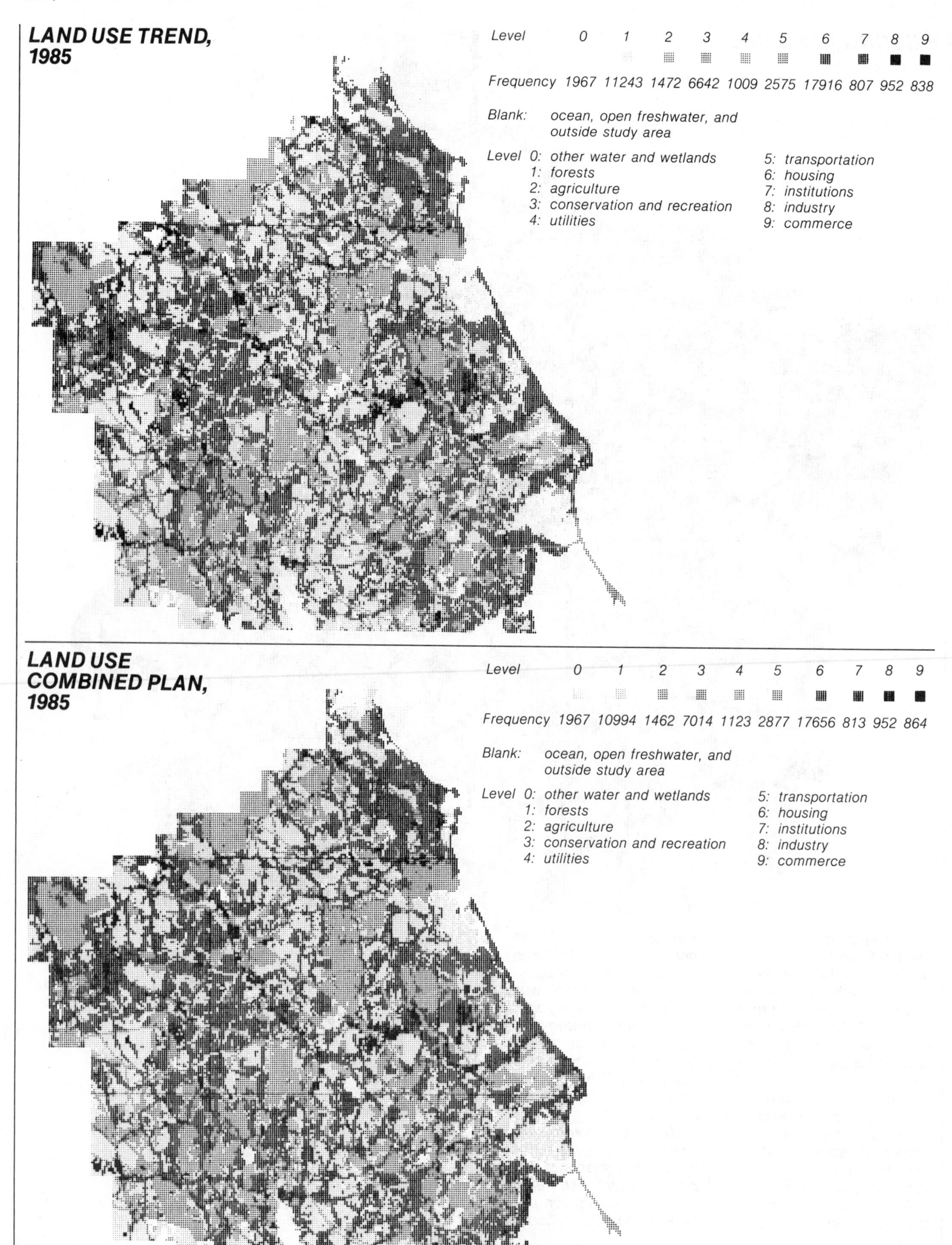
LAND USE TREND, 1985
Level 0 1 2 3 4 5 6 7 8 9
Frequency 1967 11243 1472 6642 1009 2575 17916 807 952 838
Blank: ocean, open freshwater, and outside study area
Level 0: other water and wetlands
1: forests
2: agriculture
3: conservation and recreation
4: utilities
5: transportation
6: housing
7: institutions
8: industry
9: commerce
LAND USE COMBINED PLAN, 1985
Level 0 1 2 3 4 5 6 7 8 9
Frequency 1967 10994 1462 7014 1123 2877 17656 813 952 864
Blank: ocean, open freshwater, and outside study area
Level 0: other water and wetlands
1: forests
2: agriculture
3: conservation and recreation
4: utilities
5: transportation
6: housing
7: institutions
8: industry
9: commerce

AIR QUALITY IMPACTS

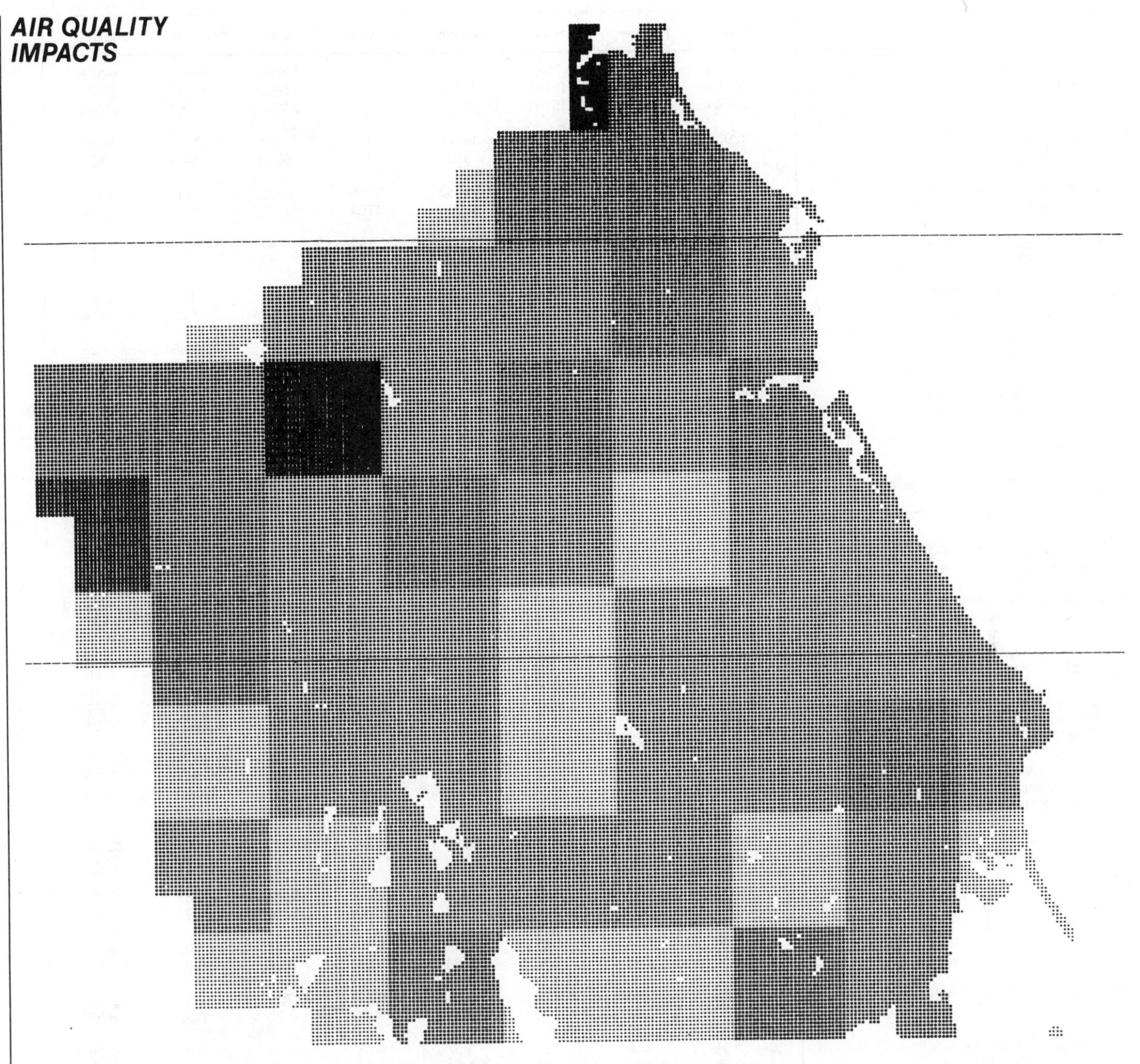

SEDIMENTATION IMPACTS

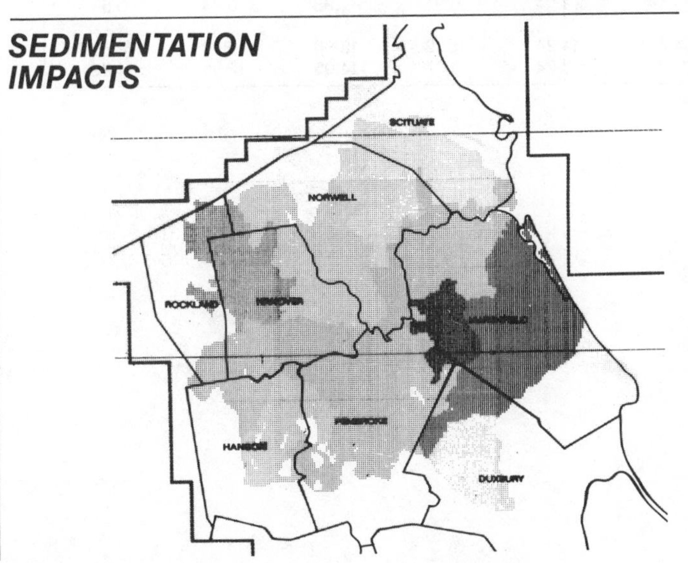

Gradations in tone indicate higher degrees of impact for the pattern shown in the land-use trend model; the darker the tone the greater the impact.

		Duxbury	Hanover	Hanson	Marshfield	Norwell	Pembroke	Rockland	Scituate
Town population by town	1975	11460	10056	6885	19139	8411	12759	14651	16367
	T	23559	23386	13766	29879	15857	24690	18157	21964
	C	20784	24474	15781	28734	11890	21378	18092	23840
New population by town	T	12089	13330	6881	10740	7446	11931	3506	5597
	C	9324	14418	8897	9595	3479	8619	3441	7473
Total housing per town (hectares)	T	3154	2885	1448	3352	2295	2562	959	2299
	C	2754	2180	1745	2906	1754	2270	1040	2519
Residentially developed land (percent)	T	49	51	36	44	42	42	36	47
	C	43	54	43	38	32	37	40	51
Water demand by town (millions gals/day)	T	1.37	2.13	0.90	2.46	1.04	0.99	3.16	2.09
	C	1.12	2.03	0.85	2.37	1.05	1.14	3.22	2.14
Diverse wildlife habitat lost (percent)	T	39	70	40	28	34	36	36	26
	C	46	69	29	35	45	25	36	29
Most remote wildlife habitat lost (percent)	T	8	100	30	15	26	56	56	2
	C	11	100	32	16	25	74	56	1
Critical resources lost by 1985 (percent)	T	62	60	13	56	47	55	86	31
	C	62	60	13	63	53	41	86	31
Significant watershed/ agricultural land lost (percent)	T	25	24	6	13	15	16	15	13
	C	10	25	6	10	6	1	17	18
Historical resources lost (hectares)	T	23	16	4	20	32	15	8	8
	C	7	18	4	8	22	1	1	14
Negative visual quality seen from roads (percent)	T	13	23	21	32	16	19	71	23
	C	10	26	19	32	16	13	78	21
New housing in excessively noisy areas (hectares)	T	118	18	30	94	55	68	18	20
	C	87	18	20	70	12	45	37	43
Acreage development costs per 1-acre single-family home (1970 dollars)	T	3863	NA	3590	3751	3886	3696	NA	3667
	C	3835	3781	3509	3686	3823	3591	3693	3661
Average development profit per 1-acre single-family home (1970 dollars)	T	8420	NA	7110	7560	8310	7360	NA	7280
	C	7740	4980	5510	6390	7350	5530	5150	5910
Increase in local tax base (millions, 1970 dollars)	T	70.05	61.88	30.53	48.65	40.82	47.15	25.18	28.92
	C	54.39	67.42	43.25	49.48	17.40	34.64	27.78	41.10
Number of school children by town	T	5522	6205	3824	7422	3986	5974	4681	6725
	C	4957	6424	6402	6854	3218	5506	4797	7111
Local capital expenditures (millions 1970 dollars)	T	7.09	8.13	4.83	8.61	5.30	7.22	6.13	8.41
	C	6.10	8.41	5.35	8.64	4,87	6.37	6.02	9.05
Local operating expenditures (millions, 1970 dollars)	T	6.73	8.02	4.77	8.61	5.24	7.12	6.13	8.41
	C	6.01	8.30	5.19	8.64	4.16	6.30	6.02	8.82
Change in local tax rate (dollars/1000 increase)	T	1.14	1.59	1.82	0.57	0.63	1.04	0.84	0.64
	C	1.26	1.57	1.64	1.04	0.91	1.98	0.44	0.53
Median income (thousands, 1970 dollars)	T	14.96	14.95	14.30	14.24	15.19	13.90	12.350	14.60
	C	14.77	15.03	15.18	13.24	15.07	14.03	12.85	14.82

COMPARATIVE ANALYSIS: TREND AND COMBINED PLAN

The chart shows a comparison of the effects that are expected to result from the Trend and Combined patterns in each of the eight towns of the area. Although the impacts of the planned pattern are considerably less, the changes brought about by urbanization in both cases are severe.

is being rapidly overrun by suburbanization. For this reason, the system is designed to deal with the issues of suburban growth management. Its allocation models distribute uses in varied patterns, according to different assumptions. For example, the 1985 trend model shown on page 50 was based on the assumption that attitudes and policies, and thus growth patterns, would remain essentially unchanged from 1975 to 1985. In other words, it is a linear projection of the pattern that existed in 1975. Each land use was allocated according to specific rules that were in accord with this general assumption.

The combined plan for 1985 was devised by combining plans developed by different teams for different towns within the subject area. The process for developing the plans and combining them is shown in the flow diagram on page 47.

The trend model serves as a baseline with which the combined plan, or any other plan or model, can be compared. The information system calculates numbers for comparison, as shown in the chart on page 52, easily and quickly. Evaluation models can then make predictions of fiscal, demographic, and environmental impacts for the trend and any other distribution of land uses one might devise. Examples of predictions for increases in sedimentation and air pollution are shown. Other evaluation models predict the effects of different patterns of land use on soils, water quality and quantity, vegetation, critical natural resources, transportation, noise, historic resources, and fiscal accounting.

Once in place, a system of this type can make many different types of predictions in responding to different issues. For all its sophistication, however, the dilemma first identified by Geddes and described with such clarity by MacKaye remains essentially unchanged. The impact of the combined plan, though less than that of the trend, is still overwhelming.

MANAGEMENT ZONES: THE EXAMPLE OF HUETAR ATLANTICA

The situation of the Huetar Atlantica region of Costa Rica is quite different from that of Metropolitan Boston, and the difference in many ways illustrates the contrast in ecological as well as in governmental makeup between the industrialized and the nonindustrialized worlds. Precisely because nonindustrialized countries are less developed economically, regional thinking holds particular promise for them. It should be possible for them to develop their use of resources in ecologically sustainable ways without first repeating all the mistakes the industrialized countries are now paying so much to undo. It *should* be possible, but it probably will not be for a very long time and for any number of social, political, and economic reasons. Still, one can hope.

Of the five regions into which the government of Costa Rica has divided the country, Huetar Atlantica, the Atlantic Coast region, is the least developed economically. The land turns upward behind a marshy coastal plain, becoming wilder and more rugged as it rises into the jagged Cordillera that forms the spine of Central America. The slopes are covered by a towering tangle of tropical rain forest, the constituents of which vary with elevation. The Center for Tropical Studies has identified and mapped five distinct rain-forest plant communities in the region, all extremely rich in both plant and animal species.[21] All of them are important in the global perspective. Rain forests have a major role in the regulation of world climate patterns, and they provide a habitat for over half of the world's wildlife species. They are also among the world's most seriously threatened biomes. Huetar Atlantica is a prime example of a conflict that is spreading through the world's rain forests and threatening to destroy them. Despite its majestic size, prodigious productivity, and mind-boggling diversity, the tropical rain forest is a fragile community. The reason stems largely from the interactions of soil and water and the ways these are affected by different types of human use.

For centuries, the human populations of the rain forests in most parts of the world were quite small, and they grew their food by shifting cultivation. Under this system, the vegetation in a small area of forest was cut down and burned, and then seeds were planted in the mixture of soil and ash. Plantings were in irregular patterns rather than in rows, and various species were mixed together rather intensively. There was no plowing. After crops had been grown for a certain period, usually two to five years, the farmers would abandon that clearing and move on to chop down trees in a new place. It was once believed that they moved on when productivity fell off because the soil had become depleted, but some recent observations suggest that weeds and insects may have reclaimed the land and made growing conditions too difficult before the soils were depleted.[22] The latter explanation may well be correct because it has often been observed that rain forests regenerate very quickly. In areas where shifting cultivation is still practiced, one can see various small, more or less circular segments of forest in various stages of succession.

Shifting cultivation in some ways replicates the natural regenerative processes of the rain forest.[23] Under natural conditions, clearings are created by floods, wind, fire, trees dying of old age, and the like. Plants that cannot grow under the dense canopy colonize these clearings, and regrowth begins. Several studies have confirmed that the process works in the same way after clearing for agriculture. Thus shifting cultivation generally seems to work in harmony with natural processes.

This is an ancient system of agriculture, long practiced in Africa and Asia as well as in Central and South America. The earliest European farmers, the Danubians, used similar practices in the oak forests near the border between present-day Holland and Germany. It is a system that can be sustained indefinitely so long as the population does not exceed the capacity of the forest to regenerate, and researchers agree that it seems to have no ill effects on the forest. When populations grow too large, however, the cycle of clearings is speeded up, and farmers find themselves returning to once-cultivated areas before the successional sequence has had time to reproduce the diverse forest of later successional stages. At this point, agriculture begins a pattern of retarding succession in the forest, soils do not have time to recover

[21] See Tosi, 1969.
[22] See Janzen, 1973.
[23] See Gomez-Pompa, et al., 1972.

CASE STUDY V

The Huetar Atlantica Region of Costa Rica

Outside its one urban area, the port city of Limon, the population of Huetar Atlantica is mostly rural and agrarian, including a great many landless and land-hungry peasants. Both through the government's land distribution program and through illegal untitled settlement, inroads are being made in the rain forests that cover most of the land. These forests, ranging from varied palm communities of the coastal strip to the towering, incredibly diverse green masses of the upper mountain slopes, vary considerably in their ability to support human development. Least resilient are the forests on red and yellow lateritic soils, which, after clearing, develop a hard crust that is extremely resistant to plant growth, both cultivated and natural. Deforestation in these areas can be, for all practical purposes, permanent. Other soils, particularly the alluvials of the lower foothills, can be productive for indefinite periods.

Despite its many limitations, much of the region is capable of supporting some economic productivity if uses and means are chosen carefully. In general, crops that incorporate some protective features of the natural ecosystems are likely to be most successful over long periods. The two vegetation profiles—natural and ideal cultivation—illustrate this principle.

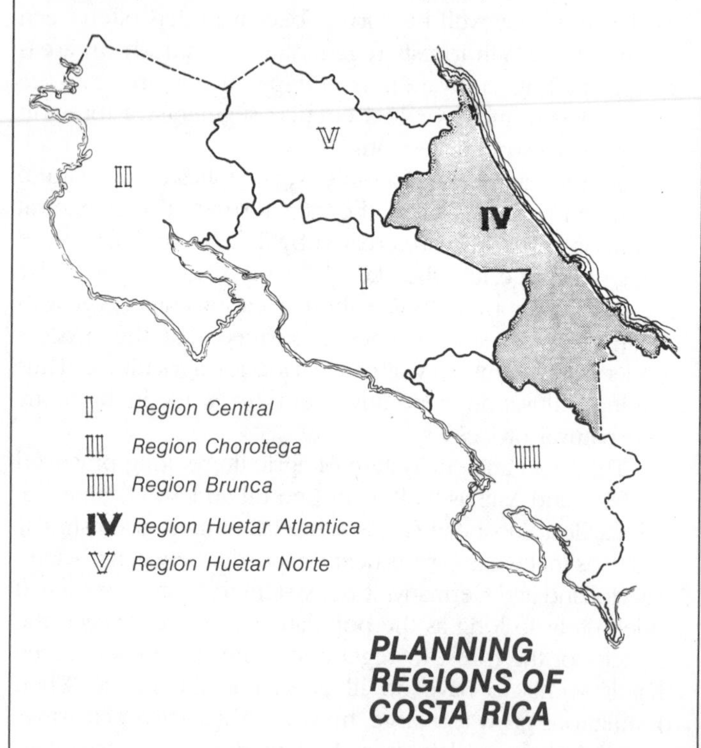

PLANNING REGIONS OF COSTA RICA

Prepared by the Regional Planning Consulting Group, California State Polytechnic University, Pomona, and OFIPLAN, the Costa Rica National Planning Agency. Project Director: Sylvia White. Consultant for Environmental Management: John T. Lyle. Plans drawn by Rosa Laveaga, research associate; profiles by Gregory de Young, research associate.

GENERAL MANAGEMENT ZONES

I: The Coastal Strip

Soils:	Regosoles
Precipitation:	< 2.8 meters/year
Temperature:	> 25°C
Vegetation:	Salt-tolerant coastal species; palms dominant
Slopes:	0–5%
Preferred uses:	Palm tree crops Fruits and vegetables Aquaculture (in lagoons) Cacao and other tree crops
Acceptable uses:	United urbanization Small-scale recreational development
Major impacts of human use:	Pollution of the ocean Destruction of coral reefs
Means of control:	Sewage treatment systems for all settlements Limited use of fertilizers, herbicides, and pesticides

II: Wet Lowlands

Soils:	Poorly drained alluvials
Precipitation:	> 4 meters/year
Temperature:	> 25°C
Vegetation:	Very humid tropical rain forest
Slopes:	0–5%
Preferred uses:	Nature reserves Limited logging Game ranches (cultivation of native species)
Acceptable uses:	Production forests Mixed agriculture where drainage can be provided
Probable impacts of human use:	Deforestation where access is provided Saturation and leaching of soils Loss of native plant and animal species Pollution of rivers and ground water
Means of control:	Building only those roads that are definitely needed Drainage works where feasible and needed Maintaining partial forest cover on cultivated lands Limiting use of fertilizers, herbicides, and pesticides to essential minimum

III. Alluvial Plains

Soils:	Alluvials, mostly well drained, some poorly drained
Precipitation:	2.8–4 meters/year
Temperature:	22.5°–25°C
Vegetation:	Humid tropical rain forest
Slopes:	0–15%
Preferred uses:	Intensive mixed agriculture (field crops, tree crops, animals) Aquaculture Limited urbanization Production forests
Acceptable uses:	Other types of agriculture
Probable impacts of human use:	Soil erosion Soil saturation where drainage deficient Flooding Pollution of rivers and ocean Loss of native plant and animal species
Means of control:	Drainage works where needed Dikes for protection of settlements from flooding Limit use of fertilizers, herbicides, and pesticides to essential minimum Ban aerial spraying Maintain partial forest cover Maintain large interconnected forest reserves

IV: Lateritic Foothills

Soils:	Laterites
Precipitation:	2.4–3.8 meters/year
Temperature:	22.5°–25°C
Native vegetation:	Very humid tropical rain forest
Slopes:	15–45%
Preferred uses:	Forest reserves Limited commercial logging
Acceptable uses:	Production forests Tree crops
Probable impacts of human use:	Irreversible leaching of soil nutrients and loss of ability to support vegetation Erosion Soil compaction where grazing allowed
Means of control:	Maintaining vegetation cover Where cultivation is deemed necessary, continuous application of necessary soil amendments Keeping roads to a minimum

V. Upper Slopes

Soils:	Laterites
Precipitation:	4.0 meters/year
Temperature:	22.5°C
Native vegetation:	Pluvial rain forest
Slopes:	40%
Preferred use:	Forest reserves
Acceptable uses:	Commercial logging Limited subsurface mining
Probable impacts of human use:	Extreme soil erosion Pollution of rivers by mining
Means of control:	Immediate reforestation of cleared areas Pollution control and rehabilitation plans required for mining permits

The system of land management zones developed for the region shows five distinct zones that are more or less homogeneous in environment and defines the most suitable uses and likely effects of development in very general terms for each.

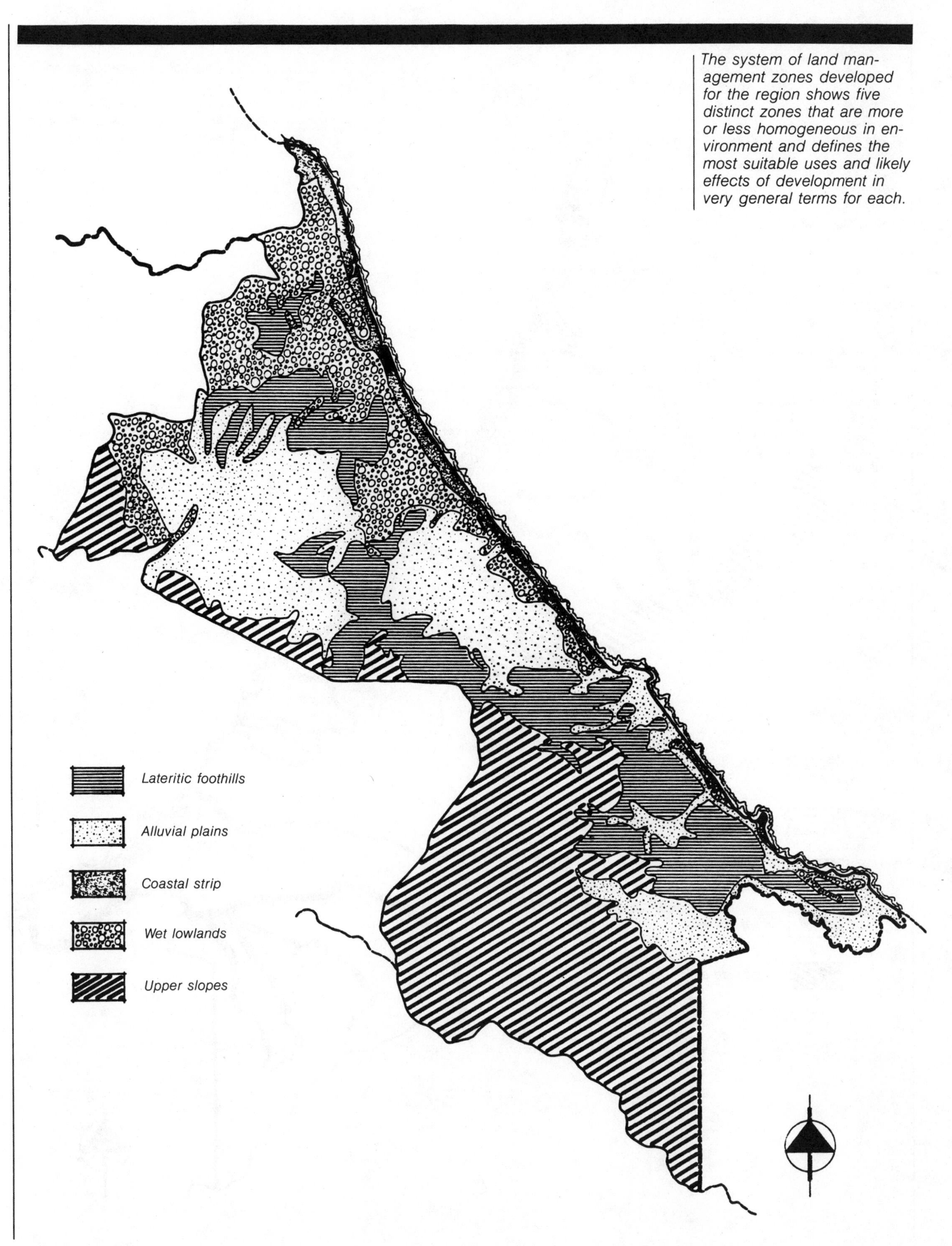

SPECIAL MANAGEMENT ZONES

The special management zones are areas with particular characteristics requiring singularly focused attention, either for purposes of protection or production.

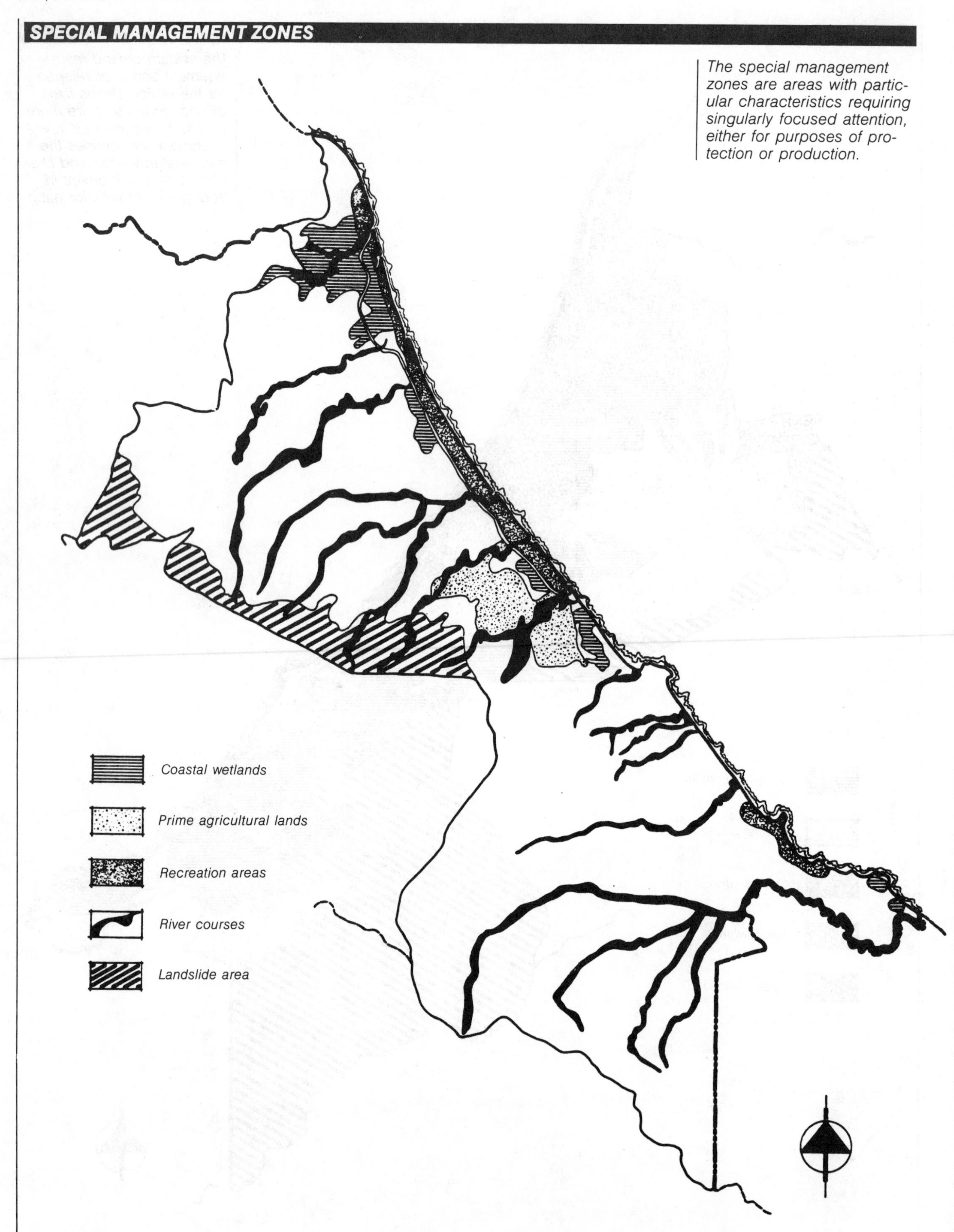

1. Coastal Wetlands

Coastal wetlands are lands near the ocean that are permanently or regularly covered by water. They are important for their very rich wildlife populations and their potential productivity. Wetlands are particularly promising locations for intensive aquaculture. Some aquaculture experiments are already being conducted, and these show great promise as a source of protein. It is important, therefore, that wetlands be protected and managed as wildlife preserves and aquaculture sites.

2. Prime Agricultural Land

One area of about 300 square kilometers within the Atlantic Region features exceptionally high-quality agricultural soils. Since these soils are rare, it is most important to protect and nourish them. It is also economically important to coax a high level of productivity from them. The area is probably a particularly appropriate one for small-scale, intensive mixed farming, combining tree crops, food crops, animal production, and possibly aquaculture.

3. Recreation Areas

The two areas in the region with considerable recreation potential are the beaches adjacent to Cahuita National Park in the southern part and the network of rivers and canals in the northern half. The major attraction of the beach areas south of Limon is a tranquil, fragile beauty that could easily be overwhelmed by heavy development. It would probably be best to encourage small-scale, dispersed, comfortable clusters of cabanas and restaurants, catering primarily to vacationing Costa Ricans and those visitors from other countries who want to experience the unique qualities of Costa Rica's natural environment.

The northern part of the coast has a potential for a somewhat more intense level of development, although the potential clientele here, as well, would probably be that relatively small number of tourists who are more interested in unique natural environments than the concentrated excitement of an Acapulco.

4. The River Courses

The major rivers play extremely important roles in the ecology of the Atlantic Region, and their functions have already been considerably upset by settlement in the area. Deforestation has led to increased sedimentation loads and to more extensive flooding and frequent changes in courses. There are indications that the river waters are badly polluted both by pesticides and herbicides from agriculture and by human wastes.

Management should concentrate on pursuing the following purposes:

1. *Control of forest clearing and each watershed*
2. *Control of agricultural pollution*
3. *Development of sewage treatment systems for riverside communities*
4. *Exploration of hydroelectric alternatives*
5. *Stabilization and control of bank areas*
6. *Flood control*
7. *Regulation of river traffic*

5. Mineral Areas*

The principal mineral resources are the copper deposits in the upper Talamanca Range. Others include possible oil deposits in the Talamanca foothills and extensive bauxite in Pococi. As these resources are exploited, they might present major environmental problems. Any surface mining will inevitably cause damage to the landscape, but it can be minimized by reshaping and replanting as soon as possible after the minerals are removed. Strict environmental planning controls must be maintained over all mining activities in this region.

6. Landslide Areas

There are several major areas of geological instability in the region. The most critical are located in the Talamanca Range and to the southwest of Guapiles and Siquirres. Because of the constant danger of sliding, future development should not be considered for these areas.

**Mineral areas not shown on the map by request of OFIPLAN.*

VEGETATION TRANSECT: NATURAL VEGETATION

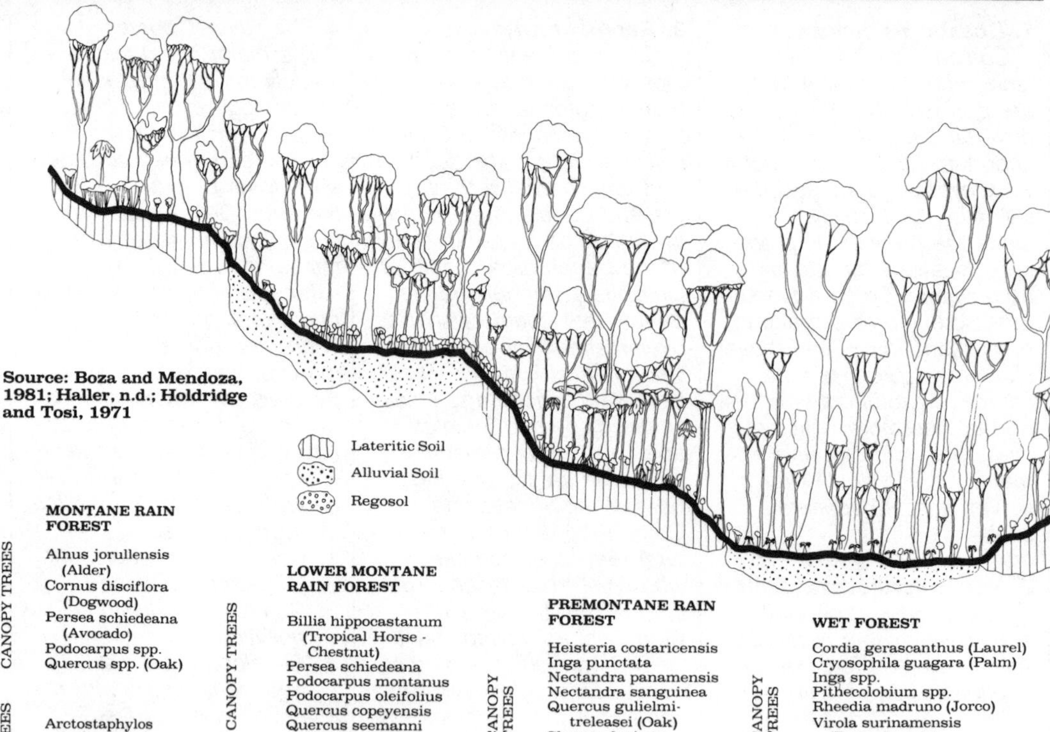

VEGETATION TRANSECT: IDEAL CROPS

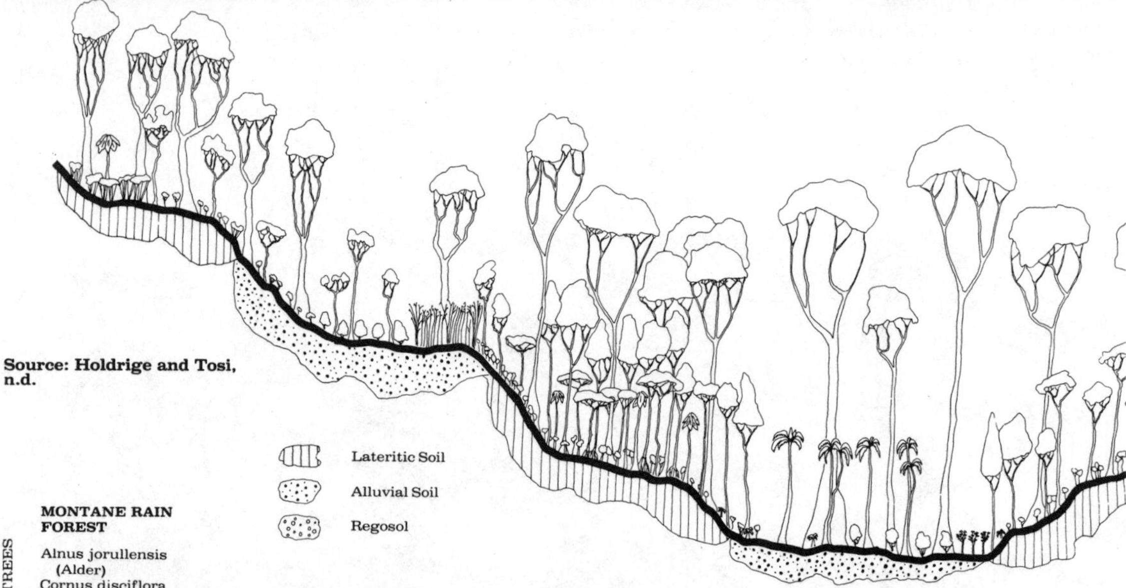

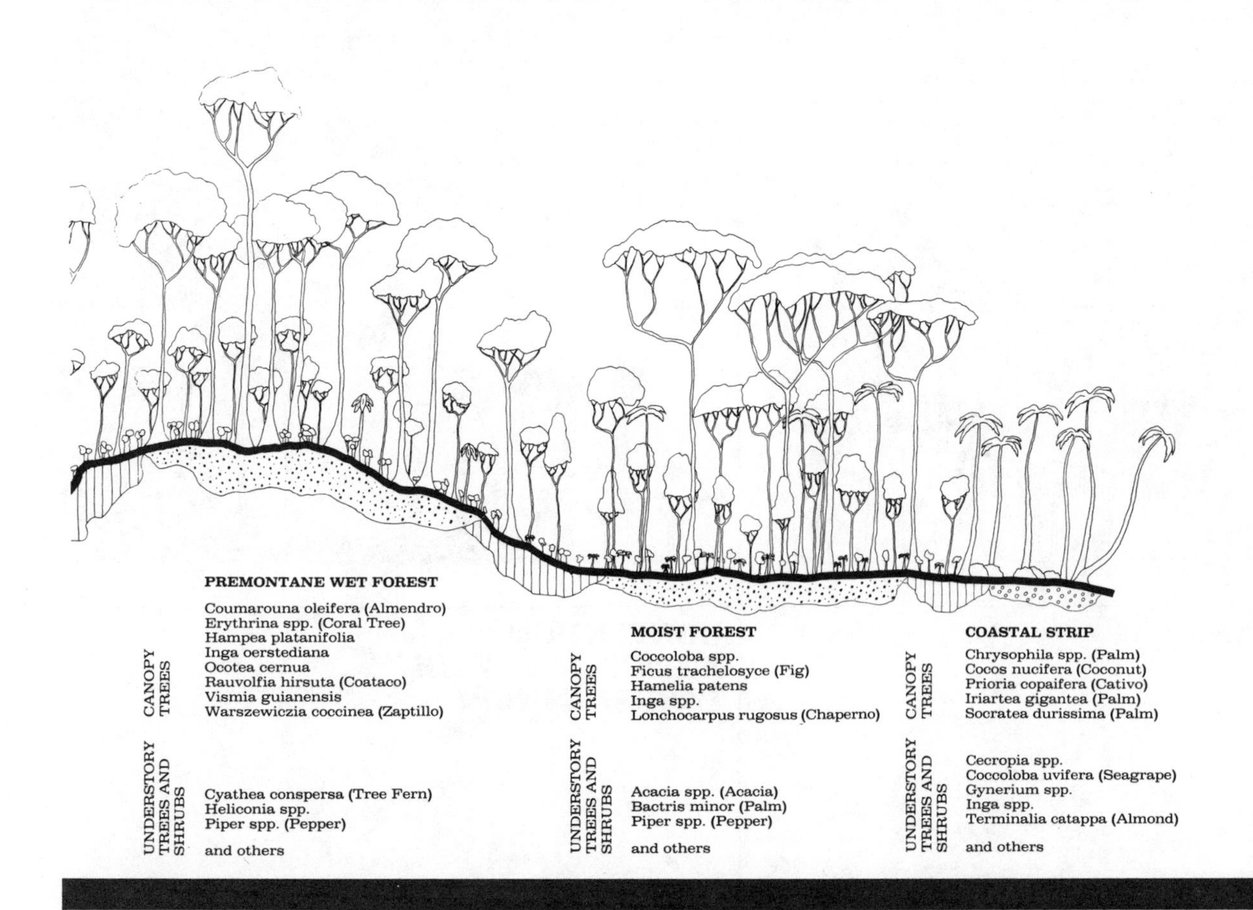
PREMONTANE WET FOREST
CANOPY TREES
Coumarouna oleifera (Almendro)
Erythrina spp. (Coral Tree)
Hampea platanifolia
Inga oerstediana
Ocotea cernua
Rauvolfia hirsuta (Coataco)
Vismia guianensis
Warszewiczia coccinea (Zaptillo)
UNDERSTORY TREES AND SHRUBS
Cyathea conspersa (Tree Fern)
Heliconia spp.
Piper spp. (Pepper)
and others
MOIST FOREST
CANOPY TREES
Coccoloba spp.
Ficus trachelosyce (Fig)
Hamelia patens
Inga spp.
Lonchocarpus rugosus (Chaperno)
UNDERSTORY TREES AND SHRUBS
Acacia spp. (Acacia)
Bactris minor (Palm)
Piper spp. (Pepper)
and others
COASTAL STRIP
CANOPY TREES
Chrysophila spp. (Palm)
Cocos nucifera (Coconut)
Prioria copaifera (Cativo)
Iriartea gigantea (Palm)
Socratea durissima (Palm)
UNDERSTORY TREES AND SHRUBS
Cecropia spp.
Coccoloba uvifera (Seagrape)
Gynerium spp.
Inga spp.
Terminalia catappa (Almond)
and others

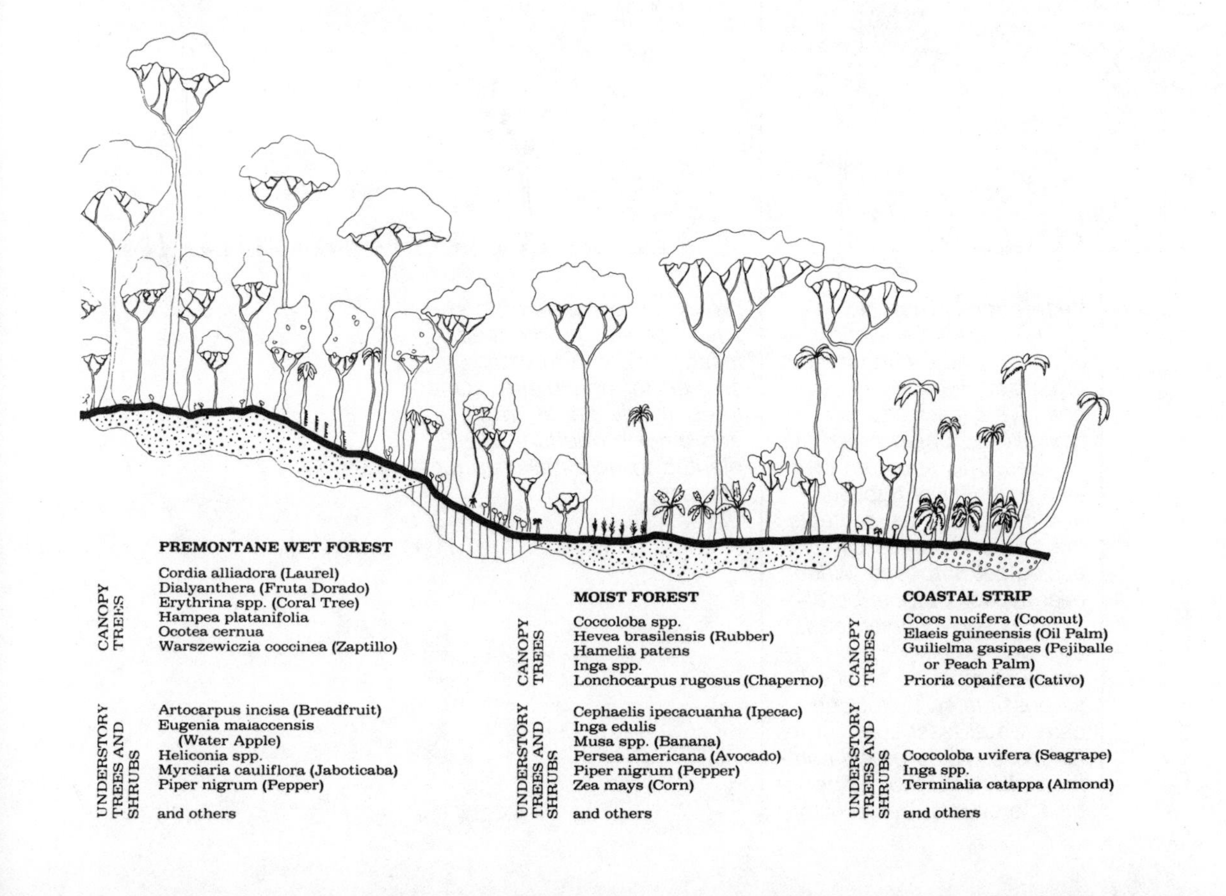
PREMONTANE WET FOREST
CANOPY TREES
Cordia alliadora (Laurel)
Dialyanthera (Fruta Dorado)
Erythrina spp. (Coral Tree)
Hampea platanifolia
Ocotea cernua
Warszewiczia coccinea (Zaptillo)
UNDERSTORY TREES AND SHRUBS
Artocarpus incisa (Breadfruit)
Eugenia maiaccensis (Water Apple)
Heliconia spp.
Myrciaria cauliflora (Jaboticaba)
Piper nigrum (Pepper)
and others
MOIST FOREST
CANOPY TREES
Coccoloba spp.
Hevea brasilensis (Rubber)
Hamelia patens
Inga spp.
Lonchocarpus rugosus (Chaperno)
UNDERSTORY TREES AND SHRUBS
Cephaelis ipecacuanha (Ipecac)
Inga edulis
Musa spp. (Banana)
Persea americana (Avocado)
Piper nigrum (Pepper)
Zea mays (Corn)
and others
COASTAL STRIP
CANOPY TREES
Cocos nucifera (Coconut)
Elaeis guineensis (Oil Palm)
Guilielma gasipaes (Pejiballe or Peach Palm)
Prioria copaifera (Cativo)
UNDERSTORY TREES AND SHRUBS
Coccoloba uvifera (Seagrape)
Inga spp.
Terminalia catappa (Almond)
and others

RAIN FOREST STRATIFICATION

STRATIFIED AGRICULTURE

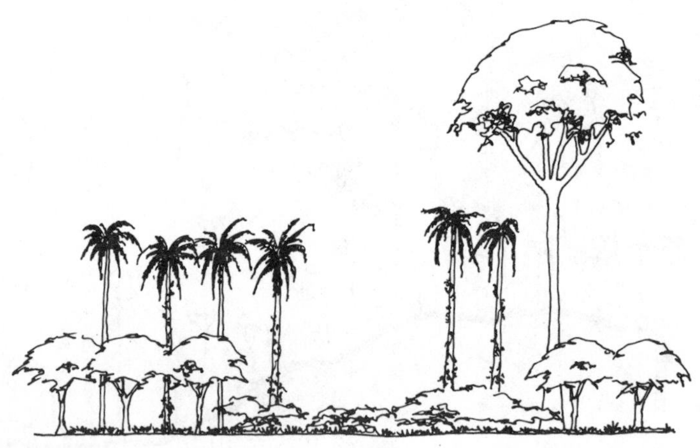

PRODUCTION FOREST

CONVENTIONAL AGRICULTURE WITH PARTIAL CANOPY

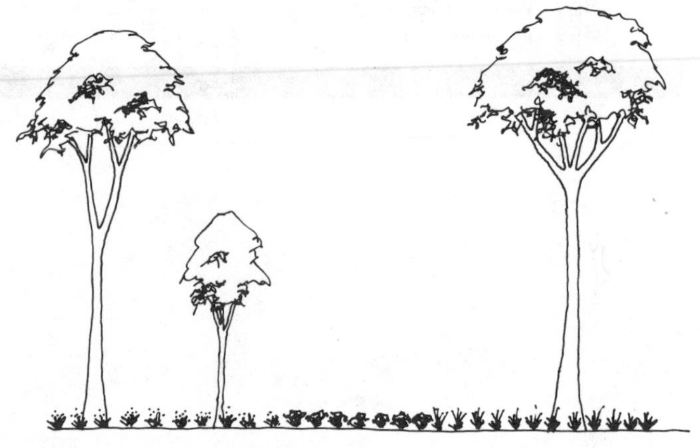

Vegetation Structure

The natural vegetation structure of the rain forest is organized vertically in several levels, each of them typically occupied by certain species. This structure creates a variety of habitats and microclimates, protects the soil, and serves other ecological purposes. Whenever the forest must be altered to serve human ends, as much of this vertical structure must be retained as possible so that it can carry on at least some of its functions. Trees grown commercially, either for timber or other products, can provide protection by assuring continued coverage with their foliage. When a forest is cleared for agriculture, some trees should be left in place and other productive trees planted to help make up for the loss.

SCATTERED RESERVES

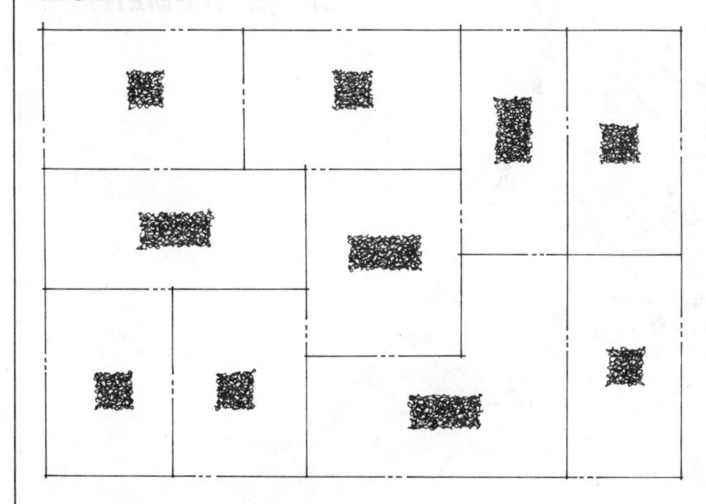

CLUSTERED RESERVES

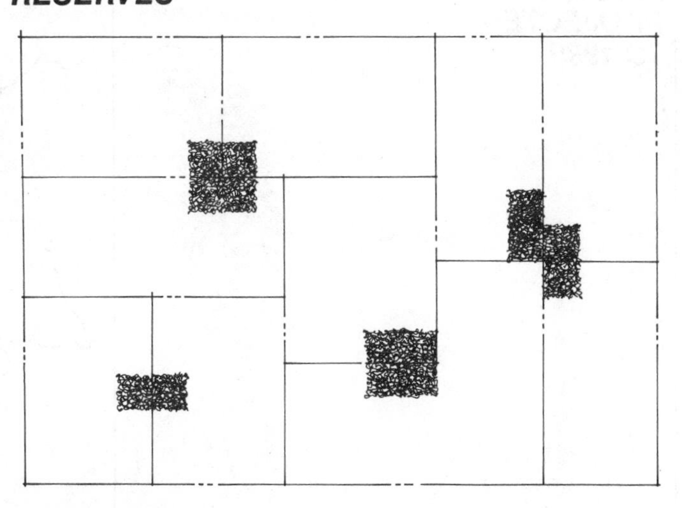

SETTLEMENTS AS ISLANDS CONNECTED BY CORRIDORS

CLUSTERED RESERVES CONNECTED BY CORRIDORS

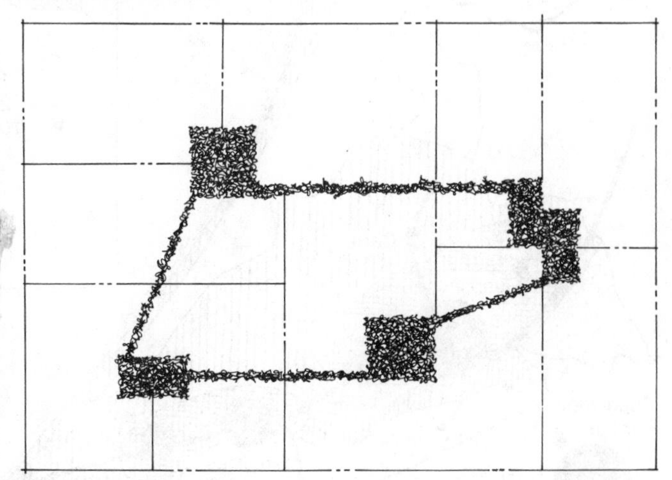

Natural Reserves

The new agricultural settlements sponsored by the government of Costa Rica have often been carved out of the rain forest. Usually, a small natural reserve has been left undisturbed in each settlement. These reserves tend to be widely scattered, in patterns something like that shown in the first diagram. With careful coordination, they might be designed to support much more diverse wildlife populations. The same total land area would enjoy greater species diversity if the reserves were larger, although fewer in number, and located closer together. Diversity would be further increased if the reserves were connected with natural corridors, preferably following rivers and streams. Ideally, settlements would then be established as islands within and surrounded by the continuous natural matrix of the larger landscape; to bring this about, however, would require more land and greater distances between settlements.

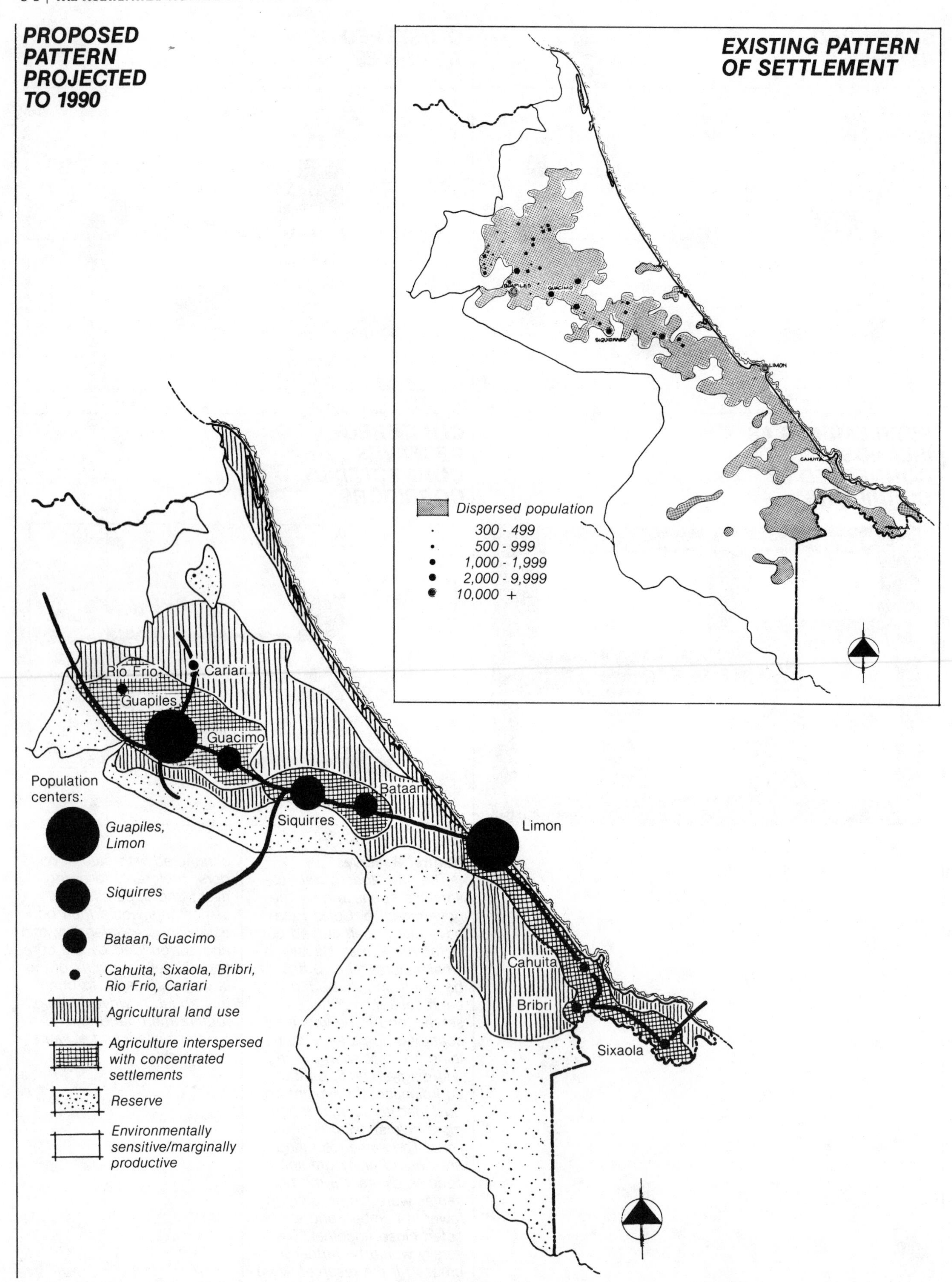
PROPOSED PATTERN PROJECTED TO 1990
EXISTING PATTERN OF SETTLEMENT
Dispersed population
300 - 499
500 - 999
1,000 - 1,999
2,000 - 9,999
10,000 +
LIMON
CAHUITA
Rio Frio
Cariari
Guapiles
Guacimo
Bataan
Siquirres
Limon
Cahuita
Bribri
Sixaola
Population centers:
Guapiles, Limon
Siquirres
Bataan, Guacimo
Cahuita, Sixaola, Bribri, Rio Frio, Cariari
Agricultural land use
Agriculture interspersed with concentrated settlements
Reserve
Environmentally sensitive/marginally productive

fully, and the capacity to support human life declines. Such a pattern seems to have led to the extinction of the Danubian culture, for example.

In Latin America, population growth has been accompanied by pressures to clear rain forests to make room for permanent cropping of the European and North American sort, and this is causing serious problems. The reason lies in the structure and function of the rain forest ecosystem, primarily in the interaction between soil and water. In the rain forest, nutrients are stored mostly in the biomass of the plant communities rather than in the soils, which are generally thin and infertile.

Lateritic soils, the reddish and yellowish types with high contents of iron and aluminum oxides, are the most common. When their vegetation cover is removed, the few nutrients they have are quickly leached out of them by the heavy rainfall. Such land can produce some crops for a few years, but yields become less each year until the soils are exhausted and can support no growth at all. Fertility can be sustained only by very heavy and expensive applications of fertilizer. When plants can no longer grow, topsoils are washed away. The erosion rates, especially on steeper slopes, are startling. Layers of aluminum and iron oxides are then exposed, and these can dry out to form a surface crust called laterite, a substance so hard that it is used in some places as a building material. The crust fends off seeds of colonizing species. Consequently, not only is the land incapable at this point of producing crops, but it also repels attempts by the forest to reestablish itself. Moreover, since most of the moisture for rainfall is provided by evapotranspiration from the trees, extensive clearing can bring about changes in climate. Decreases in rainfall of as much as 21 to 24 percent have been observed to follow widespread clearing of forests in the Amazon basin of Colombia.

At the same time, the Huetar Atlantica region is populated by poverty-stricken people who desperately need better means of feeding themselves. Costa Rica has an active and effective program of land distribution, but it has difficulty keeping up with the demand. It is common practice for landless people to invade parts of the forest and clear them to plant crops, regardless of soil quality. As the need for agricultural land grows, the government has also begun establishing settlements in the rain forest. Whether legal or illegal, agricultural settlements in most parts of Huetar Atlantica will be hard to sustain over long periods. The conflict will not be quickly resolved, of course. Hungry people are not likely to be swayed by arguments of ecological imperatives. What might help is to find ways of coaxing some levels of productivity from the land while maintaining the rain forest ecosystem.

Even within a rain forest, the environment varies, although the variations in the overall pattern of lush growth are not very obvious. Rainfall varies from high to extremely high (about 2.4 to 4 meters per year), and the soils are not all lateritic. There are sizable stretches of relatively fertile alluvials in this region that can sustain agricultural production. Since some areas are more suitable for agriculture and settlement than others, each can be used in a different way to achieve some balance of production and preservation.

Given the pressing nature of the conflicts, and the institutional base within which any plan for this region will be effected, the implementation device has to be simple and straightforward. A sophisticated information system of the sort developed for Alaska or Metropolitan Boston would be useless.

General Management Zones

The device proposed by the Cal Poly consulting team for Huetar Atlantica and illustrated on pages 54–64 is a system of General and Special Land Management Zones. There are five general zones, each characterized by a different and fairly consistent combination of soil type, plant community, temperature, and rainfall. Within each of these zones, the potentials and limitations for various crops, the likely environmental impacts, and the effective control measures are usually the same. This means that land management decisions and practices in each zone can be based on a uniform set of guidelines. These guidelines describe agricultural practices and control measures that replicate at least to some extent the functional and structural devices used by the natural ecosystem to maintain itself, which means primarily to protect the soil. They are outlined in the charts that accompany the case study. Because the resources available for environmental management are so limited, the guidelines are extremely simple, concentrating on essentials, with little attention to minor matters. A brief description of each zone will be helpful in explaining the management zone concept.

The Coastal Strip of Costa Rica is a narrow stretch along the ocean's edge, extending from the Panamanian to the Nicaraguan border. It is a beautiful, tranquil landscape, covered mostly by palm trees. Over forty species of palms are native to the area. The climate is hot and humid (average temperature of about 25°C, average rainfall of about 2.8 meters) but less so than most of the Atlantic Region. Small-scale agriculture has long been practiced in this zone and has usually been successful in the sandy soils ("regosoles" in the international parlance). Most successful have been the palm tree crops that follow the patterns of the natural plant communities. Coconut, oil palm, pejibaye, and palmito fit the environment and have great economic potential.

This basic principle—encouraging human cultivation to follow the example set by natural communities—is one that might well be adopted in all development activity in this region. It is an especially promising approach in the zone called the Lateritic Foothills. These rise from the coastal plain in fairly gentle slopes, forming the lower skirts of the mountains. Their soils are deep red and yellow laterites, and they are densely covered by very humid tropical forest. Rainfall is higher than that along the coast, ranging from 2.4 to 3.8 meters (about 95 to 150 inches). This combination of qualities is most inadaptable to clearing and settlement for the reasons explained above. For all that, the slopes are manageable and accessible, and the location is near existing population centers. The pressure for settlement is therefore enormous, and, in fact, settlement is rapidly taking place.

Any activities in this zone that remove the plant cover from the soil, even temporarily, will cause a certain amount of leaching, erosion, and fertility loss. If the cover can be quickly reestablished, however, and the soil protected, then some forms of economic productivity can be sustained. The best means of accomplishing this is by planting tree crops. Cultivated trees can protect the soil in the same way that natural vegetation does. Several of the native hardwoods that have been overly exploited in so many natural forests, but unaccountably never grown commercially, have extremely high economic value, and some grow fast enough to yield lumber in less than thirty years.[24] Thus, it seems probable that native hardwood plantations can be an ecologically compatible and productive use of the lateritic foothills. For the shorter term, food-producing trees like cacao, macadamias, almonds, and other nut and fruit trees can also be compatible. Diversity is especially important here because of the constant threat of pests that can strike at any time and destroy the entire population of a single species.

Another problem lies in the fact that tree crops are almost entirely a commodity for export, and although they will stimulate international trade and produce cash income, they will not directly feed the population of the region. For this purpose, the Alluvial Plains will have to be much more intensively cultivated. This zone is uniformly fairly flat and covered by rich alluvial soils. The natural plant community is moist tropical forest. By overlaying maps showing existing population centers and transportation routes on the management zones, we can see that, probably through a combination of good sense and trial and error, most settlement and development up to very recent times had occurred on these plains. Only when the population grew beyond the numbers that this zone could easily support did settlement begin expanding into the lateritic areas.

In fact, however, the alluvial soils can be much more intensively used than they are now to produce food. By improving farming practices—in some cases, for example, by introducing measures as simple as contour plowing and crop rotation as well as by careful choice of crops and in others by shifting from export to food crops—the productivity of the alluvial areas can be multiplied. Various mixed cropping systems also have great potential, especially those that use human labor to best effect and minimize the use of fossil-fuel-burning machinery with its very high capital and energy costs. By careful design of such mixed cropping systems, the productivity of these fertile areas could probably be increased severalfold, thus relieving much of the pressure on the more sensitive, less fertile lands.

The other two zones have very limited capacity for economic use. The Wet Lowlands feature extremely high rainfall (about 4 meters, or over 150 inches per year), very flat land (with slopes of 0 to 5 percent), and tropical wet forest cover. The soils are mostly alluvials, but because of the level of rainfall and the flat topography, they are continuously waterlogged in a great many places. Since this condition becomes even worse when the vegetation cover is removed, most forms of agriculture are not feasible here. Traveling through a part of this area, I saw cornstalks wobbling, if they had not already fallen, because the water-saturated soils could not maintain a hold on their roots.

[24] See Tosi, 1971.

This is another situation where economic activity might well echo the natural communities. Paddy rice, for example, would probably thrive under these conditions. Game ranching has considerable potential as well, since wildlife populations are prodigious and diverse. Instead of trying to raise cattle under such hostile conditions, as many farmers do, raising the native species might hold much greater promise. There are several kinds of edible wild pigs that thrive here, for example, as do alligators, which could be raised for their hides.

The Upper Slopes are even less hospitable to human beings. The rainfall there is about the same as in the wet lowlands (4 meters), but the soils are lateritic, the native vegetation is pluvial forest, and the slopes are mostly over 40 percent. These conditions combine to preclude virtually any economic land use without disastrous consequences. These areas should all be set aside as natural preserves.

The five zones, then, if they are managed with awareness of their possibilities and limitations, might support a broad range of uses without destroying their life-support capacity. They can provide food, work, and economic return for at least the number of people who live in the region now, although how much more population growth they can sustain is a question that requires more analysis. We can summarize the major suitable uses in each zone as follows:

Coastal Strip:	Palm tree crops; small-scale food production
Lateritic Foothills:	Tree crops; forest preservation
Alluvial Plains:	Intensive food production
Wet Lowlands:	Limited production of native species; forest preservation
Upper Slopes:	Forest preservation

Special Management Zones

Within this overall framework of general zones, there exist a number of smaller areas that feature particular and unique potentials and/or sensitivities and that will therefore require more specifically detailed management programs. These include the following:

Coastal Wetlands: These are highly productive and have great promise for intensive aquaculture, but they are extremely sensitive and susceptible to damage by erosion and pollution from the lands that drain into them.

Prime Agricultural Lands: These constitute a small area with the best soils in the country and therefore must be managed with extreme care.

Recreation Areas: Areas where environmental amenities are attractive and are therefore beginning to attract sizable numbers of vacationers. The temptation to overexploit them for economic return is great, and they could easily be overwhelmed and degraded.

River Courses and Their Flood Plains: Both of these areas are dangerous for settlement and critical to ecological function. Some have hydroelectric potential whereas others have transportation potential. All are in need of pollution control programs.

Mineral Areas: These are mostly copper deposits located in places where uncontrolled mining can cause serious damage.

Landslide Areas: In these, the rocks are extremely unstable and dangerous.

Planning Integration

The landscape management program given here is actually just one part of a larger regional planning effort. Besides myself, the landscape planner, the team included a regional (economic) planner, Professor Sylvia White, who directed the project; an agricultural economist; a transportation planner; and a demographic planner. The problem was to bring together the two different approaches to regional planning, the economic and the ecological, that we discussed earlier. The two are not mutually exclusive and should clearly be brought together if we are to have any chance of reconciling economic development with ecological integrity. In the example of Huetar Atlantica, the General Land Management Zones provided a framework for the other physical aspects of the plan.

The map on page 64 shows the existing population distribution. As I mentioned earlier, this tells us that settlement has historically followed the soils that could support it. Until recent years, the only inhabitants of the upper lateritic slopes were small Indian tribes, of dispersed populations, who do little clearing and live very lightly on the land almost entirely apart from the economic system. The surge of settlement by people from outside is a recent phenomenon.

Nevertheless, the agricultural and economic studies supported the ecological analysis in the contention that the alluvial plains could actually support much larger populations than they do now. Simple improvements in agricultural practices and more careful selection of crops could greatly increase crop yields, and the development of agricultural industries could provide a great many more jobs on very little land. Consequently, the most promising regional development policy for Huetar Atlantica is an increasing intensity of use in areas that are already settled along with controlled expansion of existing towns. This is the pattern shown in the development plan on page 64. The alternative approach, which would be a policy of population dispersal and colonization of new lands, would be environmentally destructive and economically unrewarding. However, there will inevitably be some exploitation of the lateritic foothills and the wet lowlands. If this can be limited to areas contiguous with development on more suitable lands, as suggested in the development plan, and if it can follow the practices recommended for each management zone, then these marginal lands can make some economic contribution without disastrous environmental deterioration.

Such an integration of economic development planning and land management planning offers great promise for the overwhelming land-use problems of the developing countries. The combination of a sensitive landscape and land-hungry, food-hungry people is unfortunately quite a common one in the more economically deprived parts of the world. Certainly, a map can do very little to solve the shortage of capital, the concentration of land ownership, the lack of technical skills, and all the other problems that bedevil efforts to use land wisely in these areas. Nevertheless, it can point in the right direction.

RIVER SYSTEMS AS REGIONAL NETWORKS

The lines of the river courses that wind through Huetar Atlantica in the map on page 58 illustrate the fact that differing regions are knitted together by the flow of water. The richest, most diverse communities of plants and animals in a region tend to cluster along these river courses and their related streams, ponds, and marshes. Human beings are attracted to the edges of rivers as well. Historically speaking, in temperate climates (although usually not in the tropics, where rivers are often disease-ridden) as new regions are settled, the land along the river bottoms is cultivated first, and the earliest towns are located on their banks. This was the pattern followed by the earliest civilizations along the Indus, the Nile, and the Tigris and Euphrates, and it was followed again in the settlement of the American South and Midwest. The river courses served as tentacles of development, which later spread to fill the spaces between. In more recent years, people have been more frequently drawn to rivers for recreation.

Given the fundamental importance of rivers and their related features in the regional landscape, it makes sense to concentrate planning attention on them. In his 1962 study of the landscape and recreational resources of Wisconsin, Phillip Lewis found that over 90 percent of the attractive, high-quality landscapes in the areas he called "environmental corridors" followed rivers.[25] Within these corridors, he identified four surface types that needed particular management: water, wetlands, flood plains, and sandy soils. His plan for recreational development and landscape conservation in Wisconsin then took the form of a network of corridors or "heritage trails" that would lace through the state.

Although river systems have long been used as a physical context for regional planning, the overriding concern in the past has invariably been economic development—the providing of power, industries, and jobs. This phenomenon has been true the world over. Noted examples include the Missouri River plan, the Volta River project, the Aswan Dam, the Karibe Dam, and the Damodar Valley project. In projects like these, rivers have rarely been considered as ecological systems, and the results have often been disastrous.

Probably the first serious mistake made in conceiving these gargantuan efforts is usually the failure to understand fully what is there before the heavy machinery goes to work. When an adequate inventory is made of a river's resources, the very magnitude and diversity of the description itself is usually enough to inspire thought and care. Such a description gives some notion of the multifarious ways in which a river, or even a small stream, serves a variety of functions and plays myriad roles in daily life as it winds through a region and knits the parts together. Most obviously, it collects the water from the land and carries it to a larger

[25] See Lewis, 1964.

stream and ultimately to the ocean. Not quite so obviously, it slowly reshapes the land it passes over, carrying soils from uplands to lowlands and carving away inches of rock each century. It also plays a pivotal role in maintaining the ground water regime since, in most places, water percolates through the river bottom and related marshes and lakes into the water-bearing rock strata beneath. Finally, it supports a rich community of plants and animals. The riparian associations growing along the river banks and in the marshes and flood plains are usually far more varied and productive than those in surrounding areas. The riches are passed on up the food chain to the highest carnivores, which also are usually abundant near rivers. As we have seen, people are attracted to the rivers because of their fertility as well as for their own recreation.

CASE STUDY: THE LOS ANGELES REGIONAL FLOOD CONTROL SYSTEM

For those rivers passing through cities, however, it often happens that their diversity of roles is reduced to a single one: flood control. The example of the rivers that flow through the City of Los Angeles is not unusual. In this dry climate, all three rivers are small and intermittent; they are vital to the regional water regime nonetheless. The earliest settlements, the Spanish pueblo and missions, were laid out near the river banks for easy supplies of irrigation water. Later, in the late nineteenth and early twentieth centuries, when the central parts of the city developed, industries were also drawn to the rivers. Though a number of observers, beginning with the earliest explorers, had pointed out the frequency of their flooding and their tendency to change course in the soft alluvial soils of the coastal plain, buildings were built right up to their banks. When the rains came, predictably there were problems. Buildings were flooded and foundations undermined, and their owners called for help. The U.S. Army Corps of Engineers was ready with a solution: concrete. By 1940, all of the rivers had been straightened and lined with concrete; once the rest of the region urbanized, the concrete channels reached to the mountains. Major earthen dams were built upstream to impound runoff water during the rainy season. This regional network is shown on page 69.

As a result, in the process of solving one problem related to the river regime—flooding—all other functions and roles of the rivers were seriously impaired. The riparian community of plants and animals was drastically reduced. Recreational opportunities were lost. Silts were trapped behind dams, cutting off the flow of eroded materials that help build beaches. The famous Southern California beaches, their source of sand replenishment gone, began to grow narrower and narrower until some of them virtually disappeared.

Ground water depletion also became a problem since there were fewer and fewer areas in the city where surface water could effectively percolate into water-bearing rock strata. This is illustrated by the section on page 70, which shows a clay lens extending under most of the coastal plain, effectively blocking the percolation of water into deeper aquifers. Only in a relatively small area called the "Forebay" can water make its way into both shallow and deep aquifers. Unfortunately, in the early years of this century, the Forebay area was developed for industrial uses and mostly paved over. Ground water recharge was seriously inhibited.

Thus what can be done with this regional river system, now become concrete channels that do nothing but convey rainwater as quickly as possible to the sea? It will not be a natural river system again, but we can reshape the system to restore much of the ecological diversity and the multiplicity of human uses. The case study on pages 69–74 illustrates some of the means for accomplishing this. These possibilities were carried out as a graduate project by William Snowden under the sponsorship of the Los Angeles County Flood Control District. The network was explored at the regional scale because it extends throughout the entire urbanized region, but there is little to be done at the plan unit, project, or site scales. The next step, as shown in the detailed designs, must therefore take place at the construction scale.

The most pressing issue concerns ground water. Two major means of ground water recharge are being established by the Flood Control District: spreading grounds and injection wells. The locations of those already developed are shown on the map on page 70. This map also shows the extent of the permeable Forebay already mentioned. The next map shows the areas most suitable for ground water recharge—those within both surface aquifer areas and Forebay. The map that follows shows proposed locations of inflatable dams—an entirely feasible technology that can retain water when it is desirable to do so and let it flow the rest of the time. Behind these dams, short sections of the channels can be reconstructed as shown in the typical design on page 71. These sections will be widened to hold water with soft bottoms for percolation. The banks will also be soft, with riparian vegetation established to stabilize them and provide wildlife habitat. Spreading grounds can also be designed as shown in the next drawing to reestablish riparian communities.

As for recreation, the U.S. Army Corps of Engineers, whose attitudes have evolved noticeably since the early flood control era, in conjunction with local park departments are developing a network of trails to interconnect entire urban regions along the edges of the channels. The map on page 72 shows existing and proposed bicycle and equestrian trails. The section drawing shows how these trails are located in relation to the channels, along with other recreational uses. It also shows the use of the low-flow channel—the smaller channel within a channel—as a canoeing run. During the dry summer season, the controlled release of small amounts of water from upstream impoundments will make such a run possible.

Such controlled releases can also provide water for irrigation. Thus, intensive agricultural plots can be established along the channel edges with water diverted to irrigate them in very much the same manner as has been done along natural rivers since Mesopotamian times. In this case, sewage water can augment the water supply by being fed through a series of treatment ponds, as shown on page 74, beforehand,

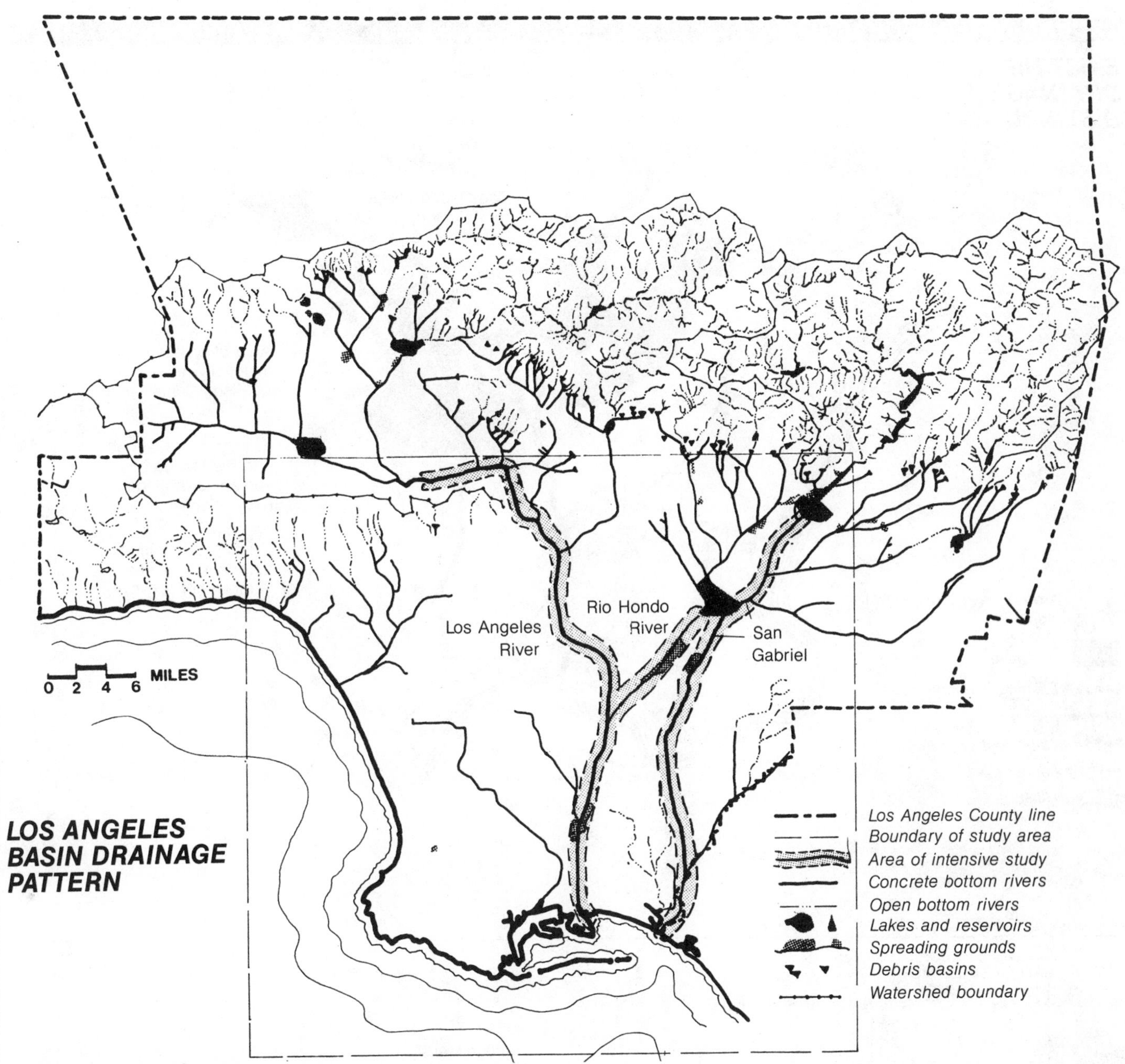

CASE STUDY VI

Multiple Use of the Los Angeles Regional Flood Control System

As Los Angeles grew in the early twentieth century, the three small rivers that flowed intermittently through the city came to be confined within concrete channels for two reasons: to prevent the disastrous floods that occurred regularly and to fix the courses of the rivers, which had tended to change frequently, thereby disrupting development along their banks. Although these two goals were achieved, it has since become obvious that the channeling made it impossible for the rivers to perform most of their other ecological roles. Chief among these is the percolation of water into underground storage. The concrete channels also ruined natural settings for potential recreational activity.

Some of the functions of natural rivers might yet be restored to this concrete flood-control system. Of the many possibilities, a few are shown here.

Prepared for the Los Angeles Flood Control District by the 606 Studio, Department of Landscape Architecture, California State Polytechnic University. William Snowden, with advisers John T. Lyle, Arthur Jokela, and Jeffrey K. Olson.

GROUND WATER RECHARGE

EXISTING DRAINAGE DISTRIBUTION

Narrows
San Gabriel Valley
(not studied)
Narrows
Area of proposed
injection wells
by Flood Control Agency
Existing San Gabriel
and Rio Hondo
spreading grounds
(very effective)
Inglewood
fault zone
Existing Dominiguez
spreading grounds
(largely ineffective)
Area of
saltwater
intrusion

Area of severe groundwater depletion
Forebays
Existing injection wells
Rivers
Water lost to ocean
Faults
Hills

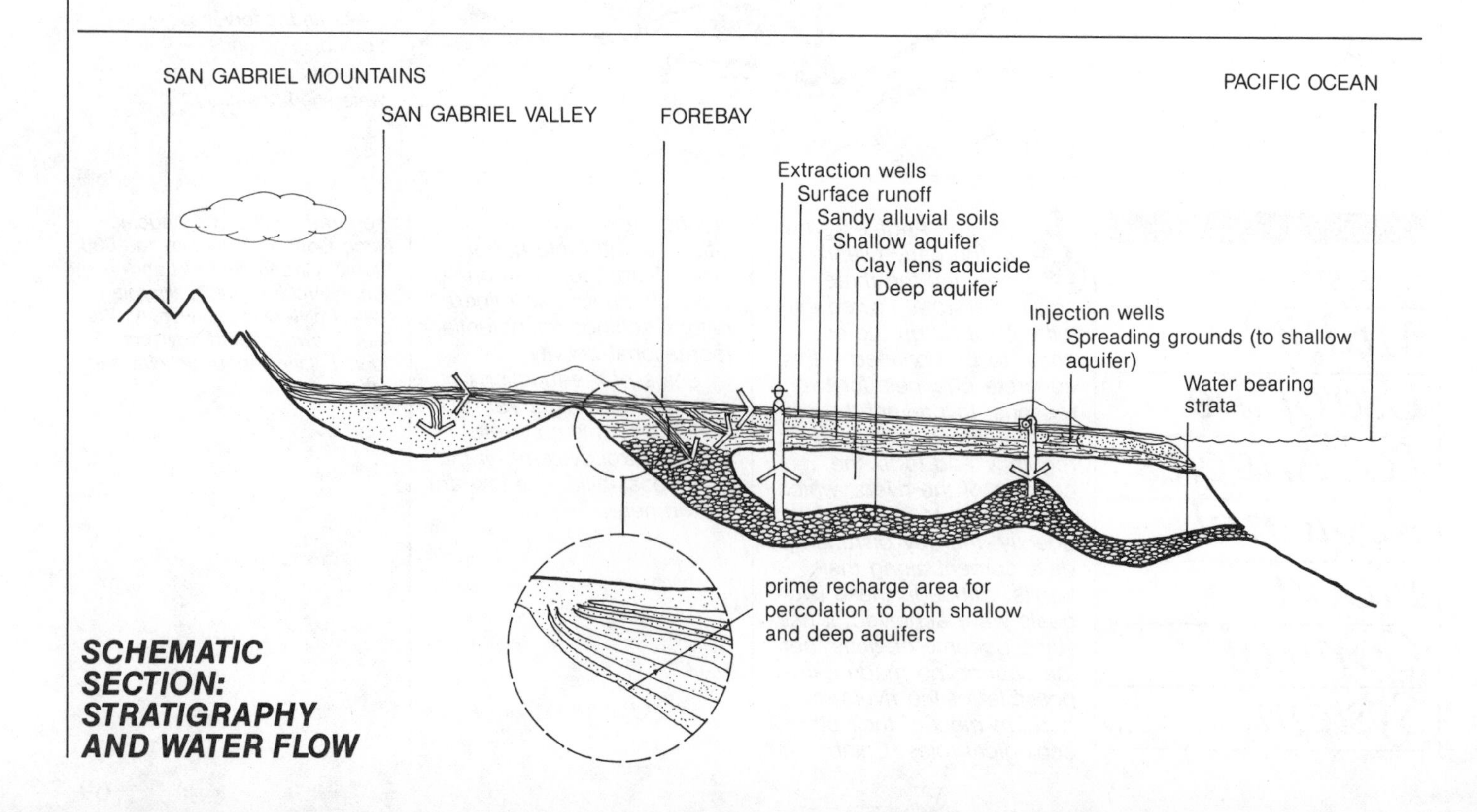

SCHEMATIC SECTION: STRATIGRAPHY AND WATER FLOW

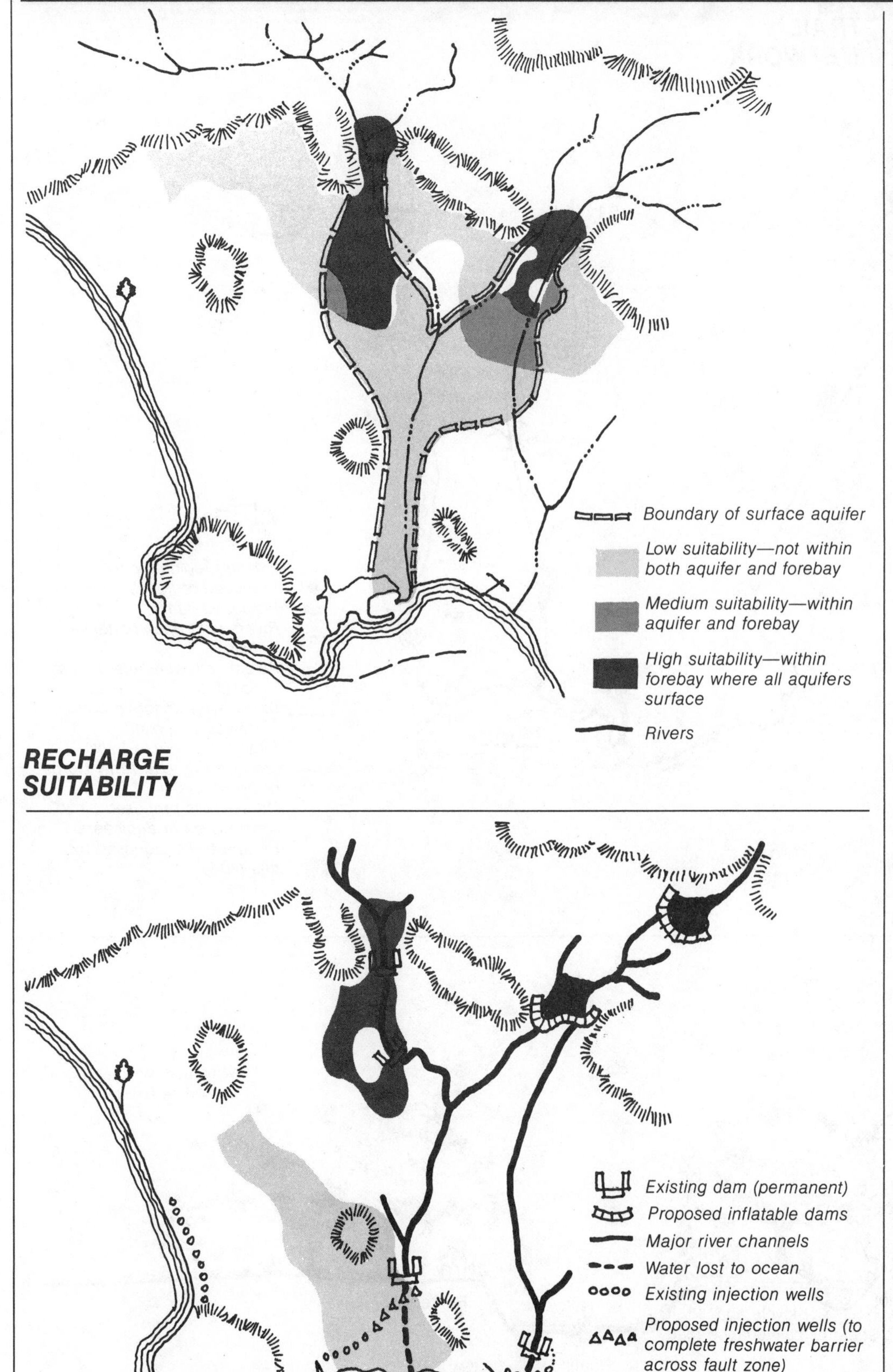

Settling ponds can provide some recharging of the underground supply, but there are not many locations where they can work effectively. Recharge of both shallow and deep aquifers can occur only within the forebay area since it alone is free from the large clay lens that underlies most of the basin and prevents water from passing through. The analysis here shows the most suitable recharge locations.

RECREATION

TRAIL NETWORK

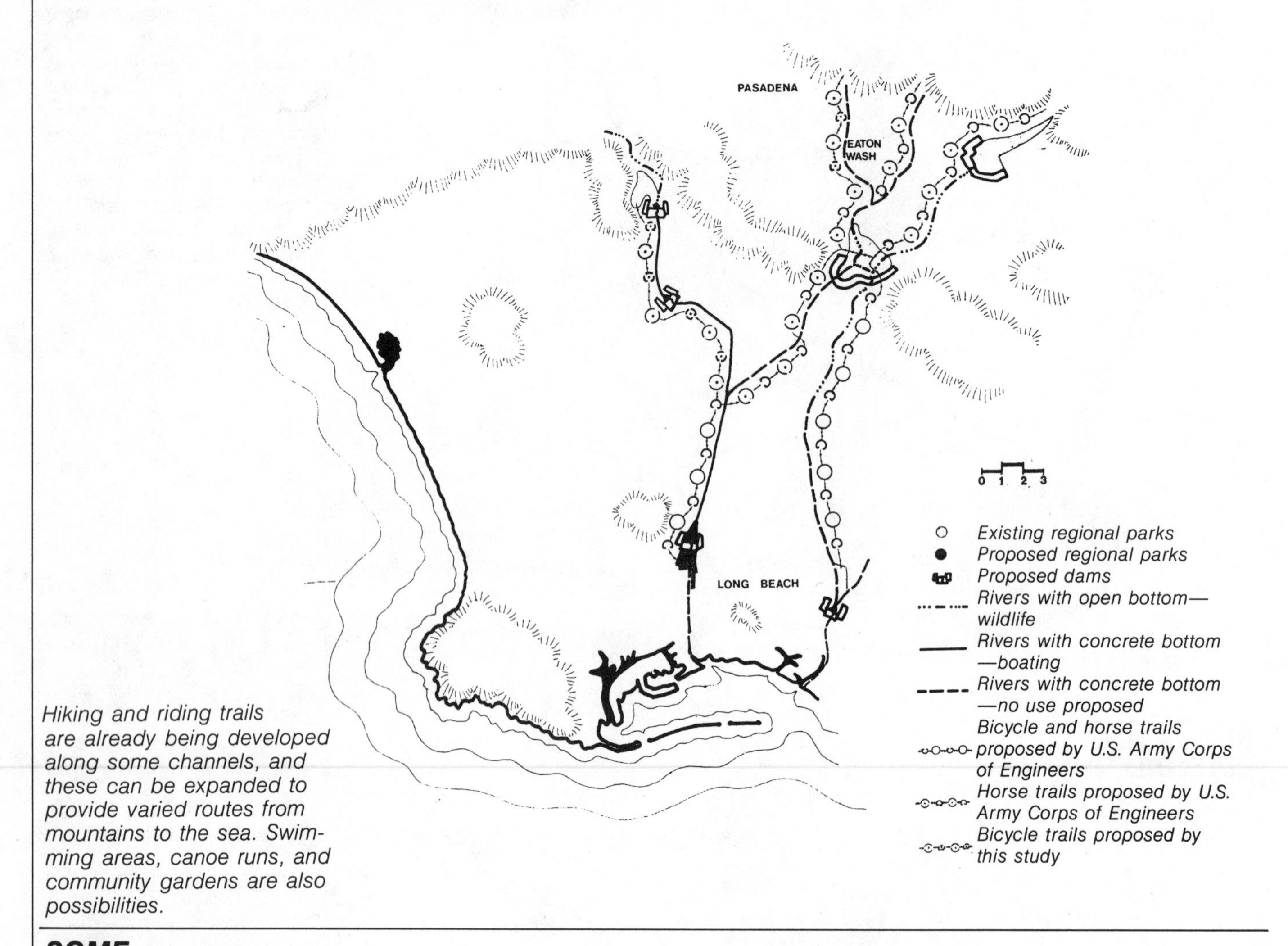

Hiking and riding trails are already being developed along some channels, and these can be expanded to provide varied routes from mountains to the sea. Swimming areas, canoe runs, and community gardens are also possibilities.

SOME RECREATIONAL USES

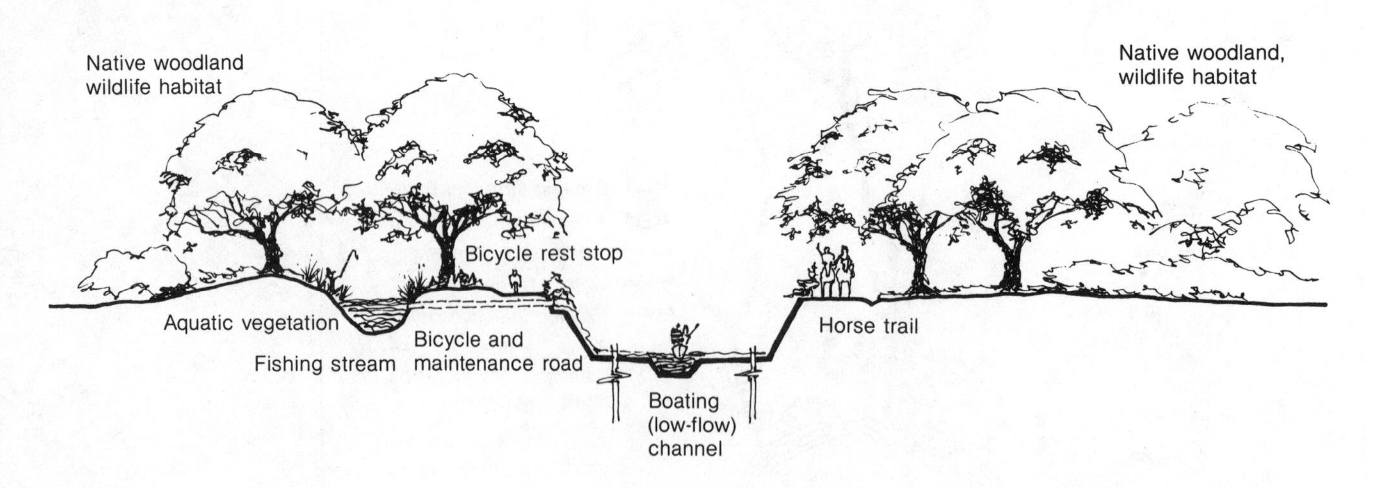

PRODUCTIVITY

BIOTIC PRODUCTION

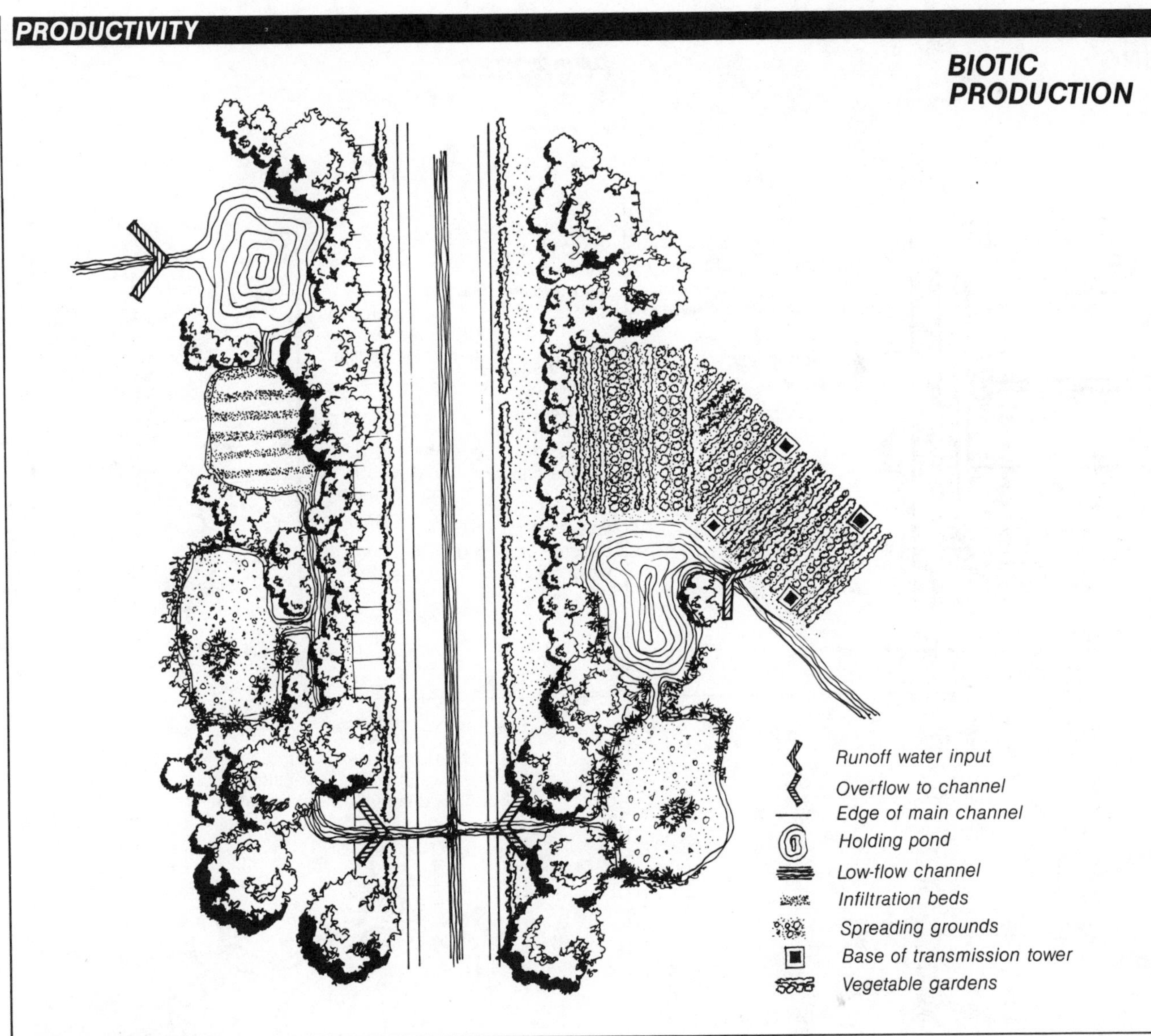

SOME PRODUCTIVE USES

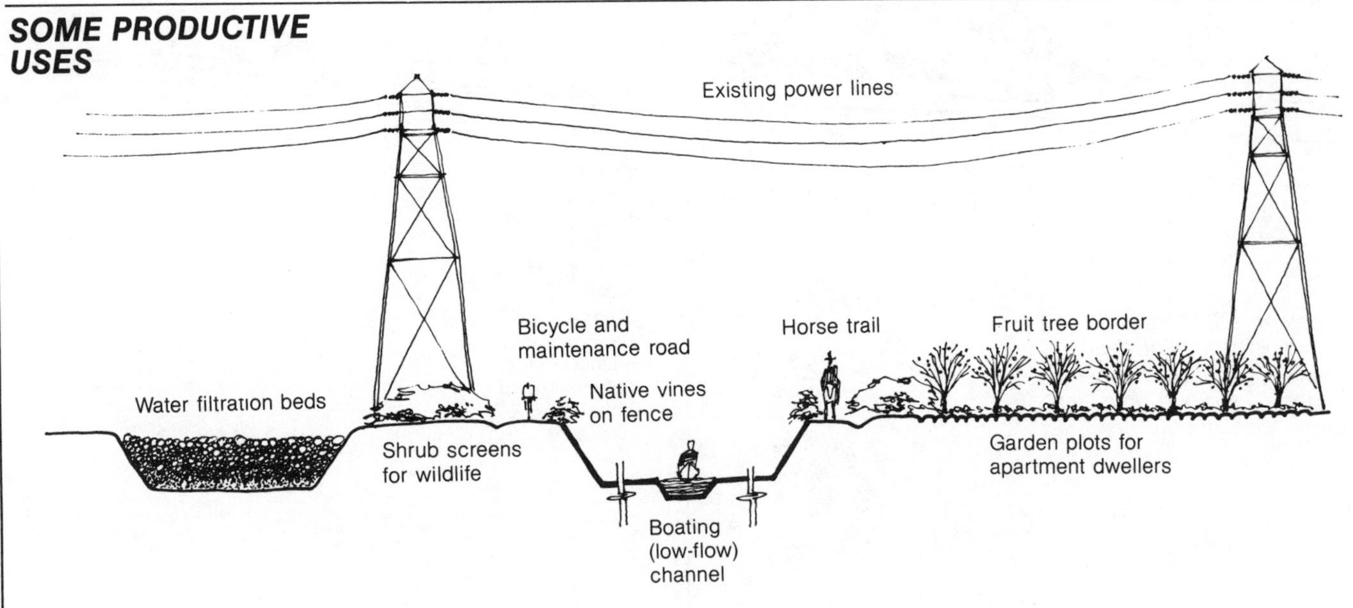

WILDLIFE HABITATS IN THE SPREADING GROUNDS

EXISTING

Flood control district policy is to remove all vegetation from streamside and levees

Low-flow channel at side of river

Water committed to series of ponds

Typical open bottom river in recharge area

Wastewater introduction

By paying attention to the principles of habitat design, the settling ponds can be shaped to support considerable wildlife populations.

PROPOSED

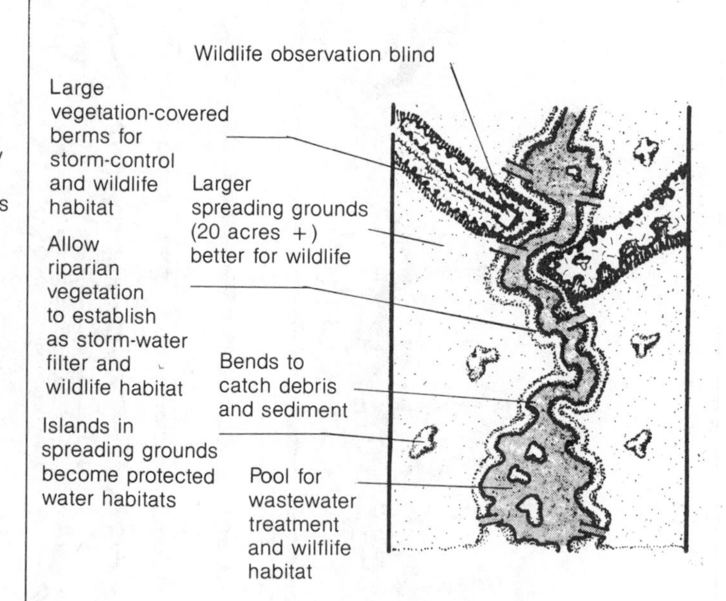

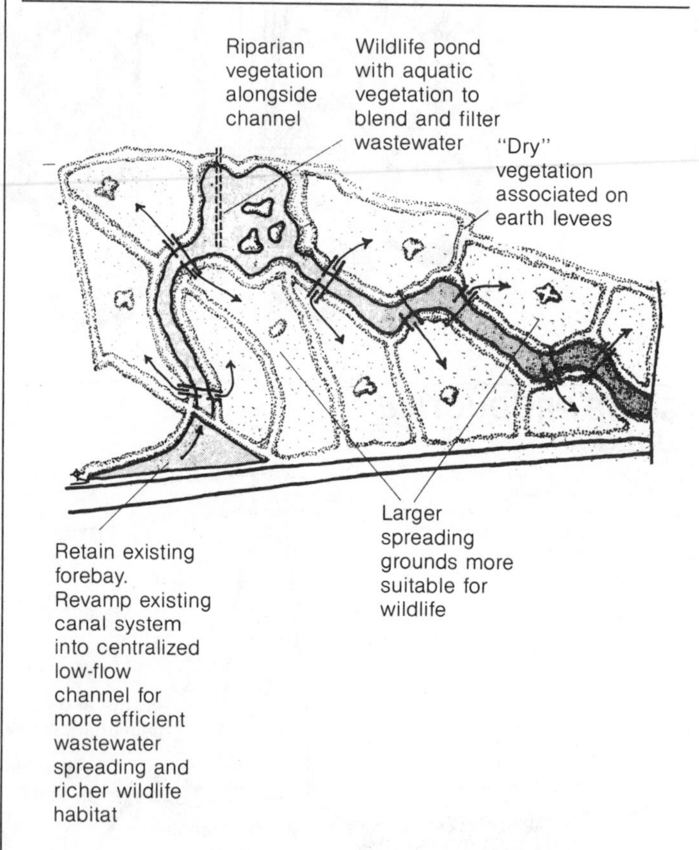

with runoff being drained into the channel and carried to the ocean.

In this way the diverse roles of rivers can be retained or restored even in heavily urbanized regions where the human influence is dominant. In fact, the Los Angeles river system could eventually be far more diverse in the activities and the living communities it supports than it was in its natural state. Here, only the first step has been taken: a coherent view of what might be.

COMMUNICATION BETWEEN LEVELS

For the levels of integration to work with a common purpose, messages have to pass between them. At the regional level, the most important messages concern goals, which are then passed to lower levels. We have mentioned that the most common means of passing down goals is the policy statement. Policy statements can take different forms and are sometimes supported by laws. This does not mean the imposing of policies by higher authority, but the shaping of policy at the regional level with full participation of the people of the region. We might propose here another law of integration: the goals at a higher level reflect the values, desires, and perceptions of the lower levels.

These case studies illustrate some of the means of passing down goals. As I mentioned, the Metropolitan Boston models can point a direction for urban growth policies as well as information and guidelines for design at the plan unit level. They can also provide bases for county-wide ordinances concerning matters as specific as roof materials.

In the Huetar Atlantica case study, the principal means of passing down goals are the establishment of management zones and related guidelines. Since putting them into practice can only be done at the ground level, ultimately by the man with a shovel, they will probably have to become more detailed and precise at each level.

Finally, for the Los Angeles river system, the communication is far less precise, though in some ways more interesting. Here, at the regional level, a coherent view of the possible has been presented, but without any attempt at exact policies or guidelines. This is another, very important role of regional design: giving an enticing glimpse of a possible future.

CHAPTER 4

Subcontinents and the Whole Earth

If regional landscape design is a rarity, design at the higher levels is rarer still. This situation is likely to change, however, as resources run short and as the tightening network of communications makes the earth ever smaller. We will be forced to think in larger increments. Critical issues that emerge only at the larger scales will have to be dealt with. Largely for want of experience, levels of integration higher than the regional have yet to take on any clear shape. Any number of levels of scale might emerge in the future, but for the present, all we can do is group the possible levels under the broad heading of the subcontinent. The size of a subcontinental unit is larger than a region and smaller than a continent; we can say little more than that.

Issues like the distribution and control of water in the water network throughout the western United States, as described in Chap. 2, can be dealt with in a consistent way only at the scale of a subcontinent. Some day, if all the conflicts are to be resolved, and if the water flow is to become both maximally beneficial and sustainable, a subcontinental plan will have to be drawn. Given the present political attitudes in the states involved, that day is probably long in the future.

AN EARLY PLAN FOR THE AMERICAN WEST

Curiously, however, a plan for this same landscape was made over a century ago, although it apparently was never thought of as such. Some discussion of that plan will be worthwhile because it will help to clarify some of the purposes and a few of the difficulties that go with subcontinental design.

The designer was Major John Wesley Powell, a man whom one does not usually think of as a shaper of landscapes or even as a planner. His place in history rests on his exploits as an explorer; he was the first to go the length of the Colorado River, although he was also a successful soldier, professor, geologist, anthropologist, and bureaucrat. Nevertheless, his most impressive achievement may have been his plan for the American West.

Powell came to know the West very well, first through his several exploring expeditions, later from living there for considerable periods. He admired the cultures of the western Indians and studied them in detail. What he came to understand was a rather simple fact that few settlers in this new landscape understood before it was too late: that this vast area west of the Rocky Mountains, which covers over forty percent of the continental United States, is different from the area to the east in fundamental ways that demand an entirely different approach to settlement and land use.

Settlers in different parts of the New World invariably brought with them images of landscape and techniques for using it derived from the lands they came from, lands which were usually quite different from those they now settled. This happened in the eastern states and the South and dramatically in the Midwest, where the failure to understand the landscape culminated eventually in the disasters of the dust bowl era.

Very early in the nation's history, the monolithic view, the failure to see that various areas of the continent were fundamentally different, was institutionalized in the survey grid. The origins of the grid are lost in historic mists, but it was less than ten years after the separation from the British Empire that the one-mile squares of the grid were dropped over all of the United States beyond the original thirteen colonies. After that, virtually all property lines were drawn on or closely related to the grid lines with the visible results mentioned in the last chapter. The American landscape, regardless of climate or topography, became a uniform checkerboard wherever it was settled.

When the various land grant acts came along, parcel sizes were based on the grid; the quarter section, one-quarter of a one-square-mile section, became the standard homestead unit. This was a reasonable size for a family farm in the East where rainfall provided adequate water for crops. But in the dusty, dry landscape of the West, where settlers came to claim land in significant numbers after the Civil War, Powell observed that it was either too large or too small—too large for irrigation farming, too small for pasture farming, which was coming to be called ranching. These were the two feasible types of agriculture in the arid West. Irrigated farms had to be located close enough to a stream or river to get a water supply. Reliable, controllable periodic watering produced dependably high yields but required a great deal of labor not just to dig but to maintain the canals. A family could not manage more than 80 or so acres, and as history has shown over and over again since the earliest irrigation works on the banks of the Tigris and Euphrates rivers over five thousand years ago, irrigation farming is really a cooperative enterprise better undertaken by communities that can share the labor. Along the Tigris and Euphrates the cooperative irrigation communities eventually evolved into the first urban civilization. In Utah, the Mormon farmers, organized in ecclesiastical communities, did very well with irrigation by applying techniques they had learned from the Indians in Arizona.

On lands too far from water for irrigation, growing crops was seldom feasible because of the low levels of rainfall. A great many settlers, trying to eke out a living on land where rainfall usually averaged less than 15 inches a year, learned this lesson very painfully and left behind barren quarter-sections rutted into dusty furrows and decorated with collapsing cabins. Cattle grazing was feasible on these dry lands but only at very low densities. An area ten to twenty times a quarter-section was needed to graze enough cattle to support a family. As a result, the cattlemen were mostly nomadic, wandering about and grazing their cattle on government land since there was no way for them to obtain enough of their own.

In hindsight, it seems obvious that a change in land grant policy was needed, but few seem to have noticed that fact, or if they noticed, commented on it. It was Powell's particular genius to see the obvious and to build upon it a design for the landscape of almost half the nation.

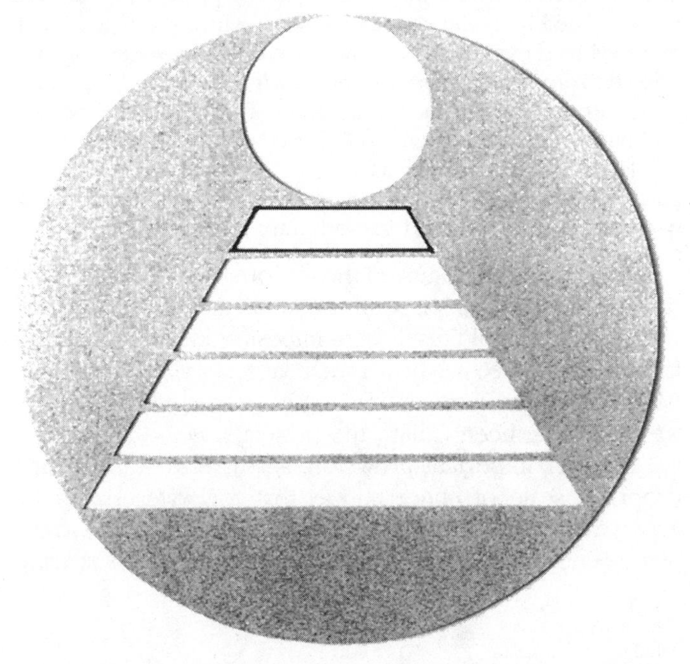

Powell's Vision

Powell submitted his report to Congress in 1878. The title is long—*Report on the Arid Regions of the United States With a More Detailed Account of the Lands of Utah*—but the text is brief. Though Powell, as a trained scientist, supported his proposals with information concerning the character of the landscape, his data were actually rather skimpy, and there was nothing like a formal method of analysis. It is altogether a fine example of what we will be calling in a later chapter the *gestalt method*. Powell did not analyze the landscape by breaking it down into pieces and putting them together again. Rather, he not only understood it as a whole but treated it as a whole. If the gestalt approach rarely works at such a large scale, it is simply because it is so difficult to see such a vast landscape as a whole. At any scale, use of the gestalt method requires a profound understanding of the land involved.

Powell saw the broad sweep of the western landscape in three geographical tiers—the lowlands, the highlands, and the valleys, hills, mesas, and mountain slopes between—and he saw it in terms of use: ". . . the irrigable lands below, the forest lands above, and the pasturage lands between."[1]

The irrigable lowlands required the most detailed consideration. They were the lands along the edges of rivers and streams, and in the flatter regions, they were fairly extensive. The limiting factor was not the amount of land but the amount of water available to irrigate it. With careful and frugal use, which Powell wholeheartedly endorsed, experience showed that an area of 80 to 100 acres could be irrigated with one cubic foot per second of water. On this basis, about 2262 of Utah's 80,000 square miles, or 2.8 percent of the territory, could be irrigated.

The minor irrigation works needed to distribute waters from the small streams could be built by individual farmers, but there were relatively few small streams, and most of the land alongside them was already taken. Larger streams required extensive networks of canals, and these were beyond the labor and capital of individuals. To provide for the cooperative efforts that were clearly needed, Powell proposed a system of irrigation districts. Any nine or more individuals could select an area of land from those described as irrigable for purposes of settlement. Each member of the group could claim up to 80 acres within the contiguous area. The group could then apply for a survey to be carried out by the Surveyor-General's office. The members of the district would develop irrigation works cooperatively, but these had to be in place within five years. As long as water was available, new settlers could claim adjacent lands in parcels up to 80 acres and join the district. Water rights would inhere in the land.

This scheme gave some assurance that settlers would have the resources to survive in the harsh conditions of the West. The most critical resource—water—would be under

[1] See Powell, 1962; p. 16.

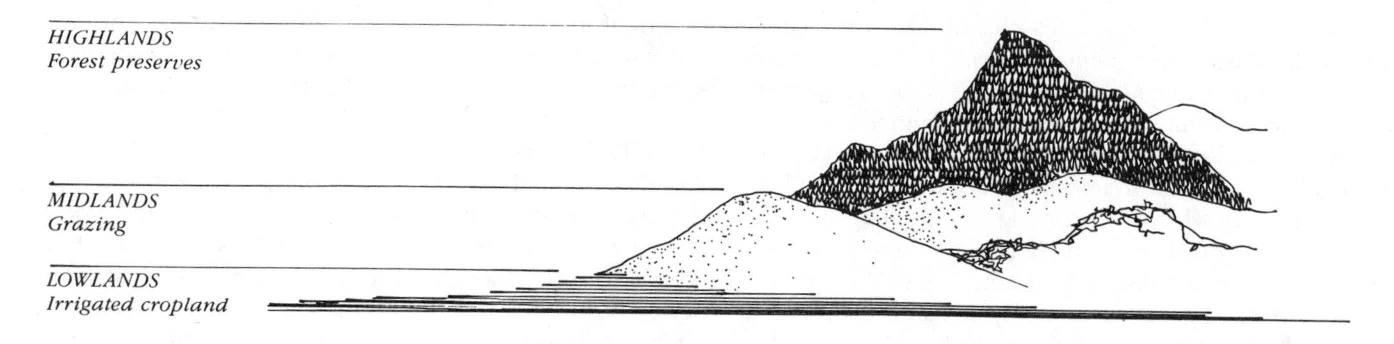

the control of those who would use it. Water speculation would be eliminated. Much of the frustration and heartbreak that attended homesteading would be avoided. For the government, the problems of devastated, abandoned homesteads would be eliminated.

The problem was that the irrigation district scheme would also mean a shift in emphasis from settlement by individuals to settlement by colonies, a major change psychologically, and a difficult one. Rugged individualism was already a tenet of the western credo, whatever the cost.

Timberlands covered about 23 percent of the western landscape, all at upper elevations, though the upper and lower limits varied with local climatic conditions. Only about half of this area was actually covered in timber, and this, according to Powell, was the result of fires set by the Indians for purposes of driving their game. The only way to restore the timberlands, he believed, was by removing the Indians. Once the burning stopped, the timber would grow again. He also believed that the timberlands should be made available, as they had not been up to that time. "Provision should be made by which the timber can be purchased by persons or companies desiring to engage in the lumber or wood business, and in such quantities as may be necessary to encourage the construction of mills, the erection of flumes, the making of roads, and other improvements necessary to the utilization of the timber for industries of the country."[2] He made no recommendations as to how this was to be accomplished.

For pasturage lands, the lands between lowlands and highlands, as for irrigated lands, Powell had proposals that were quite specific. Because the grasses were scanty—though valuable in both summer and winter—he recommended that tracts of 2560 acres, or four sections, be made available for "pasturage farms," or ranches. Each of these large ranches would need a small irrigated area of about 20 acres for growing the rancher's own food and hay for the cattle. These 20 acres would have to be within irrigable areas, of course, but the other acreage should be nonirrigable. Since irrigation was again involved, districts would again be required. Powell foresaw that without these safeguards, the first settlers along a stream would stake claims to all the land along the banks, thus gaining control over the land settled later. Pasturage districts would also have nine or more members. Their houses would be grouped as close together as possible for access to roads, schools, and other surfaces. All pastures would be used in common; there would be no fences. Powell recognized that drawing property boundaries to allow the irrigable areas and grouping of houses would require careful attention to the landscape: "... the division of these lands should be controlled by topographic features to give water fronts."[3]

Besides their recognition that the unique character of the landscape demanded a different pattern of settlement, several other features in Powell's proposals distinguish them from the views that commonly prevailed in the settling of the United States. One is the strong implication of a watershed concept. Though the word "watershed" itself is never explicitly mentioned, the end result of the proposals, had they been followed, would have been a system of watershed control. Each district, by its nature, would have been in a single watershed, and the sum of districts in a watershed would have necessarily become the next larger unit of control. Especially in a dry landscape, any logical organization that begins with the landscape is likely to end up with the watershed as the basic land unit.

Even more important is Powell's idea that the actual drawing of property lines should be done according to specific local conditions. His statement on this is instructive:

> The lands along the streams are not valuable for agricultural purposes in continuous bodies or squares, but only in irrigable tracts governed by the levels of the meandering canals which carry the water for irrigation, and it would be greatly to the advantage of every such district if the lands could be divided into parcels governed solely by the conditions under which the water could be distributed over them; *and such parceling cannot be properly done prior to the occupancy of the lands, but can only be made pari pasu* with the adoption of a system of canals; and the people settling on these lands should be allowed the privilege of dividing the lands into such tracts as may be most available for such purposes, and *they should not be hampered with the present arbitrary system* of dividing the land into arbitrary tracts.[4]

Thus Powell recognized the importance of relating subdivision lines to the shape of the landscape, foreseeing the problems that would result from imposing an arbitrary system from the perspective of a much larger scale of concern. What he proposed was a drawing of property lines at what we have here been calling the project scale.

Another important provision, again relating to the appropriate scale of concern, was that for water rights. In every case, water rights were to be attached to the land, permanently inseparable from it. Until the water was actually

[2] See Ibid., p. 40.

[3] See Ibid., p. 40.
[4] Ibid., p. 51; my italics.

in use, moreover, title was not to be bestowed. Doing so would prevent the usurpation of water rights by speculators or by users who were not inhabiting the watershed.

In basing his plan on the flow of water, Powell was following what we might call the limiting factor approach. He may or may not have been aware of the law of the minimum described by Justus Liebig just thirty-eight years earlier. Liebig had noticed that crop yields were often limited not by lack of major nutrients like phosphorus or nitrogen, but by minor nutrients for which needs were small but which were available in quantities that were even smaller. This led him to state that the growth of a plant was dependent on the amount of food "presented to it in minimum quantity." The concept was later expanded to include the whole range of inputs and conditions that an organism takes from its environment, any one of which might be limiting. The implication is clear: To raise productivity, one deals first with the limiting factor.

In the West, since the limiting factor was water, it was reasonable to concentrate first on that. There was a strong logic in Powell's taking such an approach—even a rare vision, as we have seen—but it may have been his preoccupation with the limiting factor that caused him to overlook some other important forces that were beginning to change the western landscape. Like almost all of his contemporaries, Powell saw the West as a vast reservoir of resources to be used, and he was almost exclusively concerned with food production. His treatment of the forests was cursory, ignoring the probability of rampant exploitation in the absence of strong governmental control. The experience of the man-made devastation of the eastern hardwood forests was still fresh in mind at this time. It had inspired efforts like the halting of sales of New York's forest lands by the State Forest Commission in 1872 and the campaign of the American Association for the Advancement of Science, begun in 1873, to conserve the nation's forests. Powell must have known of these movements and of the fledgling efforts in Congress to establish national forest reservations, but for some reason he did not take up their cries.

Likewise, he failed to consider the establishment of national parks and the whole idea of landscape preservation. Although a new idea, it was well-established in his day. The Yosemite Valley had been preserved by Congress for public use and enjoyment 14 years before Powell wrote his report, and the first national park, Yellowstone, had been set aside six years before that. Some of the most awe-inspiring landscapes in the world were within the area of his report, and certainly well known to him. His boats had passed under the Rainbow Bridge and through the Glen Canyon and the Grand Canyon on their epic journey down the Colorado, and Zion and Bryce canyons were in Utah, the region covered by his studies in detail. All of them went unmentioned.

Even more surprising was his failure to consider the prospect of urbanization. The West was clearly ripe for it. Thirty years earlier the gold rush had shown how sudden hordes of settlers concentrated in a few places could alter the landscape. San Francisco had made it clear what a boom town could be like and the effect it could have on the resources of the countryside. The spasms of suburbanization that would wrench Southern California were still two years from their beginnings, but the trends that would bring them on, especially the new railroads, were already visible.

These are the sorts of large patterns that take on importance at the subcontinental scale. Powell saw clearly the function of the water flow, the large pattern of rivers, valleys and watersheds, and the agricultural pattern that would fit this kind of landscape. That was a promising beginning, but the human claims on landscape extend much further. As the human influence grows, the more complex patterns of conservation, preservation, and urbanization take shape. At the subcontinental scale, we can see them in larger terms as interrelated systems that influence and complement one another. While we can readily comprehend a conurbation and its growth tendencies at the regional scale, we can really grasp the larger context of its supporting resources only at the scale of the subcontinent. Consider again the map of Southern California's watershed, which covers most of the western states. Serious consideration of the complex effects of urbanization on resources and their allocation can only be undertaken at this scale. Likewise, we can see preservation and conservation as systems that include any number of different, sometimes widely separated, but still interrelated landscapes.

With the luxury of hindsight, we can ask now: What if Powell and his associates had considered conservation, preservation, and urbanization in their proposals for the future of the American West? What if they had carefully plotted all the truly breathtaking scenic resources—not only all those that later became national parks, but also areas like Lake Tahoe and Glen Canyon—and recommended that they be set aside right away? It would not have been difficult while most of the West was still public land. Or what if they had seen the forests as a natural matrix to be managed as a protective landscape that would wrap around the productive and urban landscapes? Most of them were absorbed into the National Forest system in the later periods, of course, but with some striking omissions and discontinuities that seriously compromise their protective role. And most intriguing of all, what if Powell had developed a policy for urbanization that would balance the growth and location of cities with the limiting factor of water as he tried to do with farmland? What kind of map might he have come up with?

It would be foolish, of course, to fault Powell for not dealing with these issues, which had not even taken the form of issues as yet. His vision of the future, as we have seen, was far beyond that of anyone else at the time. The questions are nevertheless worth asking and pondering because they can give us some notion of the potential breadth of design, the knitting together of land use and resources that might yet be possible at the subcontinental scale.

The Failure of the Plan

As it turned out, Powell's proposals were never adopted anyway. As Wallace Stegner put it, "The confusing, impracticable, loophole-filled land laws remained as they were, the rectangular surveys went on projecting themselves across drainage divides and out from river and creek valleys, often concentrating all the water within miles in a single section and making it easy for the man who owned the water to engross whole duchies of dry land."[5] The bills that Powell wrote to implement his proposals were never acted on by Congress, and though a land reform movement rallied around his report, the debate settled into a long-term wrangle that would go on for decades. Nine years later, in the midst of the panic brought about by the drought of 1887, Congress did finally authorize and appropriate funds for Powell to go ahead with the survey of irrigable lands that was needed to lay the groundwork for his plan. He would first complete the topographic survey of the West, a gargantuan undertaking in itself, and then he would carry out a hydrographic survey, measuring stream flows. With the information from these two surveys, he would then plot potential storage sites, headwork locations, canal routes, and finally irrigable lands. It would take seven years and seven million dollars, he estimated, and for the time being Congress was willing.

Powell was strongly supported initially by the powerful Senator William Stewart of Nevada. It soon became clear, however, that Stewart and his colleagues saw the survey primarily as a guide for their own land speculation, planning to claim the potential water storage sites. When the Attorney General's office placed a moratorium on all land claims until Powell's survey was complete and a new hard policy in place, Stewart's enthusiasm cooled. He and his associates saw to it that Powell's budget was cut to an inadequate pittance in 1890. The grand design soon petered out, and Powell left the federal bureaucracy to devote his time to ethnology and philosophy.

There are few landscapes of subcontinental size left in the world that are still in the relatively virginal state of the American West in 1878. Nevertheless, its lesson might yet be of value to a great many partially developed areas. As with the regional scale, the most promising applications probably lie in less developed nations where most of the larger patterns of natural ecosystems are still in place and where they might still become the foundations of broadly integrated human/natural patterns. The regional scale pattern shown in the Huetar Atlantica example could be generalized to the larger scale.

In practice, most landscape policy making at the subcontinental scale occurs at national levels. The subdivision of nations into regional units linked by national policies, like the five regions of Costa Rica shown on page 54, is fairly common. Nevertheless, a serious consideration of the differing resources, structures, and capacities of the regions, the flows of energy and materials among them, and the meaning all these factors have for the subcontinent remains uncommon.

THE GLOBAL LEVEL

All pathways through the levels of scale meet at the global level. If we include its upper atmosphere, the earth is the only closed system, one with few inputs and outputs—the inescapable one. Since the first hazy images of our lovely spinning sphere came back from space, the knowledge of this fact—the superficial knowledge, not the profound comprehension of it—has become commonplace. The words that reflect that knowledge have entered our everyday vocabulary. Buckminster Fuller's vivid image of "spaceship earth" has become a cliché. On a slightly more technical level, we commonly talk about the biosphere, the thin mantle of life that covers most of the globe, with the inference that we recognize its unity. James Lovelock has presented arguments that the earth is not just metaphorically, but actually, a single living organism. We even hear references to Teilhard de Chardin's difficult notion of the noosphere—the global mantle of intellect. For Chardin, the growth of this mantle of intellect is productive of a larger consciousness, one biological in nature and the latest stage of overall evolution in the direction of ever greater complexity. He calls it the "... autocerebralization of mankind becoming the most highly concentrated expression of the reflective rebound of evolution."[6] In practical terms, it means that, for better or worse, the human intellect has taken over the earth.

Given the stage of our development in information and communications, we can probably consider the two—the biosphere and the noosphere—inseparable. Whether we want it that way or not, the network of human control has come to encompass all living things, and the responsibility is clearly ours. The global ecosystem is a human ecosystem, and the vestiges of natural landscapes, or protective landscapes, are really small patches on the larger whole, dependent for their survival, like all the rest, on human foresight and intelligence. Given the trends of the recent past, that is a sobering thought, but like the earth itself, it seems inescapable.

Nevertheless, gloom and dejection are not entirely in order. Since the early 1970s, the human race has sporadically shown signs of responding to the challenge, even occasionally of being inspired by it. Institutions for global management have begun to take shape. Since springing out of the 1972 United Nations Conference on the Human Environment,

[5] See Powell, 1962; p. xviii.

[6] See Teilhard de Chardin, 1966; p. 111.

the United Nations Environment Program (UNEP) has shown signs of growing confidence. While we often hear expressions of doubt and disappointment about its effectiveness, it is important to remember how controversial is its very existence. UNEP has little authority to do more than gather and publish information and to stimulate and coordinate projects that are generally accepted as noncontroversial. Nevertheless, its philosophy is positive and constructive, stressing as it does the interdependence of development and environmental protection. UNESCO's program on Man and the Biosphere is an equally well-intentioned complement to UNEP.

Someday, almost certainly, there will be effective environmental management at the global scale, and landscape management will certainly be a major part of it. The agency may grow out of the U.N. programs, or it may take another form altogether. Since this form will have to evolve from political realities, it is, for the time being, unpredictable. What does seem likely is that the management agency will be very limited in its powers, severely circumscribed by political necessity. Managing the global ecosystem will not be carried out by massive undertakings at the global level; rather it will be the sum total of myriad efforts undertaken at all levels of concern. This implies a common focus, a flow of information, and at least some shared goals. The laws of integration, in fact, demand the largest, most encompassing goals of all at this largest scale. Providing these goals may be the most important role of our future.

Arriving at a common focus or goals will have to be preceded by vital information. We need to know what we are working with and in terms clearer than those given by the hazy photographs, or even the recent very sharp ones, from space. Fortunately, information at the global scale is accumulating at an impressive rate. World distribution patterns of most basic resources have been mapped, at least in a general way. A number of public and private agencies are continuously making inventories of and reassessing and reevaluating global resources. The reports of the Club of Rome have received widespread attention. And when the Carter administration decided to undertake an assessment of resources to the year 2000, it was planned on a global scale.

A major step—perhaps the decisive step—toward global scale design will be the development of a unified information base: a Geographic Information System that will include and integrate the data that have already emerged and are continuously emerging from research on global resources. Some efforts, such as those of Doxiadis and Buckminster Fuller, have moved in this direction in the past, but only very recently has computer technology reached a level that makes a useful system technically possible. There are a great many questions still to be answered. Some of these are technical, such as the variables that should be included and the appropriate level of resolution. Others are institutional. Who will manage the system and where? Who will have access? How do we make sure the information is not used for competitive advantages or even for military purposes?

Experience gained with regional Geographic Information Systems gives us some notion of the form a global system might take. It will include a collection of locationally distributed information in map form. In addition to inventory maps of resources like soils and vegetation types, current data showing the status of changing patterns like desertification, urbanization, and deforestation will be available. An attribute file will maintain facts and statistics describing structure and function, cross-referenced to location. It will integrate existing global information bases, like the famous redbook that maintains counts of endangered species, and it will very likely give better focus and direction to globally oriented research efforts. It will provide the raw material for modeling the global flows of materials and energy, in a way similar to the Club of Rome effort, but with considerably more precision. Thus it will give us a means of predicting shortages, overloads, and dysfunctions well before they occur.

Unfortunately, this Global Geographic Information System will undoubtedly paint a rather bleak picture. Every aspect of the global ecosystem is seriously threatened by human activity in some way, and each threatens to break down partially or completely sometime within the twentieth or twenty-first century. Maurice Strong, while he was Director of the United Nations Environment Program, wrote that "... what we face in dealing with the environment is a series of problems that will become regional crises at different times and places if we do not act quickly enough, and that could escalate into disasters of sufficient magnitude to have political and moral implications for the whole world community."[7] Strong went on to propose six major directions of program emphasis: population control; resource conservation; economic and social progress; balance of resource distribution among rich and poor countries; mobilization of science and technology to focus on environment, resources, and population issues; and international control of ocean resources. One might argue that each of these areas involves the landscape in some way, but all the ways are far from clear. Many of them, in fact, are indirect and rather tenuous.

The stated objectives of the World Conservation Strategy—which is jointly sponsored by UNEP, the World Wildlife Fund, and the International Union of Nature and Natural Resources—are somewhat more to the point. These include:

- Maintaining essential ecological processes and life-support systems
- Preserving genetic diversity
- Keeping animal and plant resources on a sustained-yield basis

These objectives sum up nicely the reasons for conservation, but they are still very general, giving little specific direction for the next level of integration. This is a common difficulty with much of the discussion at the global scale—the fact that it deals with issues so vast that their relation to other scales of concern, ultimately to the man with a shovel, is obscure. For all the talk about the biosphere, the noosphere, and spaceship earth, it is extraordinarily difficult from the perspective of the project or the plan unit, or even from that of the regional or subcontinental scale, to see the linkages that join the issues into the global system.

[7] See Strong, 1974; p. 93.

That they are there, we rarely doubt, but until we can clearly discern the linkages, the knowledge remains an abstraction, and abstractions have little power to inspire action.

Some vagueness and uncertainty are probably inevitable for some years to come as information concerning the global ecosystem accumulates, institutions take form, and goals gradually emerge. Nevertheless, it might be educational and even useful to take a brief look at the information and the institutions that exist already and see what sorts of global goals they might suggest. It is important to understand at this point that virtually all manifestations of landscape change at the global level are cumulative; that is, they result from many actions in different places, rather than from single massive events. Every decision at every scale thus plays a part. In short, we are all involved.

For this exercise, the modes of order introduced in Chap. 1 give us a usable framework.

GLOBAL STRUCTURE

The structure of the global ecosystem includes all living things along with the soils and climates that support them. We will discuss at length in Chap. 11 the ways in which these elements interact to make every part, every species and population, important to the whole. Throughout history, human activities have progressively depleted the very foundation of this structure, the soil—a long-term trend that has accelerated with the growing populations and increasing use of technology in recent years.[8] According to the Global 2000 Report, the "... condition of soils is of central importance now, and its importance is likely to increase in the years ahead"[9] and "... significant deteriorations in soils can be anticipated virtually everywhere, including in the U.S."[10] The major agents of soil deterioration, according to the report, are the following:

1. Desertification
2. Poor drainage in relation to irrigation systems, which brings about waterlogging, salinization, and alkalization
3. Deforestation on steep slopes and in tropical areas
4. General erosion and humus loss, mostly the result of agricultural practices
5. Loss of land to urbanization and other forms of development.

Since all these agents of soil loss are manageable, the goal for design is obvious.

A great many of the living elements of the global structure are also threatened. According to the Global 2000 Report: "Largely a consequence of deforestation and the 'taming' of wild areas, the projected loss over two decades of approximately one-fifth of all species on the planet (at a minimum, roughly 500,000 species of plants and animals) is a prospective loss to the world that is literally beyond evaluation."[11] Indeed, it is soul-wrenching; no sane person could consider it acceptable.

[8] See Hyams, 1976.
[9] See Barney, 1980; p. 98.
[10] Ibid., p. 105.
[11] Ibid., p. 224.

The many arguments for preserving all species of wildlife have been eloquently stated in a number of other places,[12] and there is no reason to repeat them all here. The case for wildlife preservation, however, is overwhelming. Even on a purely practical, anthropocentric plane, the need to preserve the integrity of our supporting ecosystems and the gene pool is by itself reason enough to make wildlife preservation a first-priority goal. It is difficult enough to assess the present value or importance of any plant, animal, or community; the future value is absolutely unpredictable. There is no way of knowing what species might be useful, or even indispensable, to us in the future. Once extinct, a species cannot be recreated. The loss of any is therefore a permanent loss of a potential resource, possibly a critical resource. Since no species lives in isolation, each depending on a network of interactions with its environment and with other species, preserving species means preserving habitats and ecosystem structures. A system of preserves for Mediterranean communities was established as a result of the Man and the Biosphere program. This included samples of each habitat type. This is a small example of what is needed on a global level.

Only a global system of wildlife reserves can really assure that all species—or even most of them—will be preserved. Such a system would need a sample of every community that is large enough to assure ongoing propagation. But how do we categorize communities in such a way that all species are sure to be included? Samples of the world's eleven major terrestrial biomes clearly would not be enough. Sullivan and Shaffer suggest that Küchler's breakdown of vegetation community types probably represents a reasonable basis for a system of reserves; if a reserve were allocated for each of Küchler's 116 potential vegetation groups in the United States, then virtually all species in this country other than some endemic and localized phenomena would be included.[13] Breakdowns following this system have not been made for other countries, but if they were, we can safely estimate that over a thousand communities would be identified, and that is a great many reserves. Although sizable numbers of these reserves already exist, adequate samples of some communities will be very hard, if not impossible, to acquire. Even where they are acquired, their size will usually be minimal, and there will still be the endemics and other special cases on the outside. Moreover, a look at Küchler's map will show that some of the communities are confined to very small areas.

All of these difficulties underscore the importance of preserving landscapes in natural and near-natural states wherever we possibly can and preserving or creating new habitats wherever we can in the man-made landscape. Nevertheless, even the preservation of sizable samples of all the identified communities will not be enough. If species are to continue to evolve and new species emerge, varied populations are needed. Landscape diversity promotes genetic variability and thus speciation. The reserve system, if and when it comes into being, will assure a minimum level of species preservation, but the gene pool is not only in reserves,

[12] For example, see Ehrenfeld, 1972, and Ehrlich and Ehrlich, 1981.
[13] See Sullivan and Shaffer, 1975.

but everywhere. A reasonable goal, then, is the preservation and enhancement of vegetation and wildlife communities in the landscape wherever that can be accomplished.

GLOBAL FUNCTIONS

Concern with energy and material flows at the global level focuses primarily on inputs and outputs moving among subcontinents to create global patterns and global issues. Energy flows, for example, girdle the earth. The movement of fossil fuels from a few producing nations to the consuming nations creates difficulties at both ends, on the route from one to the other and on the output end after consumption.

Energy

The route between the inputs and the outputs are worldwide concerns. The international dangers that attend the movement of fossil fuels—especially by tanker, but also by pipeline—are matters of common knowledge. Oil spills, vulnerability to terrorism, and the threat of blockades and embargoes, and thus war, are three of the better known. On the output side, the most serious environmental effects of fossil fuel burning are also global issues. Acid rain can fall anywhere from a few hundred to a few thousand miles from the place where the sulphur or nitrogen oxides that created it were released into the air—usually by burning of oil or natural gas. Parts of Sweden, Norway, and the eastern United States and Canada have been especially affected. While the acids are now known to have deadly effects on freshwater fisheries, their effects on larger terrestrial ecosystems are still not well understood. It seems certain, however, that they do considerable damage to soils, forest trees, and agricultural crops, thus further depleting resources that are already stressed from other causes.

Another major output of fossil fuel combustion—carbon dioxide—may be the most serious threat of all to the global ecosystem. The precise effects of the buildup of carbon dioxide in the atmosphere, which is already trapping heat near the earth's surface through the greenhouse effect, are still a matter of conjecture, but it seems likely that at least some global warming will result. Over the next few decades, any changes will probably be of regional importance, but within the next century, there is a real possibility that a worldwide temperature increase could bring major shifts in climatic patterns and cause partial or complete melting of the polar ice cap.[14]

The increase in atmospheric carbon dioxide is also closely related to the loss of vegetation communities discussed earlier. Though most of the carbon dioxide rising into the atmosphere has been thought to originate with fossil fuel burning, recent research suggests that a large portion, as much as half, results from deforestation, through release from decaying vegetation and exposed humus.[15]

Whatever the relative amounts from these two major sources, studies carried out by the Brookhaven National Laboratory for the Global 2000 Report predict that carbon dioxide emissions will increase by 35 to 90 percent by 1990 if present trends continue.

[14] See SMIC, 1971.
[15] See Woodwell, 1978.

That the world is likely to run short of oil and natural gas within decades has been widely documented and has become a generally accepted fact. What has not been widely considered, however, is the possibility that the limiting factor in the burning of fossil fuels may not be the supply but the capacity of the sink. It seems likely now that the damages caused by acid rain and the threat of dangerous levels of atmospheric carbon dioxide, not to mention more localized problems like smog, would probably bring about curtailment of fossil fuel use, if such decisions could be made sensibly, even before supplies run short. Should this be the case, the use of coal would be affected as well, even though coal supplies are not expected to run short for several more centuries.

Whatever the more limiting cause may turn out to be, the shift away from reliance on fossil fuels is imminent, in fact has already begun, and the implications for the global landscape are profound. Most serious studies of the subject (for example, those of Stobaugh and Yergin[16]) recommend shifts toward what Amory Lovins called "the soft energy path."[17] Soft energy technologies are generally characterized by renewable flows (primarily sun, wind, vegetation), diversity, flexibility, and a match of scale and energy quality to end-use needs. The soft paths, in short, make use of the energy generating and conserving potentials of the landscape. It is a sustainable use of the landscape and a benign one in that damaging effects are minimal. Moreover, it is an extensive use in contrast with the intensive character of mining and processing of fossil fuels. Soft energy paths will require a great deal more land than hard energy, and much of it will have to be located in close proximity with end uses, which means mixed in with other human activities and land uses. If the scale of technologies is indeed matched with end use, then facilities will range from very small to very large. We can anticipate they will change the landscape in a number of ways, raising a plethora of new issues.

One of these is the location and siting of generating facilities, particularly of solar farms and wind farms. Since solar and wind sources tend to be dispersed, again in contrast with fossil fuel sources, such farms will be quite large, covering areas of many square miles, and quite visible. Since the incidence of both wind and solar radiation vary considerably over the earth's surface, moreover, they will be most effective only in certain locations. Desert areas, being usually both windy and sunny, are prime candidates. The Whitewater Wash case study illustrates some of the conflicts that can arise in a location that has ideal energy generation potential but which at the same time has other attributes that make it important for other uses.

Even larger—both in size and in the magnitude of potential conflicts—are the prospective biomass plantations that will grow plants as alternative energy sources. A study commissioned by the Swedish government concluded that Sweden's energy needs could be entirely met by solar energy by 2015 at acceptable cost and without major social change.[18] To achieve this goal, huge biomass plantations totaling nearly

[16] See Stobaugh and Yergin, 1979.
[17] See Lovins, 1977.
[18] See Johansson and Steen, 1978.

7.5 million acres, or roughly one acre per Swedish resident, would be needed. Solar radiation levels being low in Sweden, countries in more temperate climates could probably expect to use considerably smaller areas, but these figures do give some notion of magnitude. Though this area is less than one-eighth of that now in productive forests in Sweden, it is still immense and poses a great many problems, which Swedish reports say can be resolved by long-range planning.[19]

In practice, much of the growing of biomass might be done in dispersed plantings, many of which might serve other landscape purposes and be intermingled with other uses. Thus, many of the decisions might come down to the project, site, and construction scales. Design for harvesting and optimum use of on-site energies might eventually, and cumulatively, become a major part of the global energy flow. Some of the techniques and design methods are shown in the University Village case study. Such simple techniques for conservation as those shown in the Simon residence might play a part as well. Whether such technologies achieve global importance or not, they can unquestionably help to achieve sustainable levels of energy consumption in local areas.

In the context of global energy issues, with their gargantuan size and Byzantine complexity, the relative potentials and importance of the soft uses of the landscape are still uncertain, and will probably remain so for some time. Nevertheless, we can discern enough from the global picture described here to make out some clear goals for the shaping of the future landscape that can be passed down from the global level. These might be summed up as design for the highest reasonable level of harvesting of landscape energies and the lowest possible consumption of off-site energies.

Water

While the functioning of the hydrological cycle in its totality is global, terrestrial flows are tied to drainage basins and therefore do not aggregate into the same sort of global network formed by energy flows. Water issues, therefore, tend to be more regional or subcontinental in scope. Nevertheless, several major issues are repeated so many times in different watersheds in different parts of the earth that they tend to focus worldwide attention.

At the global level, water supply exceeds water consumption by nearly seven times. Nevertheless, many watersheds, especially in arid and semiarid zones, face the prospect of drought and of wide fluctuations in supplies during the next few decades, and these fluctuations may bring about an increase in movement of water from one basin to another. As we saw in the Los Angeles example, cities in the industrialized world tend to soak up water from rural and natural areas.

The Global 2000 Report predicts that the perceived need for more reliable water supplies, especially for irrigating food crops, along with the need for flood control and electricity, will bring a great increase in hydraulic works. As much as three times the volume of water now impounded in reservoirs may be impounded by the year 2000, a volume that amounts to 30 percent of the world's runoff. This implies vast changes in the world's landscape and an awesome degree of human control. In effect, the landscapes of whole drainage basins will be redesigned; decisions concerning both their design and management will determine the functions of these ecosystems for the foreseeable and unforeseeable future.

It is unlikely, however, that all of these hydrological works are really needed. Conservation and recycling technologies can also play major roles in meeting water requirements. As with energy, there is a need to match scale and quality with end use—an important point that is often ignored in the desire to apply the latest technology. Like soft energy technologies, however, smaller-scale technologies that use the water processing capacities of the landscape, such as those illustrated in the Aliso Creek case study, require more land area and much more careful and sensitive design of human ecosystems than more conventional engineering works. Their advantage is that they alter the landscape less dramatically and are probably sustainable for much longer periods.

Food

Food, moving as it does from one nation to another around the world and being heavily dependent on fossil fuels and their chemical derivatives, has become, like energy, a commodity of global dimensions. Some authorities consider this global network very fragile, subject to disruption and breakdown.[20] Changes in weather patterns and political relations, not to mention pests and blights, can cause food shortages almost anywhere on very short notice. Moreover, as the world population grows, the pressure on food production capacity becomes greater. At least since the beginnings of systematic economics, it has been frequently predicted that the world's population will eventually outgrow the capacity of its agricultural land. Ricardo and Adam Smith both made such predictions. The time may finally be drawing near. The ability of the world's farmlands to produce enough food is already stretched thin. Deterioration threatens productivity even further. On a global scale, as we have mentioned, the quality of soils is decreasing. A survey by the Council on Environmental Quality showed that overgrazing and overcropping, resulting in severe soil erosion, are serious problems in 43 of 69 countries surveyed.[21] Yet clearly, more and more food will be needed.

The Global 2000 Report foresaw a 90 to 100 percent increase in world food production between 1970 and 2000. This is based on U.S. Department of Agriculture studies and assumes an increase in cultivated land and higher levels of production on presently cultivated lands. The real price of food is expected to increase by 30 to 115 percent, an estimate that might well turn out to be much too low.

The projected increase in cultivated lands is only a matter of about 4 percent. Very little arable land still goes uncultivated, and most of what exists is marginal in quality, on steep slopes, or in remote locations. The heavily mech-

[19] See Emmelin and Wiman, 1978.

[20] See Brown, 1978.

[21] See Bente, 1977.

anized, large-scale character of most present agricultural practices means that productivity can be increased only by raising nutrient inputs, primarily in the form of fertilizers derived from fossil fuels. Fertilizer use in 2000 is projected to be about 2.6 times present levels, worldwide. Considering that most of this will have to be derived from petroleum and the present volatility of the global fossil fuel market, this is an unsettling prospect, even if we assume supplies will be available. Furthermore, agricultural fertilizers are the main contributors to water pollution, accounting for about 70 percent of the nitrogen that enters surface waters, for example, and their increased use will undoubtedly aggravate this problem.

Certainly, as the Global 2000 Report points out, we need to consider alternatives, and the most promising options considered so far involve smaller-scale farming, less use of fossil energy and more of human energy and skill, and the application of "intermediate" technologies that employ varied means between total hand labor and total mechanization. Generally, these involve greater diversity of crops and a higher intensity of management, usually on small plots of land. These options have particular promise for developing countries, but their importance is universal. As food prices rise, we are likely to see more agriculture of this sort being practiced inside towns and cities and near their outskirts in the industrialized countries. It is happening already. The realization that urban food supplies are not secure has profoundly affected the consciousness of a great many city dwellers, many of whom are already growing some portion of their own food supplies. For such small-scale efforts to make full use of the land's potentials, they will require careful, detailed design. Though it may be decades before such agriculture becomes widespread enough to have any real impact on the global food production picture, the technical base for it is emerging rapidly. As with the technologies for energy and water conservation and processing in the landscape, it can have a profound impact on the landscape. The University Village case study shows the form that one such intermediate-technology agricultural development might take, and the North Claremont case study shows how a suburban landscape might assume different patterns in response to different levels of demand for food production.

GLOBAL LOCATIONAL PATTERNS

It is a fact of life that resources are distributed very unevenly over the earth's surface, and an even more significant fact that populations are also unevenly spread. Moreover, the two patterns seem more or less independent; that is, concentrations of people do not necessarily correspond to concentrations of resources.

Biologically, we can subdivide the world into the following eleven biomes, or types of ecosystems:

Tundra	Tropical rain forest
Northern coniferous forest (taiga)	Tropical deciduous forest
Temperate deciduous and rain forest	Tropical scrub forest
Temperate grasslands	Tropical grassland and savanna
Chaparral	Mountains
	Desert

Most industrialized nations are located in the temperate zones whose natural vegetation is deciduous, rain forest, or grasslands. The greatest population concentrations are also in the temperate zones as a rule. Nevertheless, the biomes most threatened by pressures of overpopulation, simply because they are inherently capable of supporting only limited numbers of people, are those at the two extremes of precipitation level: the desert and the tropical rain forest. Despite their very different ecological character, the richest,

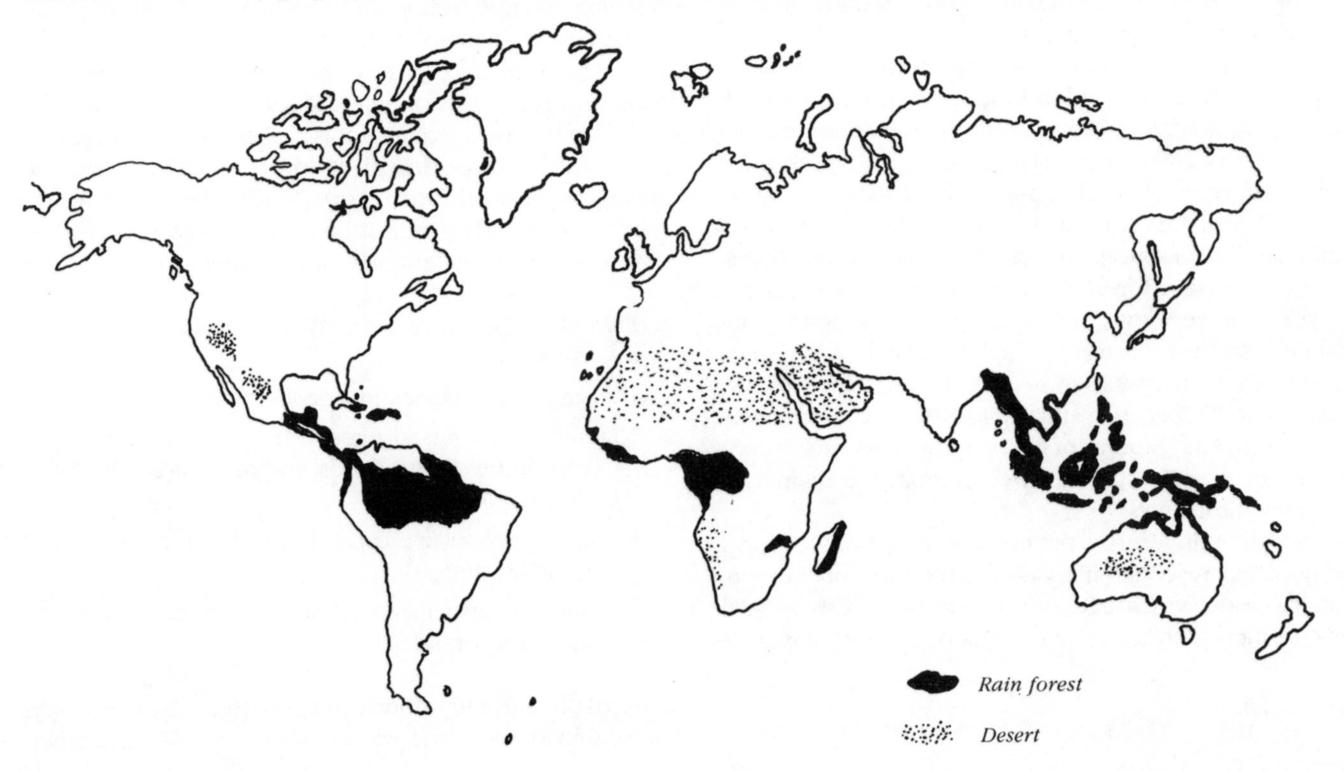

most diverse of biomes and the sparsest have common problems. In both cases, populations beyond the land's capacity try to force a living from the soil and leave it barren and unable to reestablish plant cover.

Over the last few decades, desertification has progressed like a cancer, mostly due to vegetation removal and overgrazing by animals ill-suited to the landscape. Every two years, the Sahara alone spreads over an additional area the size of Connecticut. If all of the lands where the threat of desertification is presently considered severe (mostly arid lands around the edges of present deserts, including large areas of that same American West for which Powell had better ideas) do actually become deserts, the result will be a tripling of the world's desert area. This would represent a major reduction in the global resource base and bring starvation to millions of people. It might also bring significant climatic change even to those outside the desertified regions. Quite possibly, the spread of desertification could spread to ever larger areas.

The spread of deserts can be stopped and even reversed, but it presents a massive undertaking. Land uses that maintain the soil, the plant cover, and the water regime are especially important at the desert's edge. The most effective device seems to be broad bands of trees—green walls around the desert—a massive, expensive, long-term effort. Algeria is building a green wall 715 miles wide from border to border, and some other countries have similar schemes.

Destruction of the rain forests is proceeding just as rapidly, if not more so. Once cleared of vegetation, as discussed in Chap. 3, rain forests are slow to regenerate, if they do so at all; in terms of human time spans at least, they become permanently barren. That they are nevertheless being rapidly cleared in large areas of Africa, Southeast Asia, and Central and South America is a tragedy of global proportions because the rain forests are a resource of global importance. They are by far the richest of the world's biomass in terms of species diversity.

The average net productivity of rain forests has been measured at about 8400 kilocalories per square meter per year as compared to 5980 kilocalories for temperate deciduous forests, 2800 for grasslands, and 2795 for cultivated lands.[22] Unfortunately, only limited ways have been found to make use of this productive capacity to produce food for human consumption. In fact, as we have seen, the capability of the rain forest environment to support human populations is very low, just as its ability to support plant and animal species is extremely high. Over half of all known species live in rain forests. A hectare of rain forest typically contains 50 to 70 species, compared to 10 to 20 in a temperate deciduous forest.[23] Studies of bird communities have shown a steady increase in species diversity from the temperate zones into the tropics.

Since the rain forest is the major reservoir of the global gene pool, its preservation is vital if only for that one reason. Another reason which may turn out to be at least as vital is its climatic significance. Since most of the water that falls as rain in the tropics was taken into the air through evapotranspiration of plants, reduced vegetation would mean less moisture returned to the air, and therefore less rain. At the extreme, we could find former rain forests becoming deserts, and thus the major problem areas of the world would share a common fate.

The removal of vegetation would also increase the reflectivity of the landscape. A green forest has an albedo of 7 to 15, which means that it reflects about 7 to 15 percent of incident solar radiation, whereas a desert albedo is in the range of 25 to 30 percent. Since such changes in reflectivity affect the warming and cooling of large air masses, and these in turn influence global air movements, vegetation removal is likely to have profound effects on such movements. An increase in the albedo of the tropical zones could therefore cause major shifts in global climate patterns.

What all of this means is that among all the pressing issues of the biosphere, the most urgent are those of the arid zones and the rain forests, and that these should have first priority for attention. Probably not coincidentally, these are also among the most overpopulated and poverty-stricken parts of the world. Attention directed to these environments would at the same time include most of the neediest people. Unfortunately, stopping the spread of deserts and the destruction of rain forests will not be an easy task or one quickly accomplished.

THE GLOBAL LANDSCAPE IMPERATIVE

We could go on and on. Such a brief discussion as this can do no more than scratch the surface of a vastly complex subject, but I believe it is enough to establish the fact that ecosystem design is ultimately a global undertaking—not that any particular effort necessarily includes an area of global dimensions, but that each effort takes place within a global context. The problems at the global scale are overwhelming, but just as they are brought about by aggregations of smaller acts, their solution lies in aggregations of smaller acts. Even from such a brief treatement as this, we can begin to discern the outlines of the goals that might be stated from the perspective of this highest level of integration to give direction to design at lower levels, and thus to coordinate the work done there. Within the broad outline of the three World Conservation Strategy objectives, we can list some more specific goals, among them the following:

- Preserving adequate samples of all natural communities
- Designing for species preservation and diversity
- Making the dispersed energy and water processing capabilities of the landscape more available for human use
- Using the productive capabilities of the landscape to good advantage
- Establishing sustainable ecosystems for arid zones and rain forests.

Goals like the first four can apply to landscapes everywhere. Obviously, some are in conflict with one another,

[22] See Golley, 1971.
[23] See Richards, 1952.

and they do not apply equally to all landscapes. Nor do they include all possible goals; others emerge at each scale of concern. Moreover, it might be argued that, since most landscape problems are more serious in developing countries, goals like these are more important there. But even though this may be true, it is not possible to follow one set of goals in one country and an entirely different one in another. The industrial nations will have to show the way. Moreover, even a cursory look at the global level of integration confirms the interconnectedness of all landscapes and thus the importance of common goals.

Expanding populations, dwindling resources, and the growing importance of soft technologies, which are by nature far more dispersed than the intensive technologies of the industrial era, all mean increasing competition for limited land. Meeting this challenge will require careful, sophisticated landscape design. Settlement, agriculture, production forests, and other human uses must join with wild reserves in an intricately patterned fabric that has to be woven with enormous sensitivity and skill. It cannot be woven all of a piece, however. The one general imperative that seems to summarize what we see when we survey the global ecosystem, paradoxically, is the need to act at the local level, to use wisely and in sustainable ways the life-supporting capacities of local landscapes for local purposes. Many, if not most, of the dysfunctions in the global system are brought about by moving materials and energy sources about, from one place to another, stressing one area by shortages, another by oversupply, still another with residuals. The moving about might be much reduced by seeking out the sustainable potentials of every landscape at each level of scale. In the next chapter, we will look at some ways of seeking these larger goals at the plan unit and smaller scales.

CHAPTER 5

Plan Units and Smaller Landscapes

Between the regional and plan unit scales lies an invisible dividing line. Above this line, visions are broad and abstract, and the setting of goals is paramount. Below it, vision is more precisely focused, and mechanisms for achieving goals often become preoccupying concerns. In this chapter, I will describe a variety of mechanisms applied at scales ranging from the plan unit down to actual construction. Each of them responds to goals of higher levels, and however far removed in the hierarchy, each reflects in some way the general goals that emerge at the global scale. Thus, design at the plan unit and smaller scales brings the larger visions of the larger scales down to earth.

THE PLAN UNIT LEVEL

At the plan unit scale, the most common mechanism is the establishment of specific land use. A plan unit is usually a land area with existing definable boundaries, either physiographic or political, ideally contained entirely within a single planning jurisdiction. In size it is small enough to permit accurate, detailed consideration of the attributes of the land that will determine its best use and large enough to include a diversity of uses. This is the scale of what is often called "community planning," the scale at which public participation is usually heaviest. A small town or a district of a large city might be considered as a plan unit; this, in fact, is the scale of traditional zoning. A plan unit usually covers between 5000 and 50,000 acres—an area that is small enough for an individual to know and personally relate to, but large enough to involve complex political decisions. It is also small enough to be represented on a single map at scales ranging from about 1 in.: 400 ft to 1 in.: 2000 ft. If we assume that the horizontal error can be kept to less than 2 millimeters, a not unreasonable level of accuracy, then the margin of error should be between 32 and 160 ft. This is accurate enough for land-use determinations, if not accurate enough for drawing ownership boundaries.

Where regional boundaries are defined by watersheds, plan unit boundaries are often subwatersheds.

Such a logical subdivision of the larger units into physiographic units is the ideal, and we often find it put into practice where regional planning practice is strongly established and where large areas of land are under the control of single management agencies. Most national forests, for example, are subdivided into physiographically logical plan units. In other situations, however, units are usually defined by pragmatic combinations of considerations that have more to do with the heat of the issues involved than with physiographic unity.

The Whitewater Wash Case Study: Resolving Conflicting Uses

The Whitewater Wash Corridor illustrates some typical plan unit scale concerns. It is a setting uniquely suited to energy generation. Winds whistle into it through most of the year, funneled by the deep San Gorgonio Trough to the north, and skies are usually clear and the solar rays intense. Very few landscapes in the world have such high potentials for renewable energy generation, and the location, on the edge of the vast population of the Southern California urban region, makes it especially appealing for this use.

It is a landscape subject to a number of other demands as well. Being one of the few water courses in a very dry region, it is both an important recreation area and a wildlife habitat that features several endangered species. Moreover, it lies within view of the premier resort city of Palm Springs. Thus, Whitewater Wash gives us a glimpse of some of the issues that are likely to become more common as we shift toward soft energy sources.

Here, the regional context did not provide clearly defined boundaries, and the desert landforms, heavily influenced by shifting winds as well as by sporadic rainfall, are subtle, complex, and ambiguous. The clear patterns of ridges and valleys that make most landscapes in temperate climates so readable are entirely absent. Whitewater Wash is just what

CASE STUDY VII

Whitewater Wash

Prepared for the City of Palm Springs by the 606 Studio, Department of Landscape Architecture, California State Polytechnic University, Pomona. Margaret Haskins, Dorothea Hoffman, Karen McGuire, and Rosemary Moritz, with advisors John T. Lyle, Francis Dean, Jeffrey K. Olson, and Arthur Jokela.

The winds that roar out of the mountains through the San Gorgonio trough and into Whitewater Wash create periodic problems of blowing sands, but at the same time they make this a site of unusual promise for wind generation of electrical power. Energy developers have applied for use permits to install wind generators on several parcels of land controlled by the U.S. Bureau of Land Management. The people of Palm Springs, which lies adjacent, are concerned about these plans because the generators would stand some 200 feet tall, disturbing their view of the mountains and perhaps also impeding recreational uses of the wash. The major issues are thus focused on conflicts between the use of the land for power generation and the preservation of visual and recreational values.

This project provided a land resource inventory and an analysis of each attribute of the landscape with respect to energy generation, recreation, and preservation. These analyses served as the basis for a suitability model for each competing use. These models, in turn, were used to optimize each use in one alternative. The fourth alternative represented a combination of uses with each occupying only the most suitable sites.

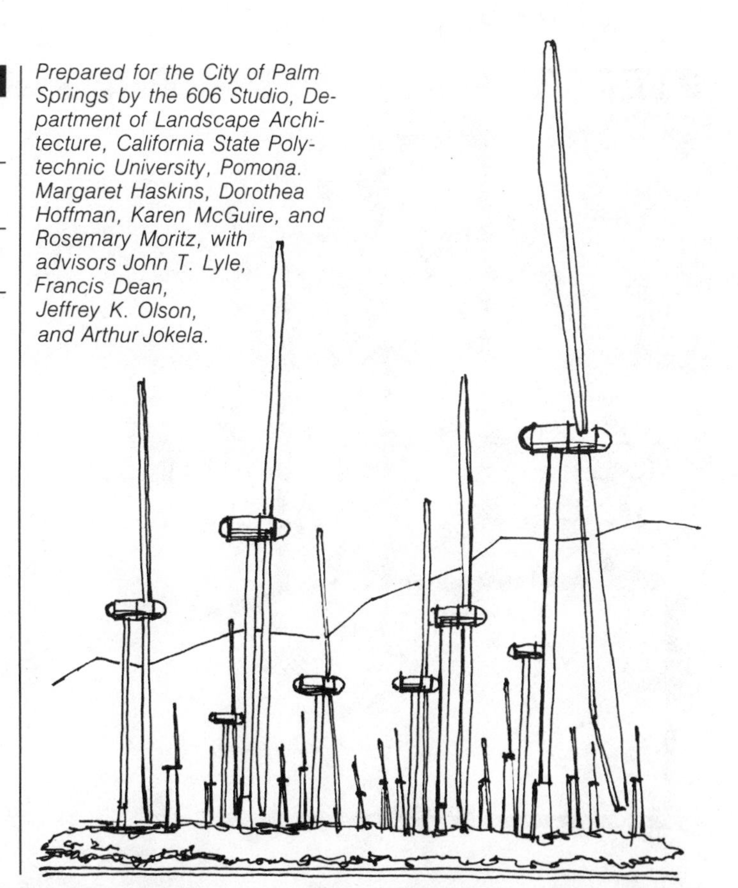

LAND RESOURCE INVENTORY

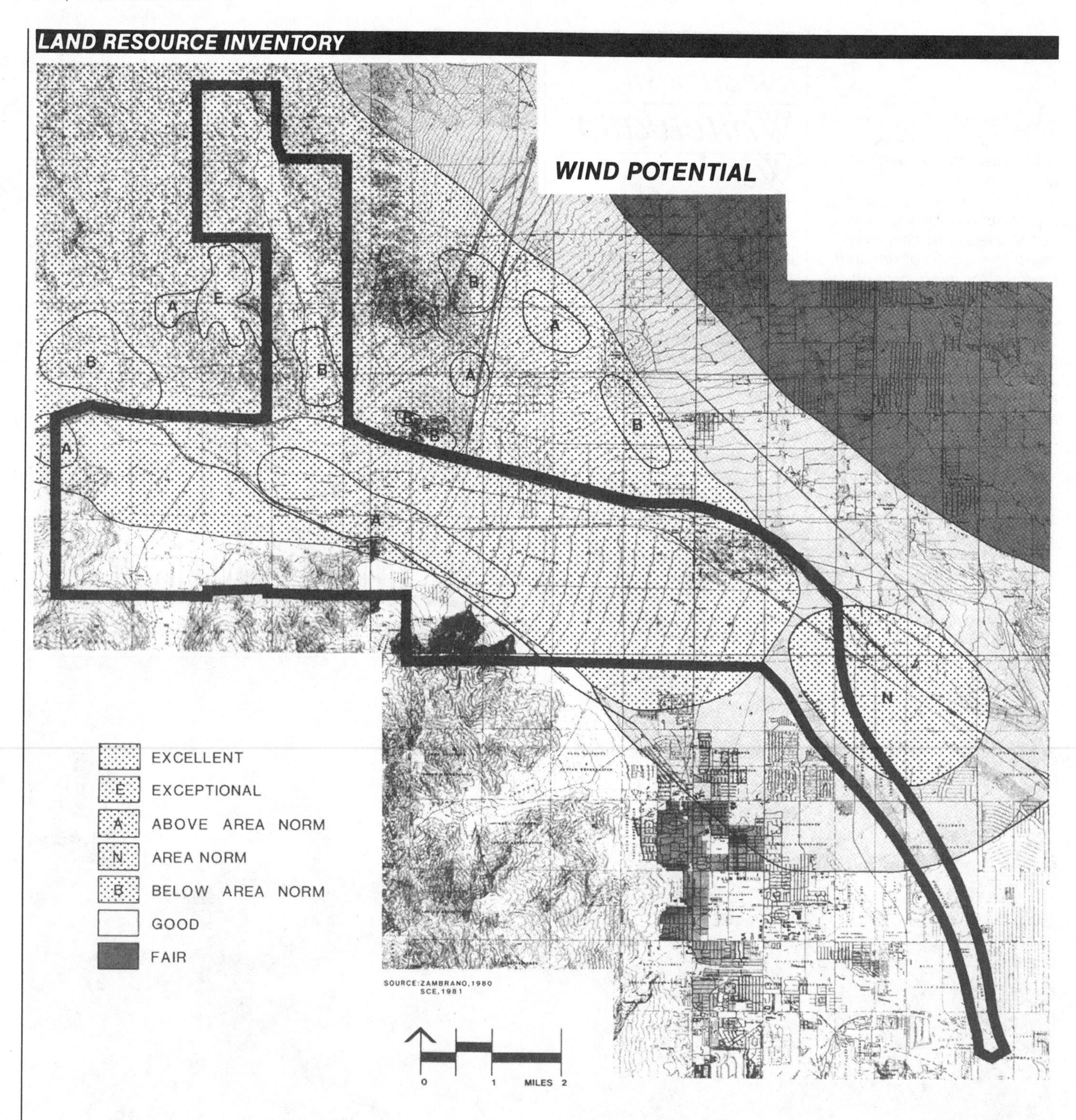

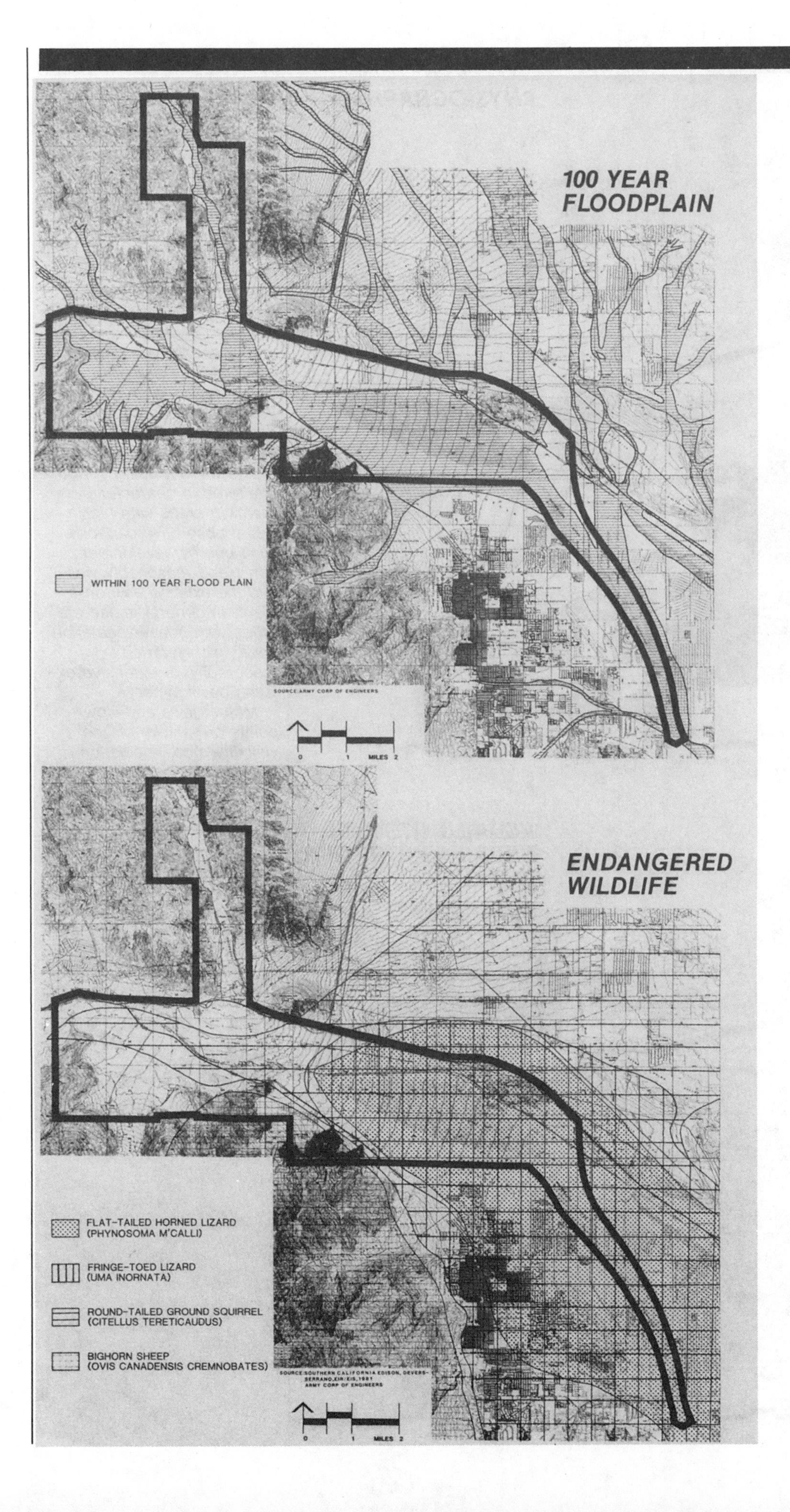

Shown here are a few representative maps from the land resource inventory for the Whitewater Wash. The wind potential map shows that virtually the entire area has excellent capability for wind energy generation, with some having exceptional or above area norm potentials.

The wildlife map shows the approximate habitat boundaries for the four endangered species found in the area. Three of these—two lizard species and one squirrel species—are located largely within the areas of high wind potential, while the fourth—the very shy big horn sheep—stays mostly within the rugged hills to the southwest.

Clearly, there is a direct conflict between wind energy generation and endangered species. Since wind turbines require fences and maintenance areas and must be linked by a network of roads, they could destroy large areas of habitat if not carefully sited.

The combination of an almost flat valley and flash flooding conditions caused by heavy rainstorms in the mountains to the north creates extensive areas subject to frequent inundation. This presents a serious problem for the construction of wind generators. While flood control structures are possible, they are particularly expensive in this area of shifting and blowing alluvial soils, and they would seriously damage wildlife habitats.

Other maps in the land resource inventory for the project, not shown here, included the following:

Existing features
Land status
Existing land use
Blowsand
Slope
Slope aspect
Geotechnical considerations
Ground water
Archaeology
Significant vegetation

LAND RESOURCE INVENTORY

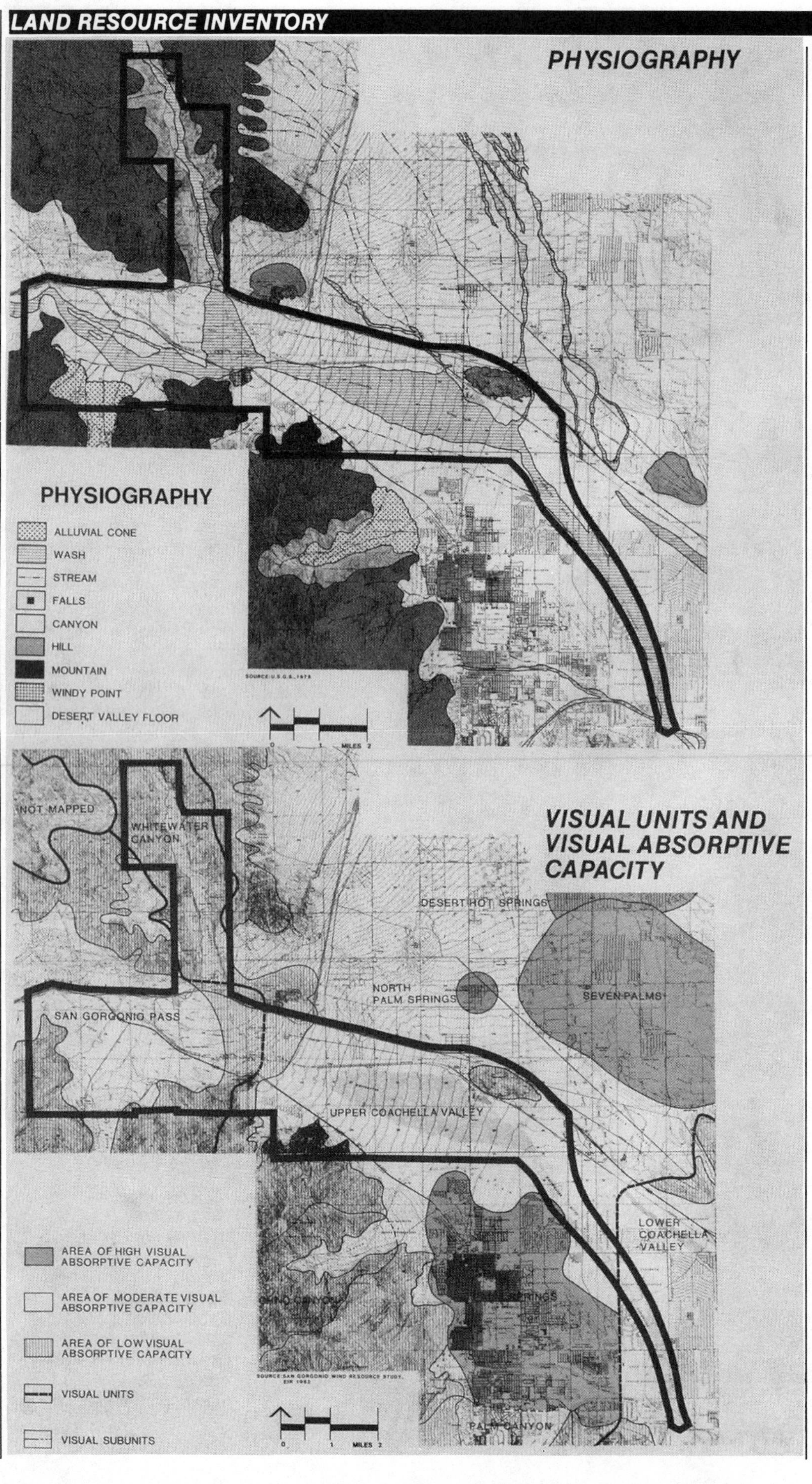

The maps shown on these two pages represent the portion of the land resource inventory devoted to visual concerns. Given the great height (100-300 feet) of the wind generators and the dramatic beauty of the surroundings, visual quality is a major consideration.

Each map shows a particular observable aspect of visual character. Visual units are landscape areas seen at one time and perceived as being unified in character. Visual absorption is the capacity of the landscape to accommodate manmade elements without being seriously altered in character by them. In areas with high visual absorptive capacity, wind energy development can blend reasonably well into the existing environment. In general, these are areas with heavy vegetation cover, highly varied topography, or extensive existing development.

Major views and viewpoints are shown on the visibility map. Population centers and highways, three of which pass through or near the area, are the most important viewpoints.

The source for the data in these maps is Wagstaff and Brady and Robert Odland Associates. San Gorgonio Wind Resource Study: *Draft Environmental Impact Report. Berkeley, California, 1982.*

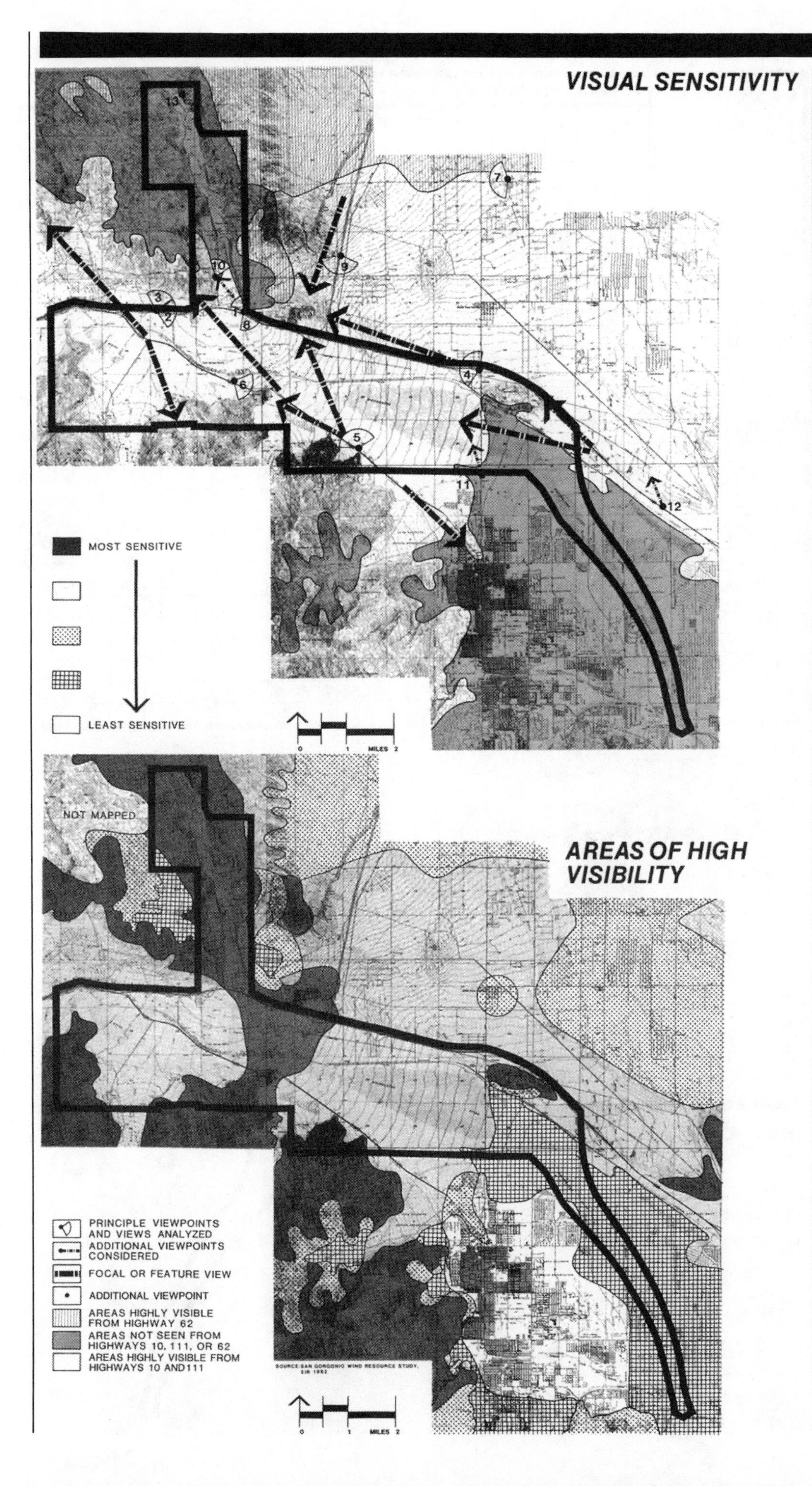

The Visual Sensitivity map is an interpretive model that uses data in the preceding two maps. Areas rated as most sensitive, for example, are those with low absorptive capacity that are seen from at least two routes. Those seen from only one place or having a greater absorptive capacity are considered somewhat less sensitive. At the other end of the scale, those with high capacity, not seen from any major viewpoint, are considered least sensitive.

SUITABILITY MODELS

Each attribute shown on the nineteen inventory maps is relevant to at least one of the three proposed types of use. Some of the attributes are relevant to all three uses. The suitability matrix is a tool for analyzing the relationship between attributes and proposed uses. The symbols shown rate each attribute for its compatibility with each use—high, moderate, low, or unrelated. Suitability models were derived from these ratings by applying rules for the aggregation of compatibility and noncompatibility characteristics.

● HIGH

⊙ MODERATE

○ LOW

☐ NOT APPLICABLE

CONCERN	ISSUE	EXISTING FEATURES: ACEC	EXISTING LAND USE: UNDEVELOPED	PERCOLATION PONDS	EXISTING DEVELOPMENT	WIND POTENTIAL: EXCELLENT	GOOD/FAIR	BLOWSAND: SOURCE AREA	DEPOSIT AREA	OUTSIDE AREAS	ASPECT: NORTH	SOUTH	EAST	WEST	RELATIVELY FLAT	SLOPE: 0-10%	11-25%	26%	GEOTECHNICAL CONSIDERATIONS: GRANITIC BEDROCK	CONGLOMERATE-FANGLOMERATE	VISIBLE FAULT TRACE	INFERRED FAULT TRACE	CONCEALED FAULT TRACE	100 YEAR FLOOD PLAIN	SIGNIFICANT VEGETATION: DUNE GRASS	RIPARIAN	OASIS	JOJOBA	CROTON AND SANDPAPER
PRESERVATION	ARCHAEOLOGY																												
	ENDANGERED WILDLIFE																												
	PHYSIOGRAPHY																												
	SIGNIFICANT VEGETATION																								●	●	●	●	●
RECREATION	ACTIVE DISPERSED	●	●	⊙	○			⊙	⊙	⊙															●	●	●	●	●
	PASSIVE DISPERSED	●	●	⊙	○			⊙	⊙	⊙															●	●	●	●	●
	ACTIVE PAD	○	●	○	○			○	○	●															⊙	●	●	⊙	⊙
	PASSIVE PAD	○	●	○	○			○	○	●															⊙	●	●	⊙	⊙
	ACTIVE TRAIL	●	●	⊙	○			⊙	⊙	●														●	●	●	●	●	●
	PASSIVE TRAIL	●	●	⊙	○			⊙	⊙	●														●	●	●	●	●	●
	ACTIVE WATER	○	⊙	●	○																			●	○	●	●	○	○
	PASSIVE WATER	○	⊙	●	○																			●	○	●	●	○	○
	SAND SAILING	○	●	○	○			⊙	⊙	●														●	⊙	○	○	○	○
	OFF ROAD VEHICLE	○	●	○	○			⊙	⊙	●																			
ENERGY	COMBINED SOLAR & WIND	○	●	○	○	●	○	○	○	●	○	●	○	●	●	●	●	○	●	○					○	○	○	○	○
	SOLAR	○	●	○	○			○	○	●	○	●	○	●	●	●	●	○	●	○					○	○	○	○	○
	WIND	○	●	○	○	●	○	○	○	●									●	○					○	○	○	○	○

The criteria were based upon two sets of factors. The first set involved limitations on building construction such as slope. The County of Riverside prohibits construction on slopes of 25 percent or greater. The second set identifies certain locations as prime because of certain favorable conditions such as wind potential. The wind potential map indicates the varying wind velocities throughout the site.

Other physical resource maps address the following concerns:

- *Aspect (the best locations for solar development)*
- *Blowsand (the limitations on solar development)*
- *Geology, Hydrology, Land Use, Soils, and Slope (the limitations on building)*

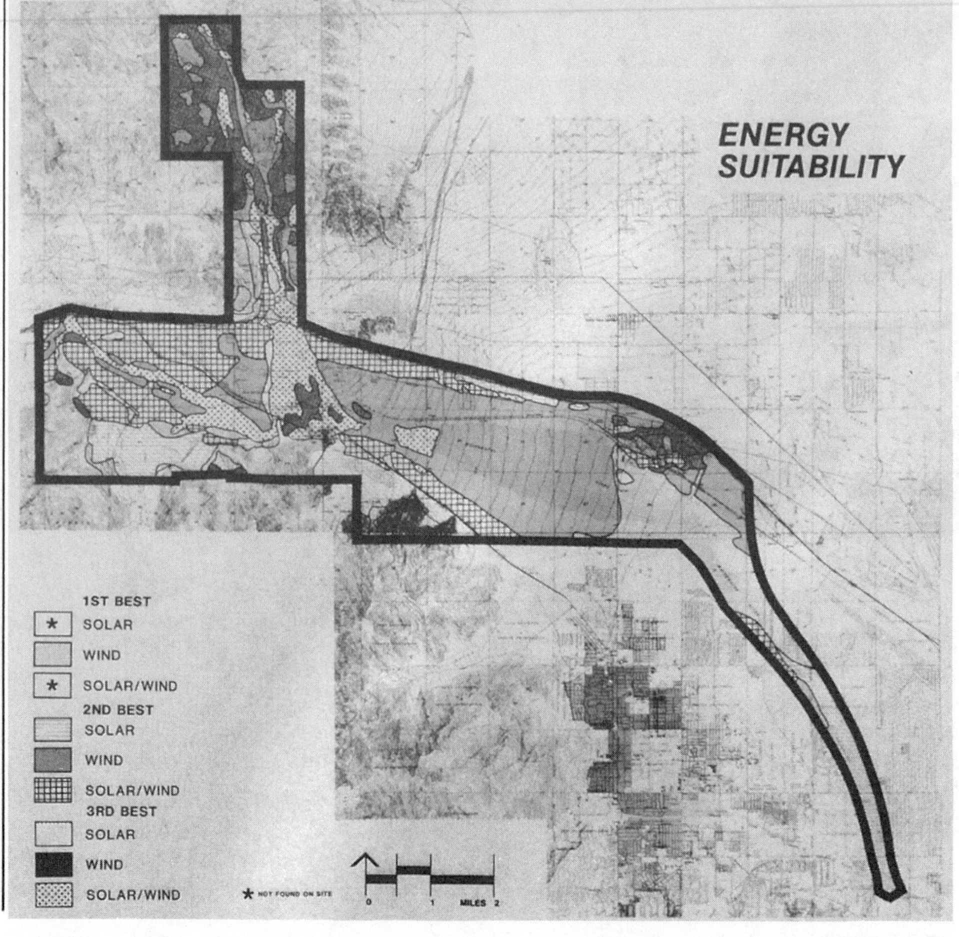

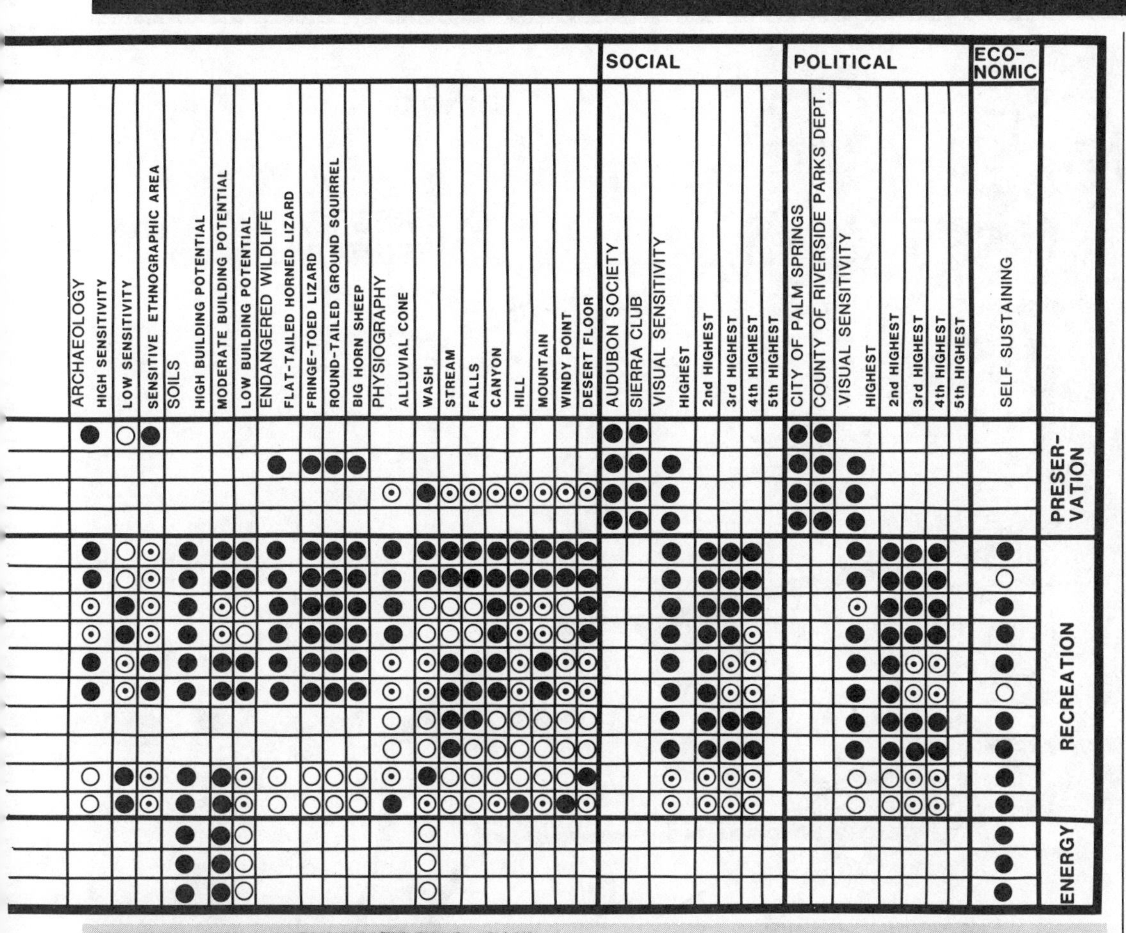

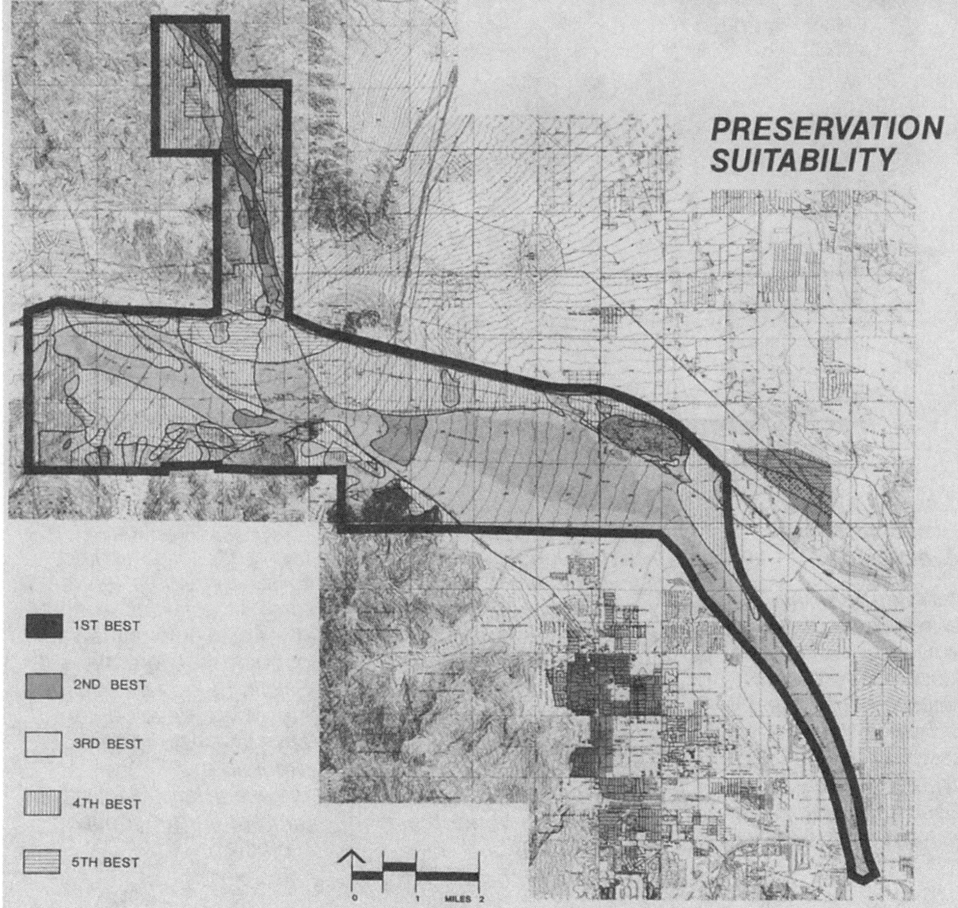

Criteria for preservation were based on two factors. The first factor involves special, rare, or endangered wildlife and vegetation. For example, both the fringe-toed lizard and flat-tailed lizard are physical resources that require special conditions for their survival. Another physical resource causing concern is the ecological balance of oasis areas, which can be easily disturbed by human intrusion. Since both their wildlife and vegetation need to be protected in order to survive, these should be set aside as special preservation areas. The Garnet Hill site is an example of such an area, world-renowned for its ventifacts (stones shaped by blowing sand) and therefore worthy of preservation.

The second factor is the presence of severe environmental hazards such as flooding. The wash area is affected more severely than any other part of the site in the event of a 100-year flood. Any structure in this area would be subject to severe damage or complete destruction.

The definition of each rating is as follows:

1. *Any four of the six preservation components, plus any "most sensitive" rating according to the visual sensitivity analysis.*
2. *Any three of the six preservation components, plus a "most sensitive" and "highly sensitive" rating according to the visual sensitivity analysis.*
3. *Any two of the six preservation components, plus a "most sensitive," "highly sensitive," and "sensitive" rating according to the visual sensitivity analysis.*
4. *Any one of the six preservation components, plus a "most sensitive," "highly sensitive," "sensitive," and "less sensitive" rating according to the visual sensitivity analysis.*
5. *Only the ratings listed in the visual sensitivity analysis.*

SUITABILITY MODELS

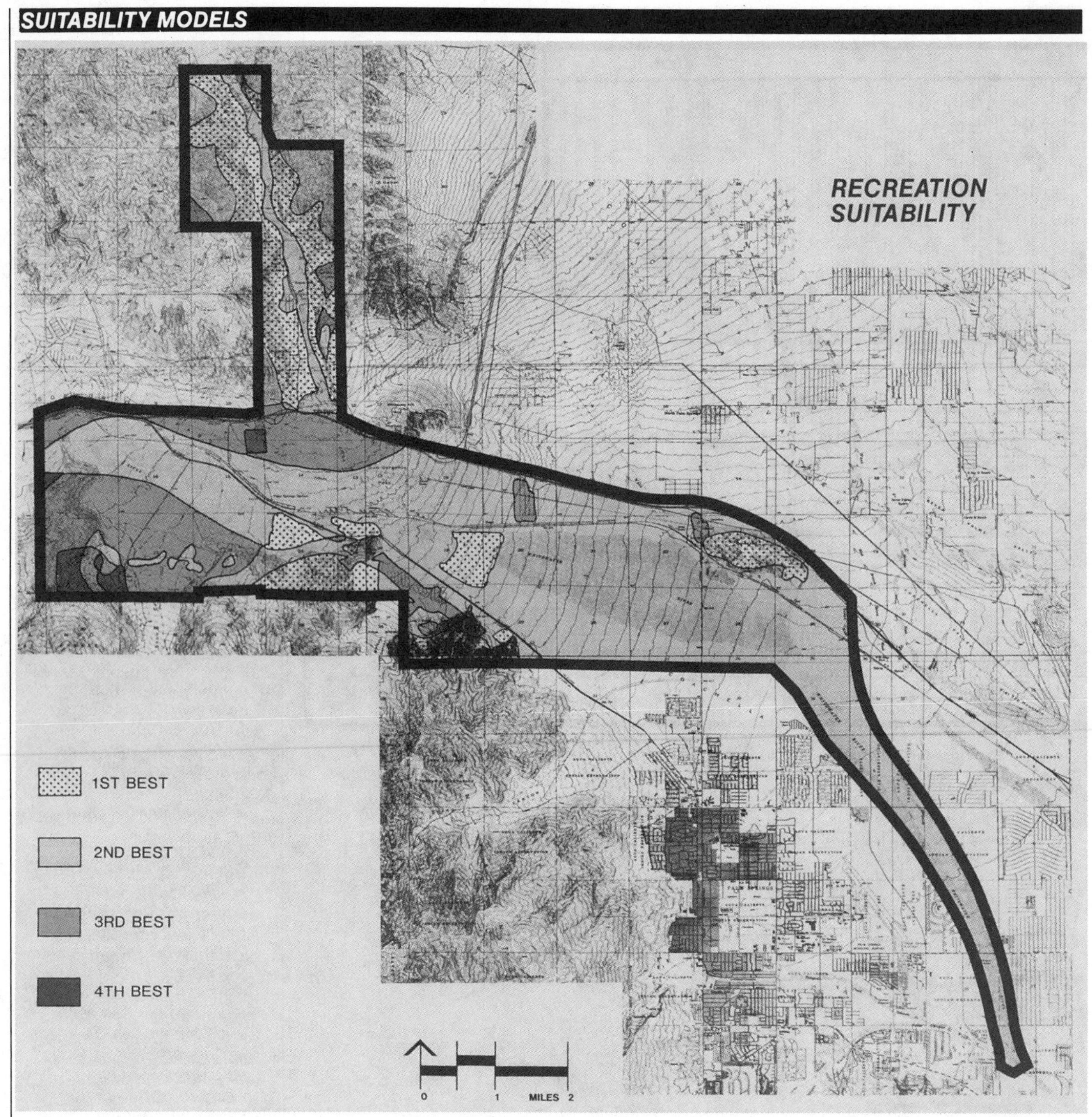

The six recreational groups are as follows:

1. *Dispersed activities (no alteration of the land required)*
2. *Water activities (a water source required)*
3. *Trails (clearing of vegetation and some grading of land required)*
4. *Pads (grading, impacting, and constructing of land required)*
5. *Off-road vehicles (ORVs) (varied terrain of hills and open space required)*
6. *Sand sailing (flat, large area with high wind potential required)*

The following are the physical resources that were reviewed for their potential contribution to recreation:

- *Physiography (hillsides for rock climbers to be indicated)*
- *Wildlife (nature trails to be established through Whitewater Canyon)*
- *Vegetation (botanical study trails to be established through Whitewater Canyon)*
- *Archaeology (artifacts to be made available for examination)*
- *100-year flood plain (the wash that is susceptible to flooding to be indicated)*
- *Soils (building of structures around Windy Point to be limited)*
- *Geotechnical considerations (construction on active faults to be limited)*
- *Existing features (structures or activities in areas of critical environmental concern to be restricted)*
- *Existing land use (building where development already exists to be prohibited)*
- *Blowsand area (activities such as target shooting restricted)*

ALTERNATIVE PLANS

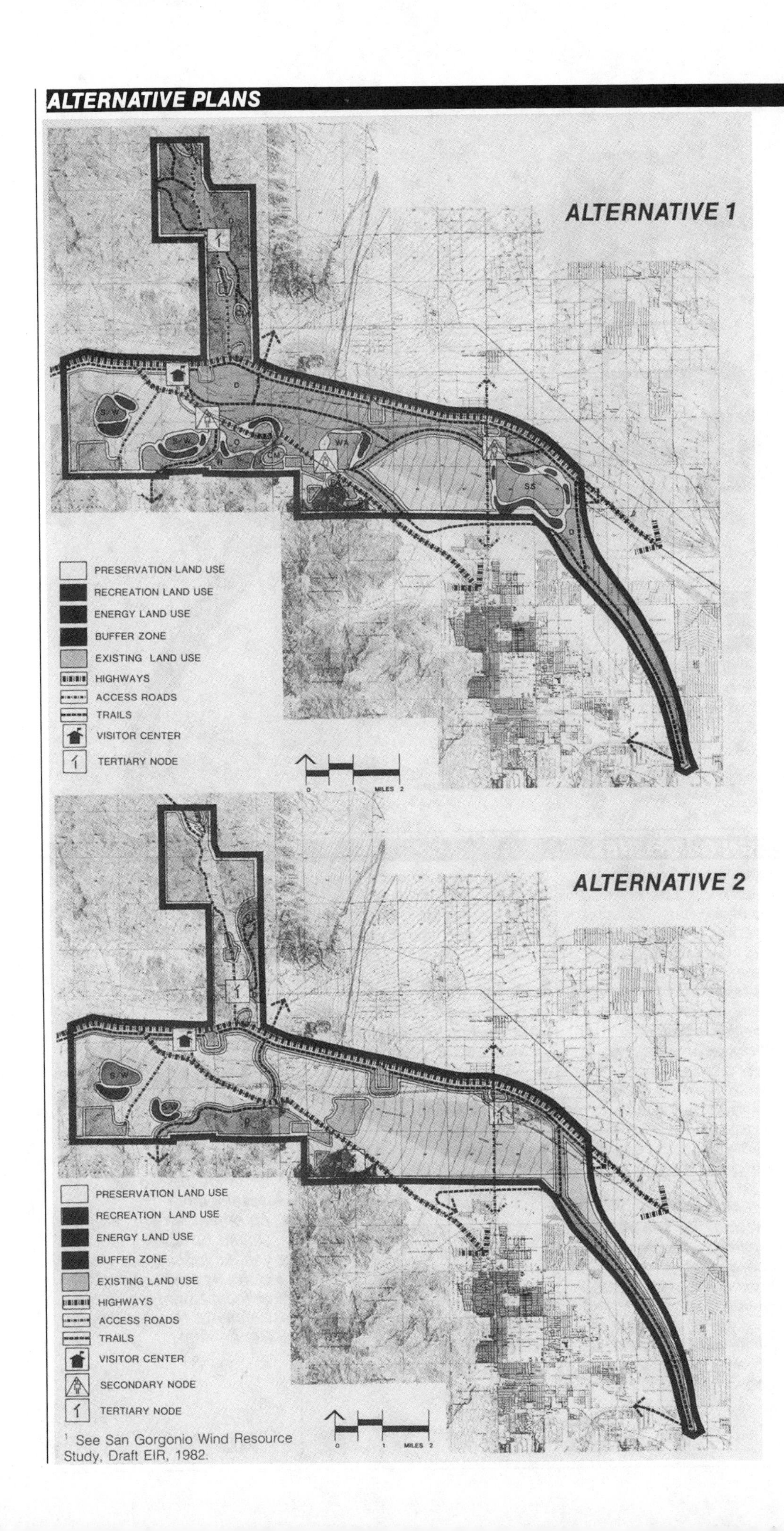

Alternative One: Preservation Emphasis

The preservation alternative emphasizes preservation issues while at the same time accommodating recreational and energy concerns.

Areas not requiring protection may be developed for recreation or energy production. In keeping with the emphasis on preservation, all recreation in the study area would be restricted to activities of a passive nature such as photography and painting.

No development of energy resources would be allowed that might infringe on sensitive areas. Furthermore, no development was to be allowed in any area within two-thirds of a mile of the highway or two miles of a residential community.[1]

Alternative Two: Recreation Emphasis

Under the recreation alternative, the places best suited for passive and active recreation would be used for those purposes. This alternative would also set aside areas for energy and preservation but only as a secondary land use. The following criteria were adopted in planning the recreational areas:

- *Areas of active recreational activities such as the use of off-road vehicles and sand sailing will be set apart by vegetation and sand dune buffers. Such containment will help preserve the rest of the desert and prevent accidents from occurring.*
- *Areas of passive recreational activities such as photography, painting, and birdwatching will serve as a transition between areas of active recreation and those marked for preservation. For instance, a passive area could be located between that assigned to sand sailing and a preserved area.*
- *A trail system will be developed to link and unify all of the recreational activities.*

[1] See San Gorgonio Wind Resource Study, Draft EIR, 1982.

ALTERNATIVE PLANS

Alternative Three: Energy Emphasis

Under the energy alternative, the areas most suitable for energy development are given priority, although other portions of the site are reserved for recreation and preservation. Possible energy systems include wind, solar, and a combination of both. The criteria for energy development addressed the following concerns:

- *Topography of the site (Areas with mainly flat terrain were preferred for both wind and solar development.)*
- *Suitability for wind and solar development (Areas where wind and solar development could be combined were given a higher priority because this type of centralization simplifies operation and maintenance.)*
- *Setback restrictions (Setbacks from scenic highways should be no less than two-thirds of a mile and from rural communities no less than two miles, not only for visual and noise abatement, but also for safety reasons.)*

Because the site is bounded by two scenic highways, its narrow, linear dimensions seriously limit the area open to development. The energy alternative considers development within the wash itself. To do so would require that the waters of the wash be diverted through some form of channelization. The impacts of such measures would be costly both in financial and ecological terms.

Keeping in mind the disadvantages, there are also opportunities in developing energy systems. Wind energy systems, which can produce 13.78 megawatts per square mile, will serve 7000 homes. The Solar One unit produces 40 megawatts per square mile and will serve 20,000 homes. Because these two natural resources are abundant and can be easily tapped, the energy alternative is worthy of close consideration.

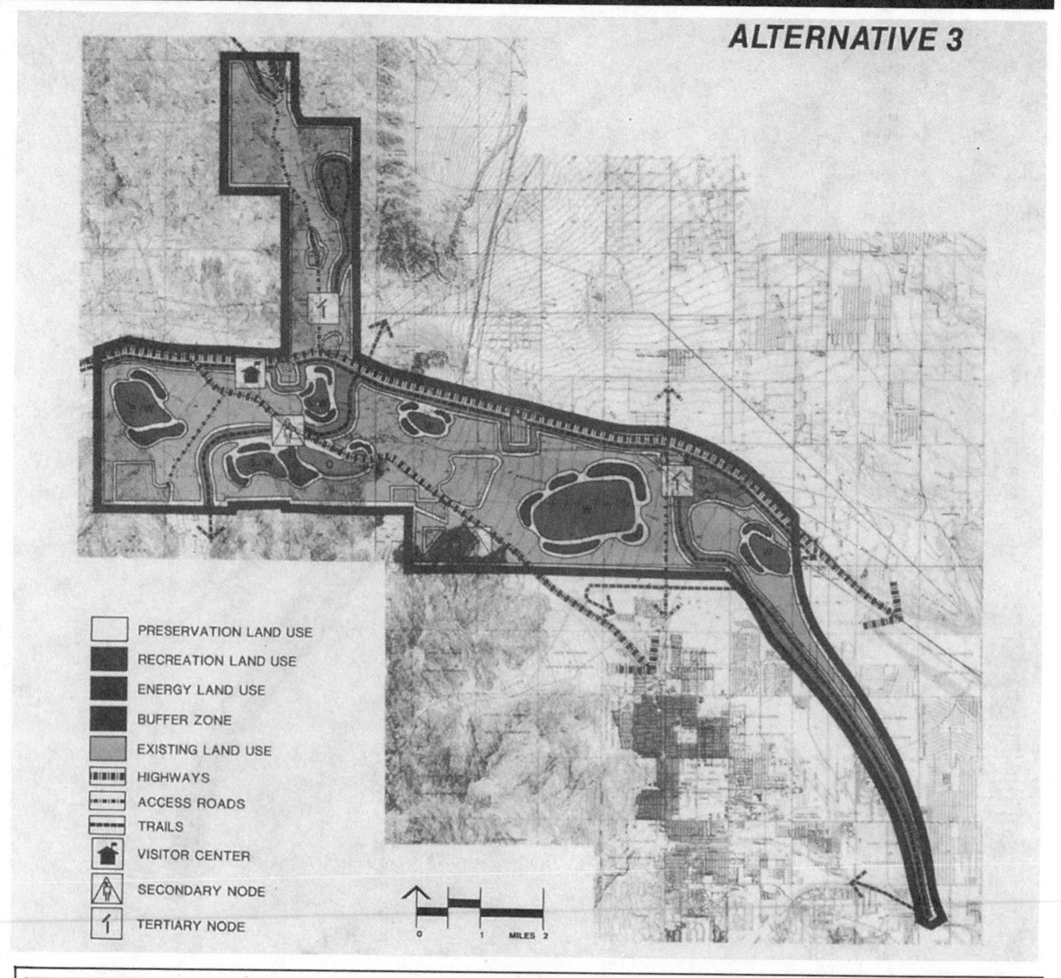

DESIGN GUIDELINES

- *Areas around access roads should be bermed and revegetated to minimize visual disturbances.*
- *Most roads should be gravel because of poor soil conditions for building and to lessen impacts on sensitive areas.*
- *All developments close to major access roads should be sited for minimal impacts on the landscape.*
- *Grading should be done during dry months and those with low wind.*
- *Immediately after construction, soil should be stabilized with vegetation.*
- *(Energy alternative only) Problems of erosion, sedimentation, ground water recharge, and periodic maintenance should be addressed prior to any flood control construction.*
- *Introduced landscape elements should exhibit properties of form, line, color, and texture that are compatible with existing natural features.*
- *Building heights should be limited to three stories so as not to disrupt the area's scenic appeal.*
- *All trail systems should be constructed on slightly compacted soils and revegetated.*
- *Sand dunes meant to serve as buffers should be hydroseeded with vegetation prior to winter rains and during months of low wind.*
- *Revegetation should be used to control erosion.*
- *All design supervision should be executed by a staff of planners and landscape architects.*

Preservation

- *Physiographic features such as ridgelines, hilltops, prominent landforms, rock outcroppings, and steep canyon walls should not be drastically altered.*
- *Archaeological exploration should conform to the specifications of the University of California at Riverside Archaeological Research Unit.*
- *Surface surveys for archaeological sites should be made before any development.*
- *An enhancement program should be sponsored by various local groups and clubs as well as the City of Palm Springs and the County of Riverside Parks Department.*

Alternative Four: Diverse Emphasis

The goal of this alternative was to accommodate all three uses—preservation, recreation, and energy—in a cohesive but diverse plan that would select the best locations for each and therefore emphasize all of them more or less equally. Remaining areas were to be developed according to their compatibility with the adjacent land use.

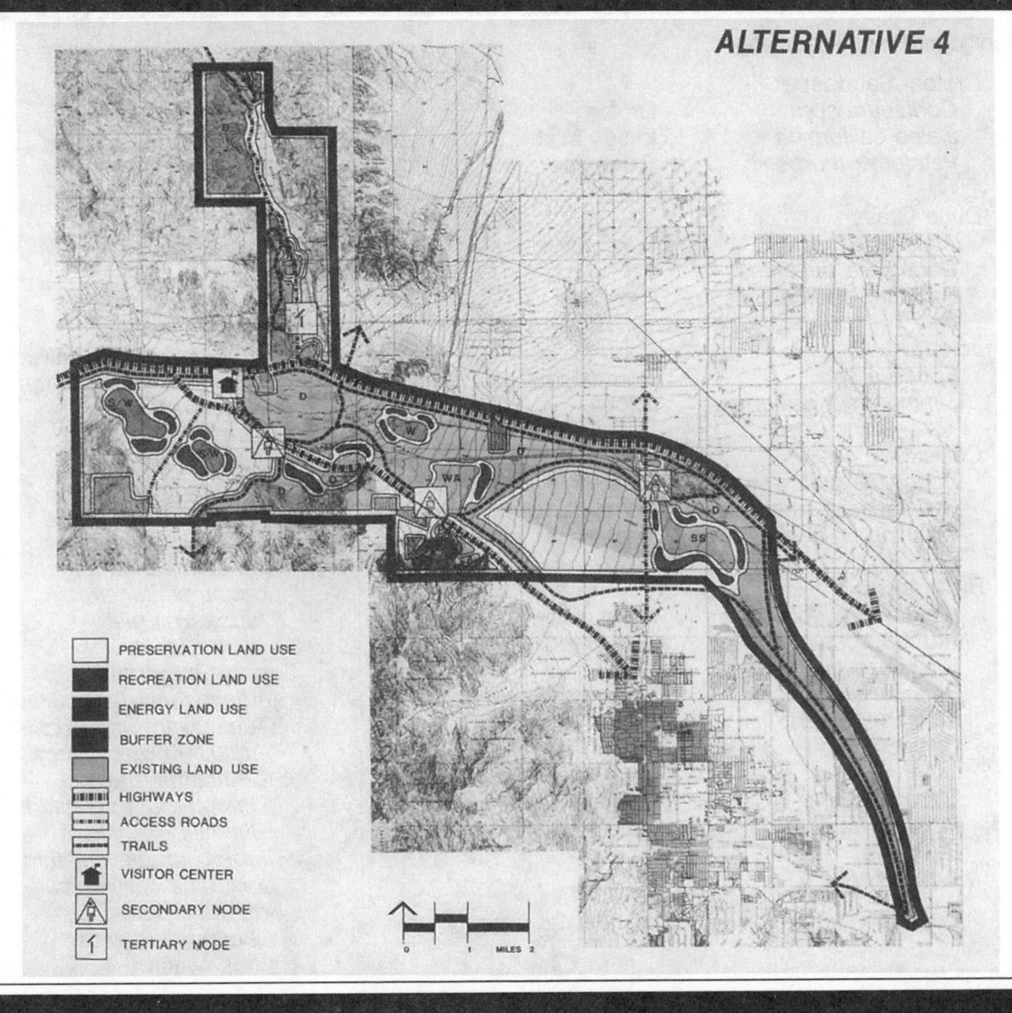

- *Physical resources and special areas should be controlled by limiting access. Only passive activities should be permitted in the following areas:*
  - *Oasis area*
  - *Riparian area*
  - *Jojoba stand*
  - *Dune grass*
  - *Croton and sandpaper area*

Also, Garnet Hill (for its ventifacts) and the Whitewater Canyon in the ACEC area should be set aside for preservation.

Recreation

- *All active recreation such as the use of off-road vehicles and sand sailing must be confined to preserve the desert and also prevent accidents.*
- *All active recreation will be provided with secondary facilities including restrooms, information centers, parking lots, and concessionaires.*
- *All visitor centers and secondary facilities will be manned stations.*
- *Concessionaires for off-road vehicles, sand sailing, paddleboats, target shooting, and food will be required to help supplement the cost of the regional park.*
- *Permits for camping and hiking (in Whitewater Canyon only) will be checked at secondary facilities to help supplement the cost of the regional park.*
- *Fees for such activities as off-road vehicles, water sports, and group activities will be collected to help supplement the cost of the regional park.*
- *A state grant should be requested to help pay for the secondary facilities at the off-road vehicle recreational site.*
- *Trail systems will link all recreational activities.*

Energy

- *No energy development should be located adjacent to recreational activities.*
- *Energy development should be restricted to small, highly concentrated areas.*
- *Wind energy development must follow setback restrictions: ⅔ mile away from roads and 2 miles away from residential areas.*
- *All energy development should have vegetated buffers for safety purposes only.*
- *The City of Palm Springs along with the County of Riverside should act as the lead agency for controls on energy development.*
- *Special consideration must be given to the visual aspects of any energy development that is to be located adjacent to the City of Palm Springs.*
- *All trails located next to any wind development should observe a setback restriction of ⅔ mile.*

PLANT LIST FOR PLANTINGS ON PUBLIC LANDS WITHIN THE PLAN UNIT

Significant Plant Associations:

Croton–Sandpaper	
Codiaeum spp.	Croton
Dalea californica	Indigo bush
Petalonyx thurberi	Sandpaper
Dune Grass	
Abronia villosa	Sand verbena
Oenothera deltoides	Evening primrose
Panicum urvilleanum	Panicum
Jojoba	
Ephedra spp.	Mormon tea
Simmondsia chinensis	Jojoba
Oasis	
Baccharis glutinosa	Mule fat
Ferocactus acanthodes	Barrel cactus
Washington filifera	Washington fan palm
Riparian	
Alnus rhombifolia	White alder
Arundo donax	Giant reed
Baccharis glutinosa	Mule fat
Baccharis sergiloides	Squaw waterweed
Chilopsis linearis	Desert willow
Chrysothamnus paniculatus	Rabbit brush
Mimulus spp.	Monkey flower
Populus fremontii	Cottonwood
Salix spp.	Willow
Typha spp.	Cattails

General Plant Communities:

Creosote Scrub	
Ambrosia dumosa	
Artemisia californica	California sage
Chilopsis linearis	Desert willow
Chrysopsis villosus var. *fastigiata*	Golden aster
Chrysothamnus paniculatus	Rabbit brush
Dalea californica	Indigo bush
Encelia farinosa	Brittlebush
Eriodictyon trichocalyx	Hairy yerba santa
Erigonium fasciculatum	Buckwheat
Jancus spp.	Wiregrass
Larrea divaricata	Creosote bush
Lepidospartum squamatum	Scalebroom
Opuntia basilaris	Beavertail cactus
Opuntia echinocarpa	Golden cholla
Sphaeralcea ambigua	Desert mallow
Yucca schidigera	Mojave yucca
Yucca whipplei	Our Lord's candle
Desert Chaparral	
Adenostoma fasciculatum	Chamise
Adenostoma sparsifolium	Ribbon wood
Arctostaphylos glauca	Big berry manzanita
Ceanothus greggii	Desert ceanothus
Cercocarpus betuloides	Mountain mahogany
Clematis pauciflora	Virgin's bower
Erigonium fasciculatum	Buckwheat
Juniperus californica	California juniper
Prunus ilicifolia	Holly-leaved cherry
Quercus turbinella	Scrub oak
Rhus ovata	Sugar bush
Sensitive Plants in the Area (Exact location unknown)	
Astragalus lentiginous var. *coachellae*	Coachella Valley locoweed
Crossosoma bigelouii	Crossosoma
Euphorbia misera	
Euphorbia platysperma	Sandmat
Monardella robinsonii	
Pholisma arenarium	
Salvia greatai	

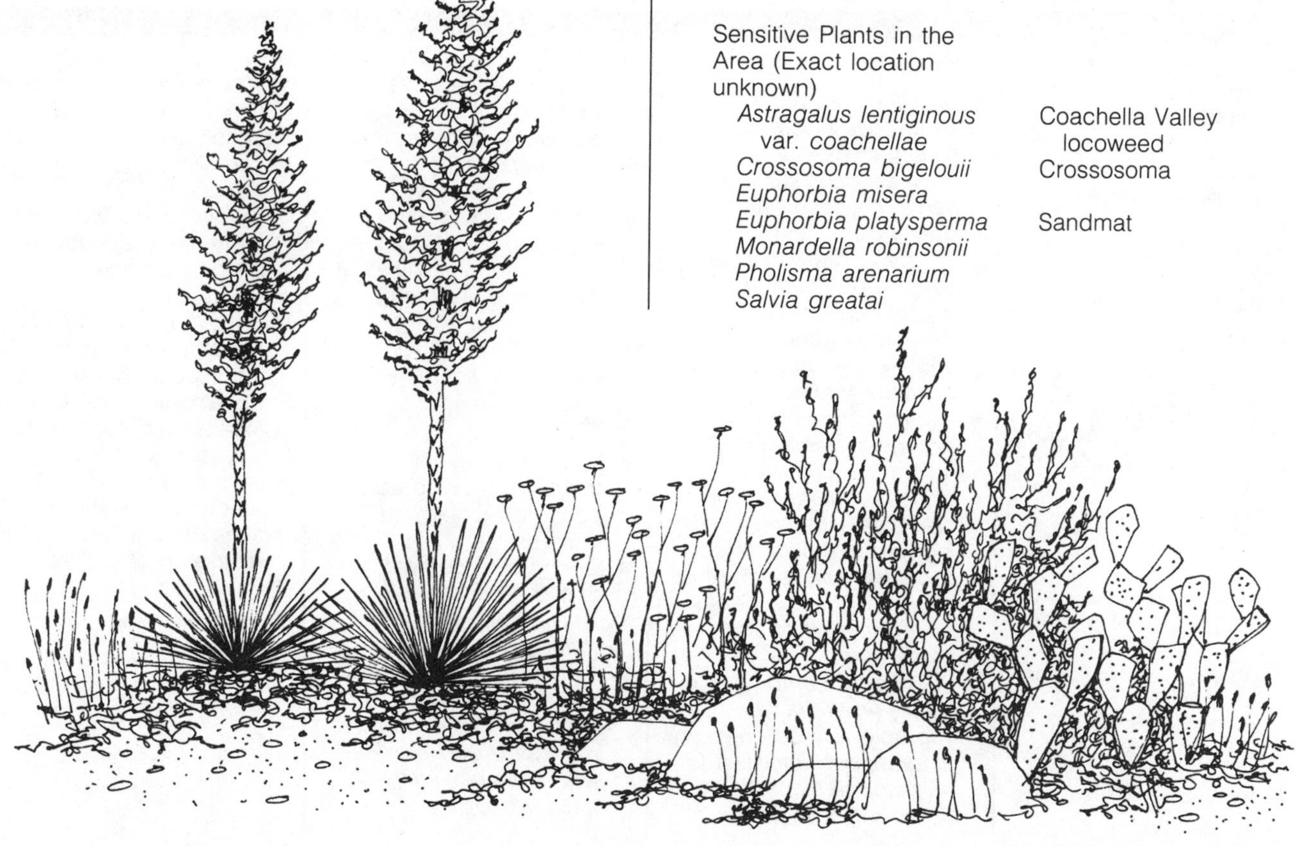

the name implies: a stretch of sand over which frothy waves of water spread after a rain in the nearby mountains, lingering until they evaporate, roll down toward the Salton Sea, or soak into the ground to replenish the aquifer. The land being relatively flat even in the wash itself, after a particularly heavy rainfall the water spreads over a broad area, the extreme being shown in the map that reveals the extent of the 100-year flood plain. The critical wash area, however, is that which is regularly covered by normal rains and which is often wet through most of the winter rainy season. Through here, water regularly flows to maintain the water regime downstream, a most important feature in an arid region, and through here water soaks into the aquifer, again a very important process in such circumstances, especially so because the aquifer provides the water supply for the sizable urban population of nearby Palm Springs. The first consideration in drawing the boundaries of the planning unit, therefore, was the extent of the wash itself. Beyond the edges of the wash itself, however, there were no clear physiographic edges to define the limits of the corridor. Consequently, the area included was determined by the issues at hand.

The U.S. Bureau of Land Management, which controls most of the land in the wash, was considering issuing permits for several wind farms in the area, and residents of the nearby resort town of Palm Springs were concerned about the threat of forests of windmills being built within their view of the rugged peaks of the San Bernardino Mountains to the north. Thus, it was necessary that the areas being considered for wind farms as well as areas where exceptional visual sensitivity and high visibility coincided with high wind potential be included in the corridor. Another consideration was the inclusion of public and quasi-public lands as shown on the land status map. The boundary on the south was determined by the city limits of Palm Springs and the edge of the very rugged foothills of Mount San Jacinto, which are mostly inaccessible. Because these boundaries were so loosely defined, and because the issues would very likely expand beyond them in the next few years, the key elements of the land resource inventory were mapped to include a far greater land area. See, for example, the wind potential map and the visual sensitivity map.

The goals imposed upon Whitewater Wash by its larger context emerge clearly from the analysis. We can summarize them as follows:

Maintain the water regime.

Use the unique wind potential to generate energy for the regional population.

Protect rare plant communities and wildlife habitats.

All of these are regional goals, although the last is also subject to federal law. But since they are, to some extent, in conflict with one another, and with the more local goals of maintaining unobstructed views and providing recreation areas, it is impossible to achieve all of them to the maximum degree. Matching the three major uses that arise from these goals—energy generation, landscape preservation, and recreation—to the landscape and reconciling the competition among them became the challenge for design.

The challenge was met by a suitability modeling process, as shown on pages 94–96. The land resource inventory described the physical features of the landscape. The potential interactions between each of these features and each of the uses were analyzed and summarized on the suitability matrix. Based on these analyses, the models were then able to show the best locations for each use, ranked by degree of suitability. Alternatives were provided by applying these models to develop different patterns of land use according to different emphases placed on each use. Thus, the competition was resolved in a rational sequence that managed to match the physical character and processes of the landscape with human values. This is only one of many ways of carrying out a suitability modeling process, and I will return to the subject later for more thorough discussion.

We have to recognize, however, that articulating land-use areas is not enough to ensure achieving any optimum combination of the goals. That can only be accomplished, in the long run, by what actually happens on the ground. Thus, following the law of scale integration once again, we look to the smaller scales for the means. The baton is passed to designers at the smaller scales through two sets of guidelines, one dealing with landform and soils and one dealing with the structure of plant communities. Both are summarized briefly on pages 98–100.

Desert landforms are especially vulnerable to disturbance by development, both visually and ecologically. If the visual quality that was such an important consideration from the beginning is to be maintained, then the forms of new development should harmonize with the natural forms of the desert landscape. This is passed down as a general goal, one that will become more specific as it is passed on to project, site, and construction levels. Ecologically, the major goals concern maintaining the water flow and stabilizing the natural landforms, primarily by planting and maintaining soil permeability.

Because plants play such critical roles in the desert in maintaining soil stability and in providing food and cover, and in some cases water, for wildlife living under stressful conditions, and because there are so few plants that thrive there, the structure of plant associations and communities becomes particularly critical. In this case, there were five specific plant associations and two general plant communities within the plan unit. The dominant species in each, as listed in the plant community structure guidelines, were the species that should be planted in conjunction with new development. We will discuss the importance of species and the interactions to shape structure in Chap. 11.

THE PROJECT LEVEL

For our present purposes, we will define a project as a landscape development operation that is under the fiscal control of a single entity, that is designed as an integral whole, and that is carried out within a finite time period, usually between two and ten years. A subdivision tract is probably the most common project scale operation, although an urban redevelopment or a recreational development like a ski resort would usually fall into this category. The size usually ranges from about 50 to 5000 acres, although a particular project can be smaller or much larger. Mapping scales are generally between 1 in.: 100 ft and 1 in.: 400 ft, which means that the maximum horizontal error should be between 8 and 32 ft, accurate enough for preliminary sketching of property lines.

Since zoning categories are often inclusive of several uses, this is also the scale at which precise land uses are often determined. For example, in a planned unit development zone that allows recreational facilities and some shopping, these would be located at the project scale.

The High Meadow Case Study

Property lines are particularly powerful tools in establishing the fit between landscape and human use. Some of their uses are illustrated in the High Meadow case study. High Meadow Ranch includes 815 acres in a rugged mountain valley studded with granite boulders in eastern San Diego County. The general use of the land was determined at larger planning scales: low-density residential with related ancillary uses. The entire tract, under a single ownership, was to be developed over a period of three to eight years, depending on economic conditions. The land was to be subdivided into large residential lots with high levels of amenity, and common recreational facilities were also to be included. Thus, High Meadow represents a more or less prototypical example of a project scale plan.

The major goal passed on from larger scales of concern, other than the low-density residential use, was to maintain the predominantly natural character of the landscape. Other goals included minimal water use and protection of wildlife populations.

Given these goals in this rugged landscape, the fit becomes all-important, and good fit requires careful analysis of the landscape. The suitability modeling process used here, as shown on pages 105–9, was based on inherent land sensitivity and impact prediction, and was quite different from that described for Whitewater Wash. From consideration of the interactions of topography, soils, plant communities, and the water regime, a definite pattern of relationships emerged, and this pattern became the guiding concept for design.

The upper slopes and the ridges are hard to reach with roads, difficult to build on, and feature important wildlife habitats. There are 324 acres, well over one-third of the project land area, that are subject to these conditions. The valley bottoms, covering about 44 acres—mostly narrow meadows with riparian areas running through—are important to the drainage pattern and for ground water recharge. Together, these two topographic forms, the highland areas—including ridges, peaks, and upper slopes—and the valley bottoms, shape the visual and ecological character of the landscape. These, it became clear, should be kept open, the highlands in their natural state and the valley bottoms to be used for recreation and common pasture. The remaining lands, about 450 acres, mostly on hillsides, were subdivided into residential lots ranging in size from 1 to 5 acres. Lot sizes were arranged to provide a relatively flat and accessible building area, with a north or south orientation for each. The ownership boundaries were made to coincide with natural features as much as possible. Ridges and drainage channels, for example, were designed to be on, or very near, lot lines, with setbacks and, in some cases, easements to protect them.

These property ownership lines were critically important because they do much to shape eventual specific form and thus strongly influence the landscape. They are also very long-lived. Once they are established, legal restrictions make them extremely difficult to change. History shows that they usually outlive not one but a long succession of buildings. Over time, moreover, varied plantings and different land management practices convert the lines into patterns that are usually much stronger than natural ones. Any air traveler can attest to their importance in the landscape. The hedgerows in northern Europe and England and the survey grid that checkerboards the American landscape are probably the extreme examples. Even the two-thousand-year-old grid of the Roman centuriatio subdivisions are still visible from the air in parts of Europe, as are the grids of their castra (military camps) in the hearts of some cities. These bold, imperial patterns form strong contrasts with the natural landforms and even with the more organized accretions of late urban growth. Property lines are as expressive in the landscape as the lines of age in a human face.

At High Meadow, we were looking for a more harmonious relationship with the natural landscape, functionally as well as visually. Especially in such rugged, steeply sloping landscapes, carefully drawn property lines can make it possible to retain the natural drainage pattern, to minimize grading, and to blend the man-made pattern with the natural one.

Looking to the site planning scale to carry on this pattern in finer detail, several devices proved especially useful. Building areas and planting areas were defined for each lot. Construction could take place only within the building area, which was selected on the basis of buildability, accessibility, visual unobtrusiveness, clearance from natural features, and distance from adjacent building areas. In this varied terrain, the standard setbacks typical of subdivision practice would not provide the flexibility needed to respond to disparate conditions.

The location of the planting areas would then be defined by each homeowner according to the siting of his house. They would cover less than half of each lot, leaving the rest to form part of the overall matrix of natural landscape.

In order to maintain the natural drainage pattern, runoff from each lot was ideally to be limited to the amount that would drain off under natural conditions. To do so required careful design of sumps, swales, and other devices.

Requirements like these, which shape the transition from the project to the site design scale, are critical in ensuring the integrity of a development. Without them, the broader patterns established at the project scale are likely to be canceled out by site design performed in a different spirit. At the same time, it is important that the site designer on each lot be allowed the greatest possible scope in responding to the precise conditions of that piece of land in combination with the needs of its owner, which he will eventually know better than anyone else. The purpose of the controls is to ensure the integrity of the whole, to establish larger goals, not to design the individual parts. The ideal is a diversity of form in siting and design within a cohesive larger framework that assures functional ecological integrity and visual continuity. Clearly, that is a delicate balance.

Usually, such controls are passed on from project to site designer through the Conditions Covenants and Restrictions, which become a legal part of the deed when each lot is sold. Being legally binding, this is a rather rigid device that requires precise language and allows little room for future adaptations. Especially for larger projects with varied topography and uses, some additional direction is often useful. Guidelines can provide a less rigid device for transmitting information to smaller scales.

Guidelines: The Woodlands Example

The site planning guidelines developed by Wallace, McHarg, Roberts and Todd [1973] for the Woodlands New Community in Texas manage to include considerable detail without being unduly restrictive. They are worthy of some attention as good examples of their kind. They begin by giving a great deal of data concerning ecological conditions and then go on to establish clear rules for the extent and locations of clearing, based on soil and vegetation types.

The main concern with respect to soils at the Woodlands is ground water recharge. All soil types are grouped into A, B, C, or D classes, depending on permeability. The A soils are the most permeable and may be cleared up to 90 percent and still provide local recharge of a 1-in. storm; B soils may be cleared up to 75 percent; and C soils, up to 50 percent. These, then, represent maximum cleared areas for each of these types. Since D soils are impermeable (mostly clays), clearance on them will make little difference and thus is not limited; high-density development is recommended.

In considering vegetation, a major concern, in addition to maintaining the natural water regime, was keeping the forested character of the landscape and its wildlife habitat. Vegetation communities were evaluated for their ability to tolerate development and related compaction and still survive, as well as for their visual importance and their ability to provide food for wildlife. Each community was then assigned a maximum percentage of clearing, with variations within each community of up to 5 percent more for open stands and 5 percent less for very large trees. The greatest percentage of clearing—95 percent—was allowed for pine stands, whereas at the other extreme only 10 percent was allowed for pure hardwood stands and the riparian community.

Other guidelines established some measure of control over grading, road rights of way, sizes of stands, and other factors, but this is enough information to give some notion of the flexibility that is possible while still assuring ecological integrity.

THE SITE LEVEL

Site design is "the art of arranging an external physical environment in complete detail."[1] It involves defining the precise location of structures, circulation routes, and activities in the landscape. The site has definite boundaries and is usually small enough to be perceived in its entirety from a single viewpoint. Sometimes, as in the cases of High Meadow or the Woodlands, the site is an integral part of a larger project for which clear goals and an integrative framework generated by design concerns have already been established. More often, we have to deal with sites that are subject to standardized and seemingly arbitrary planning restrictions like setback lines and height limits, leaving their connections with a larger context obscure.

In dealing with the larger context and with the diversity of internal concerns that include visual form, circulation and parking, orientation, convenience, economy, utilities, and a great many other matters, site design can become a complex and subtle process. I will not try to deal with it here in any comprehensive way, especially since there are several good books devoted to the subject.[2] Instead, I will discuss three fairly common types of situations in which the natural resources of a site meet and mingle with human development in three important but quite different ways. These we can call the goals of absorption, utilization, and protection.

Site Absorption: The High Meadow Example

In the absorption situation, buildings and activities are visually absorbed into the landscape, disturbing it as little as possible, keeping its basic form and cover essentially natural. Structures dig into the ground, bend around trees, often yield in their geometry, avoid confrontation. Activities find places to happen without the land's being reshaped.

An example in which the goal was absorption is the recreation center at High Meadow Ranch. The framework, you will recall, was set at the project scale. Natural hilltops

[1] See Lynch, 1971; p. 3.

[2] Ibid.

form a rugged skyline, and valley bottoms used for recreation and pasture provide a green floor. Between floor and skyline, the buildings are worked into the slopes, absorbed among trees and rocks as much as possible. The site for the recreation area includes a low hill, studded with granite outcrops and partly covered by stately spreading oaks. At its base, the hill spreads out into a narrow valley that is relatively flat, with a tiny intermittent stream running through it.

The design of the site, shown as part of the High Meadow case study on page 109, follows the larger pattern, leaving the hill as is but lacing it with hiking trails. The soccer field, riding ring, and tennis courts are located in the valley. An existing pond to one side becomes a fishing pond. Around the base of the hill are the structures: the information office (in a previously existing structure that was once a stagecoach way station), the swimming pool and changing rooms, the clubhouse, handball courts, and the amphitheater. These nestle among trees and rocks, visible but unobtrusive, deliberately located so as to be as inconspicuous as possible. Achieving such an end requires an architecturally informal arrangement with the rather loose relationships among buildings and spaces being defined in the nongeometric terms of nature. All new plantings use native materials that grow in the area. The natural setting remains clearly dominant.

Site Utilization:
The University Village Case Study

The utilization situation is quite a different one and leads to different forms. Every site has resources that vary in kind and quality, but at the very least, they include sunlight (or solar energy), soil, and some water. These and any other resources that may be present can be used to some extent to help provide the basic needs of the people who live on the site or nearby. The extent depends on the nature of the resources, of course, along with the kinds of need. Nonetheless, the possibilities of using on-site resources productively is too often ignored.

The University Village case study on pages 111–13 illustrates a situation in which the resources of a site were proposed to be used intensively to contribute to the support of its inhabitants. It was designed as a living and working environment that would satisfy the basic needs of a group of university students to the greatest extent practicable, with provisions for intensive mixed agriculture, aquaculture, solar heating, and recycling water and animal wastes. To accomplish such a complete use of on-site resources required a careful organization of the site. Buildings were oriented to the south on the hillside to take advantage of solar radiation, and dining and meeting rooms were placed at the top for the sweeping views. Citrus groves were also planted on the hillside for its drainage and optimum temperature. Vegetable gardens were situated at the base of the south slope where runoff water is available but hillside erosion is not a problem. The fish pond, which is also used for irrigation water storage, was located at a slightly higher elevation to permit its water to be released into the furrows. Pastures, pigpens, and other barnyard facilities were banished across the road as were the biological sewage treatment ponds, so that their associated odors and pests would be as far as possible from the dwelling areas. Treated sewage water is pumped by windmill back up the hill and then distributed by gravity to water the orchards. Under present health restrictions, it cannot be used on vegetable crops, although this situation may change in the future. The flow diagrams explain the use of water and energy in this system. Proper resource utilization calls for a design that will work closely with these flows, optimizing the potentials of orientation, gravity, and other physical characteristics. It is a highly functional approach, one that yields a site organization that expresses and reinforces the inner workings of the man-made ecosystem.

Resource protection also calls for close attention to landscape function, but in a rather different way. A critical resource may exist over a whole site or only a small part of it, but even in the latter case, its influence is likely to spread over a very large area, dominating a great many site decisions. Often, this influence brings about conflicts with other factors of concern.

Resource Protection:
The Madrona Marsh Case Study

The Madrona Marsh case study (pages 115–18) illustrates the point. Madrona Marsh is an 11-acre remnant of vernal marshland situated in a 160-acre parcel that so far has been undeveloped because of its oil wells. It lies in the midst of the heavily urbanized town of Torrance in the Southern California urban region. It provides the habitat for a rich wildlife population, from microscopic pond species to major predators like the rare White-tailed Kite. It is also, like San Elijo, an important stopover in the Pacific Flyway.

Since development of the area is ever more impending as more and more wells are taken out of production, environmental groups have fought long and hard to save the marsh. Even the most ardent preservationists generally concede that at least part of the 160 acres will probably be developed for urban uses, and their admission poses an interesting and important site design problem. Can the marsh possibly remain a viable wildlife habitat as urbanization closes in? How much land needs to be preserved to maintain its integrity? How should the land be organized and managed to provide the best combination of habitat value and economic use?

The site plan shown here poses some answers to these questions. The principles that guided the design probably have broad application in other situations where wildlife habitats and urbanization coexist cheek by jowl in a state of uneasy truce.

Before examining the site plan, it will be worthwhile to develop a better understanding of the clash of values involved at Madrona Marsh. The economic arguments for development are strong. The 160 acres of open land are divided into two parcels, with the marsh located on the smaller of the two, which covers 54 acres. Across the street to the west is the very successful Del Amo shopping and business center. Industrial and residential districts are on the other sides. A major traffic artery, Sepulveda Boulevard, runs along the southern edge, and a railroad spur also runs in from that side. The entire parcel is zoned for industrial use. Estimates of land value vary from a low (which is

CASE STUDY VIII

High Meadow Residential Development

The narrow valleys and rock-studded hillsides of High Meadow Ranch can accommodate about 250 single-family houses with a common recreation area on their 815 acres without the natural character of the site being disturbed. This balance can be achieved, however, only by carefully seeking out the best building sites. A sensitivity model was used to define the areas suitable for development. Analyses of geology, drainage patterns, slopes, soils, and vegetation communities permitted areas to be categorized by their degree of sensitivity (high or low) and feasibility of mitigation. The model indicated that the more sensitive areas were concentrated on the upper slopes (as a consequence of their steepness, thin soils, and wildlife habitats) and in the valleys (as a consequence of their potential flooding and dense groves of ancient oaks). This finding suggested a land-use pattern restricting structural development to the lower slopes, with the upper slopes and peaks kept in their natural state and the valleys used for recreation.

Prepared by John T. Lyle. Landscape analysis by Environmental Analysis Systems, Inc.: Kenneth Knust, Michael Steiding, and Robert Small. Prepared for New Environment Research Corporation; Raleigh Kirkendall, President.

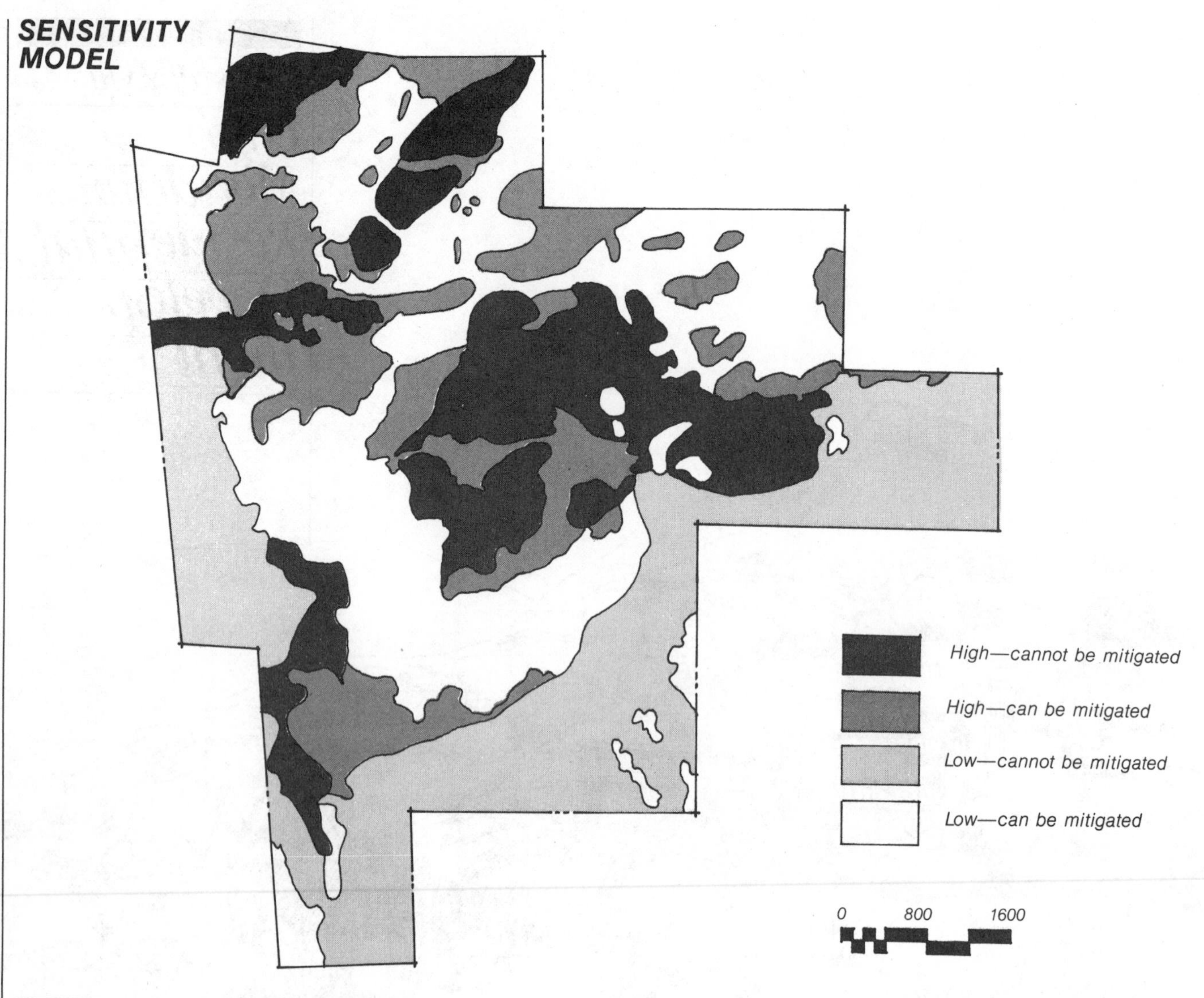
SENSITIVITY
MODEL
High—cannot be mitigated
High—can be mitigated
Low—cannot be mitigated
Low—can be mitigated
0
800
1600

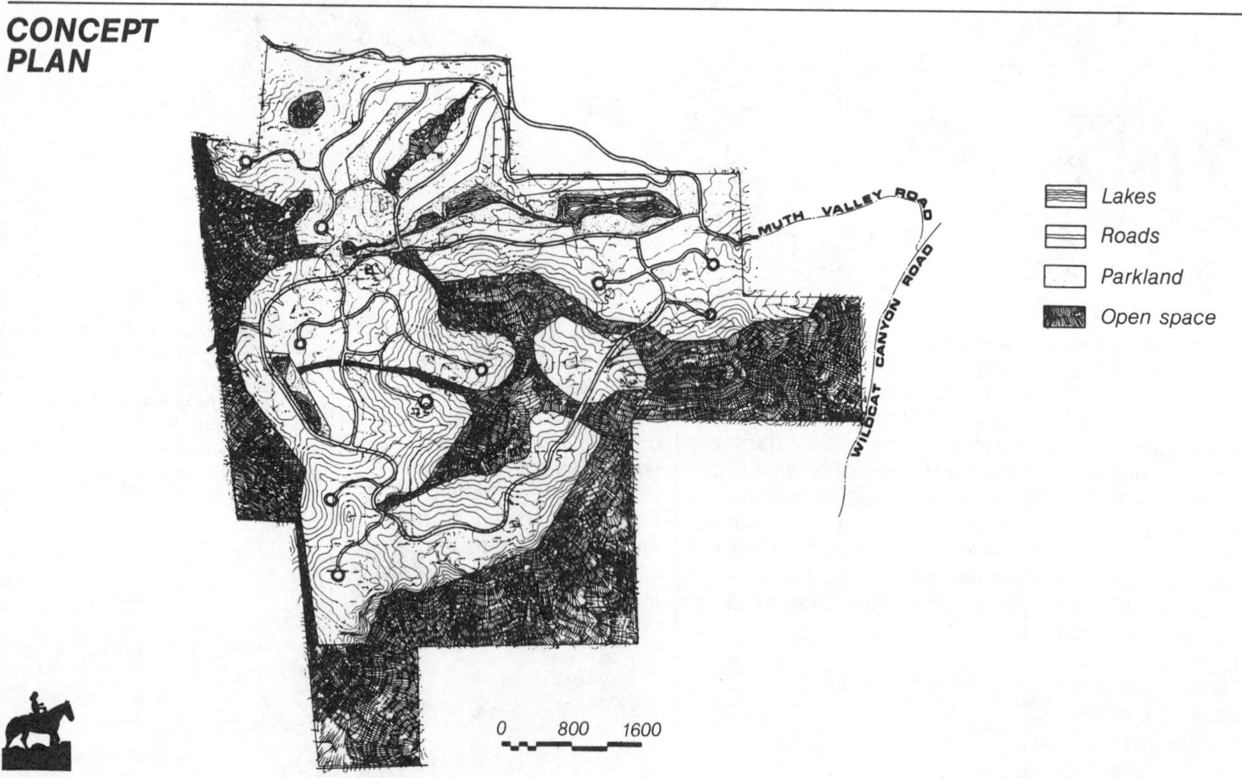
CONCEPT
PLAN
MUTH VALLEY ROAD
WILDCAT CANYON ROAD
Lakes
Roads
Parkland
Open space
0
800
1600

EVALUATION OF CONCEPT PLAN

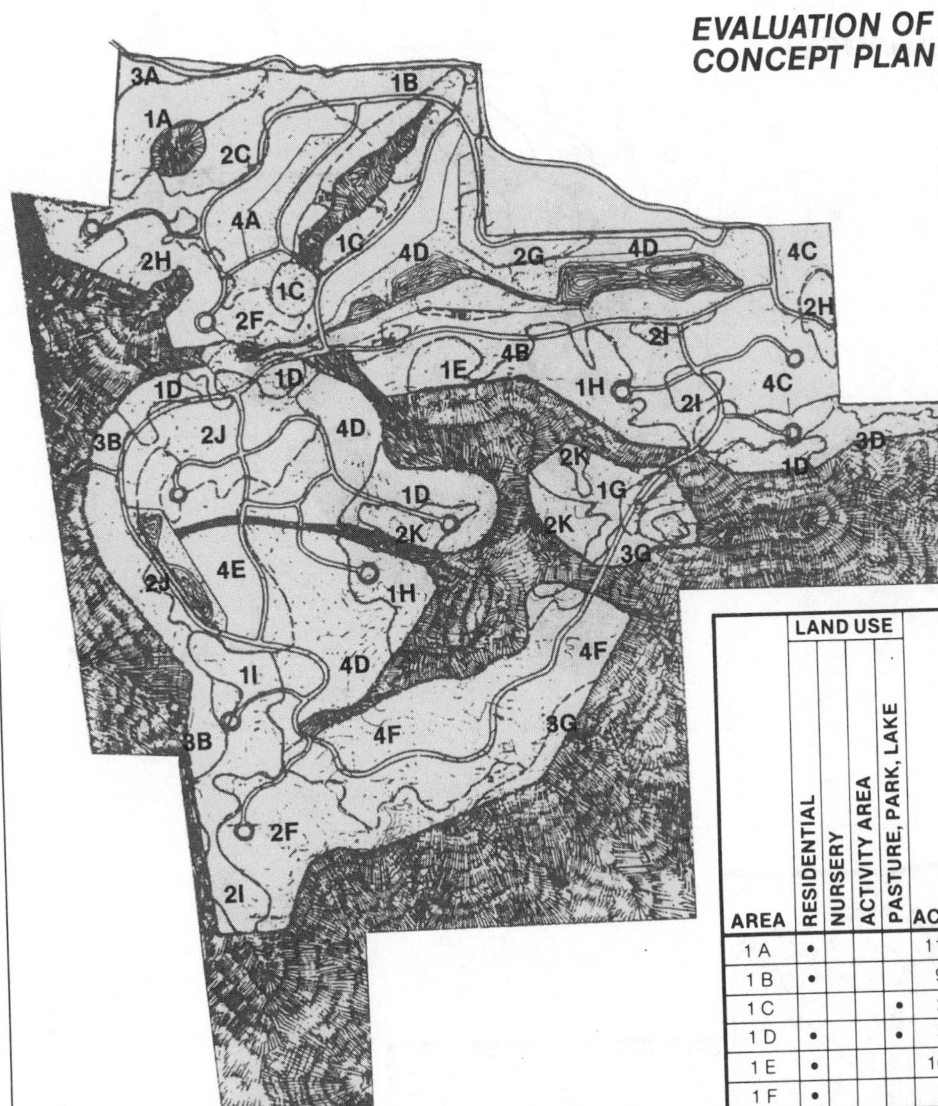

Predicted impact levels:

1 A-L *High—cannot be mitigated*

2 A-N *High—can be mitigated*

3 A-C *Low—cannot be mitigated*

4 A-F *Low—can be mitigated*

The sensitivity model was first used as a base for the development plan and then laid over the plan to identify the potential environmental impacts on each lot. The summary chart at right shows the sensitivity attributes of each land area. The chart was used to indicate the mitigation requirements to be observed by each builder and as a basis for the environmental impact report. In this way, the processes of design and impact prediction were combined.

	LAND USE							SENSITIVITIES										
AREA	RESIDENTIAL	NURSERY	ACTIVITY AREA	PASTURE, PARK, LAKE	ACRES	LOTS/NET ACRE	NUMBER OF LOTS AFFECTED	GEOLOGY	ROCK OUTCROPPINGS	HYDROLOGY	SURFACE WATER	DRAINAGE	SOILS (EROSION)	SOILS (RUNOFF)	SOILS (SEWAGE EFFLUENT)	SLOPE	VEGETATION	WILDLIFE HABITAT
1 A	•				11.01	.50	5.5	•				•	•			•	•	•
1 B	•				9.95	.65	6.5	•	•				•		•	•	•	•
1 C				•	2.64	-	0	•	•				•		•	•	•	•
1 D	•			•	8.48	.56	2.5						•	•		•	•	•
1 E	•				16.18	.54	9.0						•			•	•	•
1 F	•				3.52	.49	2.0						•			•	•	•
1 G	•				18.93	.53	10.0		•				•	•		•	•	•
1 H	•				7.74	.63	5.0		•				•			•	•	•
1 I	•				17.00	.56	9.5	•					•	•		•	•	•
2 A	•				3.48	.50	2.0	•				•	•					
2 B	•				8.03	.65	5.0			•		•						
2 C	•				8.55	.65	5.5	•	•				•	•	•	•		
2 D	•				1.69	.59	1.0	•					•			•		•
2 E	•				11.85	.65	8.0	•					•		•	•	•	
2 F			•		6.68	-	0	•		•			•				•	
2 G	•				6.28	.59	4.0		•								•	
2 H	•	•			3.52	.49	1.0						•			•		
2 I	•				11.30	.49	5.5		•				•			•	•	
2 J	•				21.06	.56	12.0				•		•	•			•	
2 K	•				12.65	.53	7.0				•		•	•				•
2 L	•				1.61	.63	1.0		•				•			•		•
2 M	•				3.78	.56	2.0	•					•				•	
2 N	•				23.04	.63	14.5	•					•				•	
3 A	•				4.62	.50	2.0	•	•				•		•	•		•
3 B	•				16.25	.56	9.0	•	•				•	•	•	•		•
3 C	•				10.04	.63	6.0	•	•				•	•	•	•		•
4 A	•		•	•	25.76	.65	13.0			•	•		•					
4 B	•			•	43.67	.59	15.5			•	•		•					
4 C	•			•	53.07	.49	23.0			•	•		•					
4 D	•				39.23	.56	22.0						•	•		•		
4 E	•			•	23.60	.56	13.0			•	•		•	•				
4 F	•				44.37	.63	28.0						•	•		•		

DEVELOPMENT PLAN

Phase	Area Use	Acres	Lots
	A Activity center	10	1
1	B Nursery	7	1
	C Residential	60	38
	D Residential	35	36
	E Residential	44	32
	F Residential	24	12
2	G Residential	74	42
	H Residential	60	35
3	I Residential	75	45
	J Residential	18	19
	Residential subtotal	390	250
	Natural preserve	320	
	Recreation	44	

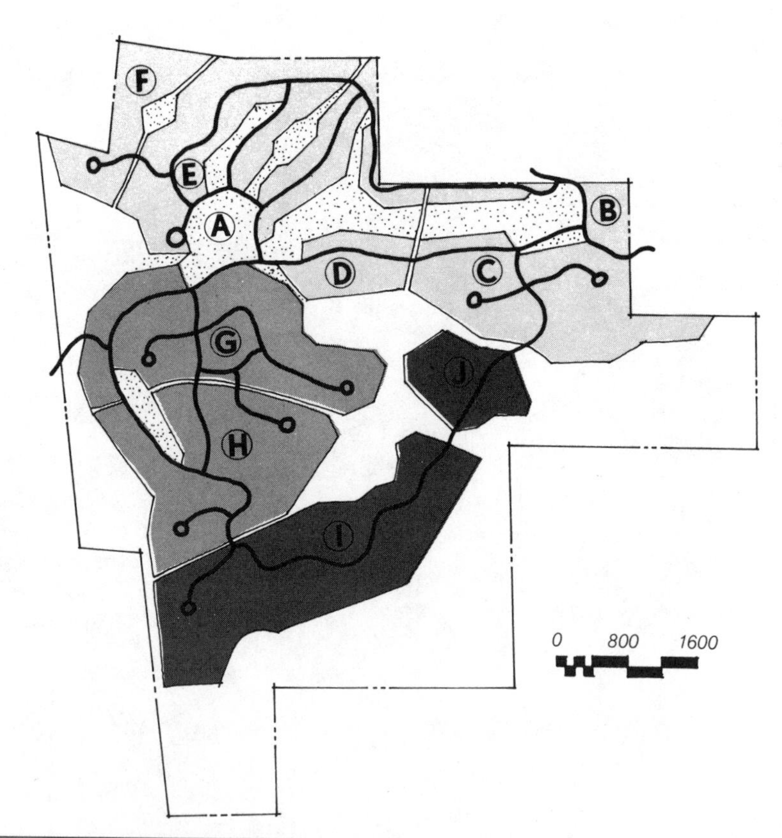

SUBDIVISION PLAN

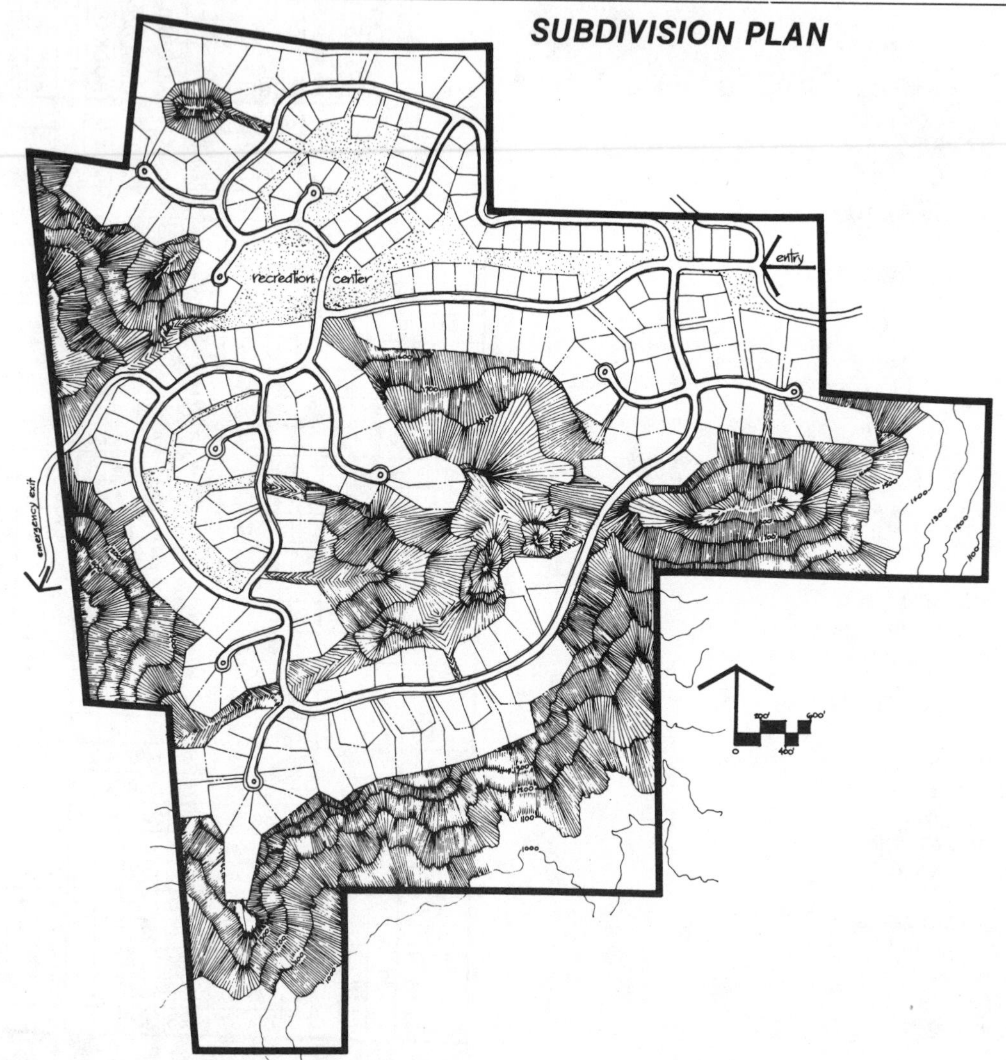

RECREATION AREA SITE PLAN

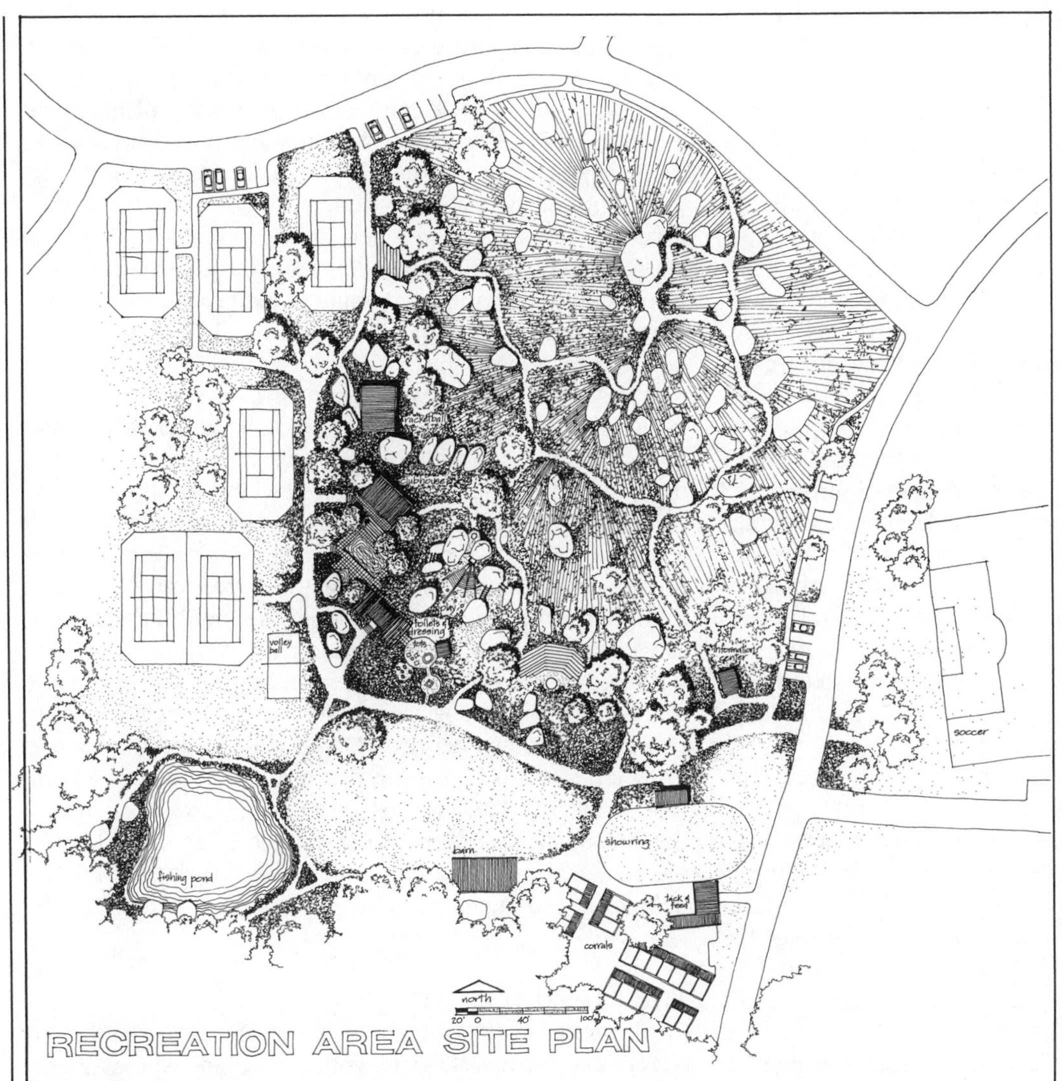

The basic conceptual relationship between development and landforms is followed in the site design of the recreation area. Valley bottoms are kept open for the flow of water to provide common recreational space, and peaks are left in their natural state. Construction is mostly restricted to the hillsides.

SITING PRINCIPLES

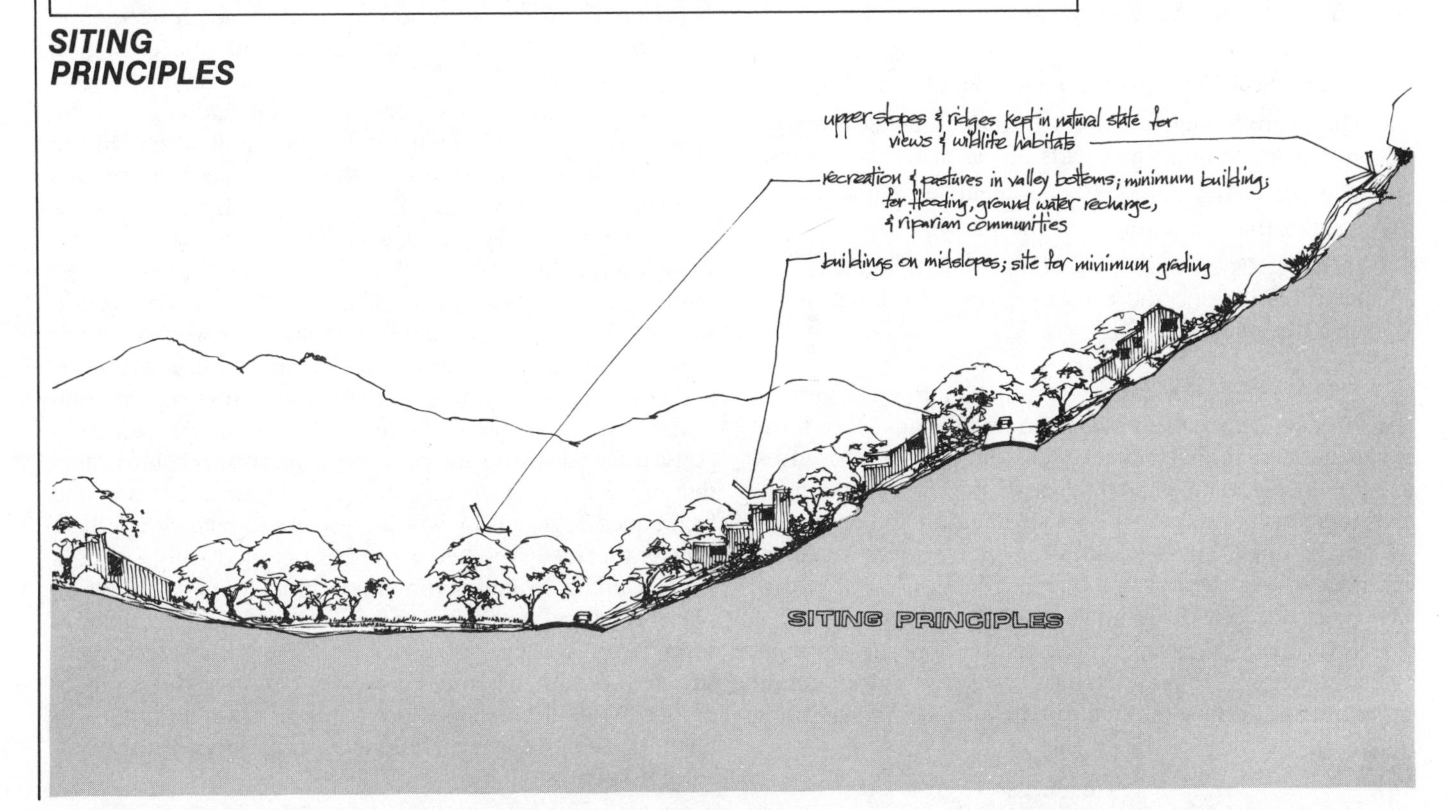

probably quite low) of about $5 million to several times that figure.

On the other side of the balance we have visual, recreational, and wildlife values. Although this is the only open land in the vicinity, it is not visually memorable, being a flat, grassy plain laced with dirt roads and patches of water. Eucalyptus trees, oil rigs, and powerlines rise sporadically here and there against the surrounding backdrop of grayish commercial and residential buildings.

Although not now used for recreation, it could be in the future, and this happens to be an important possibility. The surrounding area has a ratio of public recreation space to population of 2.47 acres per 1000 people, as compared to the accepted standard of 10 acres per 1000 people.[3]

As for its wildlife value, despite the influences of surrounding urbanization, Madrona Marsh remains a surprisingly rich example of vernal marsh habitat. The Friends of Madrona Marsh argue that it is the only one of that type remaining in the southern part of California and that freshwater habitats are always important in a dry landscape. Its species are numerous and diverse, particularly at the micro and macroscopic levels, and the trophic structure seems sound and complete. Rapid decomposition of plant materials reflects efficient nutrient cycling processes.[4]

Given such conflicting values, one would expect controversy, and, in fact, the battle over Madrona Marsh has gone on for well over a decade, with a great many twists and turns and legal ramifications. What concerns us here, however, are the site design questions. If the Madrona Marsh is eventually preserved, it seems unquestionable that the land should be intensively and efficiently used to satisfy as many needs as possible and to the highest degree.

Based on analyses, the 160 acres can be subdivided into four zones of influence. In descending order of ecological importance, these zones are as follows:

1. The Critical Natural Area, including all wetlands
2. The Primary Support Zone, including habitats of species that are important parts of the marsh ecosystem
3. The Secondary Support Zone, including habitats of species that sometimes use the marsh
4. Valuable Open Space, including some marginally related habitat and the remaining open land that drains into the marsh

These zones provide the framework for site design. Of the three alternative site plans, the first assigns all of these zones to recreational, conservation, and preservation uses. This, of course, is the ecological ideal, providing for optimum management of the marsh because it affords full control of all four zones. It also provides for a range of recreational activities much needed in the area. For economic reasons, however, it is obviously unlikely to be realized.

The second alternative presents the opposite extreme. Only the eleven acres of the first zone, the critical natural area, remain undeveloped, the remainder of the site being allotted to industrial uses. As partial compensation for the loss of the support zones, provision has been made for drainage swales to direct runoff water, with quality controls, into the marsh, with a community of riparian plants along the edges of the swales replacing some of the habitat lost in the support zones. This scheme maintains the basic functions and major habitat value of the marsh, but with a reduction in its population and diversity of species. Madrona Marsh would therefore be a far less rich environment and would contribute virtually nothing to local recreational needs. Nevertheless, should all else fail, half a loaf would be far better than none.

The third alternative—the most likely of the three to be accepted—shows the fifty-four-acre parcel maintained for conservation, preservation, and very low-key recreational uses, with the other parcel allotted to industrial use. Thus, the entire primary support zone remains intact, along with a sizable part of the secondary support zone. By some regrading and additional planting, these areas can be adapted to support larger wildlife populations than they did previously, as well as make room for human participation. Buffers are designed to protect the marsh from human intrusion while at the same time providing observation blinds for wildlife viewing. This new Madrona Marsh will also become something it never was before, an extraordinary, visually pleasing island of nature whose edges meet the city head on. Assuming the means can be found to acquire the land, the keys to the marsh's future viability will lie in the site design and later in its management.

THE CONSTRUCTION LEVEL

Construction scale design, like site design, is a complicated subject that has been widely written about elsewhere, usually in terms of esthetics and human comfort and convenience. Even those professionally concerned with the ecological character of the landscape and with maintaining the integrity of natural processes usually focus their attention on the larger scales, the regional or the plan unit levels. In the midst of larger-scale analyses, where the sweeping patterns of nature are clearer, it is easy to forget that all plans are only paper until they are converted to action on the ground. Bridging the gulf between thought and action by physically reshaping the landscape itself is the task of construction design. Thus all the levels of integration come down eventually to construction design, to the man with a shovel.

For our present purposes, we can define construction as the physical reshaping of the landscape and then envision construction design as the process of anticipating construction, that is, describing specific form and placement and the techniques or means by which these are to be accomplished. This is the case whether we are concerned with a park, a garden, an apartment complex, or agricultural fields. Form, placement, and techniques all have to be decided upon at some point, and the process of deciding is considered construction design, whether it involves conscious forethought or not. We will consider here some of the implications of form, placement, and technique in construction that affect natural systems.

[3] See Earle, 1975.
[4] Ibid.

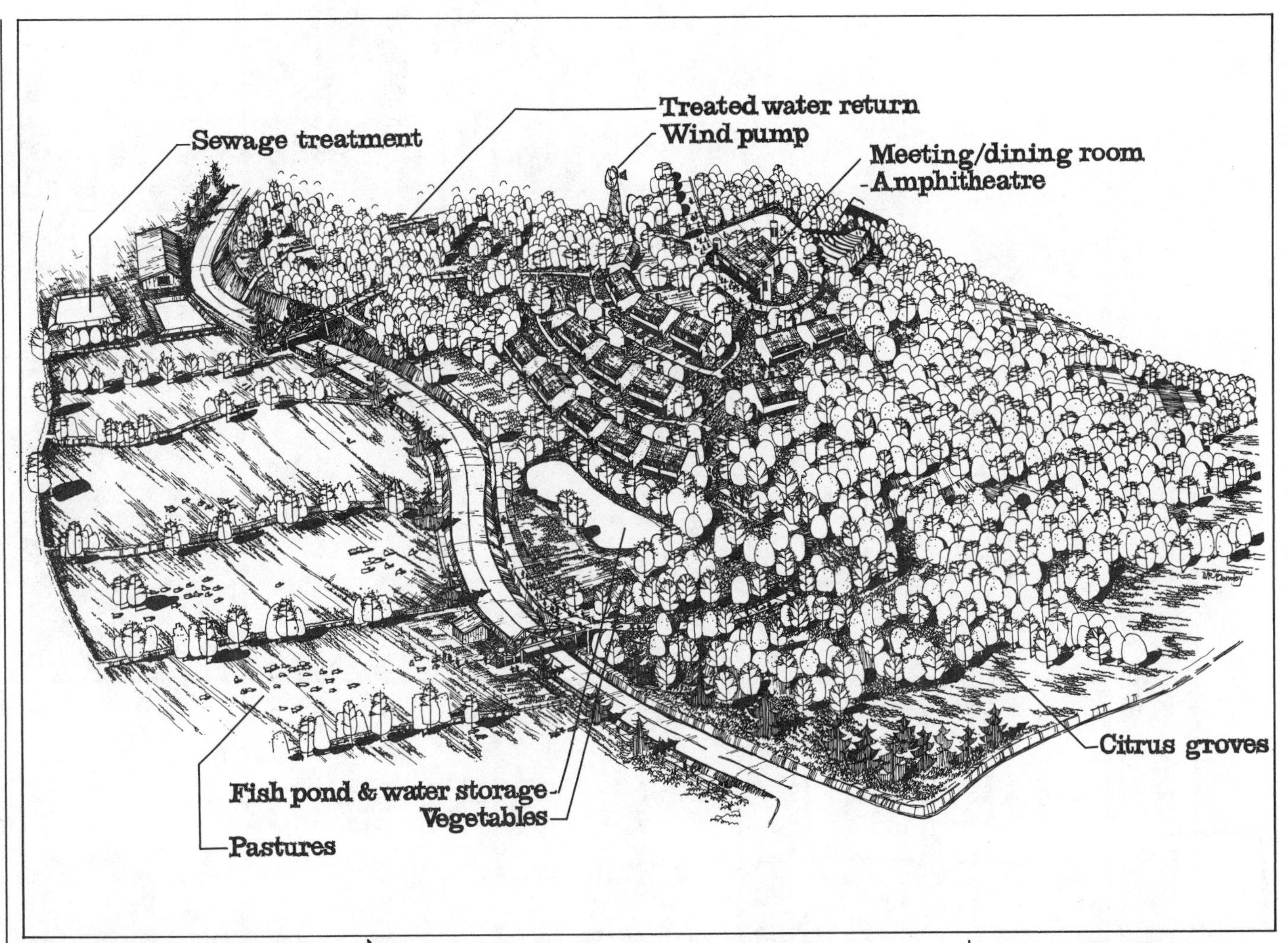

CASE STUDY IX

University Village

A small community of about 150 students could provide for most of its own food, water, and energy needs within a site of less than 100 acres. To do so would require making highly efficient and conservative use of on-site energy and water resources and a variety of appropriate recycling technologies, including biological sewage treatment and methane digestion of animal wastes. The vector flow diagrams illustrate the basic pathways of materials and energy, showing inputs, outputs, and feedback loops as projected for the initial as well as the eventual steady state of operation.

Such efficient use of on-site resources requires careful fitting of land use to landform. Buildings are sited on south-facing slopes for maximum potential solar gain. Citrus groves are also best located on slopes with solar exposure, whereas vegetable gardens and fish ponds need flat land. Sewage is carried by gravity to the greenhouse and ponds for biological treatment; the treated effluent is pumped by wind power back uphill for toilet flushing and irrigation.

The analysis showed that—although productivity in a life-support ecosystem of such design can be very high—complete self-sufficiency is probably not an economically efficient goal because it would inhibit making the best use of landscape resources. The land is not well suited for producing certain basic needs, such as cereal grains, but it is very well suited to produce greater quantities of some foods, like citrus, than the community could reasonably consume. Thus, some balance of self-support and trade is probably the most economically efficient goal.

Prepared by the 606 Studio, Department of Landscape Architecture, California State Polytechnic University. Sabine Bestier and Xantharid Virochsiri, with advisors John T. Lyle and Jeffrey Olson.

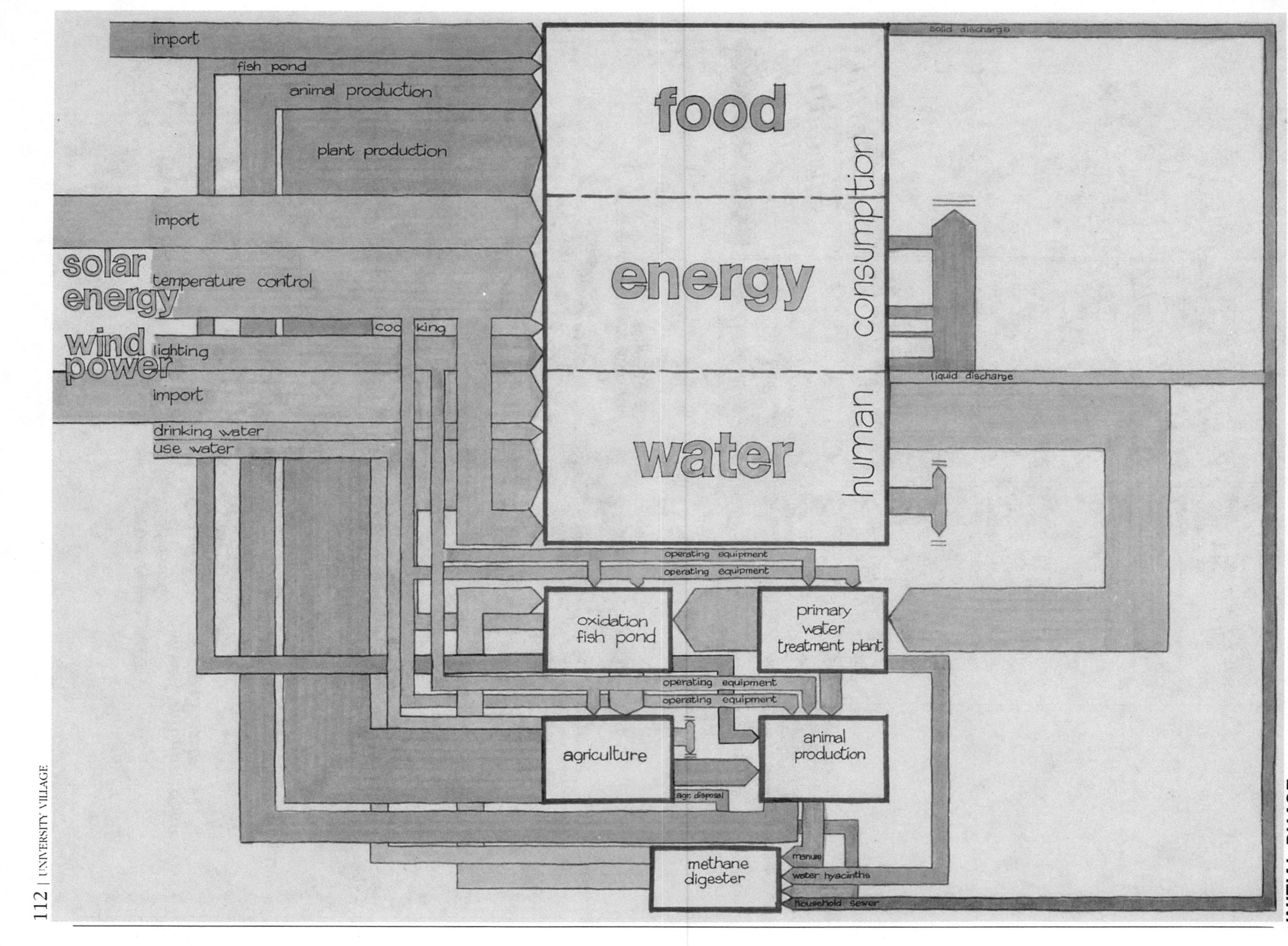

INITIAL PHASE

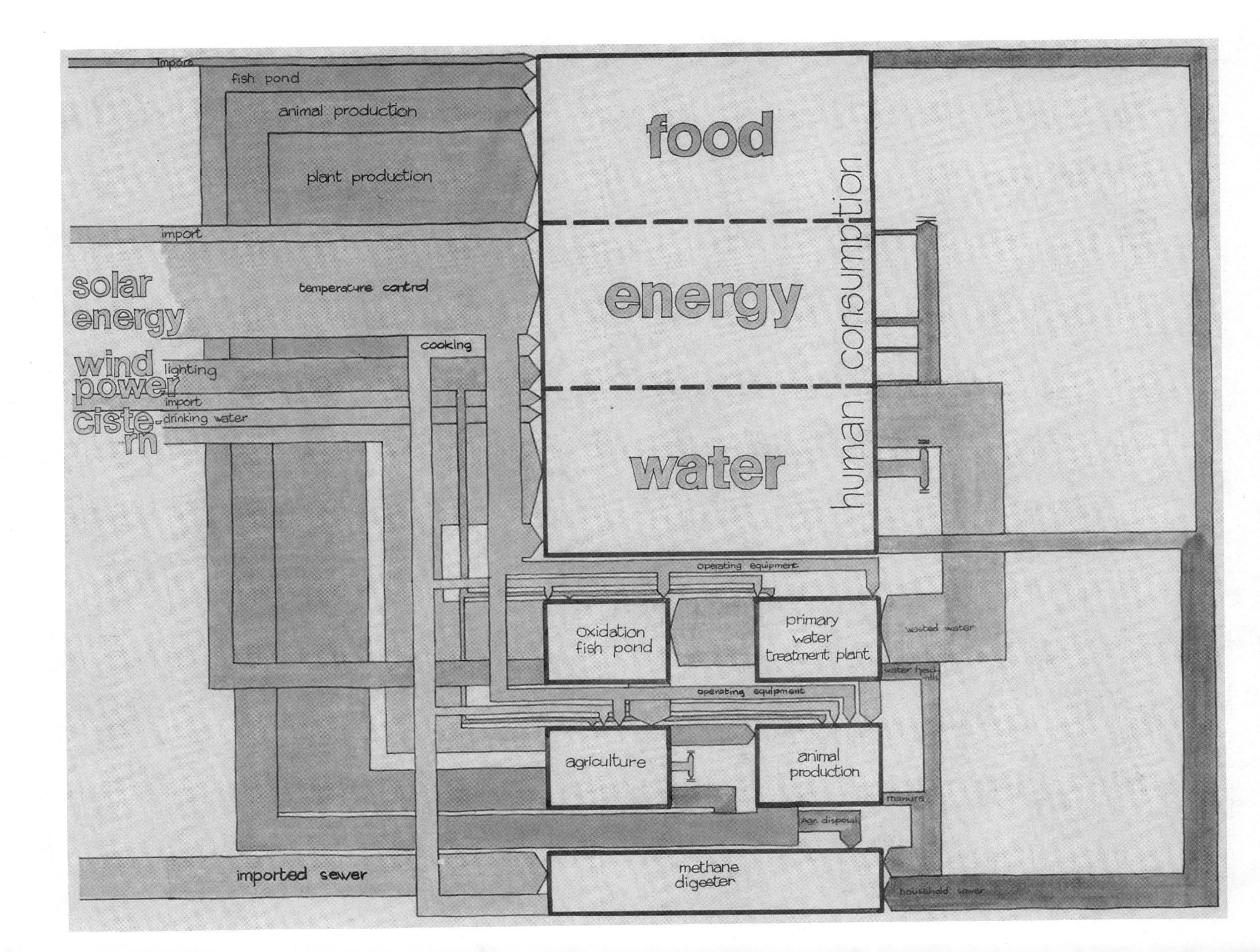
import
fish pond
animal production
plant production
import
solar energy
temperature control
wind power
lighting
import
cistern
drinking water
cooking
food
energy
water
human consumption
operating equipment
oxidation fish pond
primary water treatment plant
wasted water
water hydraulic
operating equipment
agriculture
animal production
manure
Agr. disposal
imported sewer
methane digester
household sewer

Usually, we think of form as a visual concern. Landscape forms have symbolic meanings that are a part of our culture and that are at the core of our relationships with nature. This is a profoundly important matter—the visual quality of the landscape—and one that is the major concern of most writing on landscape design. We will return to it shortly, but first we will consider form in relation to natural resources.

The specific form of the landscape has a strong influence on the workings of ecosystems. This is obviously true in nature, where we find that form is shaped by process and process in turn shaped by form, until the two are so intertwined by continuous feedback loops as to be virtually indistinguishable. In the design of human ecosystems, however, we tend to separate form and process, to think of them in somewhat different terms. What is important here is the fact that form has equal ecological importance in human ecosystems. The difference is that, for the short term at least, form is not shaped by natural process over time but by human control. By shaping the landscape, human beings determine the ecosystem function, structure, and locational relationships.

Functionalist architecture recognized the relationship between building and function and tried to reverse the relationship with the motto, "Form follows function." (By function, a combination of human activity and construction technique was meant, but the principle is essentially the same.) If in nature form and function are inseparable because of continuous feedback and there is therefore no cause/effect relationship, and if in historic architecture, function was given no choice but to follow forms that were visually derived, then in order to develop a more workable architecture, why not anticipate function and then design forms to correspond to it? It was an important idea, and it had enormous influence, but eventually broke down when it became clear that there were too many factors influencing form to allow it to follow function alone. In fact, it was such a very powerful idea that its passing has left a void in architectural theory that has proven most difficult to fill.

In landscape design, to avoid dangerously simple slogans, we might say that since form influences ecosystem function, structure, and locational patterns, along with human activity and symbolic meaning, it should be shaped accordingly. The relative significance of each of these factors will vary according to the situation at hand. The important point is that, at the site and construction levels especially, we need to consider them all, to merge them into forms that have symbolic meaning and human utility as well as ecological integrity, all joined into harmonious wholeness. Landscapes designed only to serve natural processes will often have limited meaning or utility for human beings and thus fail as human ecosystems.

Case Study: Wildlife in the Spreading Basins

An example of the influence of specific form is shown in the Los Angeles Flood Control case study. The major purpose of the spreading basins, which are clusters of rectilinear depressions located along the flood control channels, is to hold surface water until it is able to settle into underground storage. Like all standing water in this dry region, these basins have become important wildlife habitats, especially for waterfowl, both resident and migratory. With a change in their forms, they could support much larger populations. Three simple measures might achieve this goal:

1. Providing meandering banks to slow the flow of water and multiply the edge area, which is the richest habitat (we will discuss the importance of edges at some length in Chap. 12)
2. Increasing the size of each basin while keeping the total spreading area the same so as to increase the overall density of wildlife
3. Providing islands well away from the banks

The spreading grounds as redesigned here would support a far richer wildlife community than the existing basins, and entirely because of the different forms.

A fourth proposal, the planting of riparian vegetation along the banks, would further increase the habitat value and considerably expand the number of species, but it would also cause loss of water as a result of transpiration. This is why the grounds are now kept cleared of plants. Even if plants with low transpiration rates were used, the water loss would still be significant. Thus, as is usually the case, there are tradeoffs involved. Is a richer wildlife community worth the loss of a certain amount of water? Calculations will be needed to determine the amounts on each side of the ledger and thus give an adequate basis for decision. As things stand at present, the decision is likely to go against wildlife, whatever the figures. The importance of wildlife, especially in cities, is still not widely understood.

Case Study: Energy and Water in a Residential Landscape

Specific form plays an especially large role in controlling the flows of water and energy, a fact that becomes more and more significant as these resources run short. There are a great many devices for effecting such control, some of them emerging from current research and others going back to very ancient times. Carefully shaped, the landscape itself will do much of the work of energy-intensive environmental-control machinery.

The section on page 120 shows a relatively simple employment of several such devices. A hill rises fairly steeply on the south side of the house, and the hillside is stepped down in a series of short terraces to slow the downhill flow, so that it will water the plants before seeping into the ground water supply. Except in a very heavy rain, none of the flow runs off the hillside. Thus, not only is the house protected, but no burden is placed on the storm drain system. Such terracing is an ancient device, seen in places as far apart as China, Peru, and Germany. We have a vast reservoir of experience to draw upon in such matters.

The deciduous trees used to control solar radiation on the south walls of the house are also a fairly common device in very hot climates. In areas with cooler winters, even deciduous trees can cast enough shade to impede solar heating. Here they create a stratum of cool air near ground

CASE STUDY X

Madrona Marsh

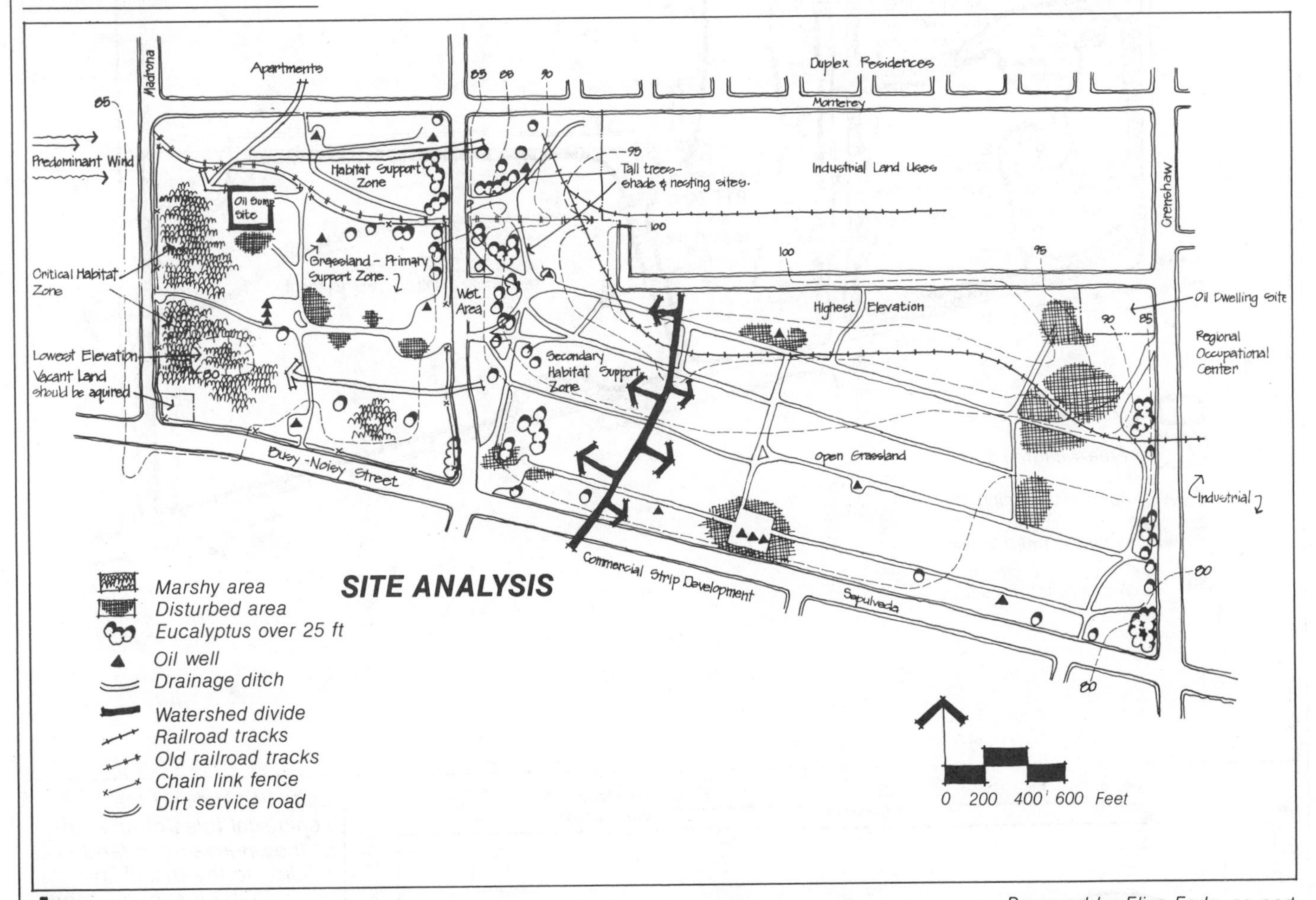

The eleven acres of Madrona Marsh, which are part of a 160-acre expanse of undeveloped land, support an extraordinary population of birds and small mammals. Since the oil wells that occupy the remaining undeveloped acres will be going out of production soon, commercial and industrial development has been proposed for the site.

Environmentalists, especially the Friends of Madrona Marsh, have struggled to preserve the entire 160 acres as open space because they are all integrally related to the marsh ecosystem and because the surrounding region badly needs recreation and green space. Such preservation is unlikely, however, because of high land values. On the other hand, the marsh probably cannot function as an effective habitat if development is permitted to extend to its edges. It is therefore necessary to inquire what a reasonable minimum land area for sustaining the functions of the marsh might be.

Prepared by Eliza Earle, as part of an M.L.A. thesis. Advisors: Mark von Wodtke, John T. Lyle, and Ronald Quinn.

MANAGEMENT ZONES

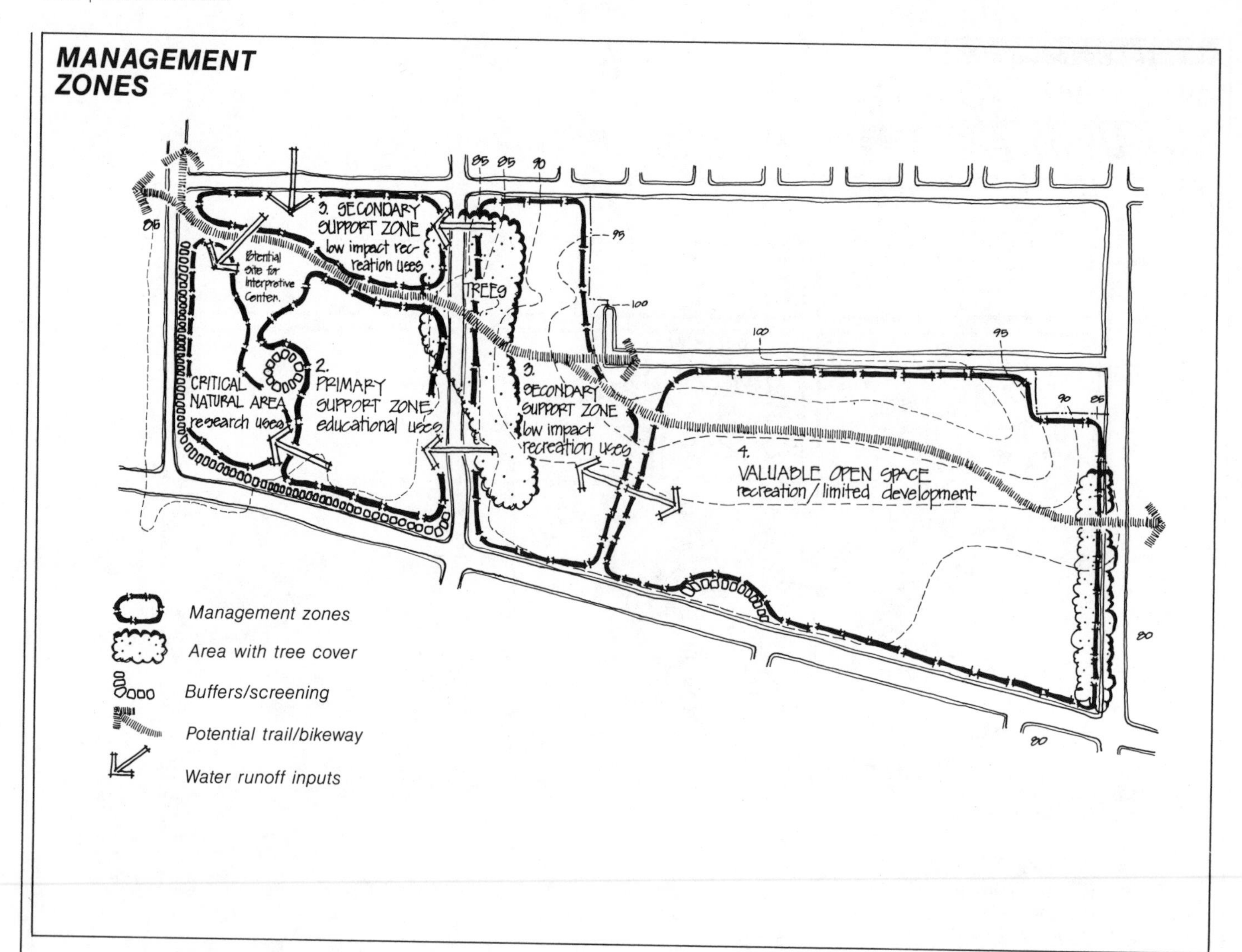

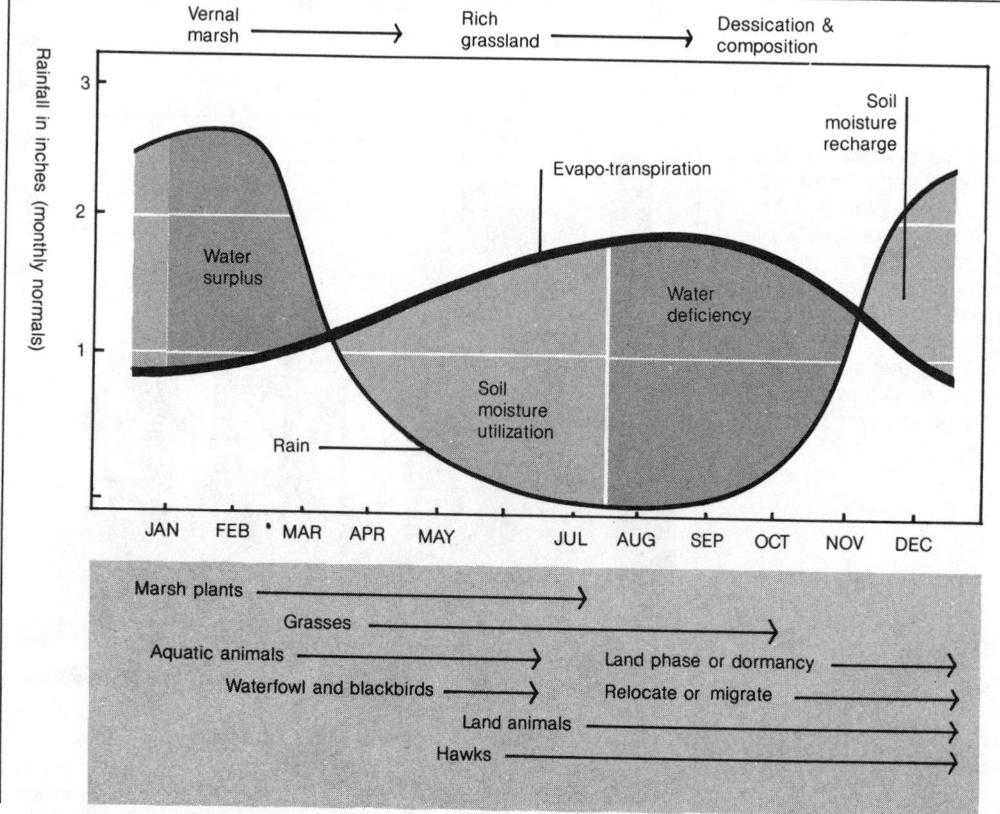

Earle first defined the ecological roles of all areas of the undeveloped land in relation to the marsh, pinpointing three support zones of varying importance. Then she developed three alternatives—one allowing the maximum level of development, one allowing no development, and a third defining the minimum land area that will assure optimal support for the marsh and leaving the rest to be developed. This study provides a realistic goal for acquisition and a basic program for future management.

ALTERNATIVE I PRESERVATION/ RECREATION AND DEVELOPMENT

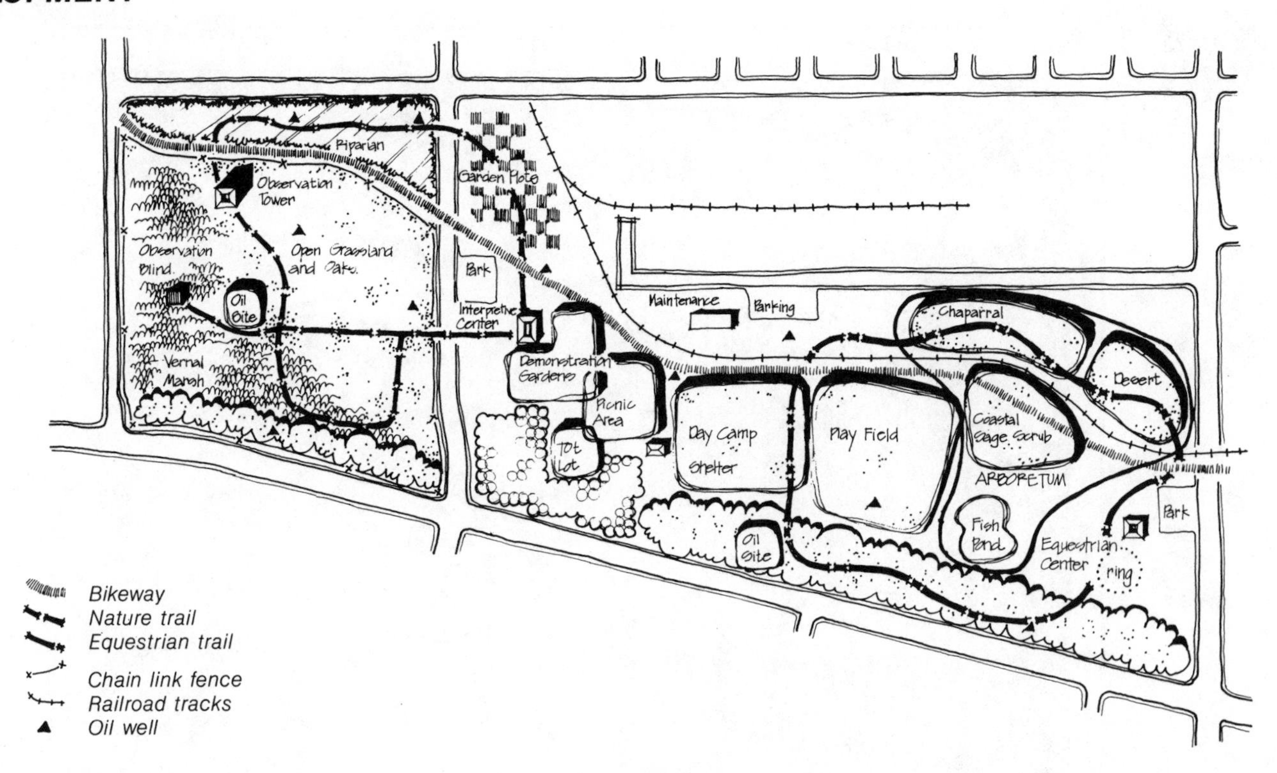

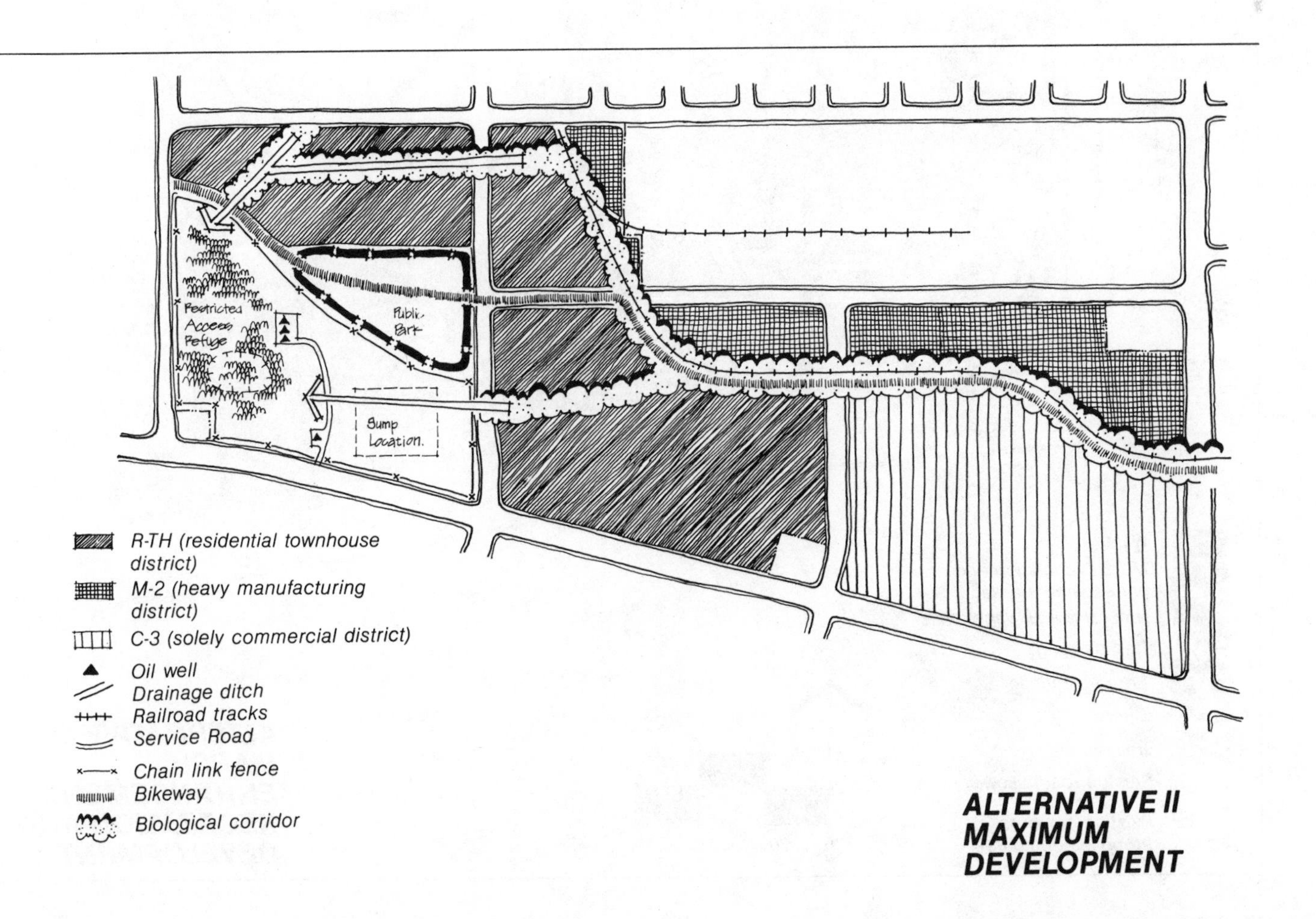

ALTERNATIVE II MAXIMUM DEVELOPMENT

Parking
Riparian Veg.
Interpretive Center
Chaparral
Free Play
Leisure Area
Fill
Observation
Blind
Water
Marsh vegetation and
enhancement area
Natural marsh vegetation
Tree plantings
Shrub plantings
Ditch
Trail
Service road/bikeway
Existing contours
Proposed contours
0
100
200
300
Feet
ALTERNATIVE III
MARSH
ENHANCEMENT,
RECREATION AND
DEVELOPMENT

CASE STUDY XI

The Simon Residence

A single-family house on a large lot presents a great many possibilities for using the life-support capacities of the landscape. If these capacities are used to the full and if electronic communications at least partly replace the energy consumed by automobile trips, this can be a resource-efficient form of settlement, although probably only for an affluent minority.

In the case of the Simon residence, the one-acre lot is divided into three zones: the protective, the productive, and the personal. The protective zone is planted in native species, the steeper areas in soil-holding ground covers. The productive areas include an avocado grove, citrus trees, a small vegetable garden, and a strawberry patch. The groves also serve to screen the house from the street and to protect it from northerly winds. The personal landscape is designed for esthetic pleasure. Here, plants are chosen for form, color, and texture rather than for their produce. As a result, they will require more care and resources than their ecological roles might justify.

The plant chart summarizes the reasons why each of the dominant species was included. As a group, they cover the spectrum of plant roles, although obviously more attention is paid to some roles than others.

Residence and landscape by John T. Lyle for Dr. and Mrs. E. J. Simon of Riverside, California.

SITE ANALYSIS

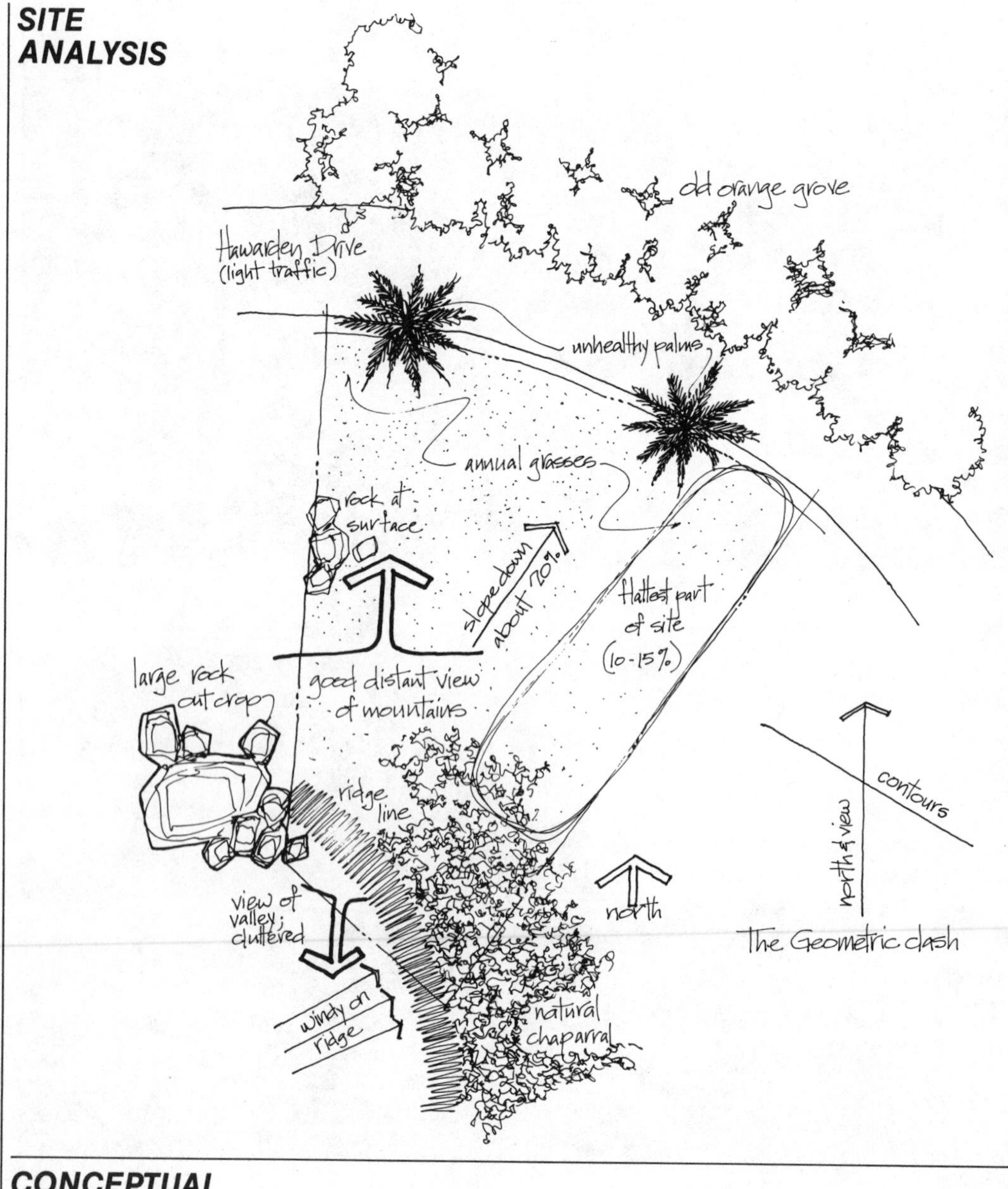

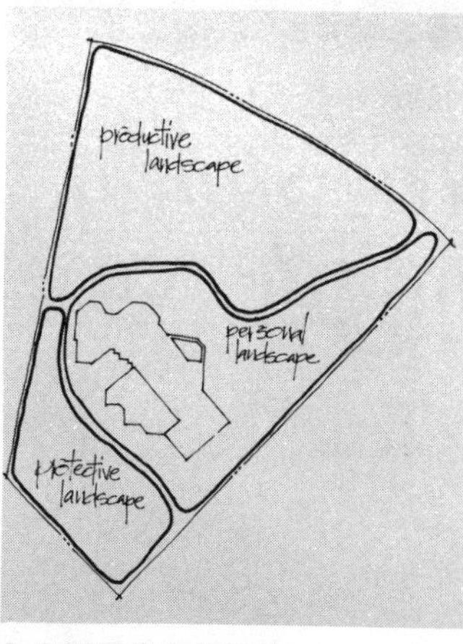

LANDSCAPE ZONES

The site analysis is an informal one which makes no attempt to break the site down into component parts or variables, but instead views the landscape as a whole or gestalt. The character of this whole is distinguished by certain visible characteristics that will strongly influence development. Important among these are the citrus grove across the street, the steeply sloping and the relatively flat areas, the rock outcrops, and the level above from which one gets dramatic views of distant mountains. Perhaps most important is the geometric clash between the northwest-southeast trend of the natural contours and the north-south orientation important for solar exposure and for northerly views. This conflict was eventually resolved architecturally by basing the plan on an octagonal grid.

The site analysis also led to a fundamental organization of the site according to productive, protective and personal goals based on existing patterns and on the chosen building site.

CONCEPTUAL SECTION

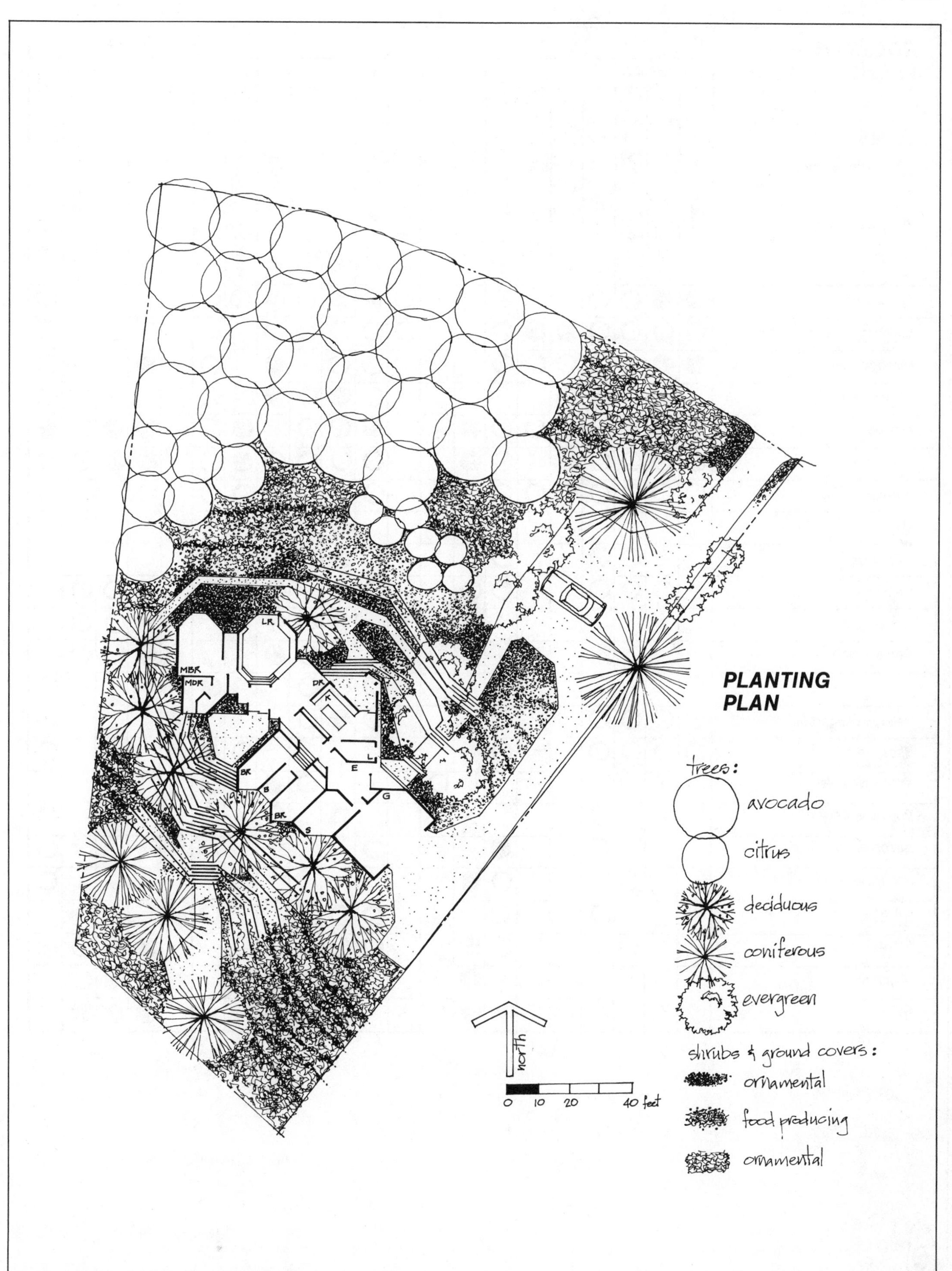
PLANTING PLAN
trees:
avocado
citrus
deciduous
coniferous
evergreen
shrubs & ground covers:
ornamental
food producing
ornamental
north
0 10 20 40 feet
LR
MBR
MDR
DR
K
L
E
G
BK
B
BR
S

ROLES OF PLANTS

● Major
○ Significant

Dominant trees, shrubs and groundcovers	Contributions																	Needs			
	Ecological control							Production			Visual				Wildlife			from humans			
	Oxygen/carbon dioxide	Filters air	Controls water flow	Filters water	Cools air	Forms still air pockets	Holds soil	Food	Fibre	Energy	Structure	Scenery	Place	Symbolic expression	Food	Cover		Water	Nutrients	Regular care	Periodic care
Pivlus pinea	●	●	○	○	○		○			○	●	●	○	●	○	○					○
Sophora japonica	○	○	○	○	●	●	○			○	○	○	○	○		○		○	○		○
Ginkgobiloba	●	●	●	○	○	○	○				○	○		●		○		●			○
Acacia dealbata			○	○	○						○	●	○	●		○					○
Avocado	●	●	●	○	○	○	○	●			●	○	○		●	○		●	●	○	●
Citrus spp.	○	○	○	○	○		○	●			○	○	○	○	●	○		●	●	○	●
Arbutus unedo	○	○	○	○	○	○	○				○			○	○	○		○			○
Cassia artemesioides							○					○	○	○							
Cistus spp.			○	○			○				○	●		○		○					
Carissa grandiflora		○	○	○			○	○			○	○		○	●	○		○	○	○	
Azelea spp.												●	○	●				●	○	○	
Camelia spp.												●	○	●				●	○	○	
	○			○			○				○	○	○			●					○
Heteromeles arbutifolia	○	○	○	○			○				○	●	○								○
Rhus integrifolia	○	○	○	○			○								○	●					○
Yucca whipplei									○				○	●		○					
Romneya coulteri				○								●		○		○					○
Baccharis pilularis				○			●				○										○
Rosmarinus officinalis			○				○	●													○
Gazania spp.				○			○				○	○				○					
Strawberries			○				○	●			○	○			○			●	●	○	●
Vegetables								●				○			●			●	●	○	
Herbs								●				○						●	●	●	

level that is drawn into the house through low ventilators and exhausted, once it is warmed, through clerestory windows above ceiling level. Devices like these require close integration of building and landscape, which is discouragingly rare in the fragmented design professions.

In the first case, the spreading basins, forms were shaped to influence the ecosystem structure, whereas in the second, they were shaped to control flows of water and energy. In both cases, they served goals established at larger scales; they provided the mechanisms, in Feibleman's terms, for realizing larger purposes.

As with the site scale, the means of passing the sense of purpose from the larger scale to the smaller vary. They may be very general or very explicit. Where sensitive environments or potentially disastrous effects are involved, they are more often explicit.

Control by Design at San Elijo

The control by design section of the San Elijo case study is an example of precise information being conveyed from the project scale to the construction scale. Here the subject is not form but technique, the means by which the form is to be achieved. Building on lands that drain into San Elijo Lagoon will seriously affect the lagoon. The stabilization of landforms without increasing runoff or erosion is especially essential to continuing management of the wetland environs. It has to be understood, however, that development of these areas will go on for years. Conditions and specific uses, and thus construction techniques, will vary. Rigid standards would unduly limit the technological resources available to future designers. Consequently, the charts shown here were developed to show the range of technical means available for each of the major developmental actions and the variations in type and degree of impact that are likely to follow. The charts are meant as guidelines for designers of these future developments and also for the review and approval of plans. Each of the techniques might be appropriate under certain conditions, but high-impact techniques like impervious membranes or retaining walls would have to be limited to situations where more environmentally benign measures would not serve the purpose. Thus, flexibility and discretion are left to future design and management in keeping with the policy of organic development. Since technology constantly changes, and since conditions vary so widely with time and place, specific forms best take shape in the context of specific situations. Construction design should be undertaken as close to the time and place of construction as is possible. Controls exerted by larger scales will ideally be no more than those needed to assure the fostering of larger purposes.

PART II:

Design Processes and Methods

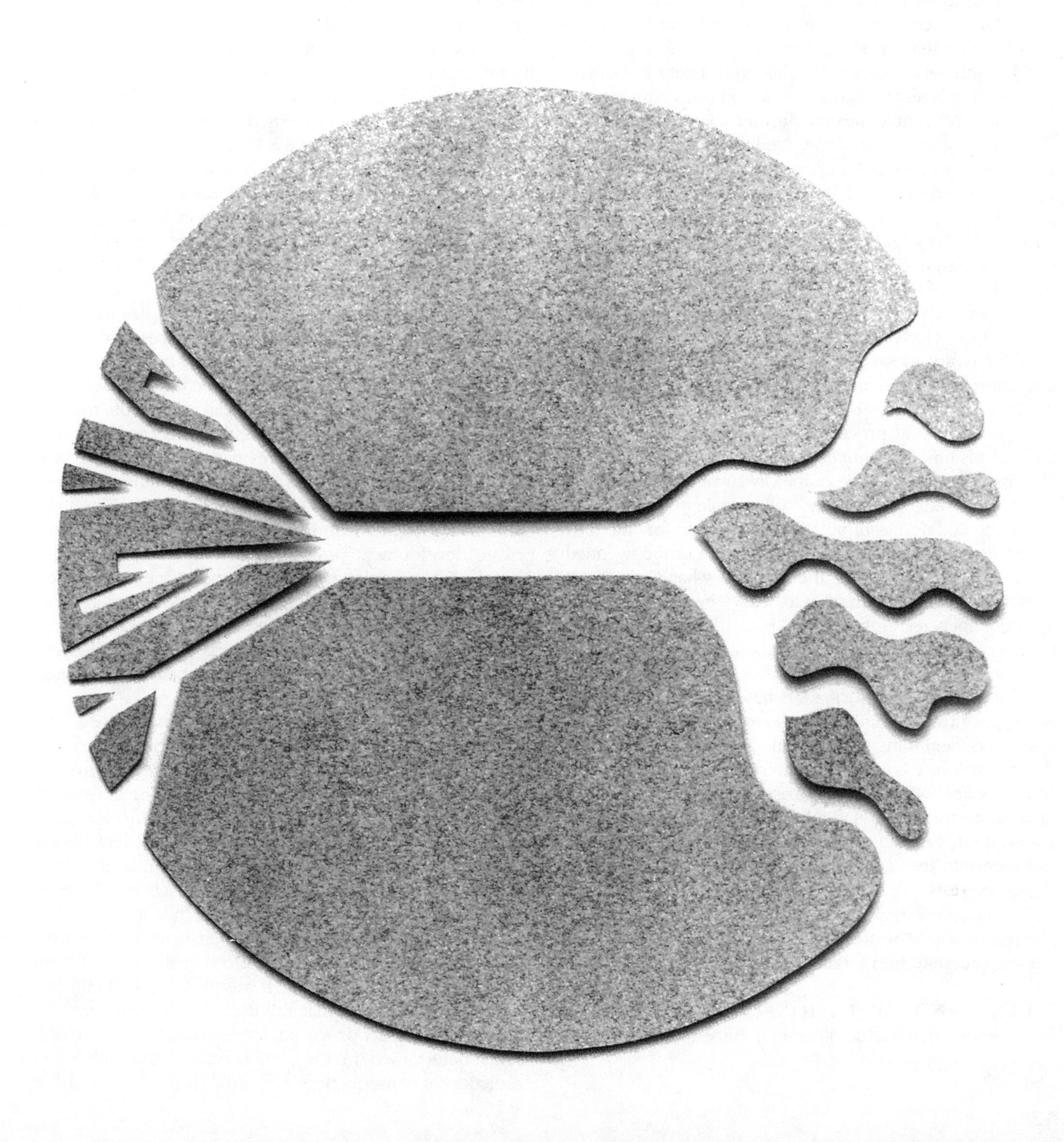

CHAPTER 6

Stages and Themes of Design

In processes of landscape design, the noosphere is joined with the biosphere. It must be abundantly clear by now that ecosystematic design is no simple matter. By attempting to shape the ecosystem in its entirety—to give actual form to ecological processes—it encompasses layers of complexity that have been beyond the scope of landscape design until very recently. Dealing with different concerns related to different integrative scales implies a multiplicity of modes and approaches that further complicates matters.

Managing such complexity requires coherent processes; it forces us to proceed in a systematic, consistent, clearly explainable sequence of steps. The very notion of design process is itself fairly new to landscape design. Until the mid-twentieth century, no one cared how a designer came up with a design. It was generally assumed to be a somewhat magical leap of intuition, and whether one liked the results or not, one did not question their origins. One simply accepted or rejected them.

This is not to say that the necessity of assembling information on which to base design proposals was only recently recognized. Frederick Law Olmstead did careful analyses of natural processes, although the knowledge of them in his time was limited. And we have seen that Patrick Geddes described the role of the city survey, which included maps describing topography, soils, geology, and climatic factors, in addition to population distribution for "the town and its extensions," during the early years of the twentieth century. These were also the years of what Carl Steinitz calls the "professionalization of data" by U.S. land management agencies.[1] John Wesley Powell also supported his proposals with analyses of the landscape, sketchy though they were.

Sometimes the information base was quite elaborate and sophisticated. Steinitz cites the example of Warren Manning, who, beginning in 1912, built a data base of 363 maps describing every conceivable physical characteristic of the United States. When the data base was complete, he drew up a master plan, showing a land-use pattern for the whole country. But, Steinitz points out, "... there was no obvious link between the data and the design."[2] The heart of the design process still remained obscure.

Some major regional design efforts were also carried out in this way. The New York regional plan and the Tennessee Valley Authority plan are examples.

THE DEMAND FOR RATIONAL PROCESS

Nonetheless, that was another era. Since mid-century, the world has changed. Since the 1960s, interest in the processes of design has grown rapidly under the influence of a number of factors whose importance seems likely to become even more drastic in the future. It will be worthwhile to spend some time examining these factors because they may tell us a good deal about what we might expect to accomplish by rational process in design.

Probably the first factor to emerge was the "environmental movement," which shed light on the enormous changes that human use of the land was bringing about in natural systems. This was the period when it became clear that we were using up our stored resources at alarming rates, disrupting the processes that support life, and radically altering natural populations, and that we were doing all these things largely by our misuse of the land. Natural systems, obviously, are enormously complex, and in order to deal with them, we needed design principles that could deal with such complexity. We needed processes that were able to apply large, diverse quantities of information.

Partly as a result of the emerging ecological awareness, challenges to land-use proposals began rising from the general population, which had been little concerned with such matters before. Federal agencies particularly felt the heat. Dams, flood control projects, and timber cutting practices became common targets of public indignation. The angry battle over Glen Canyon Dam was symbolic of the era, but private development projects were commonly challenged as well. Court battles became commonplace, and a great many cases were lost by project proponents, even those with quite sound plans, for lack of a clear demonstration of the validity of their proposals. Courts demanded rigorous logic, and the need for defensible plans became blindingly clear.

Related to these challenges was a growing insistence by some citizens' groups as well as by people in decision-making capacities for participation in the design process. They wanted to review plans as they were being developed, and they wanted to have their own ideas considered. To do so, they had to understand the process, which meant that it, and its results as well, had to be communicable.

And then came the bombshell: the National Environmental Policy Act of 1969. Whether or not it is true that the requirement of this act for an environmental impact report for projects significantly affecting the environment was only a minor afterthought, it was profoundly significant. The various state acts that followed the federal one also included the impact provision, and it changed the way land use is determined in the United States. Although some prediction of consequences was certainly implied in earlier

[1] See Steinitz, 1979; p. 15.
[2] Ibid., p. 15.

internalized ad hoc methods, and responsible landscape architects had always considered the effects of their designs on natural systems, prediction now became an explicit requirement.

All these emerging concerns and the related legislative and judicial decisions pointed to the need for clearly defined processes of design and established the criteria for their effectiveness. These included

1. A *capacity for complexity,* or the ability to use a great deal of information from a variety of sources on many different subjects from diverse disciplines
2. A *capacity for prediction,* or the ability to estimate the potential effects of a proposal on the existing environment
3. *Defensibility,* or a clear and logically correct framework to support claims
4. *Communicability,* or the ability of a proposal to be understood by the general public.

Each of these criteria emphasizes analytical aspects of design at the expense of the intuitive ones that had so long held sway. Not surprisingly, most of the experimentation with design process that was given impetus by these new demands followed suit. Some of the theorists of design process envisioned methods that would be as clear and precise, as conceptually simple, as those of science itself.

The series of steps that serves, more or less, as the standard paradigm for design process has the simplicity of scientific method, but lacks the precision. RESEARCH—ANALYSIS—SYNTHESIS—EVALUATION is a reasonable enough sequence of events, but it really tells little more than that we need to know something about the subject matter before trying to reshape it. The terms are too abstract to tell us much about what we actually do.

Perhaps more disturbing is the implied linearity in this sequence, a hint that if one follows these steps, one will inevitably arrive at the single best design. Much of the pioneering work in design process accepted this implication. In fact, the idea that there is a best way that will lead to the best plan still permeates a great deal of the thinking on the subject. Nature, it is sometimes said, will tell us what to do. As a result, the purpose of design process sometimes seems to be merely to interpret for nature.

In reality, however, nature is silent, ambivalent, contradictory. We know now that she will not tell us what to do. In any given situation, any number of different plans are possible. The recognition of diverse possibilities is the all-important element missing from the four-step paradigm and from so many other efforts to define design process. Recognizing possibilities takes creative thought, and creativity tends to be stifled by a rigid framework of logic. When we stifle creativity, we shut out a great many possibilities, and in a world that so desperately needs better solutions, that is something we cannot afford to do.

We need to dispel, once and for all, the notion that design is, or can be, a science. The very nature of the scientific method requires that the world be broken down into ever smaller parts in order to understand how it works. Science seeks to control, to ignore all of the variables save one in a situation, and then to learn something with absolute certainty about that one variable. Design necessarily deals with all of the variables simultaneously, and not only to understand but to project new forms. Moreover, scientific knowledge depends on the separation of the scientist from his subject, of man from nature, whereas a designer is necessarily an integral part of the landscape he deals with. Detached objectivity is not possible; influence is inevitable. Although using the knowledge gained by science, the purpose of design is putting things—often very diverse things—together, but never with the hope of absolute certainty. Design is ultimately an integrative activity. To quote *The Aristos* of John Fowles:

> The scientist atomizes, someone must synthesize;
> The scientist withdraws, someone must draw together;
> The scientist particularizes, someone must universalize; . . .
> The scientist turns his back on the as yet, and perhaps eternally, unverifiable, and someone must face it.

We still need intuition and imagination; the baby should not go out with the bathwater. Our real challenge is to apply both creative and analytical modes of thought to design processes.

Modes of Thinking

One hears a great deal of talk about combining analytical and creative thinking, but one hears of very few specific ways of consummating the marriage. In practice, they are rarely successfully joined because they are so fundamentally different. Research on the workings of the human brain suggests the possibility that these two modes of thought are separate functions, emanating from two separate halves of the brain. Thus, they may be not only symbolically opposite but physically opposite, each half underlying one of the two major modes of thought and consciousness.[3] The left hemisphere would appear to specialize in analytical, logical thinking, primarily in linear sequence, while the right specializes in what psychologist Robert Ornstein refers to as "holistic mentation; . . . its responsibilities demand a ready integration of many inputs at once."[4] Intuition and creativity seem to reside here.

Thus it is our analytical side that establishes and follows an orderly sequence, that organizes complex information, and that describes what is there. Its biggest shortcoming is that it cannot go *beyond* what is there. Only the creative side can intuitively grasp complex situations, can leap into the future and its possibilities, and, by imagining what might be, can pose hypotheses, questions, images, and goals. Its biggest shortcoming is that it may or may not be right and by itself has no way of knowing whether it is right or not.

Since each human brain includes both sides, and since the intricacies of design clearly require both sets of capabilities, we need to understand how to incorporate what Ornstein calls this "fundamental duality of our consciousness"[5]

[3] See Sperry, 1964.
[4] See Ornstein, 1972; p. 52.
[5] Ibid., p. 58.

into processes of design. Whether design is the task of a single mind using both of its halves in interacting harmony, or a team effort in which each member contributes primarily the efforts of a single side, it is important that the two sides work together. Without some understanding of how this cooperation can be achieved, one side or the other may well dominate the process without the designer or team being aware that this is happening; worse yet, they may negate one another's efforts in a silent war. Psychiatrist Roberto Assagioli calls it ". . . a stormy and difficult marriage . . . which sometimes ends in divorce."[6]

As a practical matter, the two very different sets of capabilities suggest a simple strategic division of responsibilities. We might say that the creative side proposes and the analytical side disposes (using the term "dispose" in the dictionary sense—to put in order or to apply to a particular end or purpose—rather than in the more common sense of disposing of—getting rid of—although the latter is also a common task of the analytical side). Proposing and disposing are alternating activities throughout the entire process of design, like the continuing interplay of creation and adaptation in natural evolution, providing the alternating current—albeit an irregular one—that charges the entire process.

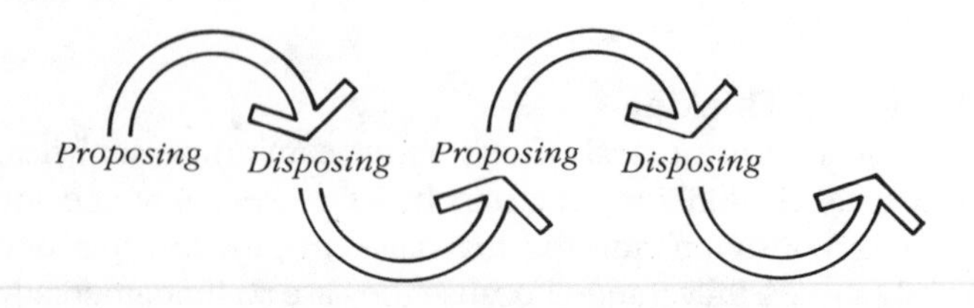

They come most dramatically into play in the later stages, when the right side proposes forms and solutions and the left side puts them in order and evaluates. Nevertheless, proposing and disposing go on all the time, even in the early analytical stages. The intuitive side might notice, for instance, that the waters of a lake are murky and thus propose to focus on water pollution as a major issue in developing a plan for its watershed.

The analytical side might then examine the facts and find that that there are heavy concentrations of nitrates in the water and the siltation rate is too fast. Water quality then becomes the issue, with emphasis to be placed on grading and the proper use of fertilizers. Or one initially thinks that a soil map might be useful, only to discover that soils in the entire region tend to be homogeneous and that preparing a soil map would take six days. Or one dimly perceives that the flow of water probably follows a definite pattern, so one gathers the data and constructs a model and finds that it does indeed work that way.

While the importance of these alternating roles in design processes is usually overlooked, it is widely recognized in some other fields. Writing of his work, psychiatrist Assagioli calls intuition "the creative advance toward reality" and assigns intellect three tasks: "the valuable and necessary function of interpreting, i.e., of translating, verbalizing in acceptable mental terms, the results of the intuition; second, to check its validity; and third, to coordinate and to include it in the body of already accepted knowledge."[7]

The scientific method has long used such a dualistic sequence, proposing by hypothesis and disposing through experimentation. Although the popular notion of science is one of analytical activity, the important discoveries invariably involve intuitive leaps that produce ideas of what could be. James Watson's story of the discovery of the double helix is one dramatic instance.[8] The analytical verification and proof then follow, usually taking far more time but certainly no more important.

The alternating cycles may be clearest in the processes of learning. Designers often notice how much the experience of design is like that of learning. Indeed, we can see design as being quite literally a learning activity. It has long been recognized that learning is commonly characterized by an ongoing cycle of freedom and discipline. Years before the roles of the two sides of the brain were well understood, Alfred North Whitehead explored these matters in some detail by stating that "discipline should be the voluntary issue of free choice, and . . . freedom should gain an enrichment of possibility as the issue of discipline" and "all mental development is composed of cycles, and of cycles of such cycles."[9] With freedom, we explore and ponder. In a new situation, if we have the freedom for it, we move from one idea or experience to another, sampling each, letting the whole sink in. Discipline, then, when we are ready for it, is undertaken to satisfy a craving for ordered knowledge that grows from these free explorations. When we have gained the ordered knowledge of the subject, then that knowledge gives us a new ability to explore and to produce, to intuit and invent in a directed way, and thus a new freedom.

Whitehead thus divides learning into three stages, which he calls the Stage of Romance, the Stage of Precision, and the Stage of Generalization. These will sound extraordinarily familiar to anyone who has survived the experience of design. More to the point, they will sound very much like what his instincts were urging while the power of convention was insisting on unremitting information gathering and analysis.

The Stage of Romance is that first stage of freedom, of apprehension, of exploratory excitement. Says Whitehead, ". . . it holds within itself unexplored connections with possibilities half-disclosed by glimpses and half-concealed by the wealth of material. . . . Romantic emotion is essentially the excitement consequent on the transition from bare facts to the first realizations of the import of their unexplored relationships."[10] For us, it is set off by the perception of a landscape with its implications of infinite complexity and a fluid set of conditions that foretell impending change. Possibilities and connections are there to be explored and sorted out, and among them are hints and suggestions of a new order that we might eventually be able to shape. This stage is dominated by the intuitive, or right, side of

[6] See Assagioli, 1971; p. 217.

[7] Ibid., p. 224.
[8] See Watson, 1968.
[9] See Whitehead, 1929; pp. 30 and 31.
[10] Ibid., pp. 17 and 18.

the brain, although there are cycles within the larger cycles so that the analytical left side plays a role here too.

In the Stage of Precision, the material of romance is put into systematic order. The left side assumes control, but not without frequent proposals from the right. Discipline and the rules of logic lead the way to ensure factual correctness. Our exploration of possibilities and connections in the Romantic Stage has revealed what our concerns are and given us a solid notion of where we are going and how to get there. Now we gather and organize the information we need, break the whole down into its parts to gain a thorough knowledge of how they really work, and put them back together in a way that formalizes that understanding. With this analytical knowledge, we gain a new freedom to explore, this time armed with the necessary information and techniques to conduct our search with a realistic sense of limits and actual potentials. This is the Stage of Generalization, in which the right side again assumes command, though still with the continuing counsel of the left. During this stage, we imagine possibilities, evaluate and compare them, and eventually arrive at a plan.

We can imagine these three stages of learning, or design, as being like the flow of a river. It begins with an array of streams, pools, and channels, flowing in varied intriguing

and mysterious ways. We know they are all interconnected, but most of the connections are hidden from view. During the Romantic Stage, we follow the network branch by branch to where it all converges in the single broad stream of the Stage of Precision. There are few choices here but to follow it along between well-defined banks until it emerges into the coastal plain and begins to spread into the diverging network of a delta, the Stage of Generalization. Here we have branches and sub-branches, pools, and wetlands—winding, twisting, turning back. But beyond, the sea looms on the horizon, and eventually that is where everything converges and becomes one.

THE SYSTEMS APPROACH

Understanding design as a learning process, then, helps us to grasp and to exert some control over the sequence of the mental attitudes involved, and perhaps most important, dispenses with the limiting notion that analysis is everything. But it still leaves us far short of any real definition of method that can fulfill the four criteria listed earlier. So we will return to the subject of rational process. For further guidance, we can turn to two fields that have been concerned with rational processes for solving complex problems since their emergence during World War II. These are the systems approach and decision theory.

The systems approach, for all its influence in planning circles, is virtually indefinable. Every book on the subject provides a different definition, and more recent works have tended to offer narrow, technical definitions. In very general terms, however, the systems approach is a logical structure for problem solving that emphasizes interrelatedness, the notion that if we alter one part of a system, we inevitably alter other parts and, ultimately, the whole. In addition to this basic idea, the systems approach includes at least four characteristics that make it especially interesting to environmental designers. These four are the consideration of problems in the largest possible context, the use of models, the role of feedback, and interdisciplinary organization.

Context

The idea of larger context is especially important for landscape design because in dealing with natural processes, as we have seen in our consideration of scale, everything is related to everything else. The systems approach, according to Churchman, is "based on the fundamental principle that all aspects of the human world should be tied together in one grand rational scheme. . . ."[11] The scales of concern give us at least the outlines of a framework for such a scheme with respect to the landscape. In terms of the systems approach, we might think of the levels of scale as systems and subsystems.

Models

As for the practice of using models, this was not entirely new to environmental designers. They had used physical, three-dimensional models for centuries. Nevertheless, the broader concept of the model as proposed by the systems approach and its vast implications were entirely new. A model in this broad sense is simply an abstract representation of reality. Its purpose, usually, is to reduce the infinite complexity of real phenomena to manageable terms. In making a model, one tries to draw out the essential characteristics of a real subject and put them together in a way that mimics the relationships that exist in reality. The model can then be used to develop a better understanding of the workings of the real phenomenon, or it can be manipulated as a stand-in for the real thing in the process of redesigning it.

The models used in systems analysis are usually mathematical, that is, the stand-ins for reality are letters and

[11] See Churchman, 1979; p. 8.

numbers. For a great many people in the field, the systems approach means categorically the use of mathematical models. For others, however, even for at least one of the founding fathers of the field, C. West Churchman, such models have lost much of their appeal. Churchman argues, correctly I believe, that a great many of the more serious issues of society are not subject to mathematical precision: "In the first place, we don't know enough to be precise, nor would attempting to be more precise help us very much."[12] Certainly, this is as often true in the shaping of ecosystems as in other fields of systems application. Usually, there are too many variables, too many unknowns, too many intangibles, too many qualitative matters like visual character. Thus, we find that other kinds of models can often be more useful.

Nevertheless, the use of models, especially several types of graphic models, which may or may not be quantified, gives us means for dealing with the infinite complexities of the landscape, especially at the larger scales. Models can represent dynamic processes as well as static forms, and they can make available the power of the computer.

Feedback

Feedback is a key concept in understanding how almost any process—natural or human—works. There are two types of feedback: positive and negative. Positive feedback is information concerning the state of a system that is used to increase or amplify change in a particular direction. Negative feedback is information that is used to dampen or decrease the rate of change. Processes of growth, change, and stability in natural systems are explained mostly by negative and positive feedback, and the same mechanisms are at work in human systems, although in these they are usually harder to understand and describe. It often happens that environmental degradation and reckless use of resources can be the result of the elimination of negative feedback. As the gargantuan Los Angeles water import system grew, for example, every protesting and objecting voice (negative feedback) was eliminated by political or economic force, and the system continued to grow until well into the nineteen-seventies despite serious and obvious problems.

As we carry through the design process, proposing and disposing, we continually test and reiterate information, models, and ideas, questioning and reshaping. After the forms imagined in the design process have been executed, the feedback loops continue, now in a real world environment, through management. Information, models, and ideas are tested and reiterated in reality. Thus feedback becomes a way of compensating for the imperfection of our predictions, which are necessarily crude and will probably never be able to take into account all causes, chance events, and the like.

Crossing Disciplines

As for interdisciplinary activity, this is not entirely new to environmental design either. Even the smallest landscape design problem—a single house or a backyard—involves information drawn from a variety of disciplines. In fact, to a practical designer, the fact that the nurturing of plants, the quality of soils, and the mechanics of structures are all considered different fields of study seems strange. He routinely works with all of them and thus automatically functions in an interdisciplinary way. At larger scales, however, more disciplines enter the picture, each of them introducing yet more complex information and techniques, and the task of assimilating and integrating them all into a coherent whole becomes very daunting indeed. A design effort at the plan unit or project scale might routinely include such diverse items as behavioral observations, attitude surveys, economic assessments, hydrological studies, and estimates of solar radiation incidence, among others. The means of fitting such an array into a logically defensible process are a major concern of the design method for such projects, involving the use of matrices, flow charts, and other devices adapted from the systems tool kit. Usually we need to consult specialists in some of these fields, and often the specialists join together in interdisciplinary teams. The organization and management of such teams are thus important aspects of the design process.

[12] Ibid., p. 20.

SYSTEMS APPLICATIONS IN DESIGN

Besides the fact that they had long been familiar in rudimentary form and were known by other names or no names at all, these systems concepts had another characteristic that made them appealing to designers and planners. This was the fact that each of them translated a mechanism that was observably at work in nature into terms that made it possible for the human intellect to put it to practical use. Thus, it became apparent that processes of design could work in ways quite similar to the other processes of nature that are its subject, suggesting a satisfying harmony of mind and matter. For this reason, the systems analysis approach found ready acceptance. The work that probably did the most to spread awareness of its potential uses in environmental design was Christopher Alexander's *Notes on the Synthesis of Form,* in which he described the application of linear programming logic to the design of a village in India, using very specific relational criteria.

Ian McHarg's *Design with Nature* in 1969 described the methodical application of map overlays to define the suitability of land for various uses. Map overlays in themselves were nothing new, but McHarg's use of matrices to analyze interactions between land variables and human activities relied on systems techniques for interdisciplinary analyses. His process, like Alexander's, featured a logical sequence of steps that led, seemingly unarguably, to the best arrangement, without any apparent need for exploring other possibilities. In this sense, both processes are examples of what decision theorists call a technical decision-making process. Such a process is generally used in situations where goals are clear, agreed upon by everyone involved, and noncontradictory, that is, where all can be substantially achieved without infringing on others. In the technical process, each step is made because it pushes the sequence of steps most effectively toward its given end or solution. There is no formal need to consider options or varied possibilities because, in the light of established goals, the most effective plan quickly becomes clear as the sequence is completed.

In the case of McHarg's method, the goals are quite clearly stated in terms of his "presumption for nature." While McHarg was widely criticized for not taking other factors into account,[13] his process is entirely justified by the clarity of his goals, however limited one might consider them. For him, minimum intrusion on natural processes is the guiding aim that makes it possible to proceed step by step to the one best solution. In practice, however, it is rare, especially when working with landscape issues at the larger scale, to find such clear agreement on goals. Usually, in fact, the planning effort is given its original impetus by sharp differences in purpose, and the whole process is enlivened by conflict from beginning to end. Although it may be possible to agree on general goals or objectives, they are still likely to be ranked quite differently by different people. A development company and a preservationist group might agree that providing housing and preserving wildlife habitats are both worthy goals, but they are not likely to agree on their relative importance.

The technical process can usually be quite effective at the smaller scales, particularly at the construction and site levels. At these scales, and often at the project level, relatively few people are involved in decisions and the major ecological goals have already been established (as suggested in Chap. 2), at higher levels. The Simon Residence, Madrona Marsh, and High Meadow case studies all show such situations. In the High Meadow example, the basic technical sequence was varied somewhat by evaluating and reiterating the concept plan several times, but the goals and direction remained the same. Reiterations of this sort are almost always essential parts of any design process in the sequence of proposal and disposal, whether they are formally expressed or not. It is important to consider and reconsider various possibilities, if only in the mind's eye or through quick sketches. As feelings and intuitions enter in and assume dominant roles, the technical process can resemble a poetic process that uses less explicit information. At the extreme, it might seem entirely nonlinear, but the basic sequence remains the same. Even a poem has to begin with an idea or observation or feeling that leads to words in a definite order.

Where the iterations occur expressly as steps of the technical process, we might illustrate this diagrammatically as follows:

Problem ⇒ *Information* ⇒ *Plan* ⇒ *Evaluation* ⇒ *Implementation*

TYPES OF RATIONAL PROCESS

In more complex situations, where different goals or priorities are involved, or where they compete or are mutually exclusive, or where the means of reaching them are not clear, or where the goals cannot be articulated at all, more complex processes are called for. In such situations, the logical sequence, not being guided or constrained by clear goals, is broken between the Stage of Precision and the Stage of Generalization, and the process is thus opened to more varied possibilities.

This brings us to the rational problem-solving paradigm adapted from the systems approach and widely used in planning. It relies heavily on the consideration of alternatives and on evaluation and feedback and can be described as a sequence of steps that proceed more or less as follows:

1. Statement of goals
2. Analysis
3. Development of alternatives
4. Comparative evaluation of alternatives by some measure of goal achievement
5. Selection of the most effective alternative or, if none proves adequate, a return to step 2
6. Implementation
7. Monitoring

If we combine implementation and monitoring into one box that we will call management, the rational sequence can be diagrammatically illustrated as follows:

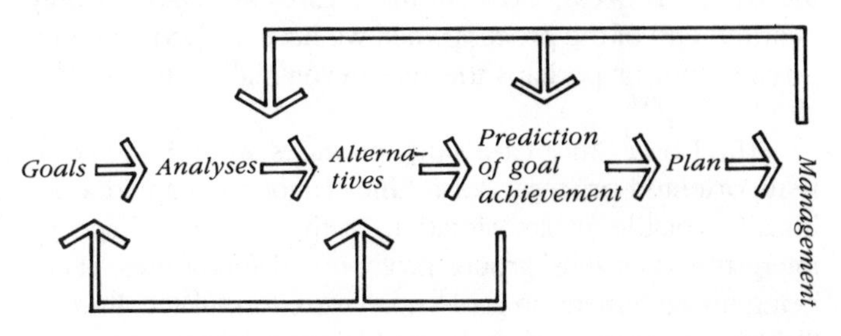

This process, sometimes called *the* rational paradigm, is often considered the major contribution of the systems approach to the environmental design fields. John Eberhard posed it as holding promise for a wide variety of architectural design problems,[14] for example, and Darwin Stuart reduced it to a three-step sequence—identifying programs, predicting effectiveness, and evaluating alternatives—with wide application in urban planning.[15]

The basic sequence of the rational paradigm is actually quite an old one, much older than the systems approach. It goes back at least as far as John Dewey's ABC steps: What is the problem? What are the alternatives? Which alternative is best? Certainly it is a sequence with universal applications. In applying it, however, we encounter certain important differences between the environmental design fields and the areas in which the systems approach has had its greatest successes. In the space and war-related industries, engineers and decision makers use this process in a rigorously quantified way. Goals are stated in terms that provide for measurement of achievement in precise quantities, often monetary. The criteria for choices usually boil down to a matter of efficiency. Although the problems may be technically complex, the quantities involved are known or measurable, and the values clear and singular.

[13] See especially Gold, 1974.

[14] See Eberhard, 1968.

[15] See Stuart, 1970.

Economic and Ecological Rationality

This state of affairs is rarely the case in landscape design. In the first place, the ecological and social factors that are the main areas of concern and the bases for choice among alternatives are hard to measure and almost impossible to quantify in terms that allow numerical comparison. According to the integrative laws, the larger goals, usually meaning ecological goals, are passed down to each scale from the next larger scale. Where all is working as it should, then, there exists a context of congruent larger goals, but there are other goals, often local ones, which are frequently in conflict among themselves and with larger ones.

In practice, it is sometimes difficult to state goals at all, beyond those clearly established at larger scales, until we know what the possibilities are. Since we are dealing here with inherently political situations involving different interests, purposes, and points of view that if not in conflict are at least incongruent, it is often far easier to articulate issues than goals. Often, then, we find ourselves beginning a design process with a set of concerns rather than a set of definite goals. We must then proceed from the concerns to an articulation of the issues and use this as a springboard for the design process. Goals in these cases are clarified only near the end of the process when we have to choose among possibilities, or perhaps they are never really articulated at all.

The Bolsa Chica case study provides an example of an issue-oriented process. Bolsa Chica Lagoon occupied a politically volatile urban situation, with a tangle of special interests and citizens' groups promoting different uses. There being no agreement on goals, none were formulated. Instead, the issues were used as bases for alternatives, and these were compared in terms of the four major areas of concern: wetland (environmental) enhancement, community enhancement, public protection, and socio-political acceptability. Lacking any way of quantifying these effects, they had to be estimated in relative terms.

In its pure form—driven by clearly stated goals, with precise quantification and efficiency as the criteria for selection—the rational paradigm corresponds to what decision theorists call economic rationality. As compared to technical rationality, which is founded on congruent sets of goals, economic rationality is founded on sets of goals that cannot all be maximally achieved by any one design. Thus, alternatives are needed to find the design that satisfies to the highest degree the most desirable set of goals. In economic terms again, this is an allocative decision.

This definition, however, is too limited in scope and too committed to quantification and to efficiency as the measure of performance to be entirely suited to the purposes of ecosystem design. Therefore, as a third mode of rational process that can deal with the broader, more varied, less precise, less determinate character of the larger landscape, I will propose what might be called ecological rationality. This mode is really a variation or expansion of economic rationality but different in that it is driven by issues—that is to say, by conflicting goals and by questions rather than by simple congruent goals—that it is exploratory in having to consider a wide range of possibilities, and that it evaluates alternatives on the basis of their predicted performance, this often being qualitatively measured in relation to, and in terms of, the natural, social, and political environment. Accepting Paul Diesing's assertion that a decision is rational when "it takes account of the possibilities and limitations of a given situation and reorganizes it so as to produce, or increase, or preserve some good."[16] Such a process is no less rational than the economic process, although in its more flexible character it works more like a natural process. It is most often applicable at the project and plan unit scales. We might diagram an ecologically rational design process as follows:

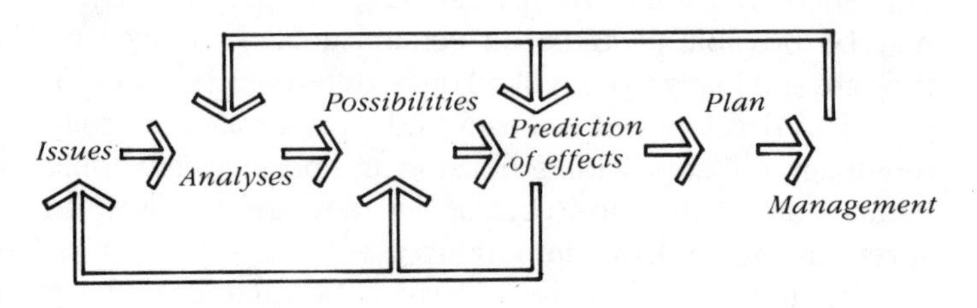

Legal Rationality

Diesing identifies three other kinds of rationality that are of interest here. These are legal, social, and political rationality. Each is appropriate for application in certain distinct types of decision-making, or design, situations.

Legal rationality depends on rules usually determined at the larger scales to define what may and may not be done at the smaller. This is one way, though often an unnecessarily rigid way, of transmitting goals from a larger to a smaller scale and ensuring that work at the smaller scale will become the instrument for realizing larger goals. Zoning ordinances are examples of mechanisms of legal rationality, as are water quality standards established by the U.S. Environmental Protection Agency. These standards are enforced by regional water quality control boards, which review all water-related projects within their jurisdictions. Thus, the regional boards take their goals from the larger (national or subcontinental) scale and their mechanisms from the smaller scales, as previously described.

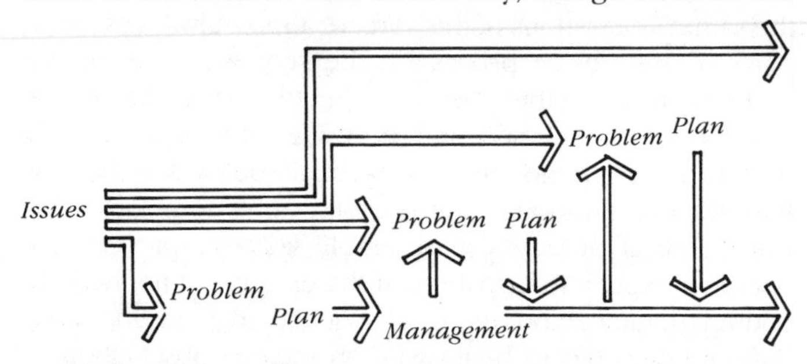

Legal requirements enter into almost every design process to some degree, of course. Grading standards, minimum pavement widths, and setback requirements are common examples. In some fields, such as highway and sewage-treatment plant design, the rules have become so thorough in their coverage of the design variables that the legal process, that is, rule following, is generally accepted as the only workable design process. Although minimum standards are

[16] See Diesing, 1962; p. 3.

assured, the disadvantages are serious. Legal rationality often defies the integrative laws by prescribing mechanisms as well as goals at higher levels, thus creating inevitable difficulties.

Diesing identifies four distinct trends associated with legalism: "(1) a trend toward complexity of distinctions and clarity of detail, such as highly technical terms, (2) a trend toward clear and distinct hierarchical differentiation . . . , (3) a trend toward uniformity, equality, and universalization where differentials are not involved, and (4) more generally, a trend toward rigidity, unchangeability, action according to rule."[17] Clearly, all these trends conflict with the ideals of creative response, regional differentiation, and adaptation to specific local conditions that are so important to the shaping of a meaningful and functional landscape. Furthermore, since laws must be precise and focus on single subjects, legal rationalism tends to work in the one-problem-one-solution way and thus to produce solutions that create other problems. We have seen this happen with water quality standards, for example.

So what can we do about the proliferation of legalism? The best answer to that question relates to the distinction between goals and standards. This distinction is important here, because while the designer at each scale looks to the next larger scale for his larger goals, ideally he looks to the next smaller scale for the mechanisms by which these goals are to be achieved. In the Aliso Creek example, although water conservation goals were suggested by a regional analysis and a water budget was established for the planning unit, the techniques for achieving that level of water use were developed at the project and site design levels. Thus, the actual ways of using water and achieving larger goals developed from the specific character of the landscape. Had the water budget and water conservation techniques been dictated by rules for the whole region, the design process for the smaller scales would have become a legalistic one, at least for matters related to water flow, and would have been severely and unnecessarily constrained. Performance controls, which establish goals, have some important advantages over prescriptive controls, which dictate the means for achieving them, but they are far more difficult to formulate effectively.

Integrative Rationality

All the rational modes described up to now not only assume our ability to see into the future with reasonable accuracy but take for granted a cohesive institutional structure. Both of these assumptions are doubtful. A host of unknowns lie in wait along the path of implementation of any plan. Changes of direction are inevitable. To make matters worse, although rational plans embrace a broad, coherent range of activities, the institutional structures are usually fragmented into any number of separate parts, or agencies, each of which can undertake only limited tasks. Wheaton wrote that comprehensive planning has usually failed because of its "definition of comprehensiveness in a world that lacks any comprehensive political power or institutions."[18]

[17] Ibid., p. 140.
[18] See Wheaton, 1967; p. 28.

In response to this incongruity between rational planning methods and their irrational context, some theorists have proposed more pragmatic paradigms. Prominent among these is the strategy that Alden and Morgan refer to as "disjointed incrementalism, or muddling through,"[19] which foregoes long-range goals and their associated values in favor of solving immediate problems. Since it is usually much easier for diverse groups to agree on such solutions than on larger goals and values, the practical results can be far greater.

And their approach brings us to a mode of rationality that Diesing calls integrative or social rationality. This mode operates incrementally and without long-range goals. Social systems develop in this way, Diesing argues, one small step at a time, as contrasted with goal-seeking technical, economic, and legal systems. Their basic trend is toward greater integration. "A system is integrated when the activity of each part fits into and completes the activity of other parts, and when in addition each part supports, confirms, and reinforces other parts by its activity."[20]

In this mode of rationality, each solution for each immediate problem contributes to social integration and divisive disagreements over goals are avoided. The difficulty, of course, is that larger purposes are likely to be lost, and without a larger purpose we may severely damage resources that may be needed in the future. Solutions to immediate problems can hamper possible solutions to future problems, but if clear, long-term goals are handed down by a higher level of integration, at least the larger ecological concerns can escape this difficulty.

Thus, incremental decisions, if they involve serious consequences, are best made with close attention to a larger framework, with conscious control, and with an understanding of possibilities. In this sense, in its consideration of possibilities and their consequences, a socially rational process is much like a series of miniature ecological processes except that each decision considers only the problem of the moment and the information associated with it. In any case, it is important to let the information base grow. Using the feedback principle, we can continually study the consequences of past decisions and search for new knowledge. Thus, information not available at first becomes available with experience and can contribute to increasingly effective solutions. This integrative process can be represented diagrammatically as follows:

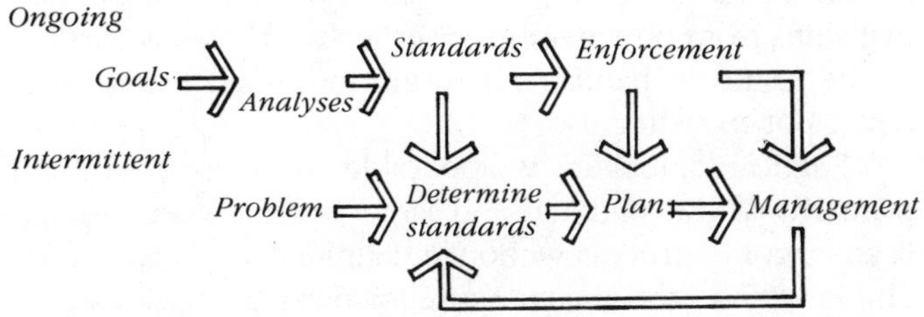

With a carefully devised beginning, a growing body of information, and conscious control at all times, this mode of rationality becomes far more than muddling through.

[19] See Alden and Morgan, 1974; p. 173.
[20] See Diesing, 1962; p. 76.

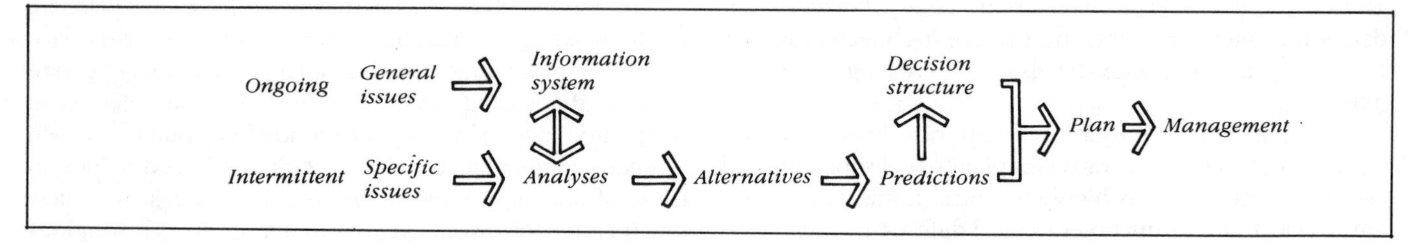

Rather, it incorporates land-use decisions with the ongoing integrative processes of a society and sidesteps disintegrating conflicts over goals and values.

This approach clearly takes time. Integrative design processes spread out decisions over long periods. We have probably all observed that the larger land-use issues usually take years, or even decades, for resolution as, one decision at a time, the integrative path is blazed. So long in term and so diffuse is the process that when a resolution finally occurs, we hardly notice it. In retrospect, it may seem as if the issues had simply melted away.

Political Rationality

Long-term though they may be, integrative methods, like the first three types, deal with only one issue or set of issues at a time. In a large, complex society, where a great many land-use issues are being dealt with at any one time, inequities and imbalances are likely to result unless these issues are guided by a common decision-making structure. And this is where political rationality enters in.

Political decisions are concerned with devising, preserving, and improving decision-making structures. Political rationality deals with "the organization of thought itself, the system of communications within which particular habits of thought are applied to materials to result in decisions."[21]

There are a great many different structures available for decisions concerning land use. In the United States, most planning decisions for privately owned lands are theoretically made by planning commissions, but the practical reality is much more complex. The movement toward public participation has tended to give private citizens a greater voice, and the technical complexity of questions concerning environmental impact have given technical specialists in staff positions decisive roles. In government agencies, similar changes have been occurring to spread the decision-making power. In general, there seems to be a long-term trend away from rigid hierarchical structures for land-use decisions and toward a more flexible organization with a leadership that shifts from one member or group to another according to the issues at hand. As a result, individual expertise is being put to better use.

Political rationality is applicable to landscape design primarily at the regional and larger scales, where design is an ongoing process without a definite beginning or end. The landscape and the structure for making decisions concerning it are being continually designed and redesigned. When it eventually comes about, design at the global level will probably be the ultimate example of political rationality in practice. Since at these larger levels, design processes and decision-making structures are inseparable, we need to understand something about the decision-making structures.

According to Diesing, a decision-making structure is composed of three basic elements:

1. Discussion relationships, which facilitate communications among members of the decision-making group, which at these levels is often very large
2. A set of beliefs and values, held more or less in common
3. Ongoing commitments and accepted courses of action

In addition to these three, a fourth element essential to decision-making structures for environmental issues is an information base. As we have seen, decisions concerning land use, if they are to have any meaning, must be informed by an understanding of the ecological processes involved. Simple data, moreover, are not enough. The information base needs to be interpreted so that reliable predictions concerning the effects of alternative plans can be made. Prediction of effects is the very core of design, especially at the larger scales. Decisions rest primarily on assessments of the future results of our actions.

This need requires a means of manipulating the information base—which we have called the Geographic Information System—to produce predictive models. Thus, for our present purposes, to Diesing's three elements of a decision-making structure, we will add one more:

4. A Geographic Information System to predict results of proposed actions

The process that a decision-making structure goes through for each of its decisions is essentially the design process. In Whitehead's sequence, Romance—Precision—Generalization, the information system provides the vehicle for the Stage of Precision.

ESSENTIAL THEMES

Looking back now at these rational processes, we find that, although they provide the logical coherence the complex issues of our age demand, they have a certain aura of make-believe. Anyone who has carried through a design process, even following the most rigorously rational paradigm, knows that the human mind does not actually work that way, and when it tries to force itself to do so, the results are uninspired. Although we need a rational framework for all the reasons discussed earlier, these rational processes all fail to account for the actual workings of the human mind. All imply linear

[21] Ibid., p. 170.

sequences of left-brain activity, whereas in reality, if appropriately complex and creative solutions are to emerge, the left and right sides should be working together in ongoing rhythms of the sort discussed earlier. While these rhythms are necessarily irregular and unpredictable, it is important to recognize as essential to the processes of design at least the larger sequence of shifting attitudes and modes of thought represented by Whitehead's Stages of Learning.

Although the specific steps described by the rational paradigms are all somewhat different, all of them embrace certain underlying themes, or broad subjects, that together provide direction. As in a symphony, each theme emerges from the fading chords of the one before it, guides the activity for a time, is explored in depth and detail with variations, and gives way to the theme that follows. Later, in all likelihood it will return, perhaps again and again. We can view each of these basic themes as having a definite place within one of Whitehead's stages and thus arrive at a general frame on which we might mount virtually any process of design.

Whitehead's Romantic Stage finds no real counterpart in the rational paradigms, which are usually described as beginning with problems, issues, or goals. These are important parts of the beginning stage, of course, but there are a great many others. During this stage, we lay the foundations for all the work that follows, describing methods, identifying participants, getting to know the land and the issues. Perhaps most important, it is then that we tune our minds and senses to the effort. This first stage of "romance" is a major part of the design process, one that takes time and shows few concrete results but that needs to be recognized. Grouping all these activities together, we will call this theme *inception.*

The next of Whitehead's stages, the Stage of Precision, is given little attention in most descriptions of rational process, which is a strange omission when we consider that for most larger-scale design efforts, this stage takes more time and research than the other two combined. In the "standard" design process described earlier, it takes half of the four steps: research and analysis. Research in this context refers to the gathering of information, and analysis is finding out what the information means for design purposes. Since this is a somewhat misleading use of the term "research," I will use instead *information*—becoming informed—and let this theme include the gathering and assembling of the needed facts. And instead of using the term "analysis," I will let the next theme, *models,* include analyzing the facts and organizing them into useful abstract representations of reality. Using the term in the broad sense described earlier, models are powerful tools for design, and all the more powerful if we think of them as conceptual constructions that we can study, reshape, and test as stand-ins for reality.

The Stage of Generalization is represented in the rational paradigm as consisting of three steps: the development of alternatives, the comparative evaluation of alternatives, and the selection of one, or some combination of two or more, as the plan. We have discussed the enormous importance of the search for possibilities. That search is an essential part of any design process, and the possibilities may be considered in any number of ways; they may or may not be then assembled into formal alternatives. We will therefore call this theme the search for *possibilities.*

In the Stage of Generalization, the cycle of creative and analytical phases emerges clearly; once we have posed possibilities, we will have to evaluate them analytically to find the best course of action among them. So the next theme, as in the rational paradigm, is comparative evaluation, evaluation based on *predictions* of performance. And from the predictions, if all goes as it should, a direction emerges, and we follow it in developing a *plan,* which might take any number of different forms and formats.

Once we have a plan, there remains the task of putting it into operation, which in the rational paradigm is usually called "implementation." At this point, the effort emerges from the Stage of Generalization into a new cycle of action and feedback—proposal and disposal—that we can best regard as an extension of design, or as a process of continuous redesign. This theme deserves the larger term of *management.*

These seven themes are—or should be—incorporated in most landscape design efforts, though with varying degrees of relative importance assigned to each. At this point, we can merge them with Whitehead's three stages to present a picture that, diagrammatically, looks like this:

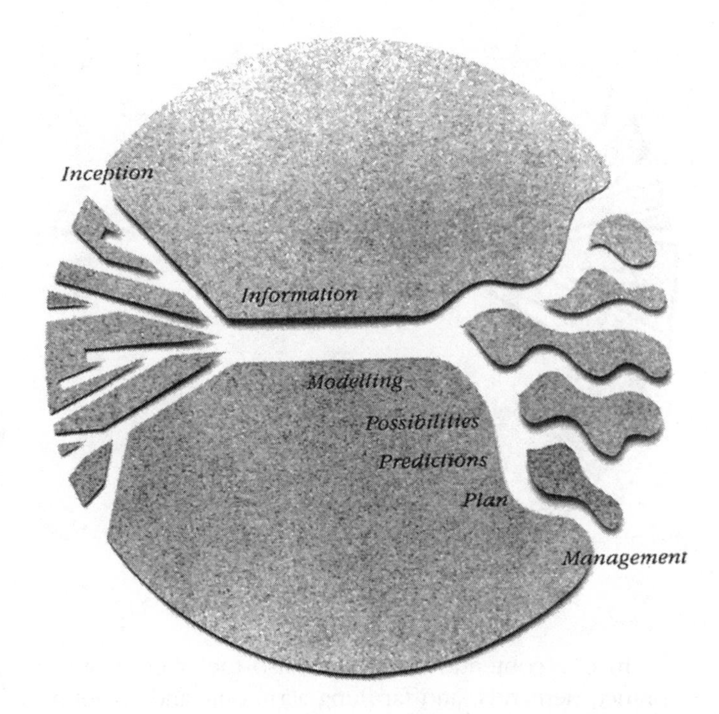

Although a rational design process usually requires all of these stages and themes, they do not necessarily have to occur in this sequence. Sometimes they are considerably separated in time. A Geographic Information System, for example, represents the Stage of Precision—information and modeling—carried out before the other stages. And themes can occur within themes. Management often involves continuous reiteration of the entire design process.

The next three chapters will be devoted to an exploration of the three stages and the themes within each stage.

CHAPTER 7

Stage of Romance

Inception

A design process begins, if all is well, in a spirit of boundless anticipation. Fragments of images of what might be flicker all around us, and myriad pathways twist ahead into the haze. A cacophony of questions sounds through our heads. It is all confusing, challenging, stimulating, intriguing, daunting, enormously exciting. It can also be disorienting, even frightening, and sometimes we are tempted to start trying to answer some of those questions right away or to choose a path haphazardly and start running along it. To do so is a big mistake, because this is a time for letting impressions sink in, for listening to people, not telling them, for questions, not answers, for dabbling and generally messing about within shadows that only slowly take form.

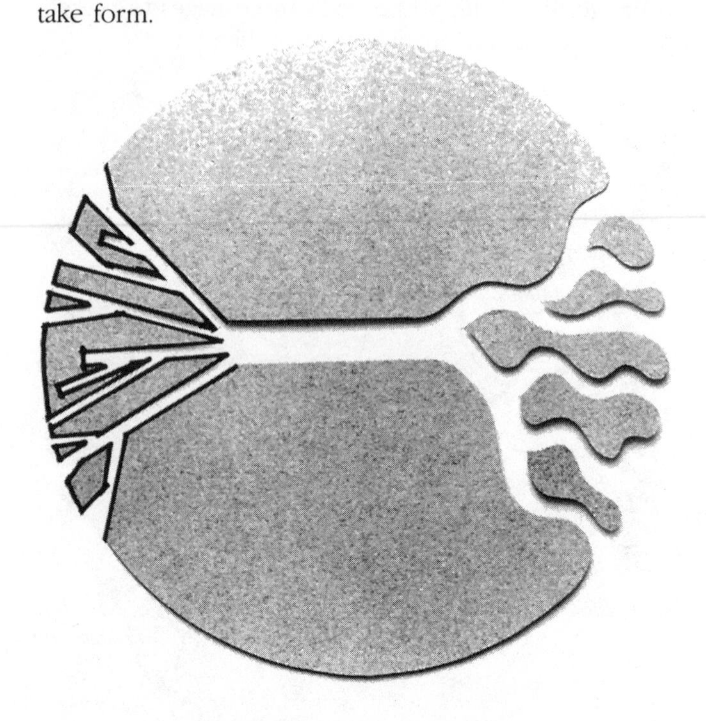

In our contentious era, it is also too often a time of conflict, demands, and far-flung acrimony, and sometimes we are tempted to start making promises, settling disputes, smoothing feathers. Again, it is much too soon for that. These conflicts usually run deep; they harbor profound implications and ramifications; they will not be settled quickly.

Such is the turmoil of the romantic stage, and it is important to let it run its course. Until it is over, the confusion and anticipation will remain, and along with them the excitement. While it is going on, we talk with those involved in the situation—again, with questions, not answers—get to know the landscape—not in scientific detail, but in form and feeling—read all the reports, letters, and whatever else may have been written about it—and think, ponder, sketch, photograph, discuss.

When the romantic stage is over, again if all goes as it should, we have a firm foundation for carrying through the rest of the process. We now know enough to focus precisely on the demanding tasks of analysis. Perhaps more important, we feel we are ready. The fundamental questions of what, why, who, and how are settled, and we can go on from there.

Answering the question *what* gives us the subject for design, tells us the matters we will be dealing with. Answering the *why* establishes the ethical purpose. Answering *who* identifies the participants, who can be many and varied. Answering *how* gives us a specific method for proceeding. These can be perplexing questions that beg for serious thought, but there is no formal logic for pursuing them. When we have studied the situation enough, discussed enough, thought enough, the answers will come.

The process of answering them is what I call the *inception* of the design effort, and this is the vital role of the romantic stage. The rest of this chapter will be devoted to the means of formulation.

WHAT?: THE SUBJECTS FOR DESIGN

The vague questions, concerns, and controversies that overshadow the beginnings of the Romantic Stage gain sharper definition as we get to know them better. At some point, they take form as goals, some passed down from larger scales and some arising from the immediate circumstances. Some of them invariably conflict with each other and thus become issues.

I described in the first chapter how the design process for San Elijo Lagoon began, with the conflict between a group of developers who wanted to fill in the wetlands for housing development and a group of conservationists who wanted to protect the wildlife. Such a conflict is a common beginning. The Madrona Marsh plan began in a similar way. In the case of the Whitewater Wash, the people of Palm Springs were concerned about the effects that wind energy development was likely to have on the appearance and ecology of their hinterland. In each situation, the initial concerns were seeds that quickly grew, as time went on, into complex branching structures of questions and issues, many of which had been germinating beneath the surface for a long time. This is the usual pattern of ecological situations. As vague concerns take shape, they become goals and issues, and from these spring the design process. Each tributary to the design stream begins with an issue, that is, with a coming forth that carries no promise of a definite conclusion. Issues are different from problems in that problems have solutions, whereas issues have only resolutions. Sometimes the two exist side by side.

As I mentioned in the last chapter, land-use issues have become far more controversial since the 1950s than they ever were before. The awareness of limits—the finite size and finite reserves of our planet—have brought resource issues to center stage. The variety of issues is unlimited, and the potential subjects of conflict are enormous. In very general terms, the major issues usually involve one or more of the following kinds of questions. One can hardly help noticing that implicit in virtually all of them are fundamental conflicts between short-term and long-term benefits, and between those for the few and those for the many.

Nature Preservation

As the profound importance of wilderness has become more and more obvious over the last half-century, more and more controversies have surfaced over the question of nature preservation. We have seen that the arguments for wilderness preservation on the global scale are urgent and compelling. On the other hand, we cannot preserve every piece of land in a natural state; land is also needed for human use. The arguments for preservation and the arguments for more active use have to be weighed in a fine balance.

Resource Conservation

Resources are by definition useful, as compared with the rest of nature, which may or may not be useful at some time in the future. David Ehrenfeld poses convincing arguments against considering present-day and possible future resources separately and pleads for what he calls "holistic conservation," or "looking at the natural world *first* as a functional whole. . . ."[1] While it is hard to disagree with that, it is also difficult to escape the conflict between resource conservation (use on a sustained-yield basis) and nature preservation (long-term nonuse) that is a century-old American tradition, dramatized in the famous Muir-Pinchot controversy and many recent public debates.

While preservationist positions are usually idealistic, conservation is often motivated to some extent by economics. Every piece of land contains resources of some kind, and every human use alters those resources to some degree. They may reside in the living community—vegetation (timber or other useful plants) or wildlife—or they may be stored in soil or rocks, in ground minerals or ground water. Or they may primarily reside in processes—solar radiation gain, the flow of water, the filtering of air, the production of oxygen. These resources are not so obvious, and since they usually have no role in the marketplace, they are assigned no economic value. As a result, they are often overlooked, an oversight that can be tragic, because ultimately they are the most fundamental resources, and in some ways the most vulnerable.

As a rule, conservation, especially of economic resources, becomes an extremely emotional issue at the very beginning. That this is so makes little sense because the resources of a piece of land usually remain largely unknown until a complete inventory has been carried out, and that does not occur until the second, or precision, stage. Futhermore, there is no inherent, irreconcilable conflict between conservation and human use as there is with preservation. In fact, conservation is itself a human use, one that usually requires management, and can often be entirely compatible with other uses. Thus it is rarely an either-or issue, but one to be resolved step-by-step as resources become known and their relationships with other needs are understood.

Protection from Hazards

Most natural hazards arise from specific conditions of the land and are therefore statistically predictable in given locations. The U.S. Army Corps of Engineers practice of plotting flood plains according to periodic probability of flooding is an example.

Among other major hazards are fires, landslides, subsidence, hurricanes, volcanoes, and tidal waves. In each case, we can identify the most hazardous areas. The question then becomes how to deal with those areas most effectively. There are three general possibilities. The first is simply to accept the risk and develop them for whatever uses make sense for other reasons. The second is to reduce the hazard by design. For example, the danger of flooding can usually be reduced by channels or dikes, and landsliding can sometimes be controlled by special foundations. Land use can be patterned to improve percolation and reduce the volume of runoff. Or we can refrain from building in the hazardous area altogether. Which of these solutions is best is a question best answered by analysis and by comparison of alternatives.

Competing Uses

In the simpler world that existed not so many years ago, land-use decisions were left to the marketplace. When government agencies got involved in land-use decisions at all, it was to determine what was the "highest and best use," which simply meant the one that would bring the greatest economic return. But now, although we still hear the term "highest and best use" far too often, it has become obvious that the use that brings the highest return is often not the one that most benefits society. In fact, the "highest and best use" may be destructive to our very sources of livelihood. The well-known loss of agricultural land to suburban growth is a case in point. It is for this reason that we turn to rational design processes for posing possible uses and predicting their effects. We need only recall the Whitewater Wash case study, where energy generation, preservation, and recreation goals competed for a given land area.

Intensity of Use

What is the volume of any particular use that any given land area can support? Nature has a variety of mechanisms for limiting use to an acceptable volume. Predators keep their prey in balance, and when a species outgrows its food supply, individuals either move elsewhere or die. There was a time when human beings were subject to these rules as well. When their populations approached the limits of their food production, the Greek city-states sent out boatloads

[1] See Ehrenfeld, 1972; p. 11.

of adventurous citizens to found new colonies, and eventually they colonized most of the edges of the Mediterranean and the Black Sea. Technology, however, has changed all that. Given their present ability to move resources from place to place, human beings have almost managed to elude the limits set by carrying capacities altogether; indeed, we find it difficult to determine how much is enough. The Aliso Creek case study, however, shows one example of use intensity; here, the water budget provided a reasonable basis for establishing the limits of urban development. Most of the other case studies involve questions of competing uses instead.

Public vs. Private

Under the U.S. Constitution, it is generally assumed that a private land owner has the right to use his land in a way that profits himself. It is likewise assumed that public lands are to be managed for public benefit. The difficulty with this distinction is that lands most suitable for public use are sometimes in private hands and those that might better be privately managed are sometimes in the public domain. This situation gives rise to issues of tenure change. What lands should be acquired by government agencies, and what lands might they better lease or trade? Whenever we consider recreation, conservation, or preservation uses for a piece of land, the question of public acquisition automatically arises.

In the United States, at least, concepts of land ownership are changing too. Courts seem increasingly willing to limit the uses of private land to protect public interests. Landowners seldom agree, of course, and this difference of opinion leads to some of the most volatile land-use issues of all.

Concentration or Dispersal

Until fossil-fuel stores were first unlocked and put to work two centuries ago, human activities were forced to concentrate in places in the landscape that, for all practical purposes, were predetermined. Once cheap, fossil-fuel-subsidized transportation had increased the options for moving ourselves and our resources about, we began to spread our activities all over the map and have continued to do so right to the present time. Planners and theorists have pleaded the case for concentration, but until the fuel shortages of the 1970s, their appeals had little effect. Now the question of concentration or dispersal is a real one wherever development is contemplated, one that needs to be considered in the light of landscape character, resources, and economics. Clustering of residential units is a form of development that has great possibilities for making better use of unimproved lands, but for it to be effective, the management of such lands must be well-defined. The North Claremont case study shows the range of options for concentration or dispersal on the fringe of an urban area, presenting four different scenarios for the future.

Other technological forces suggest the advisability of more dispersed patterns in the future. Electronic communications are making less face-to-face contact necessary in business, education, and a great many other activities. This innovation is reducing the need to move about and thus making it possible for people to live farther apart without necessarily increasing energy costs.

Visual Change

Most people relate to nature through their eyes. An astonishing number of land-use issues begin with concern for visual quality. It is only too true that the scenic character of the landscape has often been badly abused by careless development, but the concern for scenery is not always, nor necessarily, constructive. It often happens that people simply want to keep the landscape looking the way it does now. What they see now is what they want to see in the future. Often, we find the present state, however egregious it may be, interpreted as the natural state. In these situations, any proposal to build anything, or to clear or reshape the land in any way, brings instant opposition. The reaction of the people of Palm Springs to the proposal to build a cluster of wind generators within their view was not at all surprising. The generators might be quite elegant structures, might actually enhance the visual character of the landscape for some, but the opposition to the visual change is considerable.

Cultural Preservation

Most of the world's lands have been used by a succession of human societies whose patterns of use and cultural imprint were quite different. In the past, one replaced another or evolved into another with little regret for the change. Today, as we look to the past with a higher regard or merely with nostalgia, we become increasingly reluctant to obliterate historic patterns for the sake of a doubtful future. The sense of human continuity is becoming more important to us, and as a result, the preservation of cultural patterns in the landscape becomes an ever more pressing issue.

In the United States, cultural preservation issues often focus on sites significant in Indian culture and history. In other parts of the world, cultural development and its manifestations often extend much further back in time. Most of the European landscape, for example, exhibits vestiges of agricultural patterns that go back thousands of years. One study of Umbria in central Italy identified five distinct patterns of land use, the earliest of which dates from Roman times and reflects the rectilinear survey grid used by the Romans. Other patterns are survivals of the eras of the Medicis and Mussolini. Altogether, they provide a visual record of historical change, a valuable part of the Italian heritage. The planners recommended using these patterns as a compelling framework for future management.[2]

In some cases, however, the arguments for preserving historical landscapes are less convincing. Preservation of the past can then become a straitjacket for the future. Unfortunately, there is often no easy way of drawing a clean line between what is worth preserving and what is not.

A glance through the issues that initiated the case study planning efforts in this book will show that most critical land-use issues fall into one or another of these nine general areas. It will also reveal, however, that the specifics vary enormously.

[2] See Falini, et al., 1980.

Embedded in the issues we usually find goals, either competing or congruent. Thus, the pathway from issues to goals can be quite short. When issues concern competing uses, for example, there will obviously be a goal to maximize each one of them. Since it is not possible to accomplish this for all of them at once, we will obviously be faced with conflicting goals, a situation best resolved through the economic process. In this process, the goals eventually provide the basis for comparing alternatives and finally for monitoring the effectiveness of the plan. The clearer and more specific the goals are, the more useful they will be for these purposes.

As discussed in Chap. 2, certain general ecological goals are established by the larger context. As long as they are respected, these are the goals that ensure overall harmony, like the contribution of Aliso Creek to the regional waterflow system, or the place of Madrona Marsh in the Pacific Flyway. In the arena of conflicting human purposes, however, even such clear ecological goals may be doubted and questioned and transformed into issues.

As we come to understand them better, it becomes clear that there are no independent issues. Rather, the issues of a landscape usually turn out to be interrelated, and so do their resolutions. Thus, for example, at San Elijo Lagoon, in resolving the sewage treatment issue, we can provide water not only to stabilize the level of the lagoon and to allow for irrigation, thus improving wildlife habitat and aquaculture productivity, but also to create greater recreational amenity, which makes it possible to direct recreational use to suitable areas. It is important to search out such interactions early rather than to pursue single issues through to single resolutions, which are all too likely to exacerbate other issues.

WHY?: THE ETHICAL POSITION

Here may be the most difficult question of all. We are not asking "Why?" in the practical sense. The reason for a design effort is explained by the issues, as we have already discussed. Rather, we are asking "Why?" in the ethical sense. Why are we, the designers, involved? What is our ethical position? Are we, as so many profess to believe nowadays, simply the processors of information and other people's values, or do our own convictions play a basic role in shaping the plan? As pointed out in the last chapter, a great many people—and their values—enter the process at larger scales. Is the designer simply one among the multitude?

No one who has participated in landscape design is likely to hold to the notion, once so naively prevalent, that design can be an entirely objective process, free of values, dealing only with facts. We know now that even our choices of facts to deal with are value-laden, and the same goes for every choice we make. Value-free judgments are hardly possible. If we do not lay claim to an ethical position, then it will be defined implicitly by the work we do. It is thus best to state our position early and then base our actions on it.

The same moral difficulty arises in every kind of design, in planning and decision-making, even in the highly technical situations that are usually the subjects of the systems approach. West Churchman, whose thinking has so strongly influenced the latter field, has posed three possible positions for a planner (designer) of any sort.[3] The first is the "goal planner," who simply does the bidding of whoever is paying him. The goal planner's task is to determine the goals of this person or group and then recommend the course of action that will best meet those goals. Churchman points out that this is really no more than a problem-solving approach because it narrows or closes out issues. It is, in fact, very close to the value-free stance. In planning circles, Churchman's goal planner is also commonly called a "hired gun."

The second type of planner works in a similar way but draws a line beyond which he will not go. He may decide, for example, that he will not consider any actions that are illegal, or that destroy the habitat of an endangered species, or that degrade water quality, or that increase energy consumption. This type Churchman calls the "objective planner."

The third type does not serve just the group or person who pays his fees but asks the question: "Who should be served and how?" This leads to two other questions: "Who should decide and how?" and "Who should plan and how?" These questions put the design effort in the very broadest ethical perspective in that any possible means and all possible effects are open for consideration. The project at hand is cast in the larger context where concern for goals are passed down from higher levels of integration. Churchman calls the planner who works in this way the "ideal planner." He does not, of course, always achieve his ideals, but he strives to maintain his moral position.

Where larger ecological issues are involved, the question of who should be served leads us into vast expanses and awesome responsibilities. It is becoming clear that a great many nonhuman species can survive only if human beings choose to make their survival possible. Albert Schweitzer pointed the way for a new kind of ethical sense when he insisted that we are ethical only when all life is sacred to us, including the lives of plants and animals as well as those of our fellows.[4] He believed that the relations among human beings no longer provided a broad enough context for considering our ethical obligations, and everything that has been learned since about the web of life that supports us all has supported his position in a practical way. Our design decisions affect a vast array of lives, both human and nonhuman, and the effects we have on nonhuman species are usually passed through the web to return eventually like a boomerang to ourselves or others of our own species.

This is not to say that the ideal designer can always promulgate the ideal plan or even know what it is. But he can do all that is in his power to explore all the possible plans and predict their probable effects on that very large and diverse number who should be served. Most of the case studies in this book represent efforts to achieve this kind of ideal design.

For our present purposes, accepting Churchman's types, I will substitute the term "designer" for his planner, and I will also add a fourth type: the designer who is committed to a particular moral purpose and whose efforts are directed to achieving goals justified by that purpose. For example,

[3] See Churchman, 1979.
[4] See Schweitzer, 1933.

he may be committed to wilderness preservation, or to reducing resource consumption, or to equal distribution of land ownership, or to some combination of these goals, and then bend all his planning efforts to achieve them. This type is well known as the "advocate designer."

Most would agree that the position of the ideal designer is the preferred position. From the point of view of social justice and ecological integrity, his is the position most likely to lead to truly meaningful plans. Nevertheless, the circumstances of a less than ideal world ensure that goal design will remain a prevalent position. Someone has to pay for all the work done, and that someone usually feels he is entitled to some measure of control. This is especially true when landowners are paying, but it is often the case with public agencies as well. And as long as there are goal designers, there will also have to be advocate designers to represent the other side. It is not uncommon to find goal designers and advocate designers clashing on issues of development versus preservation in a process that we might call adversary design.

The ideal designer's position is probably most often assumed by those establishing goals at the larger scales, where conflicts are not so immediate. For example, the federal legislation that created the California Desert Conservation Area and provided for its management—the Federal Land Policy and Management Act of 1976—was in effect a broad mandate for ideal planning. Some quotations from the act will be instructive:

The California desert environment is a total ecosystem that is extremely fragile, easily scarred, and slowly healed.... [Sec. 601(a)(2)]

The use of all California desert resources can and should be provided for in a multiple use and sustained yield management plan to *conserve these resources for future generations, and to provide present and future use and enjoyment*.... [Sec. 601(a)(4); italics mine]

.... the public must be provided more opportunity to participate in such planning and management.... [Sec. 601(a)(6)]

Multiple use and sustained yield are well-established principles in the management of public lands, of course, and traditional watchwords of the U.S. Forest Service. The Federal Land Policy and Management Act provides new definitions, however, which are worth quoting here:

The term 'multiple use' means the management of the public lands and their various resource values so that they are utilized in the combination that will best meet the *present and future needs of the American people*...a combination of balanced and diverse resource uses that takes into account the long-term needs of future generations for renewable and nonrenewable resources ... and harmonious and coordinated management of the various resources without permanent impairment of the productivity of the land and the quality of the environment with consideration being given to the relative values of the resources and not necessarily to the combination of uses that will give the greatest economic return or the greatest unit output. [Sec. 103; italics mine]

The term 'sustained yield' means the achievement and maintenance in perpetuity of a high-level annual or regular periodic output of the various renewable resources of the public lands consistent with multiple use. [Sec. 103]

Some will be skeptical about these definitions, of course. Such noble purposes are as easy to grind out in the chambers of the Senate as they are hard to achieve on the ground. Nevertheless, they essentially convey what we mean by ideal design. And such ideals, once stated with the authority of Congress, may very well spread over a larger base.

It is important to understand, however, that defining an ethical position does not eliminate moral quandaries. All landscape design is fraught with them. Even the ideal designer has to be concerned with the possibility that the decisions of his broadly based clientele may decide on a course that his analysis tells him is ecologically wrong. What then? We can only say here that this sort of impasse seldom occurs. Surveys of attitudes toward the landscape among broad segments of the public usually show a strong disposition toward the protection of natural systems and at the same time a real interest in maintaining opportunities for human use. Still, conflicts do arise for which there is no general resolution. Each of us can only sort out the circumstances of each situation and find his own way. As it turned out, the plan for the California Desert eventually drawn in response to this Congressional mandate fell far short of the ideals expressed at the higher level.

WHO?: THE PARTICIPANTS

At this point I will add to the laws of levels of integration another principle: the larger the scale, the larger the number of people affected by, and therefore potentially involved in, the design process. Thus, the larger the scale, the more complex the question of "Who"?

For convenience, we can divide those involved in landscape issues into three groups. These are the design group, the participant group, and the client group. The design group includes those who carry out the task of design and develop the plan. The participant group includes those who are directly involved with the design group as advisors, reviewers, representatives of particular interests, and so forth. At the smaller scales—up to the project level—this usually means owners and developers. At larger scales, it is usually a more diverse group, with roles vastly expanded since 1960. The client group includes residents, users, and all those who are affected by the decisions made in the course of the design process, but who may or may not directly participate.

The Design Group

At the construction or site design scales, for working with clear and limited goals, the design group may consist of a single individual. At scales larger than these, it is likely to be a team, usually including professionals with varied expertise. Deciding what areas of expertise are needed and how they should be organized is one of the important tasks in formulating the design process.

Dissecting knowledge of the landscape into specialized disciplines is an artificial product of the academic system, effective in gaining specific detailed information, but ineffective in dealing in a practical way with larger issues. The difficulty, of course, lies in the fact that each specialist is concerned with only a piece of the picture. It was an early article of faith of the systems approach that the answer to this limitation was an interdisciplinary team that would assemble all the pieces to produce the whole picture. In practice, however, organizing successful interdisciplinary teams has turned out to be a rather complicated matter because of fundamental differences in outlook among disciplines.

Charles Kilpack has suggested that among those involved in the analytical tasks of mapping geographic variables and developing map models, there exists a continuum in attitudes toward information and its use.[5] On the far left, he placed those primarily interested in capturing and representing data as accurately as possible; on the far right were those whose chief interests lay in using it to represent relationships. If we go beyond the analytical, or precision, stage of the process, we can expand Kilpack's continuum to include on the even farther left those whose focus is generating and collecting data (the more basic scientists) and on the even farther right those primarily concerned with exploring possibilities and in shaping a plan. Then the whole continuum of planning group members looks something like the following:

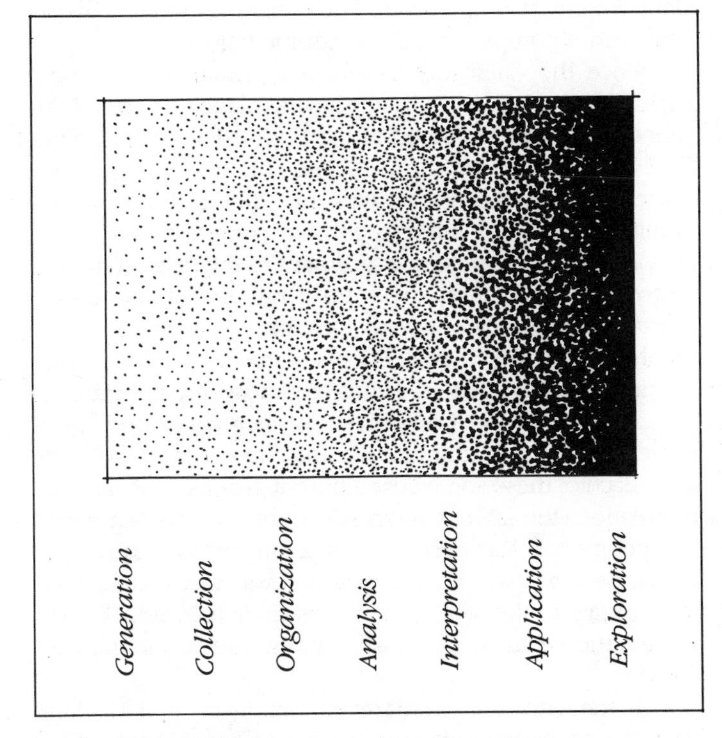

Roles With Respect to Information

Kilpack is careful to qualify his continuum as artificial and general, and my expansion of it is certainly no less so. Some specialists undoubtedly transcend such categories. The continuum is nevertheless a useful device for envisioning the composition of a design team. The people on the left will ensure the adequacy and accuracy of the information, whereas those on the right will ensure that it is well- and wisely used. The distinction is important because among the most common failings of planning efforts are faulty and inadequate information on the one hand and the lack of coherent, clearly focused use of information on the other. Reasonable distribution of team members across the continuum can make such failings less likely. The distribution described here obviously parallels the organization of brain functions described earlier.

Another question that arises in team composition concerns the roles of the various specialists. One common form of team organization is that in which a number of specialists not only contribute the expertise of their disciplines but also share more or less equally in each stage and activity of the effort. This is the kind of team often used by the U.S. Forest Service in responding to the congressional mandate for interdisciplinary planning. Typically, a team for a plan unit might include a wildlife biologist, a forester, a hydrologist, a recreation specialist, a landscape architect, and an engineer. The team leader, whose responsibility it is to carry the effort forward, might be any of these. Out of the merging and melding of ideas comes a plan that, in theory at least, represents the integration of the disciplines represented.

Another quite different format is what we might call the core-and-consultants organization. This involves a core design group (in some cases, only an individual), the members of which are not regarded as specialists in any of the disciplines involved but who do have a working knowledge of all of them and whose most important professional ability is a talent for integration; that is, a knack for putting all of the knowledge together. These are usually landscape architects and planners. Their position is in the middle right portion of the spectrum. The consultants in the organization then function as consultants usually do, providing information from their disciplines as needed, answering questions, and usually reviewing the work, but without being otherwise involved in the decision making. Their involvement is as extensive and intensive as the project requires.

Most team structures tend to be variations on one or the other of these basic formats. Each has both advantages and difficulties. The integrated team has the advantage of allowing each discipline full participation in decisions, but it has the disadvantage of lacking a central integrating force, which is especially important on the proposing side of the alternating current. The team as a whole is supposed to perform this function, and that can be very difficult for people whose professional abilities are highly specialized and hedged by disciplinary boundaries. Usually, we find the members of such teams heavily concentrated in the left-hand portion of the information orientation spectrum.

The core-and-consultants format, on the other hand, allows everyone involved to contribute in the way he knows best, but it tends to exclude specialists from decision-making responsibilities, which may also cause difficulties. It tends to favor the right side and thus fosters creativity.

Whatever the organizational format, there is still the problem of selecting the areas of specialization where information is needed. Since this depends heavily on the concerns and issues involved, it is useful to develop a matrix for representing relationships between issues and experts.

[5] See Kilpack, 1982.

Such a matrix might look something like this:

	Hydrologist	*Geologist*	*Soil Scientist*	*Economist*	*Agronomist*	*Wildlife Biologist*
Ground Water Depletion	●	●				
Agricultural Productivity	●		●	●	●	
Market for Food Products				●	●	
Loss of Wildlife Populations						●
Land Values				●		

This gives us a good indication not only of the areas of expertise needed but also of the relative magnitude of tasks for each discipline and the areas where disciplines are likely to overlap and require cooperation.

Participant groups function in a variety of relationships with design teams. The only consistent difference is that, whereas planning teams are by definition working in their professional capacity, participant groups are composed of volunteers. Sometimes the two work so closely together that there is no other meaningful distinction; at other times, participant groups do nothing more than perform casual periodic reviews.

Public Participation

In federal projects involving land use, the legal requirement for public participation is usually met by forming a Citizens Advisory Group, among whose members are representatives of all special interests involved. The Advisory Group (or participant group) for the California Desert Conservation Area Plan, a subcontinental-scale effort that covered over 25 million acres, for example, included members with special interests in energy and utilities, earth science, public affairs, the social sciences, wildlife resources, mining, state government, outdoor recreation, native Americans (Indians), archaeology, botanical resources, and environmental sciences.

The designers of the Yosemite National Park plan in the late 1970s went far beyond the Citizens Advisory Group concept and attempted to bring the entire interested public into the effort as participants, that is, to make the client group the participant group. They prepared a workbook and a planning kit and invited everyone who cared to take the trouble to devise his own plan for the park by choosing among a long list of possible actions for each of ten planning units. About 59,000 packets were mailed out and over 20,000 were returned, making this probably the most massive citizen participation effort ever undertaken. The practical value of so much participation is doubtful, however. Statistically, a random sample of about 2000 returns would yield the same results, and in fact the park service included 5000 of them in its analysis.

In highly controversial situations, participation by the public is usually much less organized and more contentious, with individuals and groups moving in and out of the scene more or less at will, promoting their own varied interests and points of views. Typical of such situations is that at work in the Bolsa Chica Lagoon case study (pages 173–79). Here, besides the landowners, three groups and six public agencies are involved, all with somewhat conflicting points of view. There is little likelihood of finding a common set of goals, and the planning process is almost certain to be long and contentious.

The client group, being amorphous and often invisible, is the most difficult to deal with. Who, in the first place, are they? The catch-all answer is: Everyone who will be affected by the planning decisions. As a matter of principle, the attitudes and values of everyone affected should be considered in the planning process, but for larger planning efforts, this can mean vast numbers of people. Especially for projects undertaken by federal agencies, the measurement of public sentiments has become extremely elaborate in recent years.

The public participation program of the California Desert Conservation Area Plan gives some idea of the magnitude of effort that can be involved. Deserts raise strong emotions in those who have any interest in them at all, and even before beginning the program, the U.S. Bureau of Land Management planning group knew that they would have to deal with a complex tangle of competing values.

Since the California Desert is a resource of national importance—Congress already having made that clear—the client group was identified as the population of the United States. Opinions were thus solicited nationwide. Of course, some groups had greater stakes than others. There were a number of special interest groups, composed mostly of people with a compelling attachment to the desert (such as motorcycle associations who use it for recreation) as well as various preservationist organizations. Even more closely involved were the actual residents of the desert, those living within the study area, and thus the most directly affected. At a slight remove, but still very much interested, were those living in communities around the edges of the study area. Each of these somewhat different groups—the national population, the special interest groups, the residents, and the people on the periphery—was invited to express its views, and each was approached in a way that was deemed appropriate to the level and kind of their interests. We will discuss the means of expression and the results in the next chapter.

In the end, the problem of identifying and working with the client group is one that has to be worked out for each planning effort individually. Usually, it will be practical considerations that limit participation. Time and money are always in short supply. Theoretically, for areas like Yosemite or the California Desert, one could argue that the clientele is global. One could even argue that it includes generations yet unborn. But clearly, consulting a global clientele, not to mention one yet unborn, would present serious—not to say insurmountable—difficulties. Thus, consultation usually extends only as far as time and money allow, and the clientele excluded by these limits must be content to be represented by the planning and participant groups.

HOW?: FROM PROCESS TO METHOD

Here we shall return briefly to the question of method and move from generalities to specifics. Most design efforts fit within the overall framework of one of the types of rational process described in the last chapter. That is, a design process will fall into a pattern of technical, economic, ecological, legal, social, or political rationality, or sometimes into a pattern that is a variation or combination of these.

The character of the issues and especially of the goals suggests which of these types will be most appropriate. If the goals are clear and congruent—that is, mutually achievable—then a technical process is appropriate. If the goals are so precise as to include the means as well as the ends, then a legal, or rule-following, process may be unavoidable. If the goals are reasonably clear and quantified but not mutually achievable, then the economic method should prevail. Goals that are large, general, imprecise, diverse in character, and finally immeasurable usually suggest an ecological process. If the goals are very indefinite indeed, it will then be necessary to devise some special process to suit the situation, one that allows us to proceed with a measure of uncertainty in the hope that clearer goals and objectives will emerge from later analyses. It is even possible for the process itself to remain a broader quest, given direction by the exploration of issues rather than by any drive toward objectives. It is also not unusual for landscape design at the project and plan unit scales to combine some characteristics of both the economic and integrative processes.

Within the general framework of rational process, we need the more specific guidance of a project plan that defines tasks, schedules, and a detailed logical structure. Usually, it is represented as a work-flow diagram. The difficulty, of course, lies in the fact that we usually don't know enough about the work that will be required to plan it with total confidence. Discoveries made along the way are likely to influence what we must do. That is why the PERT and CPM charts commonly used in aerospace projects, with their very specific task definitions and split-second timing, are rarely applicable in landscape design. Work-flow diagrams vary greatly with the nature of a project and are thus far more flexible. We can best view them as sets of working assumptions, subject to change as work progresses, with certain exceptions. One thing that never changes, for example, is the basic logical structure, which assures the conceptual validity of the project from one step to another. (We have already discussed the several reasons why a defensible logic is *essential* to support design, especially at the larger scales.)

Examples of project plan diagrams appear in a number of the case studies, but I should make it clear that in each case the diagram is meant only to give some idea of the process for that case study alone. There are no universal processes and no universal diagrams.

CHAPTER 8

The Stage of Precision

Information and Models

As the meandering streams of the Romantic Stage merge into the single broad channel of the Stage of Precision, the early excitement of infinite possibilities gives way to a sense of directed purpose. Any rational consideration of the future requires a well-ordered knowledge of both past and present.

We have already divided the Stage of Precision into an information theme and a modeling theme. In the first, we described the landscape in terms of its component parts, and in the second, we reassembled the parts into models that allow us to see it in a new light. Modeling does not necessarily follow immediately after information gathering. The two are sometimes well-separated in time. It often happens, especially at the regional scale, that an information system assembled to describe a landscape is used to generate models only as the need arises, in an ongoing sequence of iteration and reiteration.

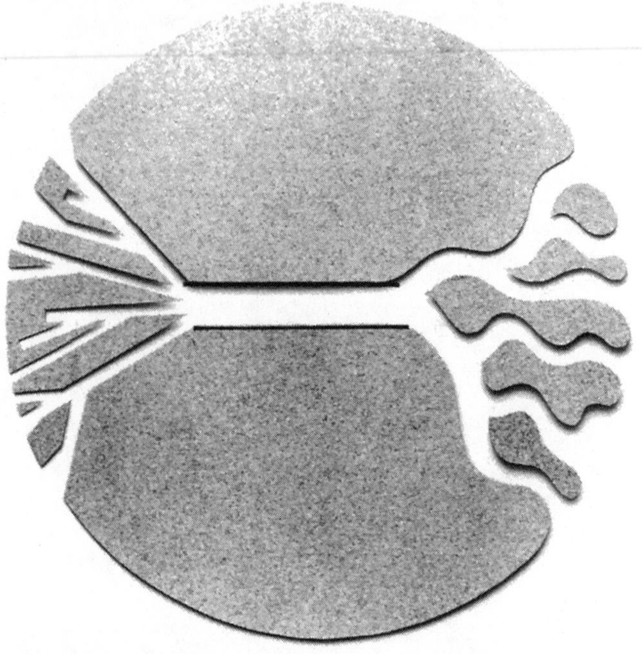

INFORMATION

First, we will consider the need for information, a term that has become increasingly loaded with diverse connotations since the onset of cybernetics in the 1950s. In the environmental arena as in other political arenas, especially since the passage of NEPA, information has become power.[1] Decisions with far-reaching implications are required by law to be made on the basis of sound information. Stewart Marquis puts it in the cybernetic context:

> Information provides the capacity for controlling work behavior. Obviously, where information is built into work-structures and work-systems the constraint is direct. Additional information about the structure and behavior of components better enables man to design ones that constrain work behavior in line with his wishes. Both kinds of information help him to answer the fundamental questions: What is controllable? and How do we control it?[2]

For our present purposes, we might divide the information useful for answering these questions into two general categories: information that describes the physical composition of the landscape and information that describes the multitude of immaterial factors (using the word "immaterial" in the sense of "nonphysical" rather than the sense of "irrelevant") that influence its future. These are two essentially different types of information and require entirely different means of collection. Before going on to explore them, it will be worthwhile to discuss briefly some general principles of information collection.

First is the principle of economy. In our data-conscious society, information overload is a common problem. It can also be a very costly problem. Information being expensive to obtain, and budgets for landscape design and planning usually being very limited, it is important to collect only that which is likely to influence decisions. Moreover, the initial waste of collecting too much information will usually be multiplied down the line. Organizing and using it are also expensive, of course, but perhaps more to the point, it is usually irrelevant information that creates confusion and lack of direction. Even after a collection of information has been organized, it can be difficult to sort out what really pertains to the issues at hand and what does not.

Here again we find an intriguing parallel of design with natural evolution. For a natural ecosystem, the separation of useful information from that which is false or irrelevant (noise in the parlance of information theory) is crucial to evolutionary adaptation.

As a way of beginning, we can look to the issues to tell us what information is needed. If agricultural uses figure among the critical issues, then obviously we will need to know about soil fertility and rainfall. If water supply is an issue, then we will need information about water sources and uses.

Nevertheless, it is not quite such a simple matter as just following the leads of the given issues. Information often uncovers new issues. When we study a landscape in detail, it is not at all uncommon to find conditions that drastically alter our initial assumptions. The discovery of

[1] See Jokela, 1979.

[2] See Marquis, 1974; p. 34.

the snail darter habitat at the site of the proposed Tellico Dam is probably the classic case, but there are a great many less dramatic examples. Careful study of the High Meadow Ranch site turned up the remains of an old Indian settlement, the preservation of which became an issue.

The principle of economy applies to the level of precision of information as well as to its quantity. To be useful, information must be accurate enough for its purpose. It is necessary to remember that precision is also expensive and therefore has to be related to actual need, and more particularly, to the scale of concern. Pragmatically, the level of precision of information should fit the character of the decision-making process. In this respect, we often have problems with data drawn from scientific investigation. For scientists, precision is nearly always critical; in fact, for them, any question of excessive precision is almost unheard of. As a result, we often find ourselves trying to generalize from enormously detailed, minutely specific scientific data. Recasting such information in a larger mold is a risky task for a nonspecialist.

Another principle that we should bear in mind is the famous Shannon theorem that the value of information is in direct proportion to its unexpectedness. The theorem underscores the fact that all information is not equally valuable and suggests that, where money limitations do not permit us to gather all the information we might want, it might be most productive to pursue directions likely to yield the unexpected. The unexpected can cause us to recast our preconceptions of a landscape and its potentials in a radical way. One need only recall the upheaval caused by the snail darter.

THE EXISTING LANDSCAPE AND ITS PHYSICAL COMPOSITION

We might think of a collection of information concerning a physical landscape as a description of the existing ecosystem that comprises the three modes of ecological order discussed in the first chapter: structure, function, and location.

Since structure involves the full range of biotic and abiotic elements in the landscape and their interactions, information concerning structure is more or less a catalogue of these elements: geological and soil composition, plant and animal communities, micro and macro climate. As we can see in countless environmental impact reports, however, such a catalogue can grow to several volumes, with the significant material buried in the mass. It is thus important, in line with the principle of economy, to search for the major elements and those that have particular significance for our purposes. To do so does not necessarily reduce the effort needed for information gathering. It often happens that it takes more research to determine the elements and relationships that have critical importance for a design than it does to simply list whatever is there.

The major elements we can consider to be those that are present in the greatest abundance and that have the largest number of interactions with other elements. In classifying plant communities, the major species are called *dominants,* and communities are usually named after the most dominant, like the alder-ash forest and the sagebrush steppe. Animal populations usually are closely correlated with plant populations. If we know what plants are present in a particular landscape, we can usually figure out what animals are there.

Significance of the various elements for design purposes may be more difficult to determine. The significant elements and combinations of elements are those necessary to maintain the integrity of the ecosystem, but these need not be the dominant species. They can include those that are important in the larger context, such as rare and endangered species.

In determining significance, it is often useful to consider the structure in terms of its substructures, among them the trophic organization and the vertical layering. Trophic structure is always significant, and we have been seen its overwhelming importance in the case of San Elijo Lagoon. There, significance could be traced to the pivotal role of the marshgrasses and to one species in particular. Although it would have been foolish to deny the importance of other structural components of the lagoon, this was clearly the one that warranted special attention in design.

In the rain forest, by contrast, where trophic substructures are little understood and a prodigious diversity holds sway, it seems unlikely that we could sort out particular groups of elements as playing pivotal roles. In this case, a critical substructure, one that we need to understand much better, is that of the forest's vertical layering. We will return to the subject of various substructures and their importance for design in Chap. 11.

DESCRIBING LANDSCAPE FUNCTION

As mentioned in the first chapter, the functional dynamics of the landscape, meaning primarily the flow of energy and materials, are the processes that drive and operate the landscape as a system. An understanding of their workings is important in dealing with issues of resource consumption and environmental degradation.

The principle of economy applies once again here. In most landscapes, there are limited numbers of processes that play central, critical roles in the shaping and working of the system. With careful analysis, it will become clear what these processes are. In San Elijo Lagoon, the key process was the energetic one of tidal flushing. In the Huetar Atlantica region of Costa Rica, the key processes were the dynamic relationship between water and soil and the exchange of nutrients between soil and biomass. Once we understand the latter more fully, we will have a good beginning for working with the rain forest ecosystem.

To identify the key processes, we usually have to give some attention to the flows of energy, water, and the major nutrients. When functional difficulties disrupt these flows, usually a lack of sufficient supplies is involved, or materials are collecting where they cannot be assimilated harmlessly. For this reason, it is important to describe at least the sources and the sinks—the inputs and outputs—of energy, water, nutrients, or any other materials that may be especially important to the landscape in question. Virtually any element can be a limiting factor or accumulate in unhealthy or dangerous quantities.

The more critical the flows, the more complete descriptions usually needed. Chap. 13 will discuss the means of describing the functional ecosystem processes at various levels of detail.

LOCATIONAL INFORMATION: FOUR TYPES OF INVENTORY

The component of the information base that usually receives the most attention is the one that describes existing locational distributions—usually called the *land resource inventory*. Such an inventory has been considered a necessary beginning for a landscape planning effort of any size at least since Geddes described the need for "regional surveys" in the early years of the century. From Geddes' time on, quite sophisticated resource inventories were developed, probably the most famous and complete of which was Warren Manning's epic 363-map inventory of the United States.[3]

Techniques for recording and storing a land resource inventory vary with the scale and purpose of the project at hand. In general, the larger the scale of the effort, the more complete and formally organized the resource inventory needs to be. Ranging from the rather casual inventories typical of smaller-scale design to the computer-stored geographic information systems appropriate for regional and larger efforts, we can describe four distinctly different types of inventories.

The Gestalt Inventory

At the construction and site design scales, we deal with landscapes small enough to be perceived and understood in their entireties. Since we can readily see the differences in landscape character from one place to another, there is rarely any need to break the whole down into layers or component variables for later recombination, as is required for the other types of inventories. We can conveniently deal with the whole by applying what Lewis Hopkins has called the "gestalt method," a *gestalt* being a whole that cannot be assembled from its parts. In a *gestalt resource inventory*, then, we identify the varying capacities of the site to support varying human use by careful viewing and map those areas with more or less uniform capacities on the basis of these judgments. We also identify specific features and qualities that will influence the design.

The inventory for the Madrona Marsh case study established three distinct zones on the basis of such field observations: the critical habitat zone, the primary support zone, and the secondary support zone. Since the entire purpose of this design was to maintain the integrity of the wildlife habitat, the inventory focused on those characteristics directly related to habitat quality. A gestalt inventory is also shown with the Simon Residence case study.

The gestalt resource inventory usually goes hand in hand with technical methods. Both are applicable at smaller scales and in situations involving relatively few people, and both rely heavily on the expertise of the designer because a great many judgments and decisions are made, more or less implicitly, by him alone.

When we deal with larger landscapes, it becomes impossible for any one person or even a group of people to know the area as a whole with both the intimate detail and holistic comprehension of a good gestalt inventory. Even if such knowledge were possible, the gestalt approach would rarely be applicable at larger scales because of the pressing need for high visibility of method and explicit rationality.

The Short-term Inventory

The stage of precision for efforts at scales larger than that of site design usually involves breaking down the site into its component variables, analyzing each with respect to its capacity for supporting the intended uses, and then reassembling them all in a way appropriate for a suitability model. Such models, which present an aggregated distribution of capacities in a landscape, are usually prepared by overlaying inventory maps. In order to lend themselves to overlay techniques, the maps need to have certain common characteristics.

Most basic among these requirements are a common scale and tickmarks to ensure the accurate placement of each overlay. Maps should also be approximately equally accurate, since the accuracy of a model produced by an overlay process is no greater than that of the least accurate map used in creating it. Moreover, they should all be drawn using a common graphic medium that will facilitate the overlay technique to be applied.

When inventories of this kind are intended for use in just one design, as is usually the case at project and plan unit scales, they are usually hand-drawn on paper or acetate. Suitability models are then developed by "handicraft" techniques. After the plan has been formulated, they will probably not be consulted again for design purposes, though they will probably remain useful for management purposes. This

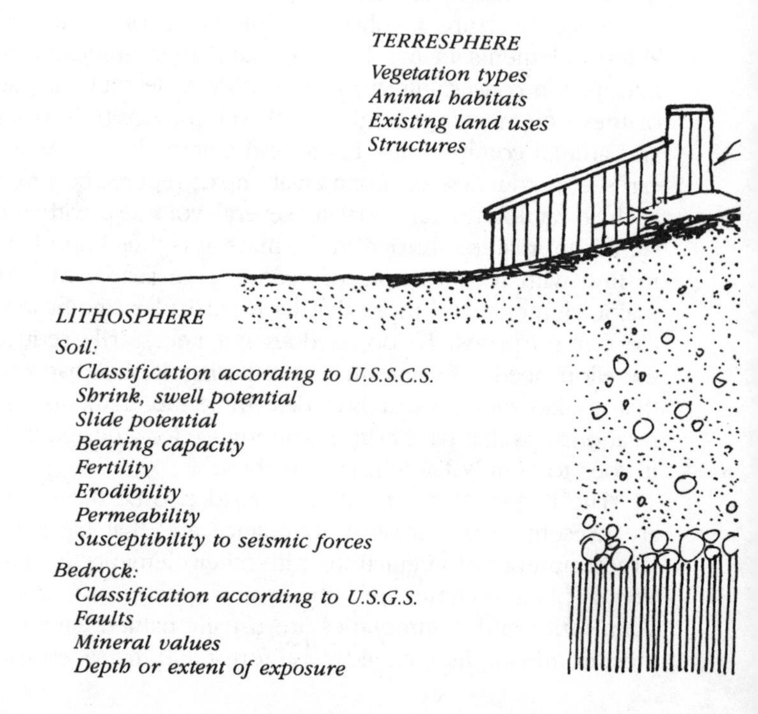

[3] See Steinitz, 1981.

is why we call them "short-term inventories." It is this characteristic that distinguishes them from both the special resource inventory and the geographic information system, both of which are prepared for long-term use and for a variety of purposes.

Manual overlay models associated with short-term inventories are rather laborious and time-consuming in comparison with the computer-aided techniques commonly used for special resource inventories and geographic information systems. For short-term use, however, the investment needed to compile an inventory for computer-aided use is often not justified. Examples of short-term inventories (at the project and planning unit scales, respectively) are shown in the Wild Animal Park and Whitewater Wash case studies. These, like most inventories, are organized as collections of maps, all at the same scale, with each map describing the locational distribution of one particular category of landscape features, such as topography, vegetation, or human use. In the terminology that has become fairly standard since computers came into widespread use, these categories (soil, slope, vegetation, etc.) are called *variables,* and specific classes within each variable (sandy loam, 20 to 30%, oak-hickory forest, etc.) are called *attributes.* For example, the variable of vegetation type might include three plant communities—say a northern flood plain forest, a conifer bog, and an oak-hickory forest. On the vegetation map, areas covered by any one of these would be shown to possess that specific attribute. Other vegetation characteristics might be mapped as separate variables. For example, timber density is commonly included in inventories for heavily forested landscapes.

The land resource inventory, then, describes what is on, over, and under the land. Every variable that might have some effect on human use or be affected by human use is included. We can envision the inventory as a series of sections cut horizontally through the landscape, in this manner:

In recent years, the very rapid development of remote-sensing techniques and related improvements in air-photo interpretation have brought about enormous improvement in the quality of resource inventories. A great many government agencies at all levels routinely collect geographic information. Unfortunately, there is little effective coordination among them, and we often find information that is inconsistent in format, contradictory, or drawn at different scales. As a result, assembling a resource inventory, rather than being an original research effort, is usually a matter of collecting existing data, checking it, reconciling it to a consistent scale and format, and finally organizing it. The variables usually included in a resource inventory are shown on the sketch. It is usually important to consider each of these, even though some may well turn out not to have any significant influence on the results.

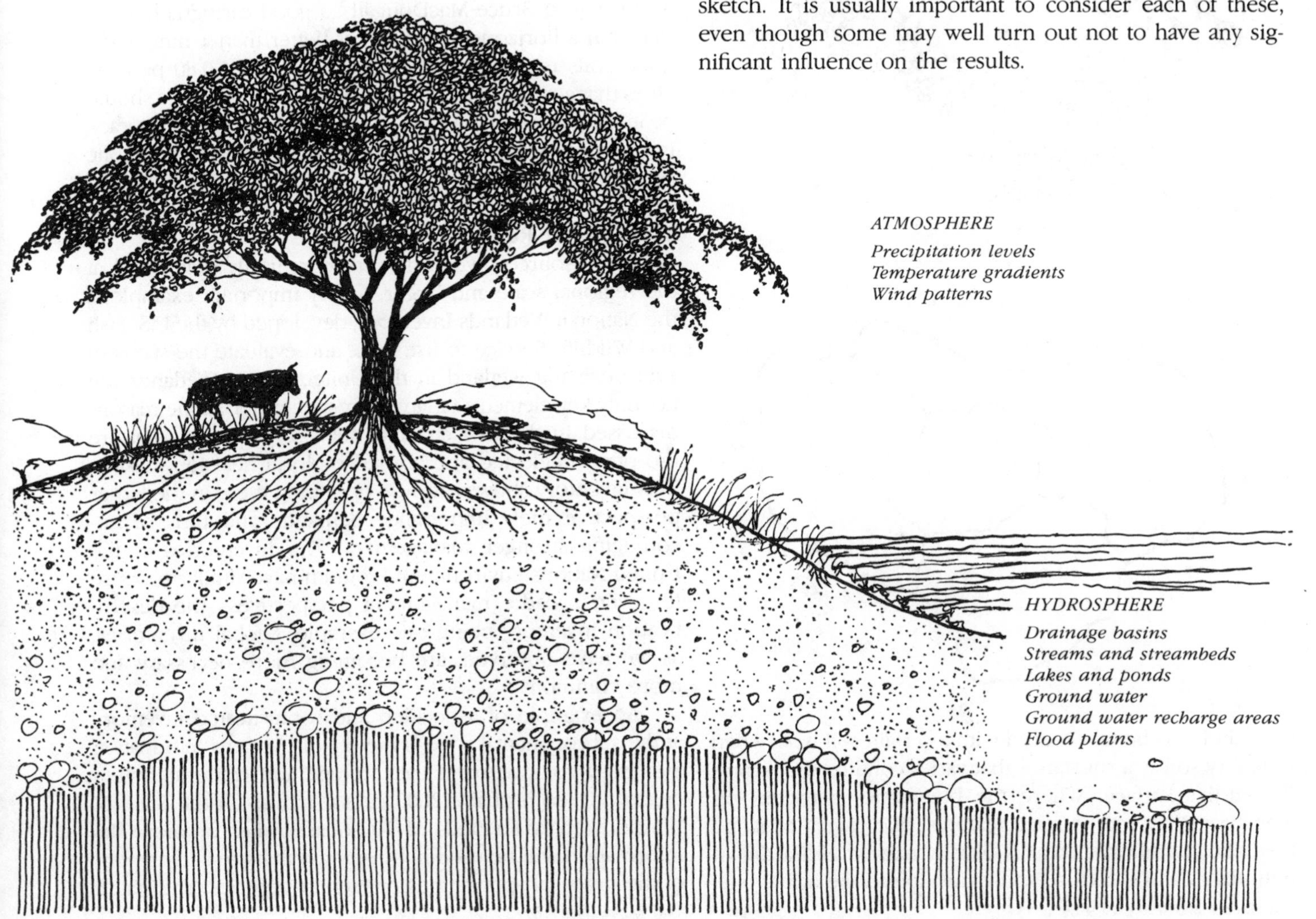

found a significance in the diversity and integrating potential of the land resource inventory that transcends its purely utilitarian role. Rodiek and Wilen express concern that so many consider inventories as "nothing more than an elaborated scientific exercise," and assert that, in fact, "inventory and classification activities are prerequisite to all conceptual thought."[4]

We might indeed view the inventory as a translation of reality into conceptual terms, in which case it is important to recognize that, in return for the clarity and manageability gained, much is lost in translation. No inventory can perfectly replicate nature.

Consider, for example, the assumption of hard boundaries. To make maps useful, we have to draw edges around geographic areas even though we know quite well that they do not actually exist in the landscape. Nature rarely draws hard lines. We draw an edge around an area, for example,

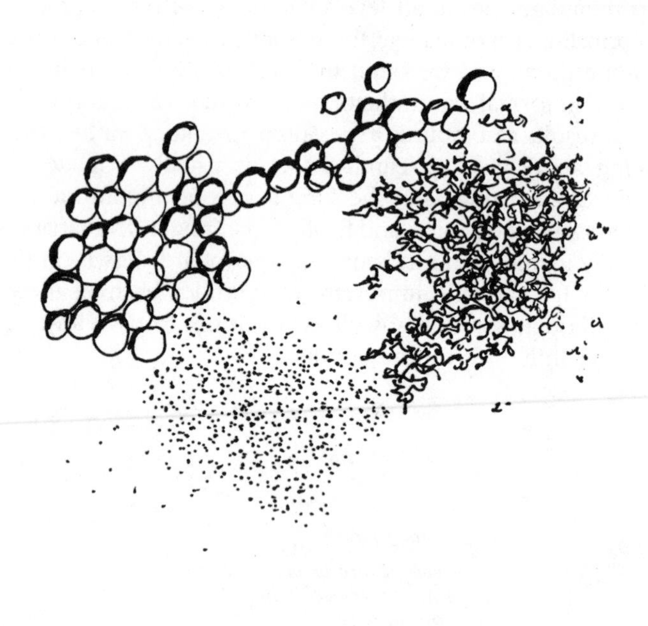

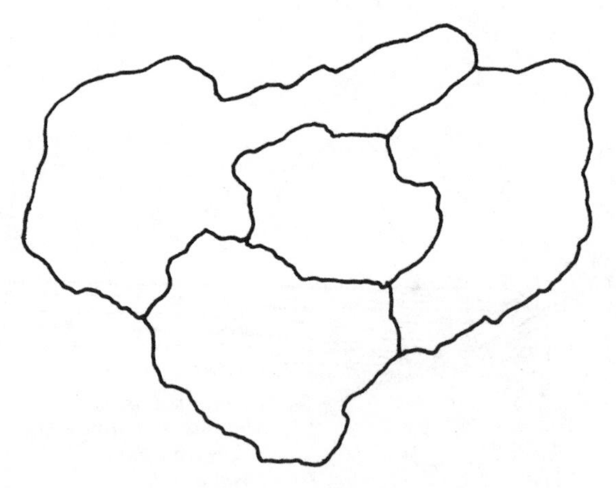

and label it as being covered with a certain soil type, when in reality soil is a substance that varies continuously across the earth's surface. Where the boundaries between soil types coincide with surface features, like ridges or slope toes, the transitional zone is usually narrow and a drawn boundary can be a close approximation to the actual one. In other cases, however, transitions are usually very gradual and edge lines largely arbitrary. Slope maps have even less precise boundaries. Again, it is an unavoidable distortion of reality, made necessary by the fact that rational thinking requires clear distinctions, but it is a distortion nonetheless and one that we should remain conscious of. When the design process is finished, decisions will have to be put into practice in a landscape that doggedly remains a continuum.

A second assumption is that of homogeneity. We assume that an attribute is essentially the same throughout the whole area within the boundaries we assign when, in reality, there are usually any number of variations. For example, the composition of soil will vary throughout any boundaried area according to the myriad slight differences found in any specific location. Again, rationality makes a demand for clarity, but we need to be aware of the distortion involved. These assumptions are not alone in introducing inaccuracies into the inventory. Further inaccuracies are caused by drawing and redrawing and from the great error magnification that results from interpretations at small scale. The accuracy of maps is normally measured in terms of allowable horizontal error for boundaries and percentage error for homogeneity. In theory, every map in an inventory should state these figures as an indication of its reliability, but in practice, they are hard to obtain. Source maps rarely include them. According to Bruce MacDougall,[5] a good cartographer can maintain a horizontal accuracy of better than 1 mm, and a good soils map has a purity factor of about 0.80 percent. Lines that actually represent broad transitional zones should be indicated as such on maps to show their relative inaccuracy; in practice, we rarely know how wide a transitional zone actually may be.

The Special Resource Inventory

Special resource inventories are commonly developed at the regional scale and larger. A very important example is the National Wetlands Inventory, developed by the U.S. Fish and Wildlife Service to list, map, and evaluate the status of every existing wetland in the United States. Wetlands are not only key elements of larger ecosystems for all the reasons discussed in the first chapter, but also resources that are too often neglected and abused by local decision makers. This inventory shows promise of putting them in the larger perspective their importance demands and thus of encouraging stronger management. The inventories of the United Nations Environment Program have similar promise for even broader management perspectives—the desertification inventory being one example. Global inventories, as we have observed, are becoming more necessary and more common.

An example of a detailed and sophisticated inventory at the subcontinental scale is the Illinois Streams Information System, a computerized data storage and retrieval system describing each of the 2000 or so streams in the State of Illinois with watersheds of 10 square miles or more. Its information will be used in three ways that are fairly typical of the practical applications of such inventories. The state's

[4] See Rodiek and Wilen, 1979–80; p. 13.

[5] See MacDougall, 1977.

Department of Conservation will use it in reviewing permit applications (it receives over 2000 in a year) and in developing stream management policies, and the Illinois Environmental Protection Agency will use it as a basis for developing water quality standards specific to each stream.

The streams are categorized as branches, creeks, drainages, forks, sangamos, sloughs, rivers, ditches, or tributaries. For each of these, a great deal of data is recorded, including various locational codes, physical characteristics, biological characteristics, cultural characteristics, public-use data, and existing intrusions. For specific station points, the data include station location, physical characteristics, biological data, water quality and quantity, and a great deal of additional descriptive data.

An important characteristic of this system is its rapid and efficient retrieval of a great variety of data, which is made possible by its use of data-base management software. It is hierarchically structured and open-ended to provide these qualities.[6] If, for example, a developer were to apply for a permit to build a paper factory or a marina in a particular location on a river, it would be a simple matter to determine the existing water quality and rate of flow in that reach of the river. Or, if the state wanted to find a site for a new regional park, the system could quickly list all of the publicly owned lands on rivers' edges with their highway access, forest cover, and specific fish populations.

The Geographic Information System

The focus on a single type of resource is what especially distinguishes the special resource inventory from the fourth type of inventory, the Geographic Information System. Such a system commonly includes two separate data files. One of these is a range of variables that are mapped, as with the handicraft inventory, at a common scale in a common format. This file, which we might call the "locational file," describes the geographic distribution of specific characteristics. The second data file, which we might call the "attribute file," stores data describing the attributes of each location. For example, the locational file might tell us that a particular plant community exists in a certain place, and the attribute file might list the plant and animal species in that community and give its rate of productivity, sensitive features, and other quantities and qualities. Thus, a Geographic Information System can include descriptions of structure and function as well as location. The two files are designed to be used interactively; that is, they are cross-referenced to one another. It is therefore possible to develop various types of land suitability models by analyzing the behavior of each attribute under the stress of land-use changes and then aggregating these in a way that is conceptually very similar to the overlay technique, but technically far more sophisticated.

Using a computer has several important advantages over handicraft methods. It stores vast amounts of information in readily retrievable and usable form. It is more efficient and capable of producing far quicker results in situations involving complex interactions and large land areas. It is especially useful when the same information base is to be used to produce a number of different models, either for a single planning effort or for a variety of efforts over a period of time.

Once the data have been encoded and the modeling program is operational, models can be generated very rapidly. On the other hand, since encoding data and developing a program can be very time-consuming and thus expensive, computer systems are not often used for one-time applications.

Speed also makes it possible for computerized resource information systems to carry out much more complex operations than handicraft methods can. Manual overlay techniques are effectively limited to situations that place equal weights, or levels of importance, on each variable, whereas the computer can weight different variables differently. For a model to determine suitability for agricultural use, for example, the key variables might be soil fertility, level of rainfall, and land slope. Should we consider soil fertility to be 50 percent more important than rainfall and, at the same time, judge a rainfall level of less than 25 inches per year or a slope of over 30 percent as being entirely unsuitable for agricultural use, the computer could handle these complications easily, whereas they would be most cumbersome to deal with manually. The computerized information system can also be linked with simulation models dealing with variations in quantities like runoff, flood levels, or water quality to produce quantified predictive models.

Information systems are primarily used by public agencies concerned with land use and by private corporations with very large land holdings—that is, in situations where political rationality is the dominant mode. To be effective, such a system must be thoroughly integrated with its user group and, since these tend to be large, institutional, and more or less inflexible, serious problems often arise. In the 1970s, a number of systems were developed but little used because the information they produced was not in a form or at a level of resolution that conformed to existing practices. Adaptation is needed on both sides. The powerful potentials of computerized information systems suggest that planning agencies should adapt themselves to make better use of the available information. In fields where they are already in widespread use, information systems have exerted profound influences on both organizational and decision-making processes.

At the same time, technology is developing means to provide a broader range of more specifically useful information. Charles Kilpack draws a distinction between a true Geographic Information System, which "manipulates data about these points, lines and areas to retrieve data for ad hoc inquiries and analyses. . . ," and computer graphic programming systems, which provide for analyses and modeling of projects on an essentially one-by-one basis. The former type of operation, which includes a great deal of record and boundary recall and the like and performs the same basic functions as the special resource inventory, he calls "routine," and the latter type, which includes land-use analyses and modeling, he calls "nonroutine." For a system to function as a true Geographic Information System, it needs to be capable of both types of operation.

Although this ideal is seldom fully achieved, a number

[6] See Hopkins, et al., 1980.

of very large G.I.S.s are now in use in various agencies. The Canadian Geographic Information System, the Minnesota Land Management Information System, and the systems used by the states of Maryland and Kentucky are noteworthy examples. Portions of the system developed for the State of Alaska are shown as a case study on pages 151–57, along with examples of its application.

Though an information system can be developed to a high level of complexity and sophistication, the underlying concepts are actually quite simple. There are two basically different formats for computer mapping: the polygon and the grid. Each has certain advantages and disadvantages.

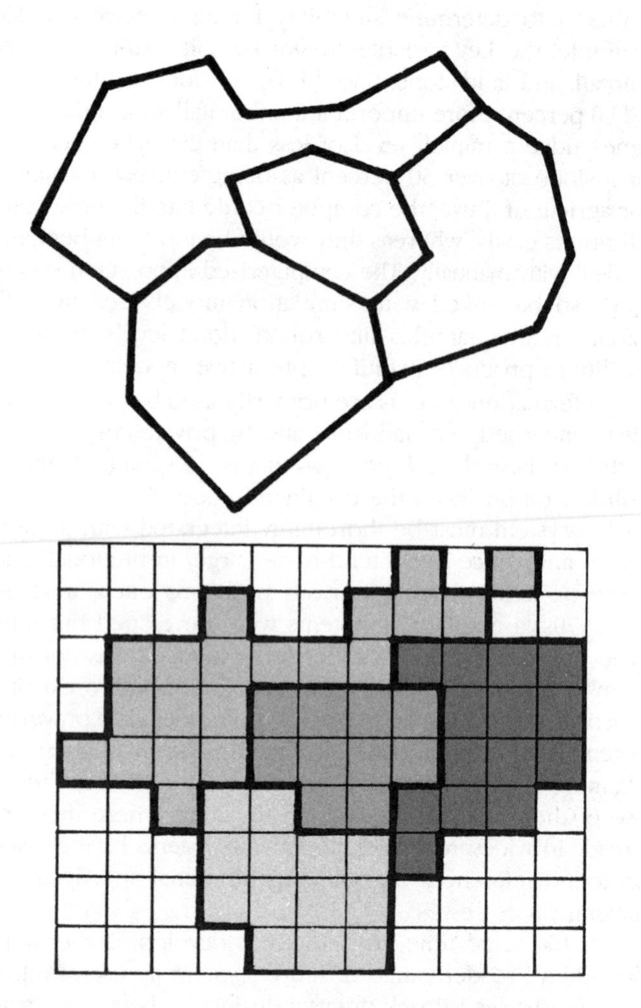

Polygon maps, examples of which are shown in the Alaska case study on pages 153 to 154, represent the boundaries of areas and linear elements with lines drawn by a plotter. The maps are encoded by a digitizer that traces the lines of the original map. If we wanted to encode a soils map, for example, we would first establish a reference point, which would probably be one corner of the map, and then begin tracing the boundaries for each soil type. The digitizer stores the boundary lines on tape as a series of X-Y coordinates. If we begin by tracing areas having Hanford series sandy loam soils, say, then the boundaries of all those areas become a data record. We then go on to trace the boundaries of other soil types, the boundaries of each becoming a new record. When all the boundaries have been encoded, the records collectively become a data file, which can be printed by the line printer as a map. Since its coordinate points are extremely close together, this map replicates the original from which the data were encoded. Soil and vegetation maps in polygon form are shown in the Alaska case study on pages 153 and 154.

The major difficulties with polygon maps arise when we try to combine, or overlay, them to get composites of different variables. Boundaries for each map are different, of course, and the mathematical operations required to overlay them and account for every tiny sliver of space become extremely complex, although programs that are able to accomplish this do exist.[7]

The grid technique avoids this problem by standardizing boundary lines at the edges of grid cells. A grid network is laid over the original map, and the appropriate data are encoded for each cell of the grid. Unfortunately, this introduces a problem of accuracy.

If we follow the rule of predominant type—that is, encode the cell to feature the attribute that occupies more than half of it—then we increase the positioning error of the original map by half the width of the cell. The size of the cell obviously reflects the accuracy of the encoded data. The smaller the cell, the more accurate the data, but, at the same time, the greater the time needed for encoding, the larger the amount of storage needed in the computer, and the greater the amount of computer time needed for modeling computations and printing out. Here again, we might recall the principle of useful levels of precision. It is tempting to establish a very small grid-cell size for the sake of accuracy, but it is wasteful to provide more accuracy than is really useful. Remember that a major use of information systems is for the development of suitability models and that such models are usually general indications of overall patterns rather than of precise representations. In practice, at the larger scales where these systems are normally used, even very large grid cells can yield meaningful patterns; in fact, grid cells as large as 1 kilometer square are not uncommon.

Maps encoded by the grid cell technique may be printed using a plotter or any of several types of printer. Each printed character represents one grid cell. Different character symbols can be used to represent different attributes, and characters can be overprinted to create shades from light to dark. The maps shown with the San Dieguito Lagoon case study were printed by a standard line printer. Such a printer is fast, inexpensive, and widely available, but it introduces distortion. Although grid cells are square, its own characters are not, being in a ratio of five high to three wide. One solution to this problem is the use of a rectangular grid cell whose proportions match those of the characters. Distortion can be avoided also by using a plotter to print grid images, like those in the Alaska and Nigeria case studies, which are products of the electrostatic plotter. But this requires expensive equipment. It should be noted that among the several types of printers capable of printing square cells, dot-matrix printers have great potential for producing highly sophisticated maps at low cost but with rather poor quality.

[7] See, for example, Dangermond, 1982.

CASE STUDY XII

The Alaska Geographic Information System

The Alaska Resource Mapping and Evaluation Program was initiated by the state's Department of Natural Resources. Eventually, the program will have created a statewide data bank of commonly formatted and automated natural resource and cultural data for use in systematic resource management and land-use planning.

The program's systematic approach is based on mapping increments or modules defined by the USGS 1:250,000 topographic quadrangles. The first quadrangles mapped included Anchorage, Talkeetna, Talkeetna Mountains, and Tyonek. Their development served to define and document the procedures, specifications, and format to be used in subsequent mapping.

Data were rectified and registered to the USGS quadrangles on the basis of the topographic, hydrologic, and interpretive information they contained and by referring to complementary 1:250,000-scale Landsat imagery.

The existing mapped data originally appeared in various formats and scales and was supported by many different types of documentation. As part of the data base design, both the maps and supporting documentation were systematically reviewed, classified, and organized for efficient automation and subsequent application. Required data that were not available were photointerpreted from aerial photography and Landsat imagery.

Data were mapped in modules corresponding to the USGS quadrangles, thus expanding the data base one module at a time. The modules are shown overlaid on the state. Data were stored as lines, points, and polygons. These were later converted into a grid-cell format—which is more flexible and efficient, and thus less expensive, for modeling purposes—by superimposing a uniform grid of forty-acre cells on the polygon coordinate files and assigning each cell a designation corresponding to that of the predominant polygon within it.

One unusual feature of the Alaska data files is their use of "integrated terrain units" for storing natural resource data. An integrated terrain unit is an area that is homogeneous with respect to the major resource variables: landform, slope, geological structure, soil, and vegetation types. These were defined by an overlay process before being encoded. Since their attribute boundaries usually tend to coincide, the boundary lines of all could be adjusted to a common line, thus eliminating the presence of "slivers" between areas that could logically be presumed to have common boundaries. As illustrated, data were encoded in such a way that every variable could be retrieved or used in modeling processes separately or terrain units retrieved or used as integrated wholes.

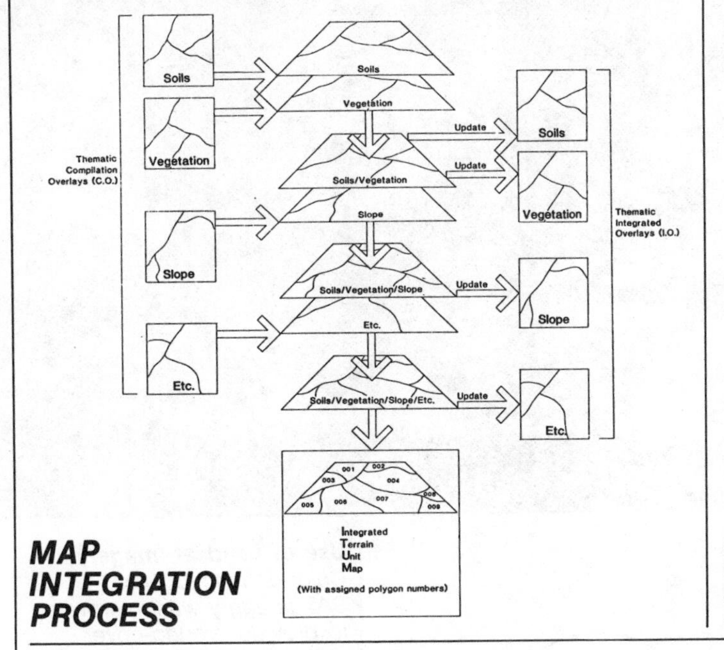

MAP INTEGRATION PROCESS

Prepared by Environmental Systems Research Institute, Redlands, California, for the State of Alaska.

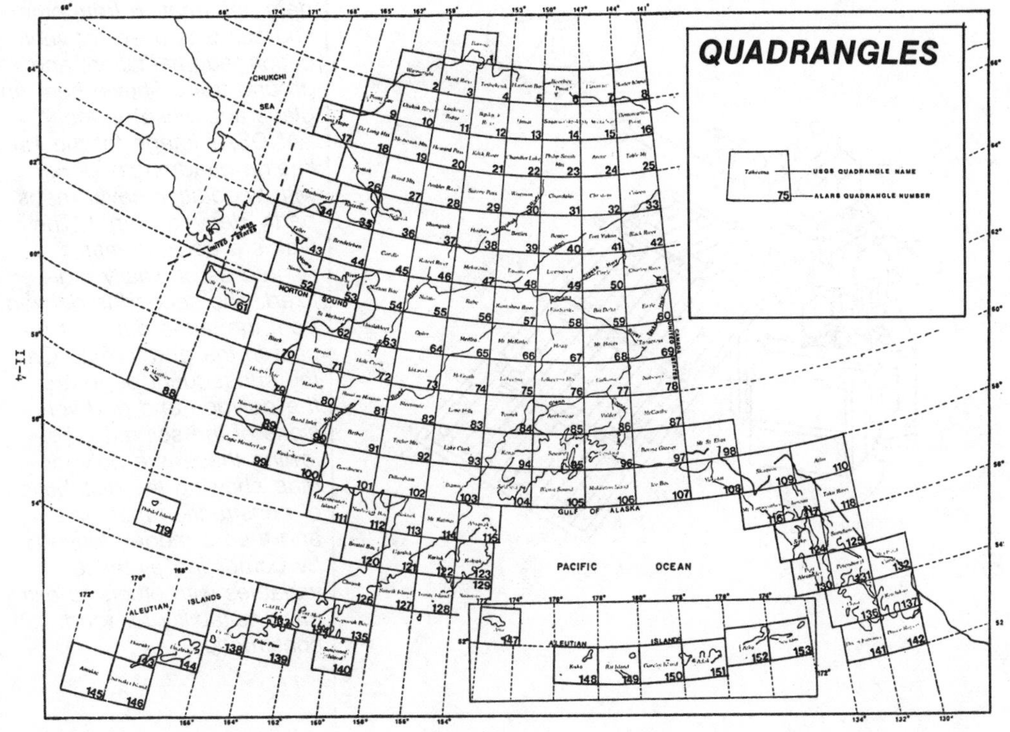

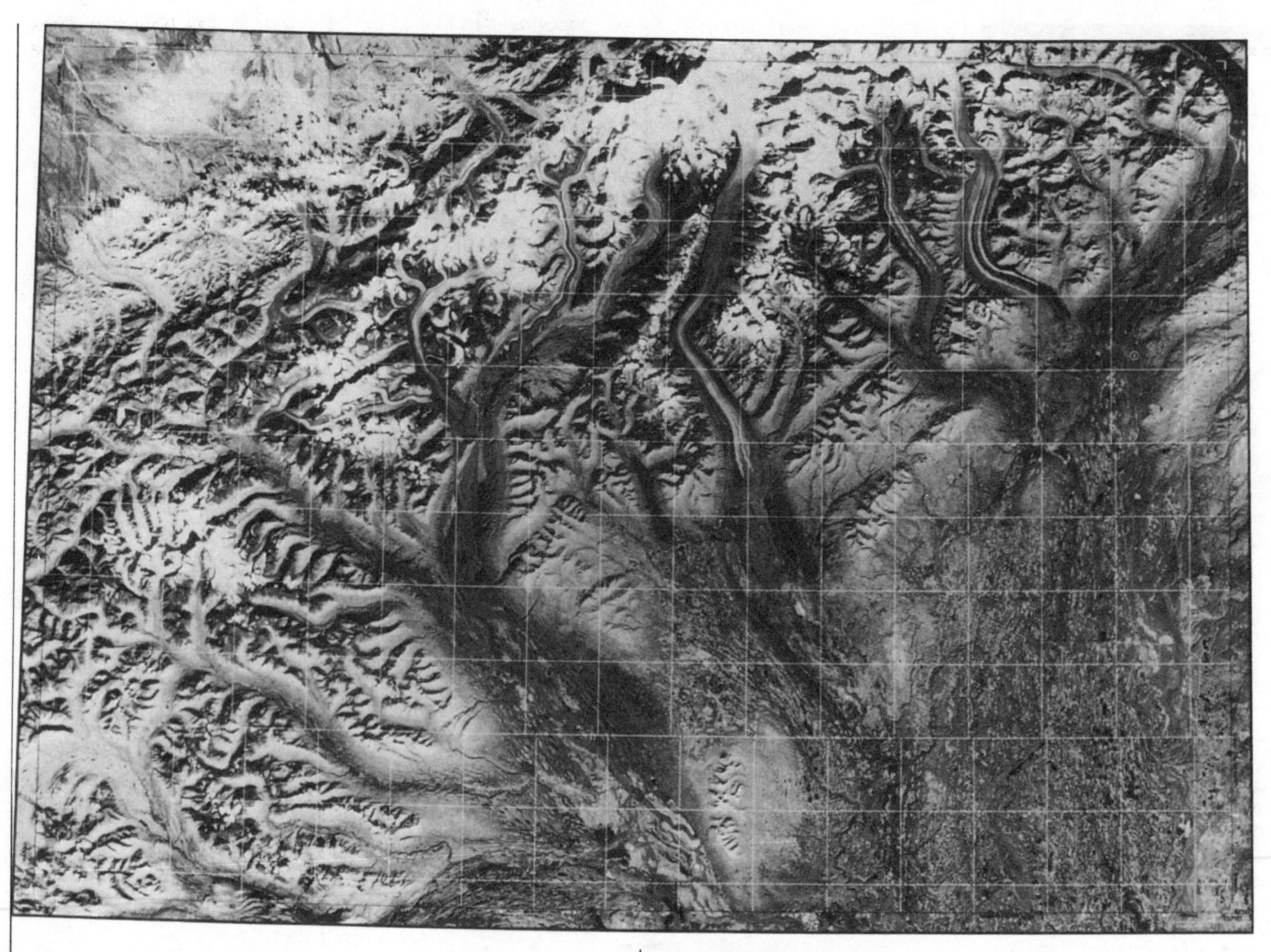

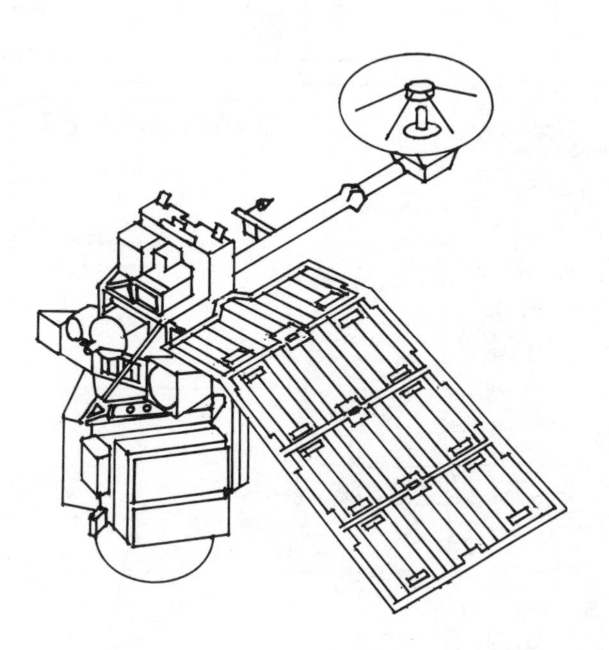

Use of Landsat Imagery
Color-enhanced LANDSAT imagery was interpreted to provide ground-cover data. Information from other sources was used for verification, augmentation, and ground truth. Shown here (in black and white) is the LANDSAT image for the Talkeetna quadrangle along with vegetation-cover maps partly derived from it. The first is a polygon map printed as originally digitized. The second is derived from the same data but recast in the grid format. On the subsequent page are the polygon and grid versions of the soil map. Following these is a polygon map showing the distribution of construction materials, and then a model derived by combining all these variables with others to estimate suitability for road construction.

PRIMARY VEGETATION

POLYGON MAP

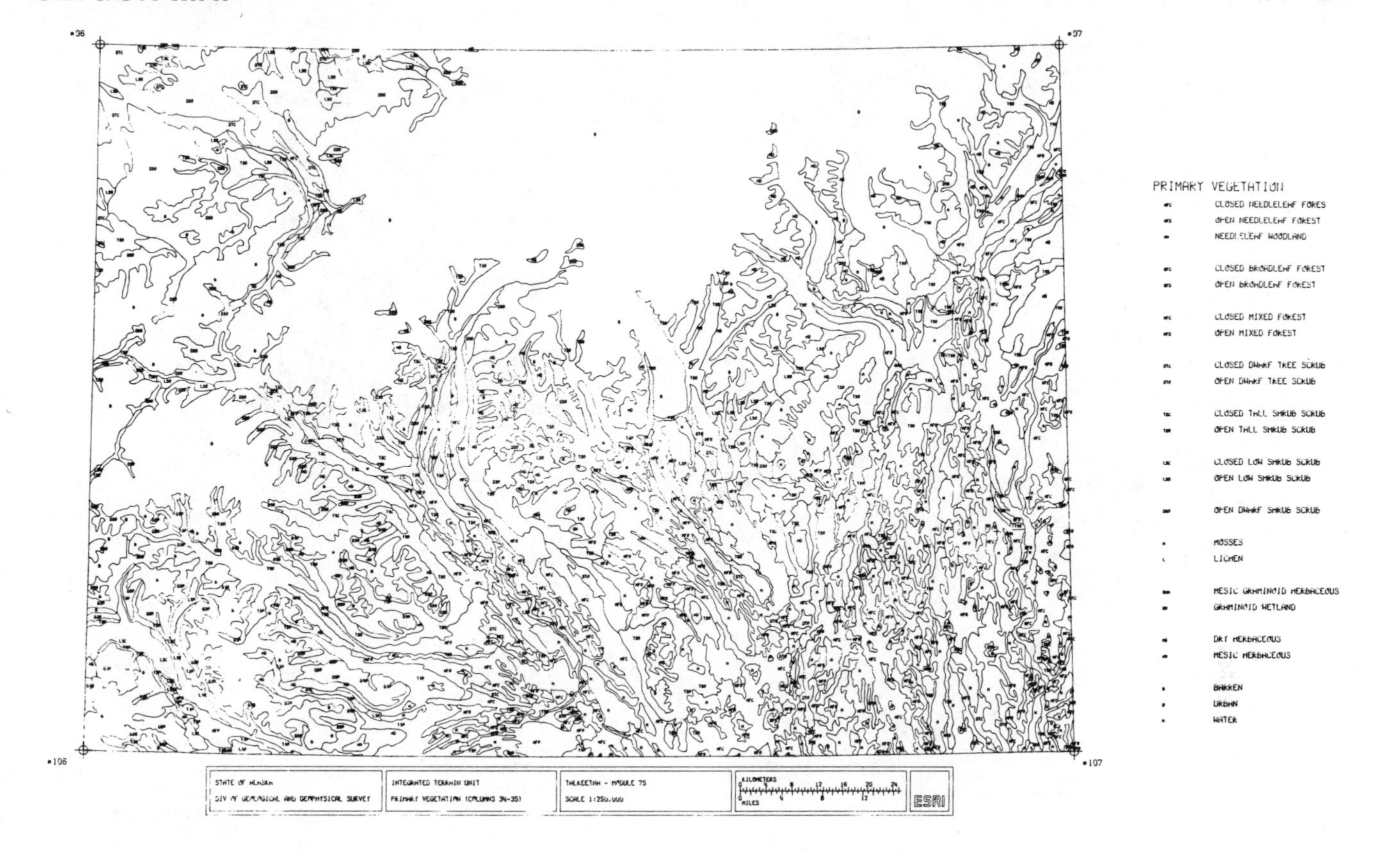

GRID MAP

PRIMARY SOILS

POLYGON MAP

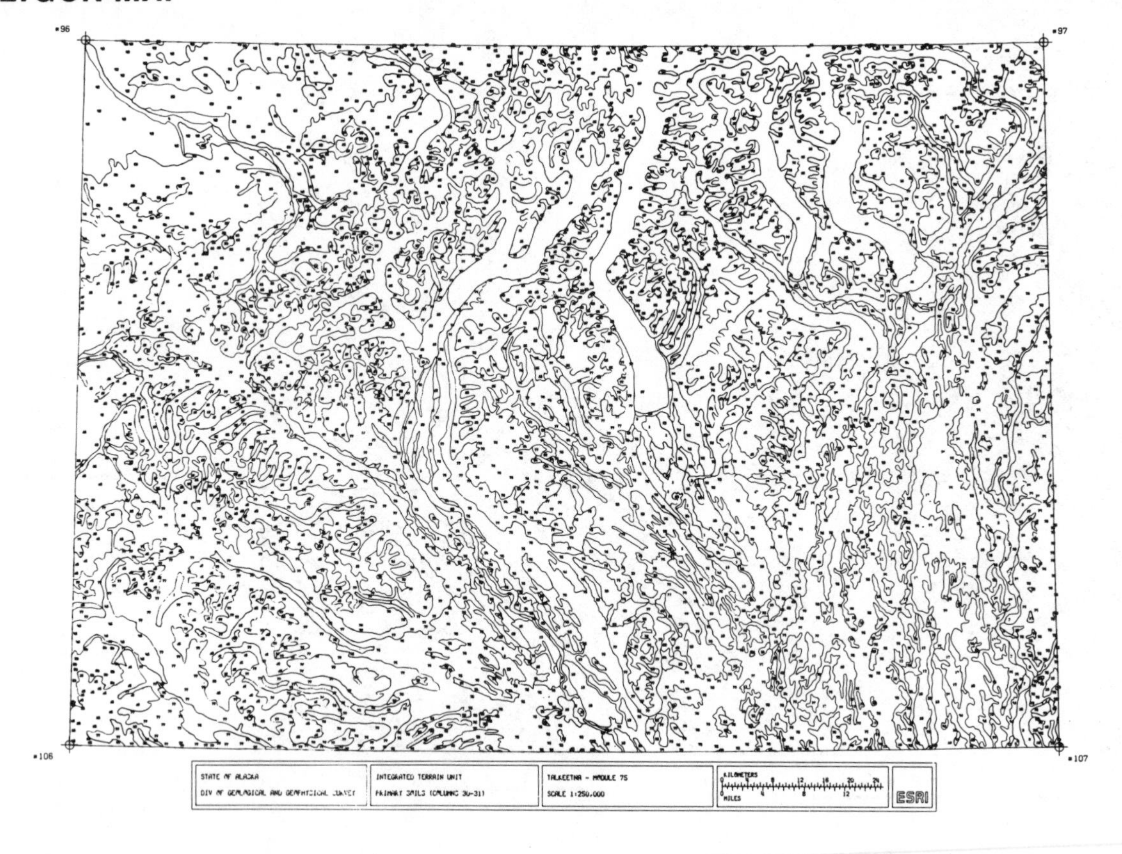

GRID MAP

CONSTRUCTION MATERIALS

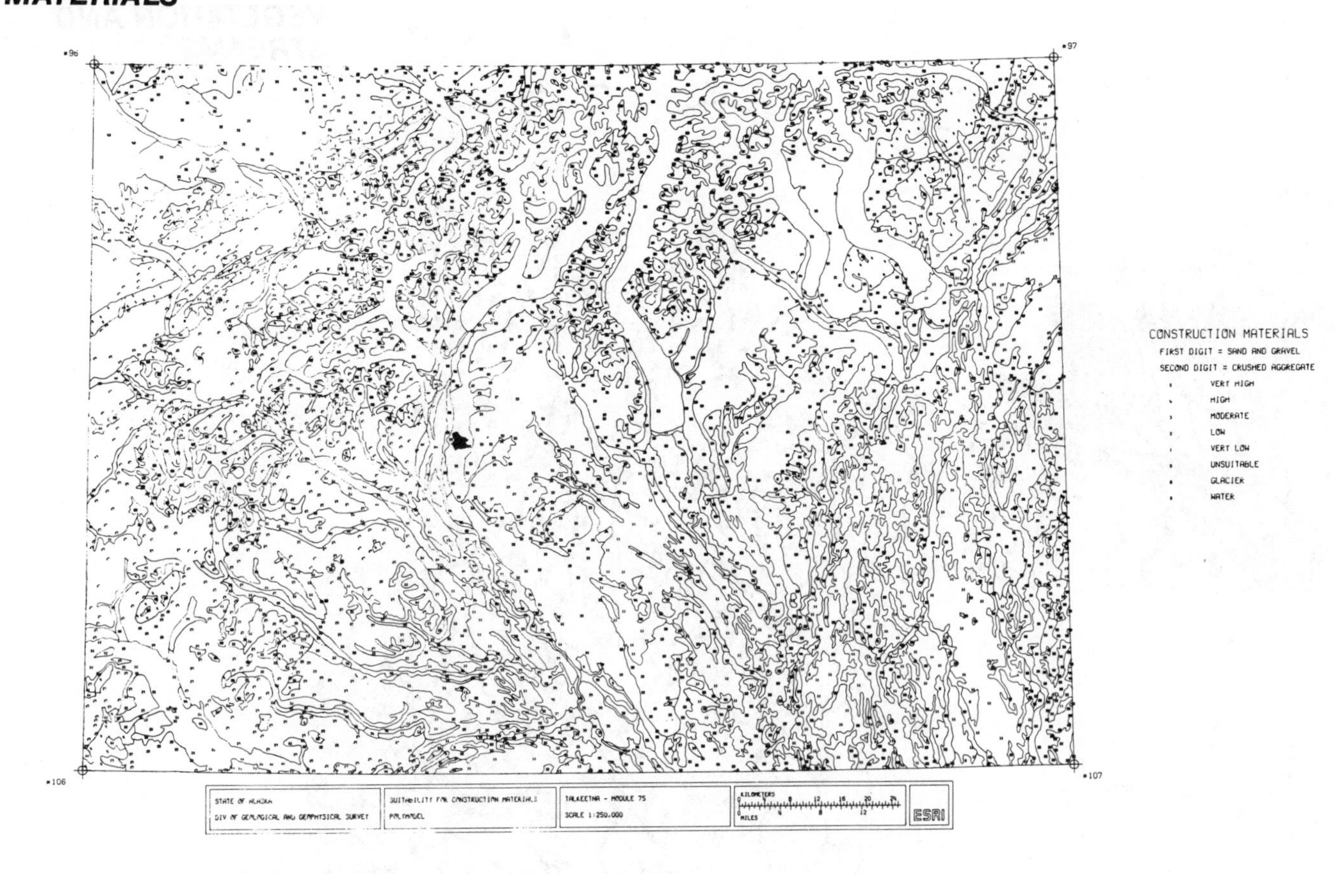

LAND CAPABILITY/ SUITABILITY: ROADS

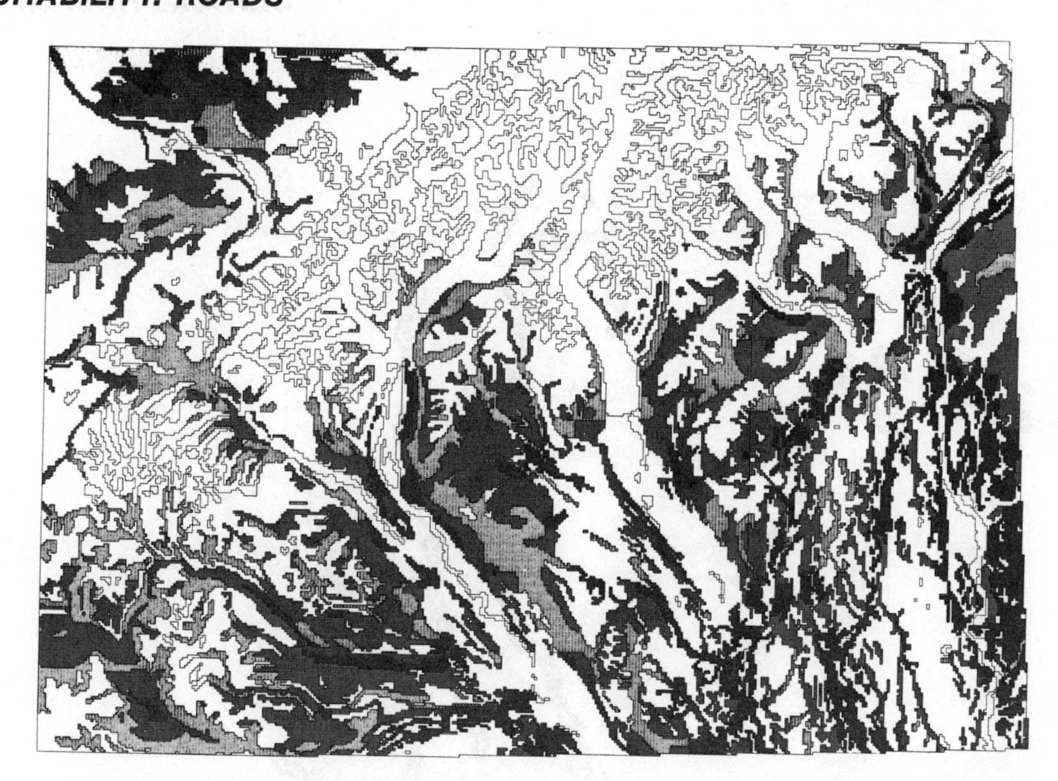

THE NORTH SLOPE

VEGETATION AND STREAMS

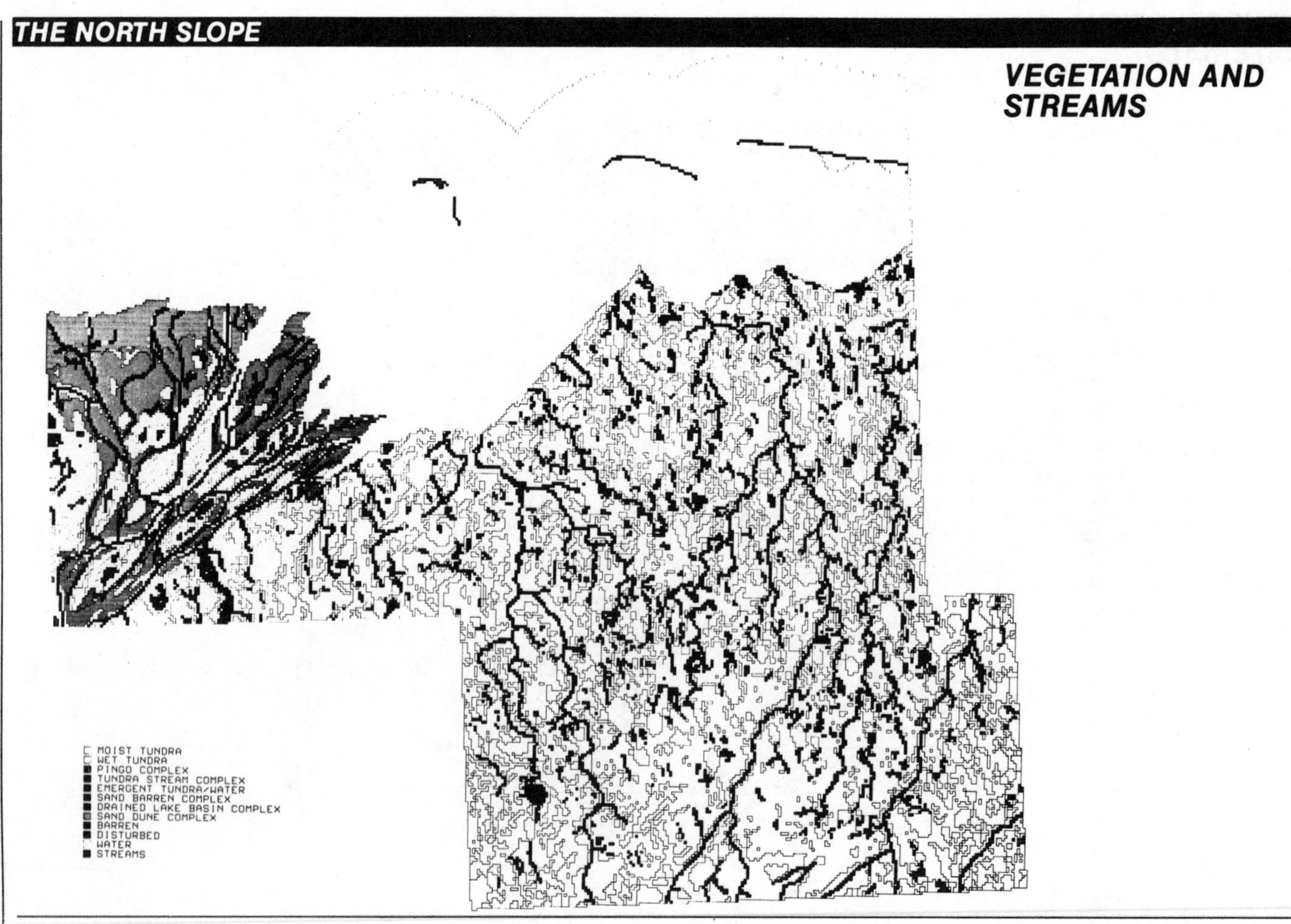

POLAR BEAR CRITICAL HABITAT

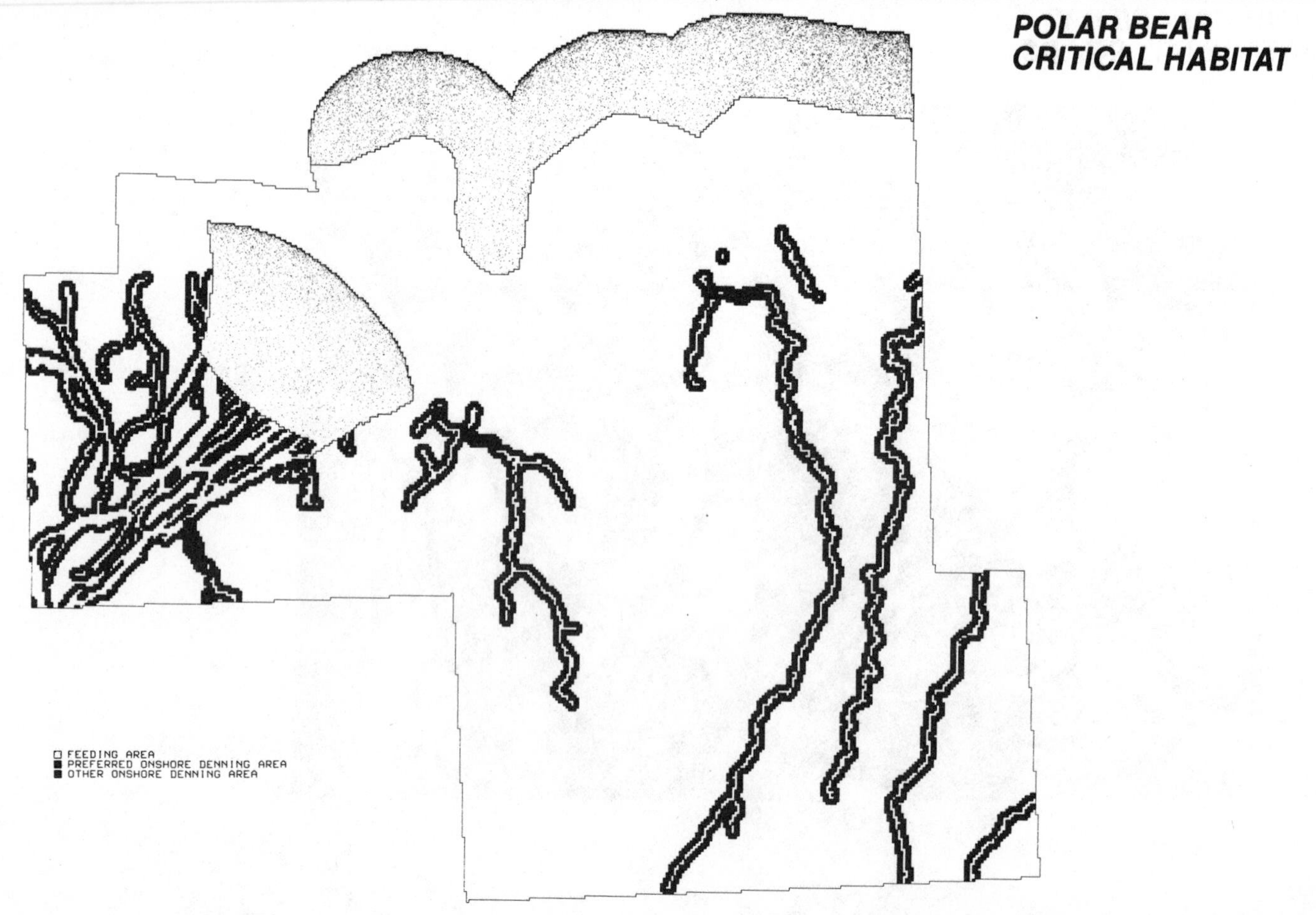

BIOLOGICAL IMPORTANCE

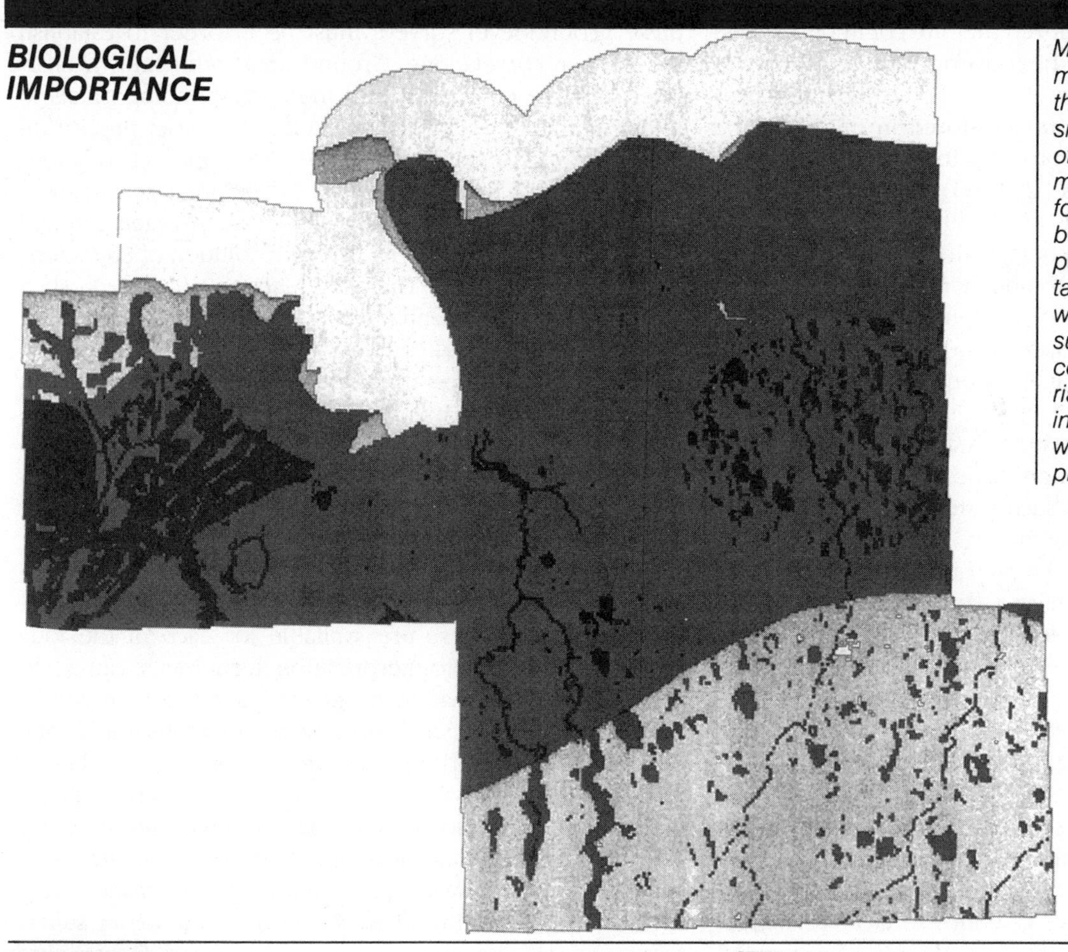

More specific data and models were developed for the North Slope at the next small USGS mapping scale of 1:63,385. Typical of data maps at this scale are those for vegetation patterns, polar bear habitats, and areas of particular biological importance, as shown here. These were used in deriving the suitability model for pipeline construction. By their criteria, very little land, and none in continuous linear patterns, was judged suitable for pipelines.

PIPELINE CONSTRUCTION SUITABILITY

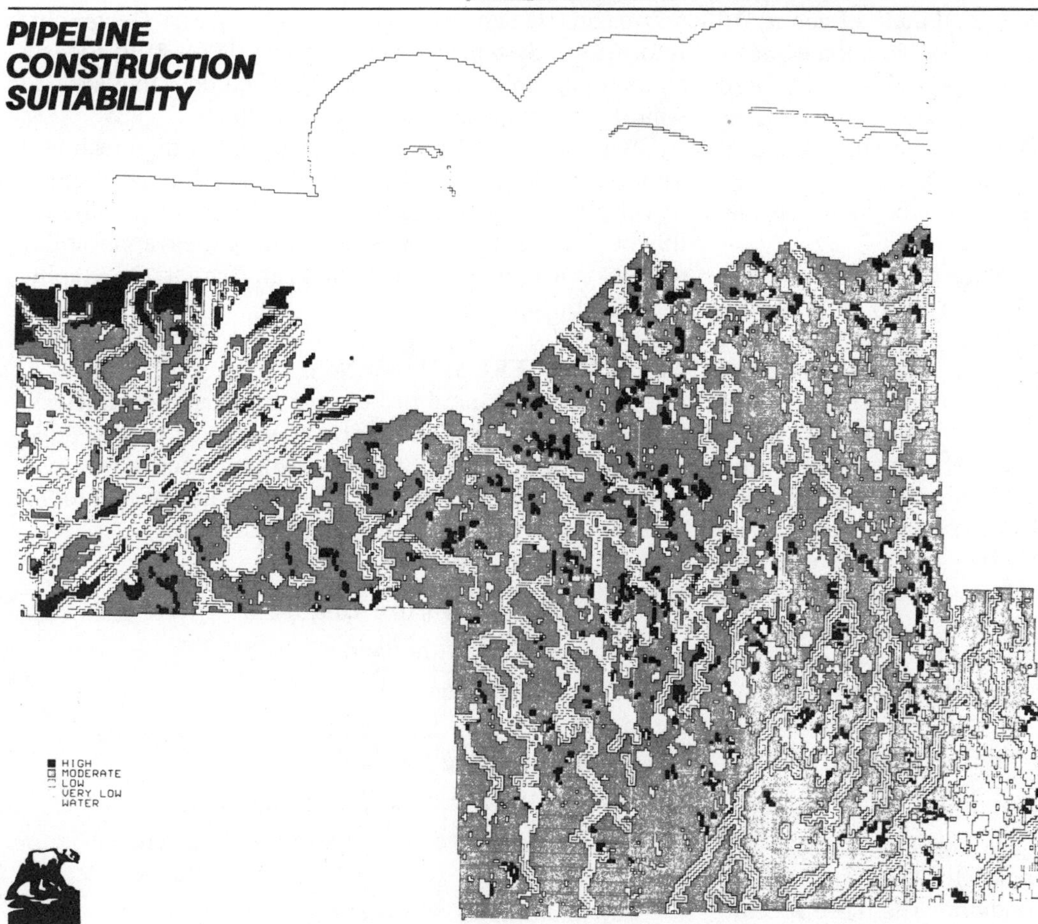

In practice, since data for most Geographic Information Systems are encoded by digitizer, they are commonly stored in polygon form, which can be converted automatically into grid cells of virtually any size. A system of variable cell sizes is easy to achieve when data are stored in this way. The Alaska system, for example, provides for maps at two scales, one for the whole of Alaska (1:250,000) and one for regional and plan unit scale analyses (1:63,360).

Rapidly improving technology in the field argues for increasing employment of computer resource information systems. The development of computer graphic software that makes microcomputers usable for spatial analysis has reduced the costs of such analysis enormously. Moreover, the ability to connect widely distributed microcomputers to large Geographic Information Systems stored in mainframe computers by telephone lines has great potential for expanding the utility of such systems, allowing, for example, a designer in his own office to "down-load" a selected segment of data for use at the plan unit or project scales. Thus, the tendency toward ever larger, more integrated information systems may ultimately benefit much lower-scale levels as well.

Use of LANDSAT Imagery

Since imagery provided by the LANDSAT program is likely to become more and more useful, it too should stimulate the use of computers. The first earth resources survey satellite was launched in 1972, and three others have been launched since. These satellites orbit the earth every 18 days in such a way as to record images of its entire surface, other than the north and south poles, from an altitude of about 570 miles. Each image covers an area that is 185 kilometers, or about 115 miles, square. The multispectral scanners used by LANDSATs 1, 2, and 3 record information in square units called *pixels,* each of which is 80 meters, or about 262 feet, square. The fourth satellite, LANDSAT D, launched in 1982 and orbiting 483 miles high, also carries the more recently developed imaging Thematic Mapper.

LANDSAT scanners produce imagery by measuring the solar energy reflected from each unit (pixel) of the earth's surface. Reflectance is measured along four bands of the electromagnetic spectrum: red, green, and two that are near infrared. This information, recorded in digital form, can be converted into pictorial patterns that look much like aerial photographs. The narrower bands and 30-meter resolution of the Thematic Mapper allow for finer distinctions than those provided by data from the Multispectral Scanner. The Mapper has two bands in the green range, for example, instead of just one, thus making it possible to distinguish between similar plant communities.

The imagery available from the EROS Data Center in Sioux Falls, South Dakota, takes three basic forms. First are the digital tapes, which can be processed by computer and combined with data from other sources, using grid or polygon mapping. For the tapes to be useful, however, the measurements of reflectance must be translated into meaningful information, such as vegetation types or land uses. Ground samples must be taken to determine what level of reflectance indicates what ground cover condition. The more specific the information needed from the tapes, the more extensive these ground-level surveys must be in order to establish the proper correlations. Ground areas with known characteristics are compared with the corresponding pixels to establish the correlations. As is often the case, precise information can become very expensive. Once correlations are established, however, images can be processed to show any number of land variables, among them vegetation and land use and water patterns. With a resolution of 80 meters, or in the case of the Thematic Mapper, 30 meters, such information is sufficiently precise for use at the plan unit and larger scales and can easily be integrated into an information system, a feat that the Minnesota Land Management Information System mentioned earlier has already accomplished successfully. Nevertheless, obtaining data interpretations accurate enough for practical use has been a serious problem. With their finer resolution, data from the Thematic Mapper will probably go far in helping solve it.

The second form in which LANDSAT imagery is available is the black-and-white photograph. Derived from digital reflectance data, these are available for each of the four spectral bands. Photointerpretation techniques can yield considerable information on ground coverage conditions, but because of their small size, the photographs are usually helpful only for work at the regional and larger scales.

The third form is the color-enhanced image, a photograph with color applied to certain areas to single them out and provide a more readable image. Serviceable in the same ways as black-and-white photographs, these images have the advantage of being easier to interpret and better suited for making design presentations. The Alaska Geographic Information System case study (pages 151–57) includes a typical color-enhanced map (reproduced here in black and white) describing ground cover conditions.

When LANDSAT imagery benefits from further technical improvements and users better understand its capabilities, it will probably be more widely employed. Especially will this be true of digital tapes by virtue of their applicability to personal computers as well as the larger Geographic Information Systems.

THE IMMATERIAL PRESENCE

Aside from its physical presence, a landscape is enmeshed in a network of invisible factors—economic, social, political, historical, and attitudinal—whose interactions will strongly influence its future. Particularly important among these in most design efforts are the following:

- Demands for particular land uses
- Market values of the land
- Political jurisdictions
- Existing laws, ordinances, zoning, and other legal factors affecting land use
- Historical and cultural events and influences
- Socio-economic composition of the resident population
- Public attitudes

Most of these are standard planning considerations that are quite adequately dealt with in other places. Two of them, however—historical influences and public attitudes—are both important enough and little explored enough to deserve some attention here.

In our preoccupation with natural processes, it is easy to forget that few of the world's landscapes are still wholly natural, that most of them have been shaped and reshaped by human activity. In such landscapes, the forms are mostly man-made; the spirits of those who shaped them are still there, sometimes pervasively, overwhelming there. In joining past and future, design must heed their lingering voices.

Patterns made by human beings are less easily explained and catalogued than those of nature. Culture enters the equation. People impose their mystical beliefs, social attitudes, technologies, esthetic perceptions, and economic ambitions on their native landscape and then further complicate matters by transporting them around the world, inflicting these same beliefs, attitudes, perceptions, technologies, and ambitions on landscapes very different from those of their origins. North American and African landscapes, for example, were reshaped in a European image. That image faded in the process of evolving native cultures, but its imprint is still very much there.

If we take the time to sort out historical trends, we can come to understand these landscapes in terms of cultural evolution. McHarg, for one, recommends the study of ethnographic history.[8]

Diverse Perceptions

The landscape that exists as a set of memories, images, conceptions, attitudes, and values in the minds of all those who see it, use it, know it, or even merely know about it—in the minds, that is, of what we have called the clientele—may be very different from the physical reality. All these immaterial existences have highly perceptible influences on the material existence, determining not only how it is used and viewed, but how decisions are made about it. A large part of the landscape design process lies in bringing the material and immaterial existences of a landscape into a viable congruence, that is, in realizing the images and values of the clientele insofar as this is compatible with the physical reality. D. W. Meinig has identified various ways in which people view a landscape, some of which match rather closely those that emerge from the kind of objective examination of the physical landscape that we have just discussed, but some of which do not.[9] These include seeing the landscape as:

Nature	Wealth
Habitat	Ideology
System	History
Artifact	Place
Problem	Esthetic

[8] See McHarg, 1981.
[9] See Meinig, 1976.

Any landscape reflects all of these, of course, although perhaps only one or a few of them in any one mind. Thus, bringing the material and immaterial landscapes into a viable congruence means dealing with a multiplicity of images, uses, and values. To do so, we need to know more or less exactly what they are. A description of the immaterial landscape, then, becomes an essential part of the information base.

There are various ways of gathering information about the immaterial landscape. If it is a well-traveled place, a long history of representations in words and pictures may already exist. To define the most admired scenes in the Yosemite Valley, Frissell, et al., assembled paintings, photographs, and documents from over the last century.[10] For most landscapes, contemporary images and values are more relevant. The social sciences provide numerous techniques for ferreting out images and measuring attitudes and values.

Largely as a result of the widespread insistence on the application of democratic principles to land-use decisions and the influence of such attitudes on federal legislation, the most extensive investigations of the immaterial landscape have usually been undertaken by the U.S. government. We have already mentioned the client group for the California Desert Conservation Area Plan—the *entire* population of the United States in addition to the more closely affected subgroups, to wit, special interest organizations, area residents, and peripheral residents. In assembling its picture of the immaterial landscape of the California Desert, the Bureau of Land Management used virtually every device in the social scientists' tool kit to solicit the views of these disparte clienteles, among them:

Fifteen public meetings held over two years to discuss various possibilities

National and local radio and television spots describing the planning program and requesting comments

Information mailings to a list of over 8000 people

Meetings of planning staff members with special interest groups, each of which prepared position papers to represent their views

Polls conducted at local, state, and national levels

Personal interviews with over 600 residents of the planning area and with officials of all cities and countries involved

Mailing the alternatives and Environmental Impact Report to a broad range of people for comments

Such an elaborate program is unusual. For most planning projects, no more than one or two of these techniques are needed to provide an adequate assessment of values. But as the planners of the California project had anticipated, their efforts uncovered a great many different perceptions of the value of the desert environment. Their summary of these included the following:[11]

[10] See Frissell, et al., 1980.
[11] See Bureau of Land Management, 1980.

The California desert as a natural ENVIRONMENT

The California desert as a RECORD of man's existence

The California desert as a HOME and traditional way of life

The California desert as a PROVIDER of food, energy, and materials

The California desert as an OPEN SPACE

The California desert as a PLAYGROUND and an escape from civilization

The California desert as a PUBLIC RESOURCE.

Most respondents valued the California desert for more than one of these qualities, and some expressed all of them, but the emphases placed on their relative importance varied widely enough to make the task of summarizing and incorporating them in the process extremely difficult.

In earlier chapters, we discussed several more selective means of gaining the views of the clientele. In the San Elijo case, questionnaires were used, and for Bolsa Chica several participating groups were taken to represent the larger public. Whatever the form of information collection, the means should always fit a project's scale and purpose.

MODELS

No matter how much economy and restraint are exercised in gathering information, we will inevitably end up with masses of data too unwieldy to manage and difficult to make real sense of. We obviously cannot remember all of it, and even major items are likely to get lost. We therefore need a means not just of organizing the data but of putting it together in a way that will lead us directly to the shaping of a plan. This is where models enter the picture.

We have already seen that no matter how arcane and complex some models may appear, no model is basically anything more than an abstract representation of reality. Its purposes are to provide comprehension, prediction, and control. A model usually incorporates those features and relationships of a system that are relevant to the issues at hand and attempts to exclude all others. It gives us a subject to work with by reducing enormous quantities of information to manageable proportions.

Although for many, the term model almost invariably means a model describing a process in terms of the variables and constants of mathematical formulae, for our purposes, such models represent just one among many useful types. Sketches, diagrams, charts, and maps can also describe relationships essential to the reality of a landscape and thereby serve the same function.

Taken in this sense, model-making becomes the pivotal step in the design process. As the model takes shape, seemingly unrelated bits of information begin to fit together, and a different reality begins to emerge, one revealing form and process. This is the springboard to design.

Ira Lowry conveniently grouped models in three classes in "ascending order of difficulty": descriptive models, predictive models, and planning models.[12] Though Lowry was referring to mathematical models for computer processing, the terms apply just as well to other types. *Descriptive models* simply draw on the information base to replicate the existing features of an existing environment or process. An example is provided by the two flow diagrams on page 165, which exhibit the relevant features of the flow of water into and out of Lake Elsinore (under the two extreme conditions of a wet year and a dry year), including the quantities of water involved at critical points. Probably more important, they give us an overall concept of how the whole system works. With rates and quantities known, it becomes an easy matter to formulate a series of mathematical models that will describe the flow of water quantitatively. Eventually, as design proceeds, such models will have to be consulted to determine specific dimensions. At this early stage, however, the pictorial flow model is far more useful because it relates the whole system to the landscape.

The sketches on page 167 represent a rather loose version of the second type: the *predictive model.* They tell us how—in the analyst's best judgment—the water flow system will probably behave if the overflow channel is redesigned in a certain way. Again, the flows for a dry year and a wet year are illustrated. For predictive models, as Lowry points out, "an understanding of the relationship between form and process becomes crucial."[13] The designer is interested in knowing and demonstrating the set of conditions that will come about if some specified change is made in the environment. To do so, he has to postulate a causal sequence, drawing on both specific information about the subject environment and more general information from the scientific literature. This task is usually an interdisciplinary one that requires the knowledge and skills of any number of specialists.

Another type of predictive model that is particularly useful for landscape design, especially at larger scales, is the type that we have been calling, in generic terms, the suitability model. Suitability models deal with locational patterns, attempting to estimate—usually in general but sometimes in very specific terms—the ability of a landscape to support human use. Examples are shown with most of the case studies. (Their types and derivations will be explored in detail in Chaps. 14 and 15.)

Lowry's third type of model—the *planning model*—amounts to a mechanization of what we have called *the* rational planning process. After linear programming techniques have developed and evaluated alternatives, the highest scoring alternative is chosen. Though rarely applied in practice, such a level of mechanization is at least theoretically possible for planning, but it can never deal with the full range of intangibles that come into play in the landscape. Modeling, so far from being able to transcend the predictive level, cannot even handle many aspects of the intrinsically tangible in a precisely analytical way. We must therefore draw a firm line between the analytically derived models that bring the Stage of Precision to its peak and the exploration of possibilities that marks the succeeding Stage of Generalization.

[12] See Lowry, 1965.

[13] Ibid., p. 159.

CHAPTER 9

The Stage of Generalization

Possibilities, Predictions, Plan

The confining banks of the river of design begin to spread out into innumerable channels in the rich mud of the delta. Here, as alternatives emerge, the design process finds its most fertile ground for insights, ideas, even visions. Here, the "return to romanticism," in Whitehead's words, becomes possible, but now with the advantage of a fund of solid information and reliable working models.

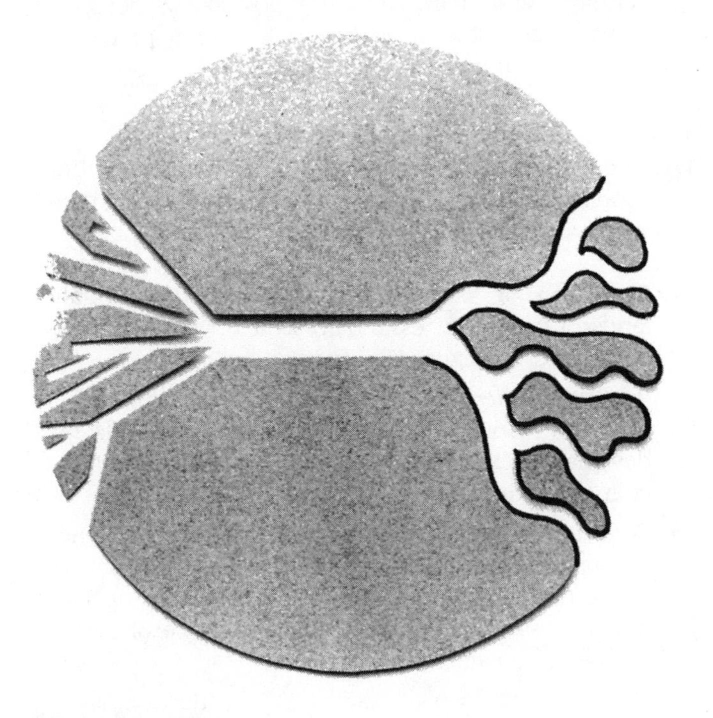

POSSIBILITIES

The entire design process is pregnant with possibilities. Some of them emerge in the earlier stages as fragments of plans, or even whole plans, come to be suggested by the issues, by this or that piece of information, or, especially, by the promptings of suitability models. We must keep every possibility in mind as we go along, allowing no idea that might contribute to an eventual resolution to be disregarded. Nevertheless, we do not come to the time for seriously considering all the possibilities until the Stage of Generalization, when modeling, in particular, has provided a newly comprehensible, multi-dimensional view of the landscape. According to Jerome Bruner, "... intuitive thinking rests on familiarity with the domain of knowledge involved and with its structure, which makes it possible for the thinker to leap about, skipping steps and employing shortcuts. . . ."[1]

[1] See Bruner, 1965.

Putting intuition to work, far from providing a means of avoiding the need to acquire knowledge, requires a particularly solid basis of it if it is to have any value.

Nonetheless, the irrational is not to be ruled out. This is the time for brain storming, for pushing aside assumptions, for wandering over the site yet once again, for sifting through photographs, for juggling variations and permutations, for turning and twisting every idea to see it in a new perspective, for endless hours of discussion, for staring at walls. Intuition works at its own pace. Insights appear and connections fall into place only when they are good and ready.

Natural processes of design and redesign follow a strategy of ongoing trial and error, posing myriad possibilities seemingly at random, just as millions of seeds try thousands of environments before a few find exactly the right conditions to sprout and grow. Designers, however, have the option of working more efficiently (with less cost) by substituting imagined possibilities for real events and critical evaluation for real trials.

If intuitions emerge as concepts relevant to parts rather than to the whole, we are faced with the task of assembly. One way to deal with it is to develop a set of ideas germane to each part.

Case Study:
The Lake Elsinore Floodway

In developing the plan for the Lake Elsinore Floodway, the landscape was broken down into specific problem areas, a major one being channel design. Within this problem area, there were three specific, closely related subareas: water flow pattern, channel section, and adjacent land use. Several possible forms for each of these are illustrated.

We might deal with the overflow of water, for example, in any of three different ways. The first is simply a channel that carries the required 2000 cubic feet per second away when the lake rises above a specified level. This solution contributes little besides protection to the town. Second, we might dredge a forebay to form a mouth for the channel, thus bringing part of the lake to the edge of the town and providing a measure of waterfront amenity. The third possibility provides for a deeper kinship between town and lake by creating a series of ponds along the floodway. Whenever the lake is low, these ponds can be kept filled by feeding treated waste water or ground water into them from upstream (as shown), waters that will gradually flow down to the lake. When the lake rises, it will flow into the ponds from the other direction, and when it rises to flood level, the ponds will carry flood waters away at the required rate.

The channel sections show eight quite different design possibilities, each offering a distinct combination of protection, amenity, and recreational space. The land-use sections, in turn, offer a number of ways for developing recreational and commercial activities and wildlife habitats in relation to the channel.

By looking at each of the problem and subproblem areas from differing points of view, the designer developed a kit of alternatives that could be combined in various ways to shape the final plan. The plan that grew out of this process is shown on page 169.

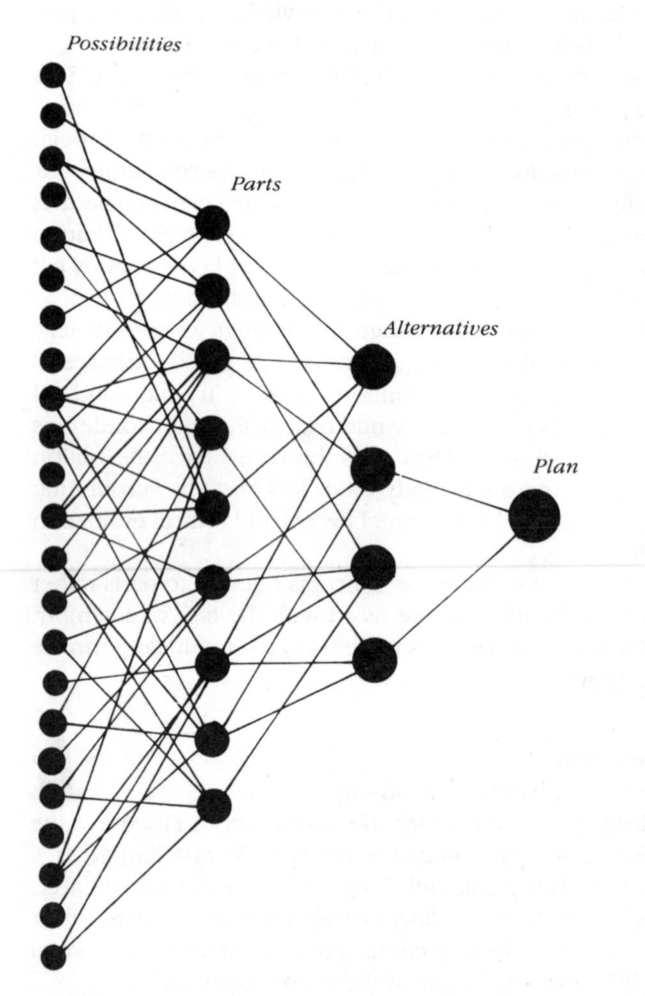

Alternatives

How many alternatives? Enough to include all promising possibilities, but not too many to be manageable. At the construction and site scales, we may not have to consider alternatives formally at all. At larger scales, a reasonable number seems to emerge for each project, usually between three and ten. Sometimes, for very complex projects, the number may be considerably greater, even thirty or forty. In such cases, the choices are usually made in stages. Thirty alternatives might be narrowed to ten, for example, and then to three and finally to one, with considerable revising at each stage.

Besides their mere number, a great many theoretical questions attend the shaping of alternatives. These questions are important because wrong choices can invalidate the whole process. Whenever controversy besets a plan, we often hear that all the possibilities were not properly considered or that decisions were rigged to favor or disfavor certain alternatives, and often these charges are all too true.

One important question concerns the factors that distinguish alternatives from one another. Probably the most common are those involving differences in degree. Federal projects, for example, commonly offer three or four alternatives: one featuring minimal or no development, one featuring maximum resource exploitation or development, and one or two that fall in between. In situations where the principal issue is preservation or development, this is usually a workable way of devising alternatives, but it presents serious difficulties since it tends to create a bias in favor of the in-between or balanced alternative. In fact, we find that any design process that sets up choices in this way almost automatically tends to rule out extremes, especially when the choices are to be made by groups or committees: "Moderation" and "balance" are almost always perceived as eminently desirable or at least safe, and this perception can overwhelm more rational comparisons.

THE CALIFORNIA DESERT PLAN

The alternatives presented for the California Desert Conservation Area are a case in point. There were four of them, described as follows:[2]

1. *No action,* or "continuing management of the public lands of the desert essentially as it is." (A no-action alternative is required by federal law.)
2. *Protection alternative,* which "favors protection of land and resources, through wilderness and related designations to preserve the desert's plant, wildlife, cultural, and historic resources."
3. *Balanced alternative,* which "considers the demands and uses that people want, along with management of the desert resources under the principles of multiple-use, sustained yield, and maintenance of environmental quality."
4. *Use alternative,* which "favors use, production, and development, including recreational access and mineral development."

The clear choice, of course, was the balanced alternative. Other difficulties present themselves, however, in reducing a vast land area with myriad possible uses to only four alternatives. These concern the grouping of uses. Facilities for energy generation, for example, proved to be a major issue. All such facilities, including wind, solar, and coal- and gas-fired plants, were included in the "use alternative" because obviously all represent intensive uses. Nevertheless, they are quite different in their locational requirements and their effects on the environment. Strong arguments could be made for locating wind and solar facilities in the desert, for example, but not coal- or gas-fired plants. Grouping them all together was an arbitrary decision based on unexamined assumptions. Moreover, any number of promising possibilities

[2] See Bureau of Land Management, 1980.

CASE STUDY XIII

Lake Elsinore Floodway

A The town of Lake Elsinore surrounds a large recreation lake in Southern California. Its downtown core, which is economically rather stagnant, stands adjacent to an old, undeveloped floodway. Because of recent severe flooding, the U.S. Army Corps of Engineers is committed to reconstruction of the floodway to carry away flood waters. The proposals in this project show how the floodway can be shaped as an important visual and recreational amenity to create a unique urban environment and serve as a catalyst for the revitalization of the downtown area.

First, the regional water flow system is described for wet and dry years, and possible forms for the floodway are presented in terms of its role in this larger system. The chosen design shows how the floodway can be maintained as a series of recreational ponds, kept full during dry periods by pumping water from underground and letting it overflow into the lake to help stabilize the lake level. When the lake overflows during storms, the flow is reversed to carry water away. This is made possible by a careful design of the bottom, as shown in the profiles.

Prepared for the City of Lake Elsinore by the 606 Studio, Department of Landscape Architecture, California State Polytechnic University, Pomona. Ikuo Sano, with advisors John T. Lyle, Francis Dean, Jeffrey K. Olson, and Arthur Jokela.

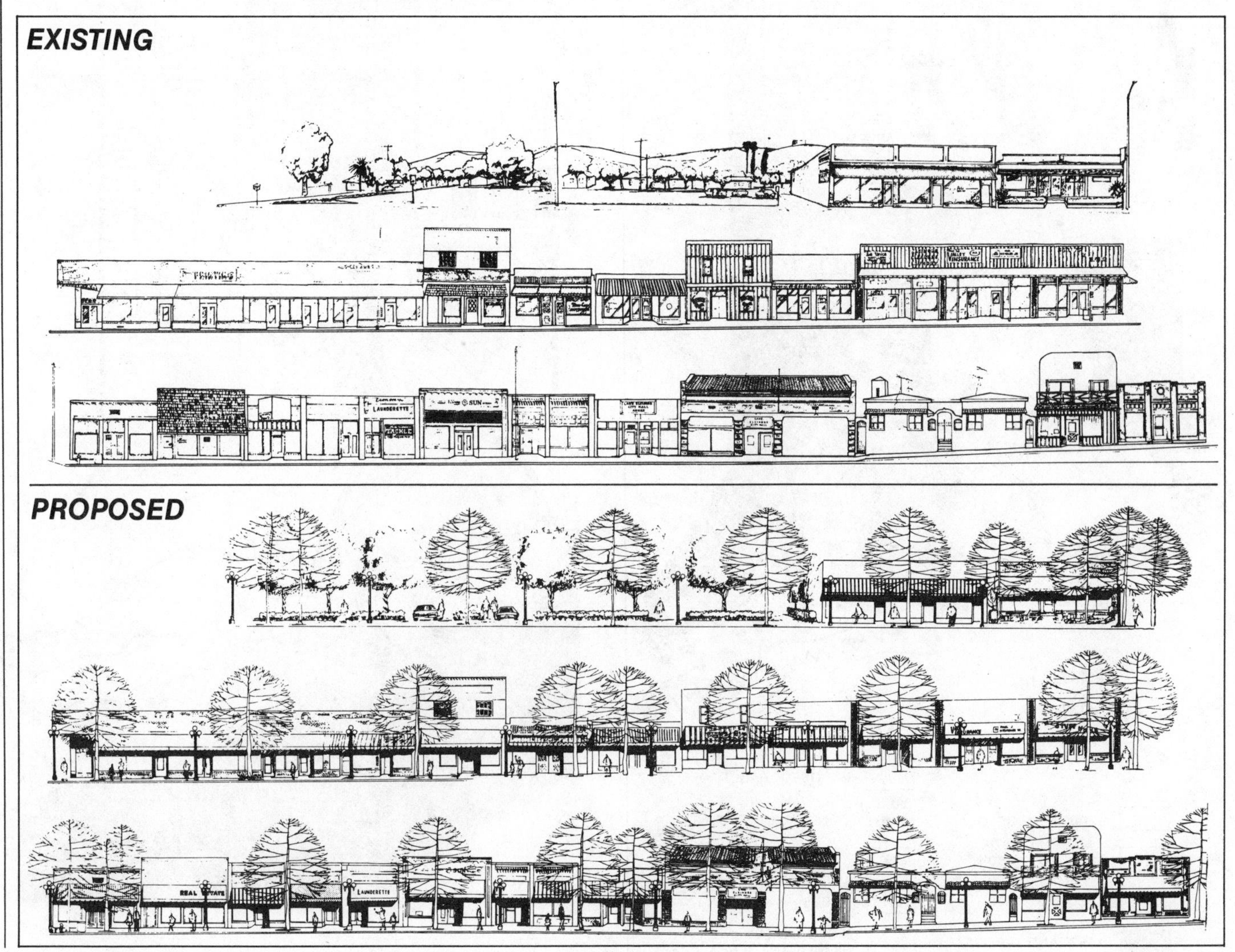

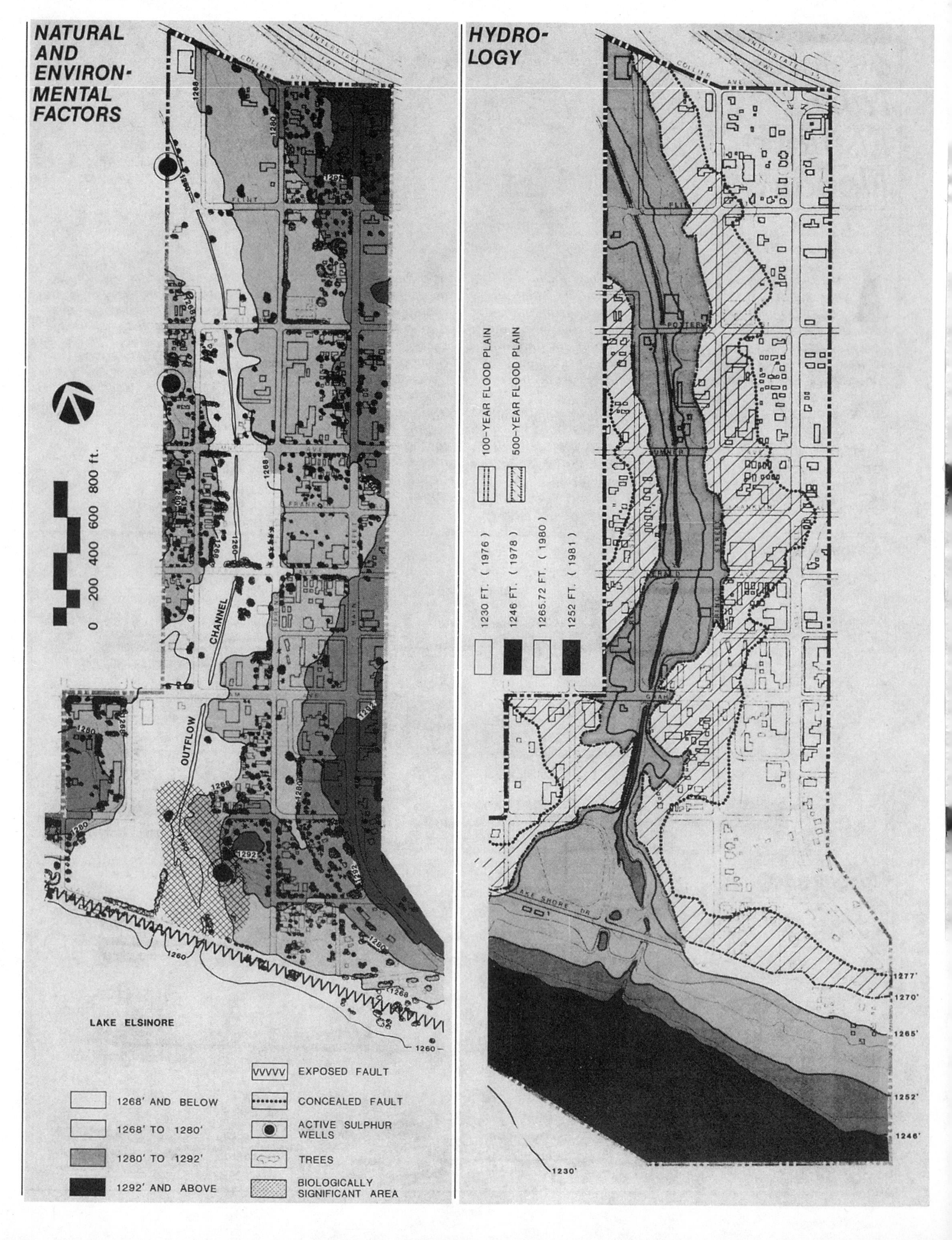
NATURAL AND ENVIRONMENTAL FACTORS
0 200 400 600 800 ft.
OUTFLOW
CHANNEL
LAKE ELSINORE
VVVVV EXPOSED FAULT
1268' AND BELOW
CONCEALED FAULT
1268' TO 1280'
ACTIVE SULPHUR WELLS
1280' TO 1292'
TREES
1292' AND ABOVE
BIOLOGICALLY SIGNIFICANT AREA
HYDROLOGY
100-YEAR FLOOD PLAIN
500-YEAR FLOOD PLAIN
1230 FT. (1976)
1246 FT. (1978)
1265.72 FT. (1980)
1252 FT. (1981)
1277'
1270'
1265'
1252'
1246'
1230'

DRY YEAR

11.8 inches/yr.
San Jacinto Basin
Railroad Canyon Reservoir
Evaporation, 4.5 ft. 12,000 ac. ft.
0 to 10,000 ac. ft.
Temescal Wash
Elsinore Basin runoff
Wasserman Canyon
for lake level stabilization
Lake Elsinore 2750 acres 7200 ac. ft.
Downtown area
Elev. equals 1236 ft.
Variable
Spring
Ground water
1,000,000 ac. ft.
70%
30%
Recharge
1670 ac. ft.
Runoff
City water supply
0.8 mgd
Irrigation
Evaporation
Wastewater treatment plant
3260 ac. ft
From Colorado River

WET YEAR

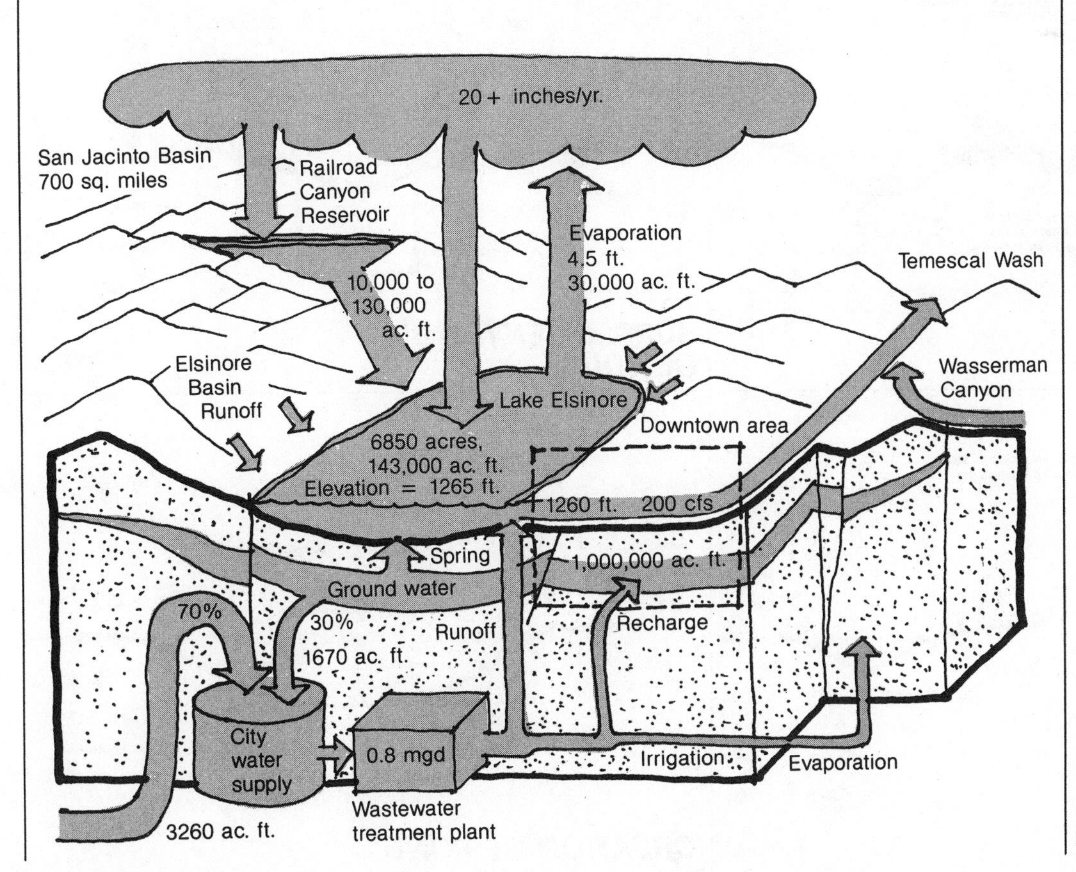

Lake levels vary enormously between wet and dry years, as shown in the hydrology map. Flooding was extensive in 1980, when the lake level rose to 1265.72 feet during heavy storms. The water flow diagrams show volumes of water during typical wet and dry years. The floodway can help stabilize these extreme variations by carrying out overflows and regulating inflows from ground water supplies. Some evaporation loss will be involved, but this will be minimal. It is important that the floodway be redesigned with regard to its role in the regional water flow system, as shown in the two models of water flow.

FLOODWAY DESIGN POSSIBILITIES

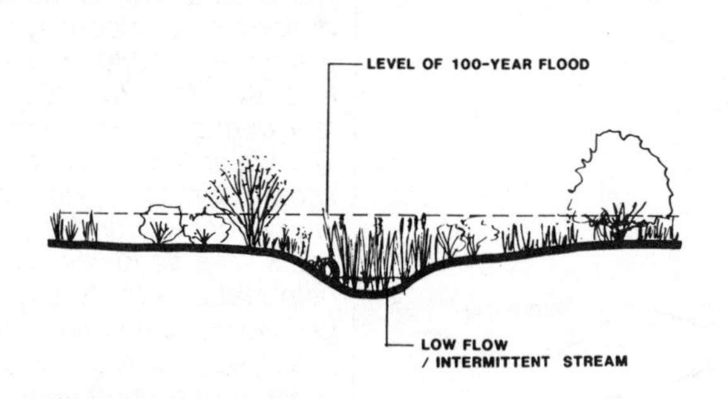

EXISTING FLOOD PLAIN

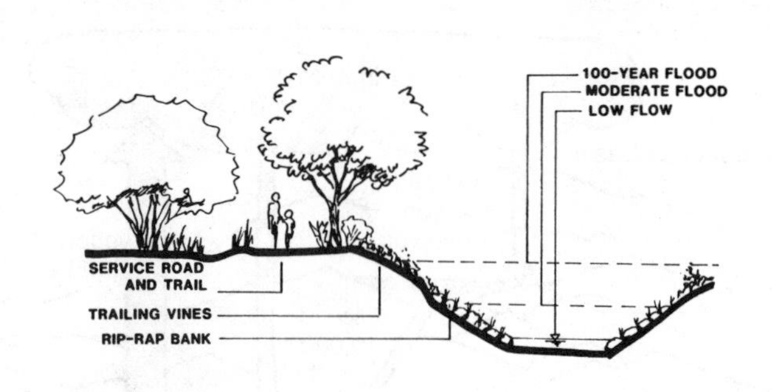

ENHANCEMENT OF RIP-RAP

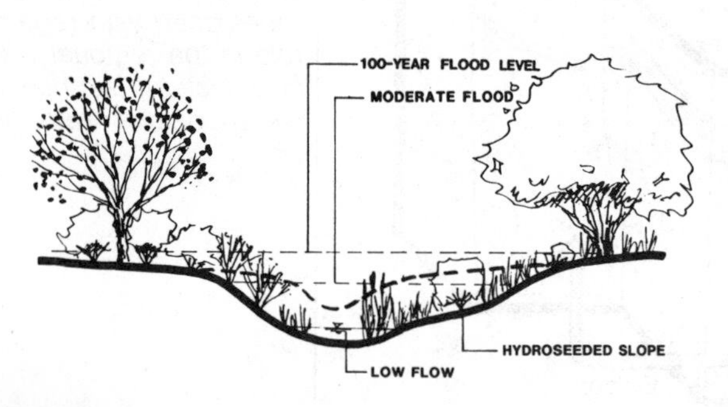

CHANNEL CLEARING AND ENLARGING

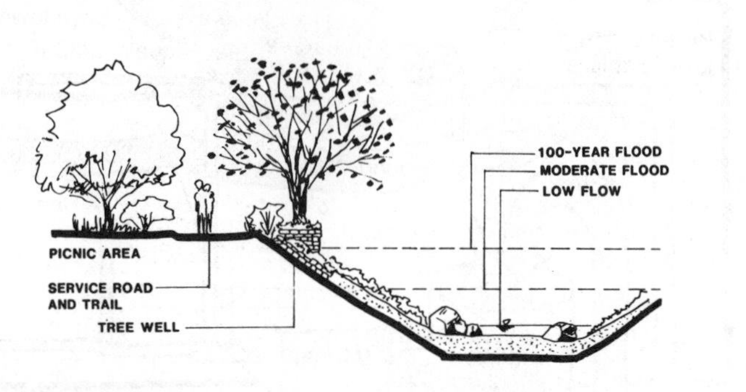

ENHANCEMENT OF CONCRETE CHANNEL

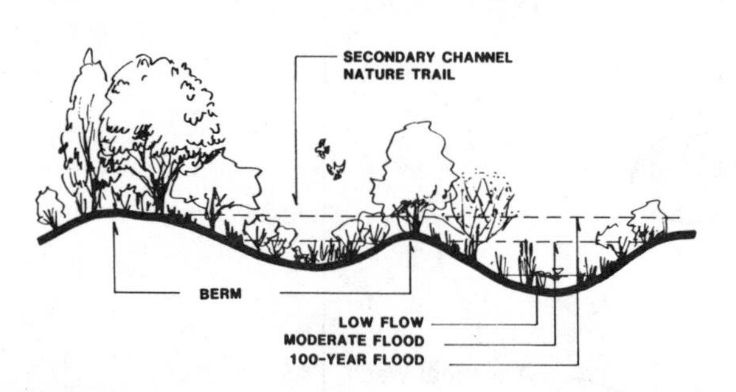

CREATION OF SECONDARY CHANNEL

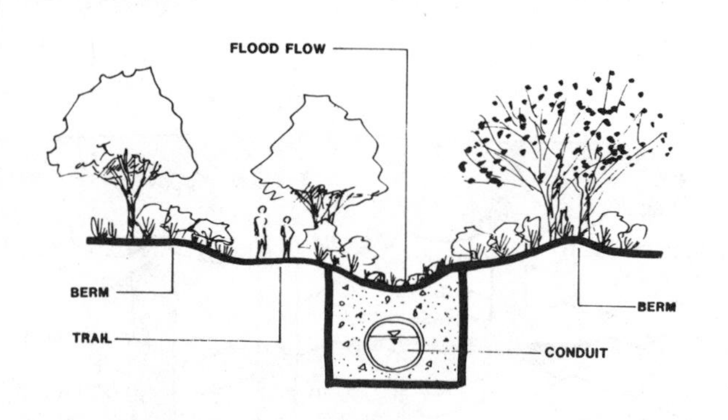

SUPPLEMENTED BY CONDUIT

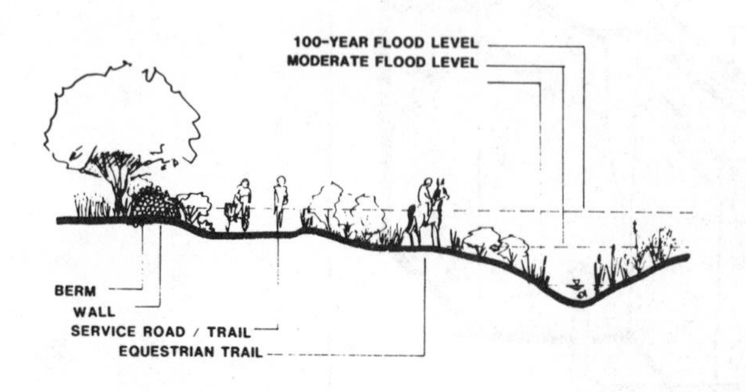

CONSTRUCTION OF STONE FLOODWALL

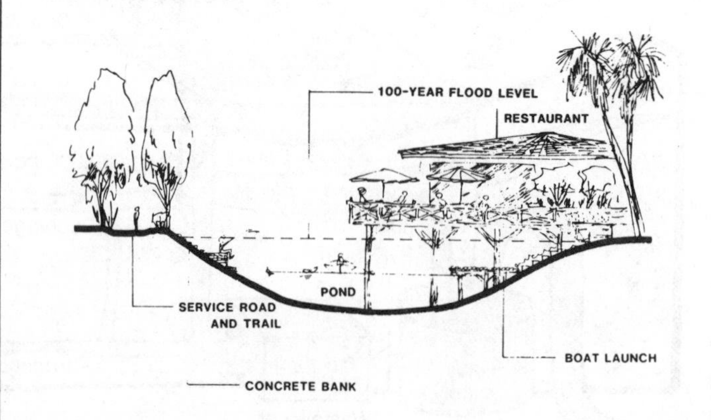

CREATION OF POND

WATER FLOW MANAGEMENT

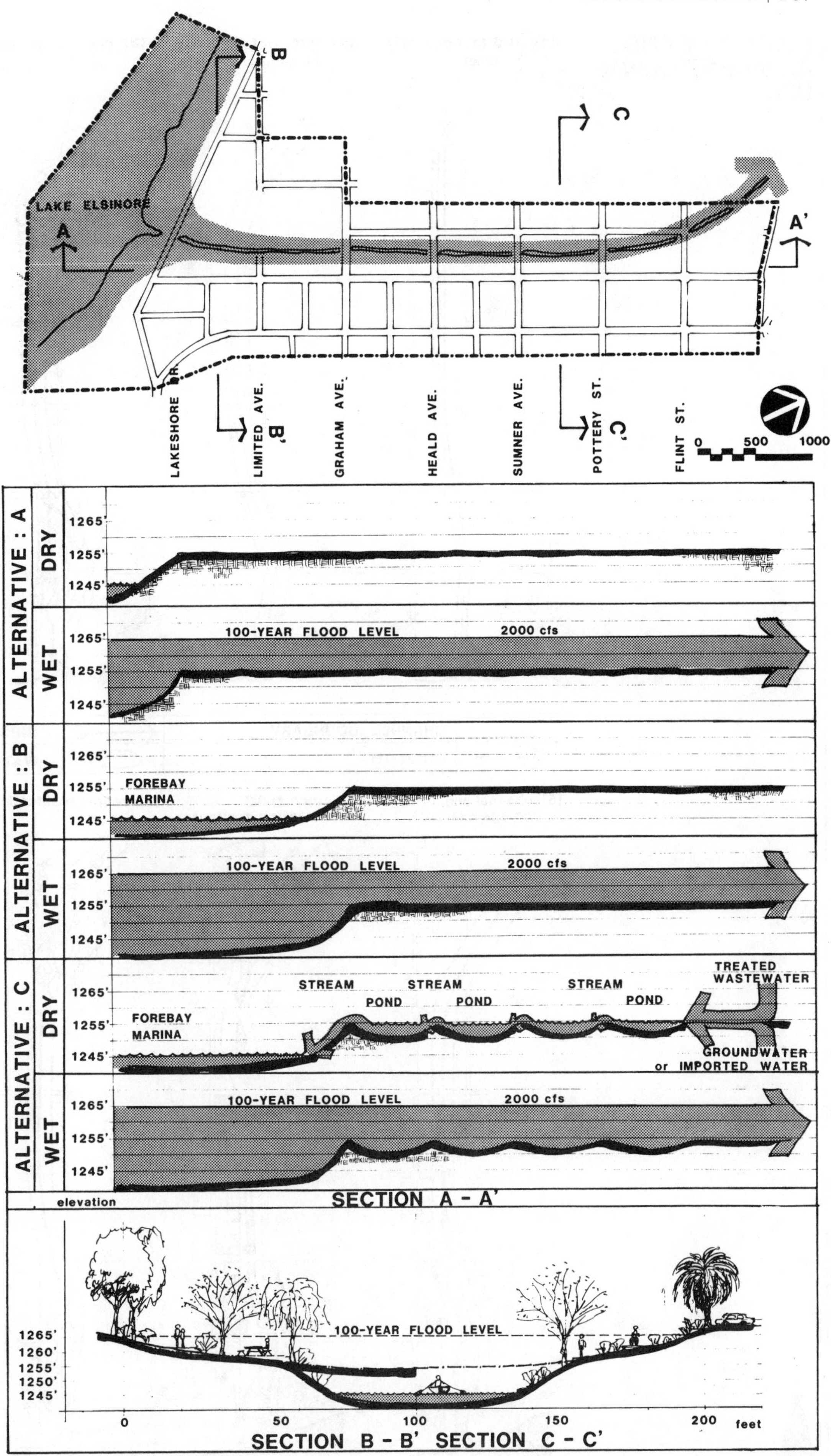

FLOODWAY AND ADJACENT LAND USE

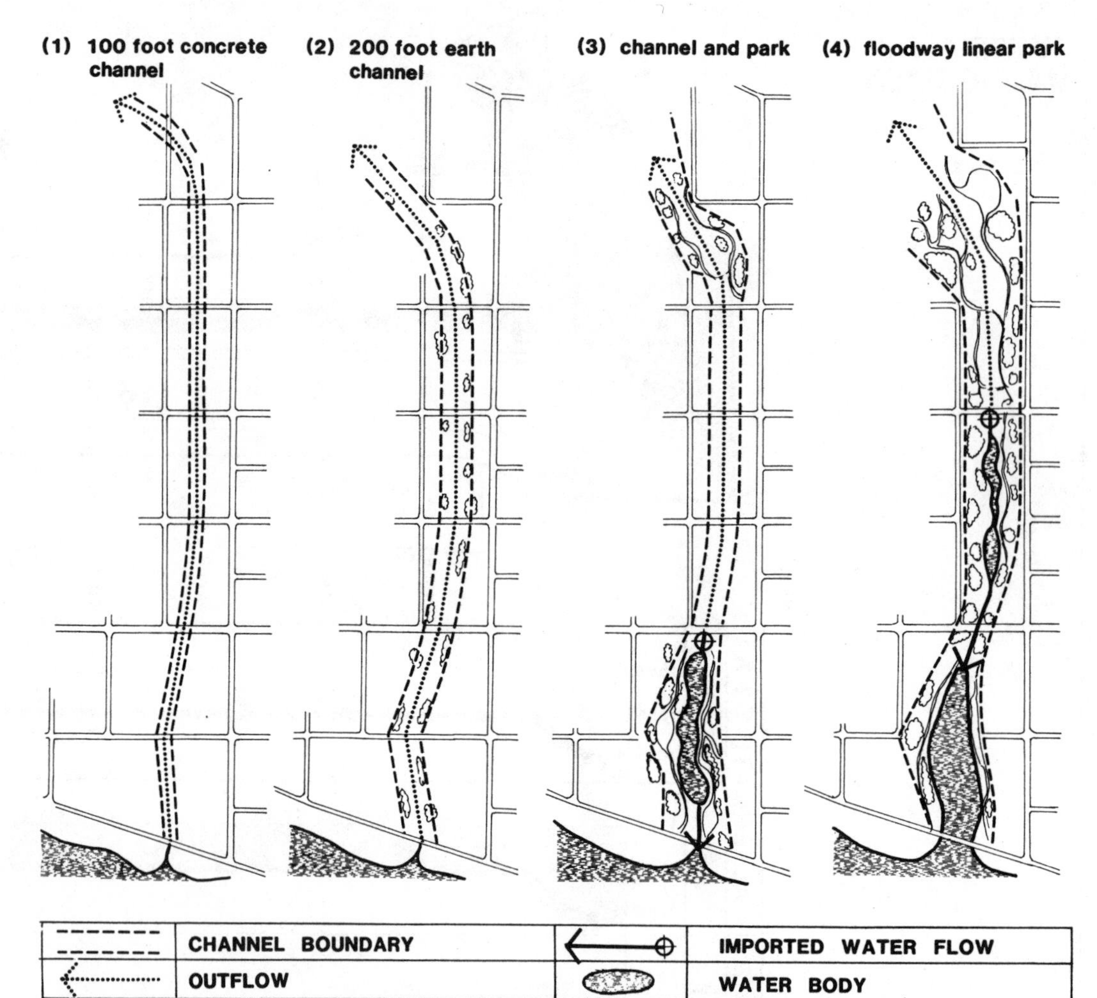

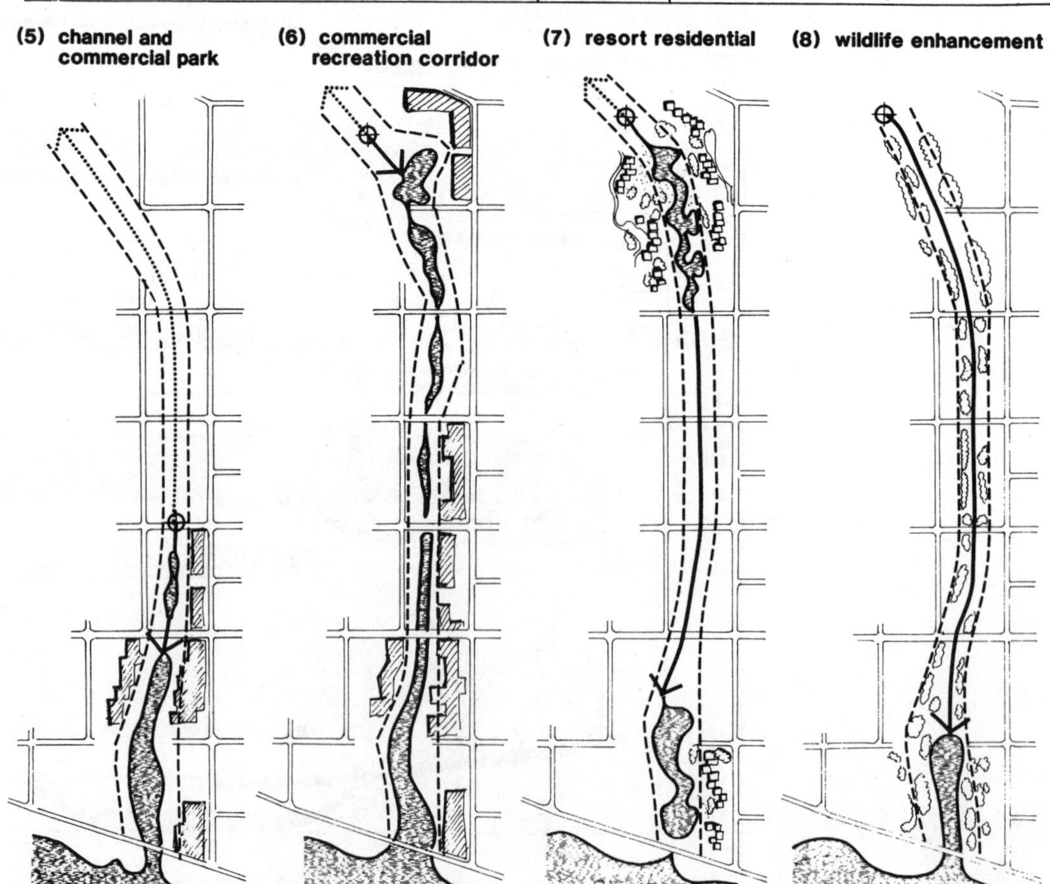

FLOODWAY PLAN

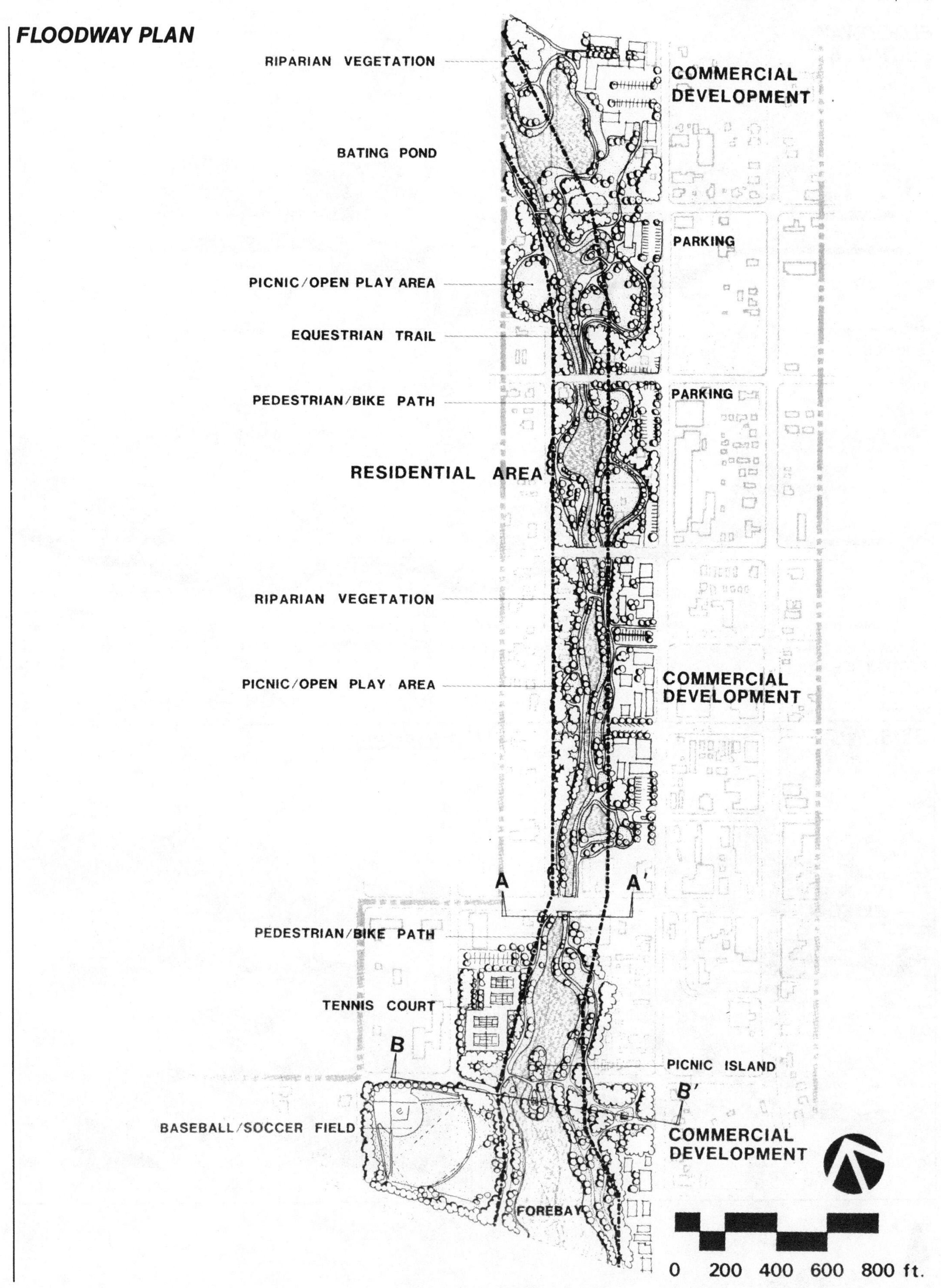

FLOODWAY SECTIONS

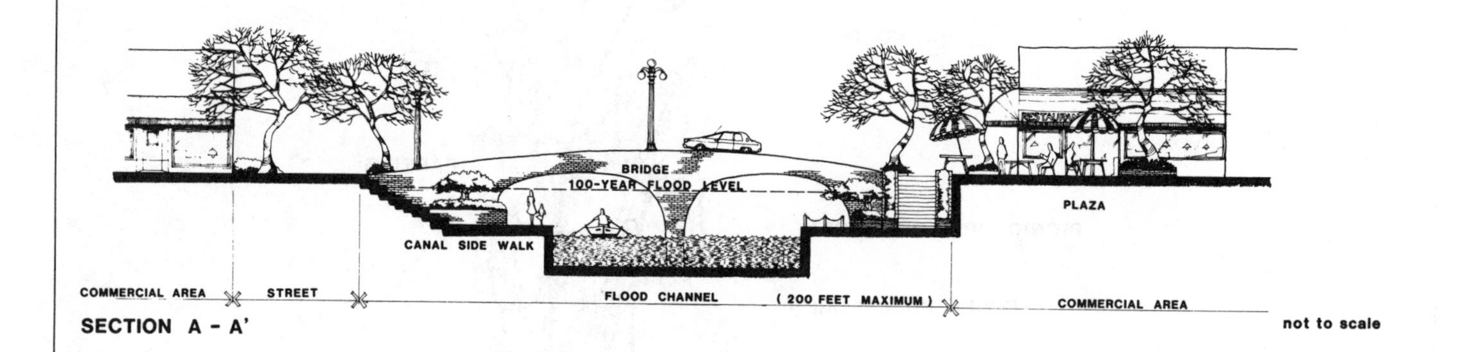

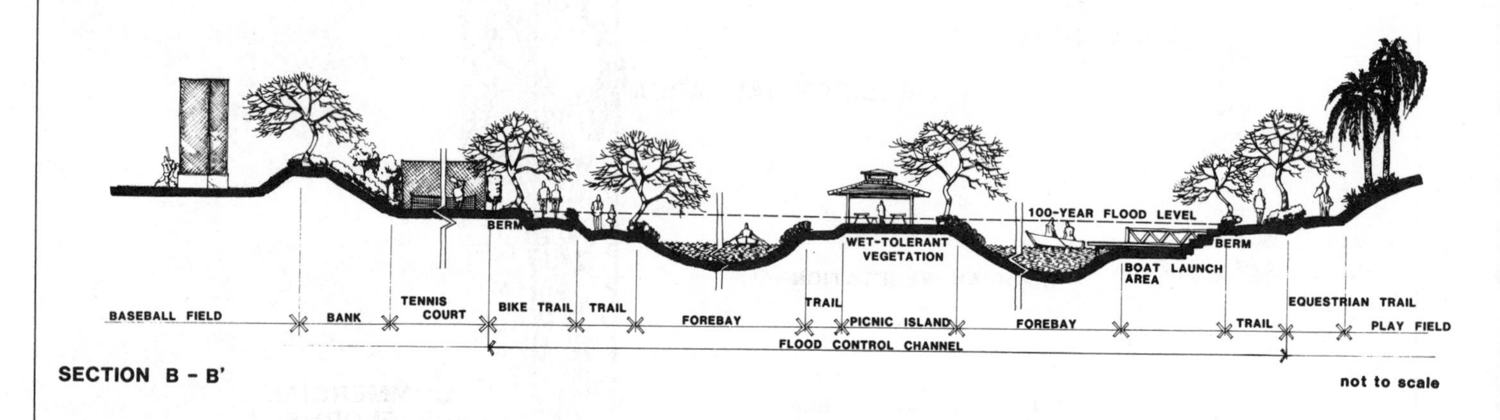

EXISTING

PROPOSED

were rejected out of hand because they did not fit preconceptions. For example, settlements in the desert were not seriously considered, even though the permanent presence of small numbers of people would certainly discourage many of the depredations that are so hard to prevent in a landscape where there is no one to observe what is going on. Settlements, even very small ones, were grouped under the heading of intensive use. That there were a great many other examples of arbitrary groupings in all of the alternatives was probably inevitable, given the early assumption that there should be a limited number of plans to choose from.

What this tells us is that the issues and the vast landscape were both far too complex, and the possibilities far too numerous, for the method adopted to derive alternatives. If it had not been assumed that all issues could be subsumed in a range from most to least use, an analysis of specific compatibilities and suitabilities would probably have led to entirely different patterns. Combinations of some very intensive uses with some very protective ones might well have emerged. More alternatives with more possible combinations would probably have presented themselves, and these could have been gradually narrowed to reach a solution. The important point here is that the shaping of alternatives should not be overly simplified, and it should not rest too heavily on unquestioned assumptions. In other words, more steps are needed between possibilities and alternative plans.

The Yosemite National Park process allowed for broader consideration of a diverse range of possibilities by bringing the public into the process more completely. The Yosemite workbook, mailed to 59,000 people and completed by 20,000, included not only four general alternatives but listed the specific possible actions that comprised each and asked participants to indicate their approval or disapproval of these. Space was also provided for participants to suggest additional possibilities. Thus, broadly based participatory design and the open-ended consideration of possibilities were carried to what are probably their outer limits.

Another basis for differentiating alternatives, one related to the degree of use intensity, is the optimization of different uses. This approach is most useful whenever competition among uses is a major issue. In the Whitewater Wash case study, for example, the key issue was competition for space by three uses: recreation, solar and wind energy generation, and wildlife preservation. Each of the alternatives presented optimized one of these three uses within acceptable limits of environmental change, and the three extremes provided the basis for exploring the means. Any number of combinations were possible, and any number of factors concerning needs were permitted to enter the decision process at this point. The plan on page 99 is the one that was eventually recommended.

Alternatives can also be thematic; that is, they can differ in basic kind or character. This approach usually produces alternatives that are fundamentally unlike in land use, in visual form, in tenure pattern, or in all three.

The Bolsa Chica Case Study

The Bolsa Chica alternatives, for example, are quite different in kind. The site is a coastal wetland in the midst of intensive coastal development in Orange County, California. It now features a sprawling tangle of oil wells, but within a few years drilling will cease and a new pattern of uses will take over. The wetlands, which are much like those of San Elijo Lagoon discussed at length earlier, support a prodigious and diverse bird population that includes five endangered species. Environmental groups and the State Department of Fish and Game are anxious to protect these populations while the landowner, an oil company, is understandably interested in realizing a profit on the development of its land. A third interest group, boating enthusiasts in the area, wants to develop a marina on the site.

The issues and the politics are extremely complex. The landscape architectural firm of EDAW Inc. carried out a study of the situation for several citizens' groups and devised a set of over thirty alternatives. The Cal Poly 606 Studio was then commissioned by the City of Huntington Beach, which adjoins the marsh, to analyze the situation further and reduce the alternatives to the smallest number that would do justice to all the varied interests. The purpose was not to devise a plan, but to summarize and clarify a most complex situation and to impart form to a muddle of controversy. The decision-making process in this case was not an economic one but rather an integrative approach, to be taken one small step at a time. The alternatives were intended to focus this process.

The first alternative envisions a regional park that would include a wetland preserve. The second makes room for a residential development on the edge of the lagoon, again combined with a wetland preserve. The third adds a small marina between the residential area and the wetlands. The fourth (called 3B here because it is really a variation on the third) relocates the coastal highway inland to allow larger boats to use the marina. Finally, the last reduces the wetland to minimal size and surrounds it with residential development, crowding a multiplicity of uses, including a marina, into the area. Somewhere among these alternatives, some combination or permutation, or combination of permutations, will evolve, but it is much too early to say what the final use pattern will be.

Also based on differences in kind, but stemming from different kinds of assumptions, is the scenario approach to developing alternatives. Scenarios are based on assumptions of possible futures. We imagine a set of conditions that might exist in the future and then project a development pattern to respond to that set of conditions. Scenarios are used not so much for developing plans as for studying options for the future.

The North Claremont Case Study

An example of the scenario approach is the North Claremont case study, presented on pages 181–83. Claremont is a small city on the edge of the Southern California urban region at the base of the San Gabriel Mountains. Until the 1950s, as the historical development maps show, it was a small town surrounded by citrus groves, for which the soil and climate are ideally suited. Since then, suburban growth has replaced most of the groves. Some of the more concerned citizens of Claremont, to try to preserve what is left of the rural atmosphere, succeeded in having much of the northern portion of the city rezoned for one-acre residential lots. Others were not sure this was a good solution. What kind of pattern would the one-acre zoning produce? What other options were there? What might the future bring?

Anne Nelson, who carried out the study, first considered a scenario that would retain the orange groves. This proved economically infeasible. The groves could not produce enough to justify their presence on this valuable land. Next she considered what was likely to occur under the present one-acre zoning in the normal course of events. This scenario, Plan A, which is shown on page 183, came as a shock to local residents who had somehow been tranquilized by illusions of rolling, pastoral greens in their rural zone.

Next, Nelson considered the other extreme, a sudden turn of events that would require optimal use of the resource potential of the land. Plan B is based on a scenario of widespread food shortage. Under this scenario, development coincides with the shortage, and available productive land is converted to maximum food production. Intensive small-scale agriculture to serve the surrounding suburbs becomes the primary use, with only a few small areas set aside for housing so that farmers can live near their fields.

Plans C and D combine residential and agricultural uses in a somewhat different way. Plan C is based on an assumption of plentiful food, moderate increases in energy prices, and a paramount need for visual open space. Agriculture is preserved in visible areas primarily for its esthetic, historical, and educational values. Most agricultural activity takes place in scenic corridors of citrus trees and "hobby farms." Most residences are on single-family lots that share centrally located strips of green landscape featuring allotment gardens.

Plan D follows a scenario of rising, though not drastically rising, food and energy costs in conjunction with stable land prices. From this assumption comes a productive and complementary combination of dwellings and urban agriculture. Most houses are either in tight clusters surrounded by intensively productive groves and farms or in individual plots devoted to polycultural farming. The tracts vary in size from 5 to 50 acres and are to be farmed by groups of varying sizes. Overall density in Plans C and D is about the same as in Plan A.

The open area shown to the east of the developed portions in all four plans is a flood plain reserved for ground water recharge. In Plan A, it would remain in its present state, fenced off and inaccessible. Plans C and D would allow for some recreational use, with hiking and riding trails located so as not to interfere with the recharge function, whereas Plan B would put it to intensive agricultural use.

It is not very likely, of course, that any of these scenarios, other than A, will be adopted. The purpose here is simply to examine the past in order to cast some light on the future. The scenario approach, like any exercise in futurism, depends on a knowledge of the past.

Some Ground Rules

Clearly, the shaping of alternatives involves a virtually infinite number of choices. Without some general principles to govern their logic, the choices can easily become arbitrary and seriously compromise the rationality of the whole process. The major principles involve questions of inclusiveness and feasibility.

It seems obvious that every reasonable and feasible possibility should have a place in some alternative, but how do we determine what is reasonable and feasible? Do we include, for instance, possibilities that may be economically viable but ecologically destructive? In such circumstances, it is necessary to invoke the ethical position of the designer. The goal designer includes all possibilities that might serve the goals of those who are paying him, but none that are unlikely to do so. The objective designer includes only those that do not transgress certain boundaries that he has established. For example, he may determine to consider no possibility that destroys the habitat of an endangered species, degrades water quality, or eliminates jobs. The ideal designer considers all possibilities that might serve the interests of his larger clientele, that is, everyone who might be in any way affected.

If we agree to include all reasonable and feasible possibilities, then clearly any excluded possibility has been judged unreasonable or not feasible. If a possibility is included in more than one alternative, the likelihood of its adoption is multiplied. One that is included in all alternatives is endorsed as virtually indispensable.

The question of feasibility deserves some further discussion. Especially in projects required to prepare environmental impact reports (under the provisions of either the National Environmental Policy Act or various state bills), "straw man" alternatives have become fairly common. The obvious purpose of these is to direct a decision in favor of an alternative preferred by the proponent. I remember one plan for a residential subdivision that posed as alternatives a regional park, which would have involved an enormous cost to the county for land acquisition, and a motorcycle racing course. Such alternatives clearly degrade the rationality, and even the seriousness, of the design process.

PREDICTIONS

At this point, possibilities begin to narrow toward a plan. The form this plan takes will depend on the circumstances, the type of process, and the scale of concern. At the construction scale, it may be a set of design drawings. At the plan unit scale, it may be a land-use plan. Following a political process at the regional or larger scales, it may take the form of a policy statement. Following an integrative process, it may be no more than a single small step that everyone involved can agree on.

In practice, choices among alternatives are often made in an ad hoc, intuitive way, but rationality requires that decisions be made on logical grounds. Following the posing of possibilities, we revert once again to the analytical mode in order to predict the impact of each alternative as a basis for our comparison.

The economic process model calls for alternatives to be compared in terms of goal achievement—that is, to what degree does each alternative satisfy the goals established

CASE STUDY XIV

Bolsa Chica Lagoon

Bolsa Chica is one of the few coastal wetlands left undeveloped in Orange and Los Angeles counties in Southern California. Oil drilling has been conducted in and around the wetlands since the 1920s but is due to be phased out within a few years. The uses of the area after that are the subject of intense conflicts involving the landowner (Signal Oil Company), the State Coastal Commission, the State Department of Fish and Game, the County of Orange in which the lagoon is located, the City of Huntington Beach, which is considering annexing it, the U.S. Fish and Wildlife Service, the U.S. Army Corps of Engineers, and several active citizens' groups.

Given its intensity and complexity, the conflict will not soon be resolved. It will take years to arrive at any general agreement on future use. The intention of this design effort was not to devise a plan for adoption but simply to define and clarify the issues in terms of alternatives and their likely consequences. The five alternatives include the stated desires of all the parties to the controversy except the landowner, Signal Oil, which did not wish to participate. The opinions matrix shows the uses favored by each group. The evaluation matrix shows, in qualitative terms, the likely consequences of each of the development actions proposed in each alternative. Although there is clearly no agreed upon alternative here, the issues are concisely stated, and the stage is set for the next step.

Prepared for the City of Huntington Beach by the 606 Studio, Department of Landscape Architecture, California State Polytechnic University, Pomona. James Theodore Bench and Jeffrey Breakstone, with advisors John T. Lyle, Jeffrey K. Olson, and Arthur Jokela. Prepared for the City of Huntington Beach.

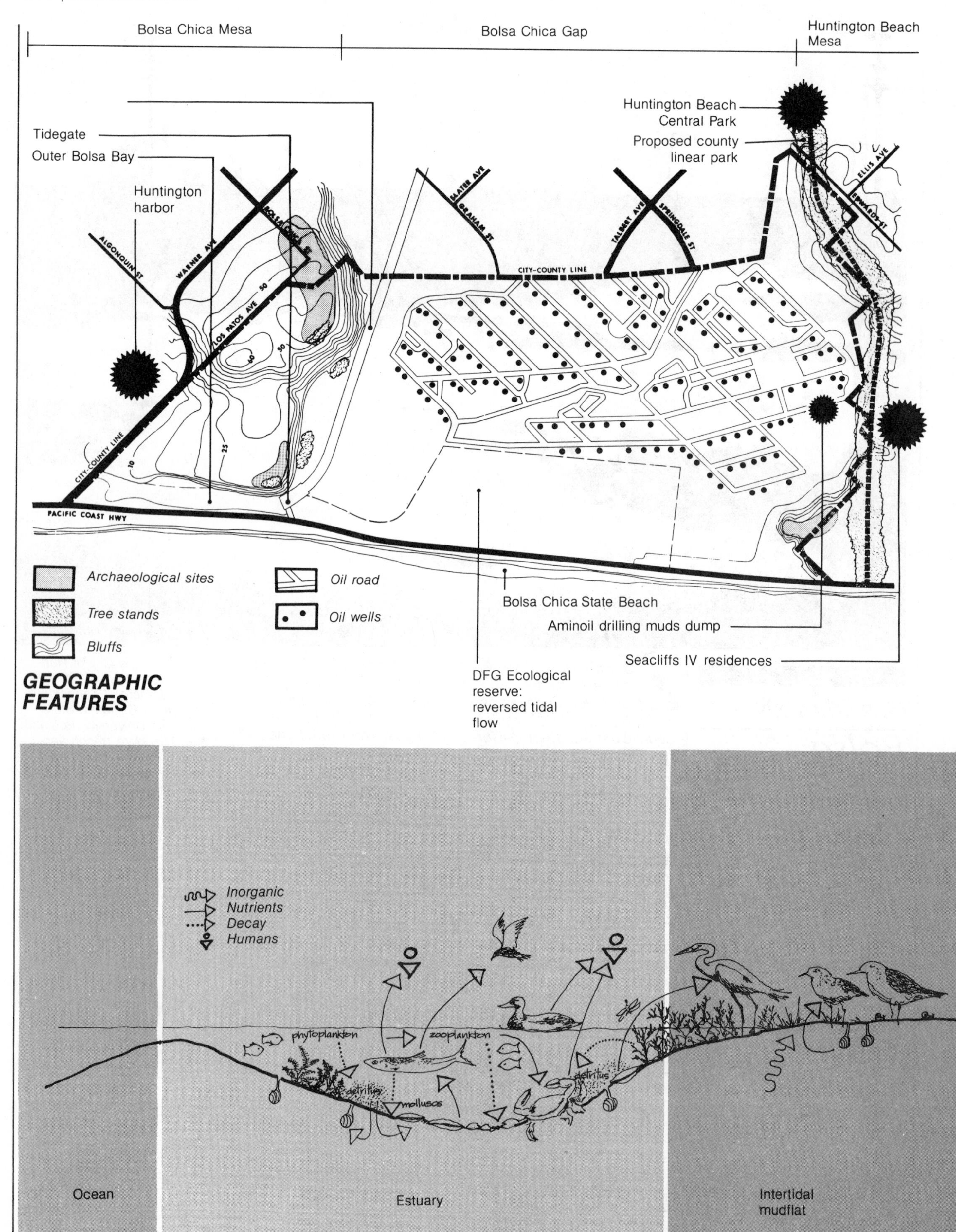
Bolsa Chica Mesa
Bolsa Chica Gap
Huntington Beach Mesa
Tidegate
Outer Bolsa Bay
Huntington harbor
Huntington Beach Central Park
Proposed county linear park
ALGONQUIN ST
WARNER AVE
BOLSA CHICA ST
LOS PATOS AVE
SLATER AVE
GRAHAM ST
TALBERT AVE
SPRINGDALE ST
ELLIS AVE
EDWARDS ST
CITY-COUNTY LINE
PACIFIC COAST HWY
Archaeological sites
Tree stands
Bluffs
Oil road
Oil wells
Bolsa Chica State Beach
Aminoil drilling muds dump
Seacliffs IV residences
DFG Ecological reserve: reversed tidal flow
GEOGRAPHIC FEATURES
Inorganic
Nutrients
Decay
Humans
phytoplankton
zooplankton
detritus
molluscs
Ocean
Estuary
Intertidal mudflat

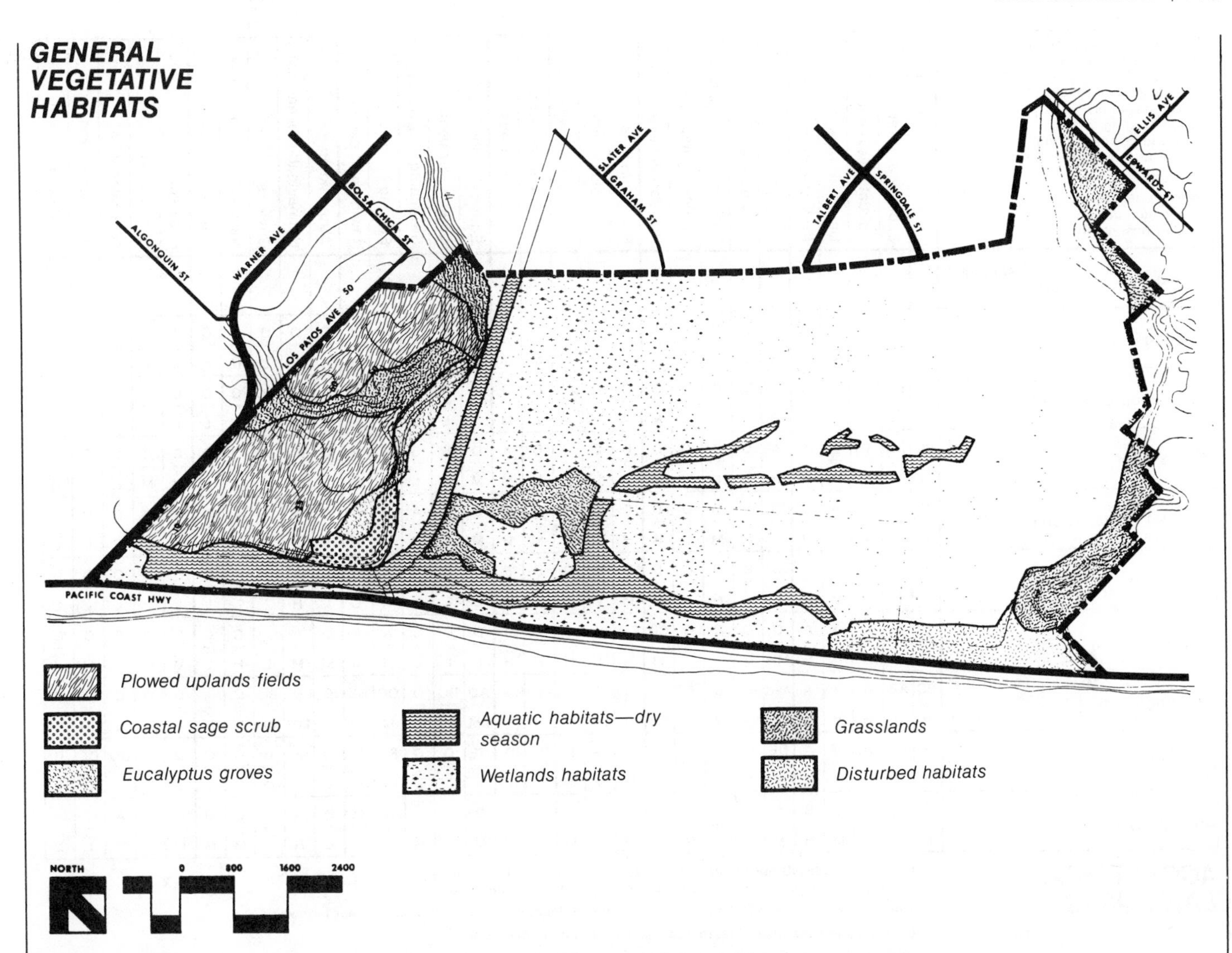
GENERAL
VEGETATIVE
HABITATS
ALGONQUIN ST
WARNER AVE
BOLSA CHICA ST
LOS PATOS AVE
SLATER AVE
GRAHAM ST
TALBERT AVE
SPRINGDALE ST
ELLIS AVE
EDWARDS ST
PACIFIC COAST HWY
Plowed uplands fields
Coastal sage scrub
Eucalyptus groves
Aquatic habitats—dry season
Wetlands habitats
Grasslands
Disturbed habitats
NORTH
0
800
1600
2400

ESTUARINE
ZONES
insects
Salt marsh

This land use is acceptable: **A** Throughout Bolsa Chica **B** In wetlands only **C** In degraded wetland* **D** On mesa only **E** On existing facilities (oil roads, mud dump) **O** Not an acceptable use. This use is of: **H** High desirability **M** Medium desirability **L** Low desirability	Tidal marsh restoration	Ecological reserve	Education exhibits	Unpaved trails	Protect natural scenery	Vista points	Aquaculture	Tertiary sewage treatment	Oil drilling mud dump	Recreation center	Picnic grounds	Open play fields	Overnight camping	Paved parking area	Paved bike trails	Equestrian facilities	Quiet water beaches	Swimming and wading	Fishing	Rowing and canoes	Water skiing	Nonnavigable ocean entry	Culverts	Navigable ocean access	Marina	Residences	Commercial	PCH relocation	Utility extension	Road construction
Signal — No response offered																														
Boating	B / M	B / M	B / M	B / M	A / M	A / H	B / L	B / L	C / L	A / H	A / H	O / L	O / L	C / M	A / M	D / L	C / M	C / M	B / H	B / H	O / L	O / L	O / L	C / H	C / H	A / L	A / L	A / H	A / M	A / M
Amigos de Bolsa Chica	BC / H	BC / H	D / H	D / H	A / H	D / H	BC / L	O / L	O / L	D / M	D / M	O / L	2*	O / L	D / M	D / M	O / L	O / L	O / L	O / L	O / L	BC / H	BC	O / L	O / L	D / L	O / L	O / L	O / L	O / L
Sierra Club	BC / H	BC / H	D / M	B / H	A / H	A / H	O / L	O / L	O / L	D / M	D / M	O / L	O / L	D / M	A / M	D / M	B / L	B / L	B / L	B / M	O / L	BC / M	BC / H	O / KL	O / L	D / M	D / M	O / L	D / M	D / M
Coastal Commission	B / H	A / H	A / M	A / H	D / H	D / H	B / M	O / M	D / L	D / H	D / H	D / M	D / M	D / M	D / H	D / M	A / H	A / H	A / M	A / L	A / L	A / H	A / H	A / L	A / L	D / L	D / L	O / L	D / L	D / L
DFG Staff	A / H	A / H	A / H	E / M	A / M	D / L	O / L	O / H	O / L	D / M	D / M	D / L	D / L	D / L	E / M	O / L	O / L	3*	A / H	O / L	O / L	BC / H	4*	— / L	O / L	D / L	D / L	O / L	D / L	D / L
Orange County	C / H	B / H	A / H	A / H	A / H	D / MH	C / M	O	O	CD / M	CD / M	D / L	C / M	CD / H	CD / H	D / L	C / L	C / L	A / H	A / M	O / M	— / M	— / L	C / H	C / L	D / M	D / M	C / M	C / H	C / H
Huntington Beach	BC / H	AB / H	A / H	A / M	A / H	ABC / H	BC / LM	OBC / M	O / L	CD / ML	ACD / H	D / LM	ACD / H	ACD / H	—	AD / LM	BC / M	O	OC / M	BC / MH	O / L	B / MH	BC / MH	C / M	— / H	D / M	D / M	O / L	AD / LM	AD / LM
USFWS	B / H	AB / H	DE / H	DE / M	A / H	DE / M	D / L	B / M	O / L	D / M	D / M	D / M	D / M	DE / L	D / L	DE / L	O / L	O / L	BE / H	O / L	O / L	B / L	B / M	O / L	O / L	D / L	D / L	DO / L	DE / L	D / L
Army Corps	C / H	BC / H	BC / H	C / M	B / H	A / M	O / L	O / L	E / M	D / M	C / M	D / L	D / M	D / L	D / L	D / M	— / M	— / M	B / H	O / M	O / L	A / H	—	O / M	O / M	D / M	O / L	A / H	— / L	— / L

ACCEPTABLE LAND USES

* Neither the USFWS nor the Army COE recognize "degraded" wetlands. The Amigos prefer to use the term "restorable" rather than degraded.

2* The Amigos recommend that any campgrounds be located on existing oil pads near highway.

3* DFG Staff like the idea of restricted wading for clam hunters only.

4* DFG Staff says O.K., depending on the location and reason for use.

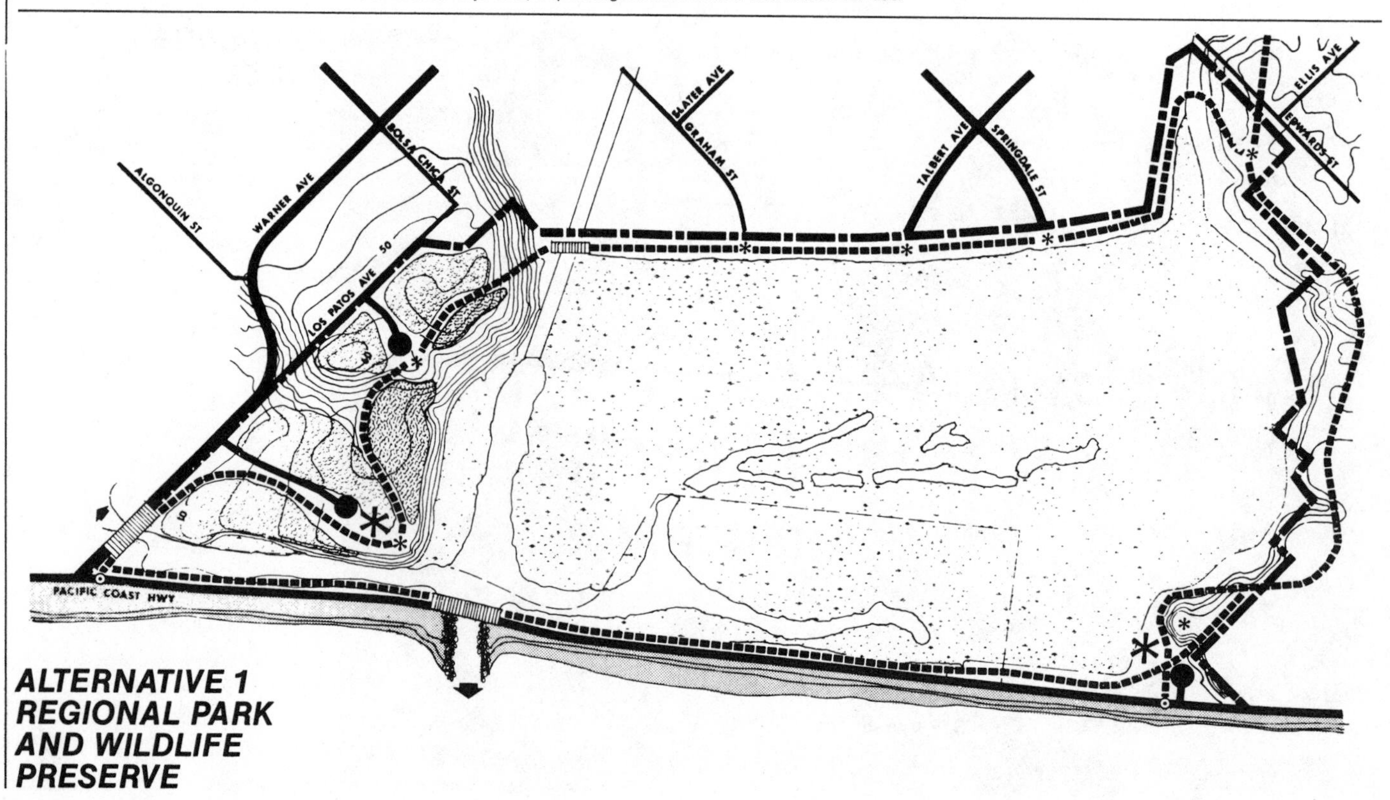

ALTERNATIVE 1 REGIONAL PARK AND WILDLIFE PRESERVE

ALTERNATIVE 2

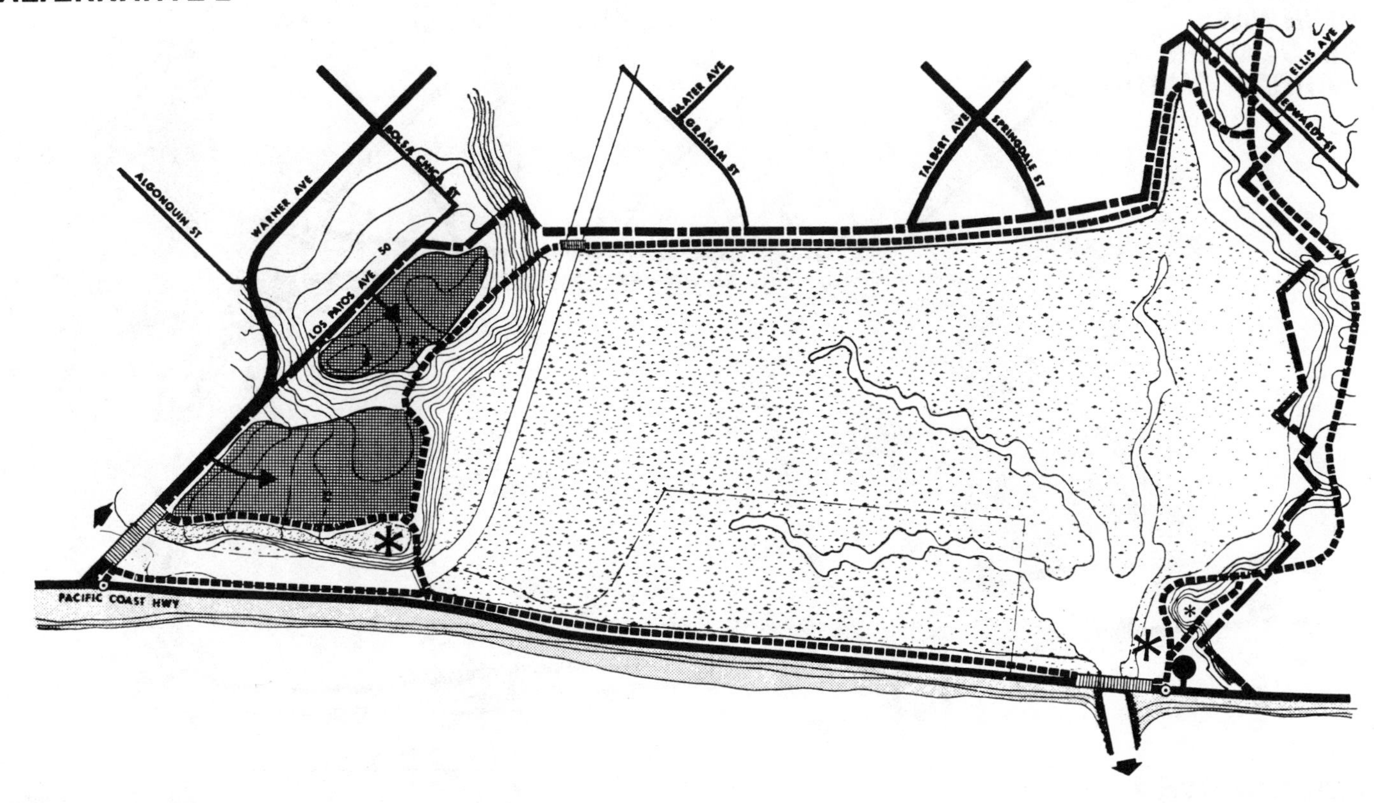

ALTERNATIVE 3 MINI-MARINA/ WETLAND PRESERVE

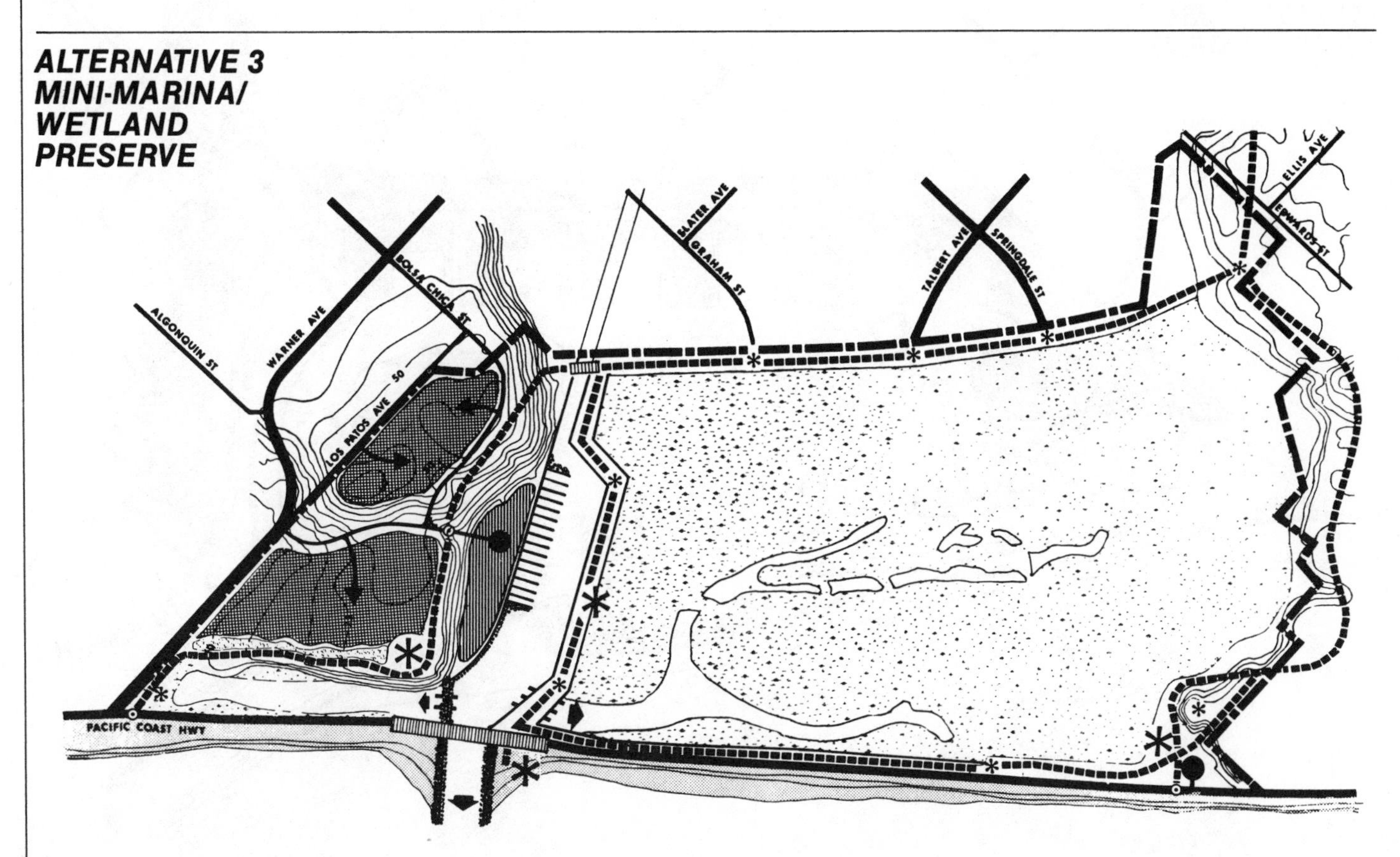

Bluffs
Tidal wetlands
Public beaches
Regional park

Residential
Commercial

Tidal access
Seaward jetty
Culvert
Bridge

Major road
Surface street
Buffer zone
Paved trails

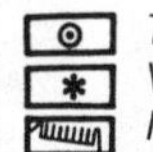
Trail crossing
Viewpoint
Marina

Overnight camp
Visitors' information center

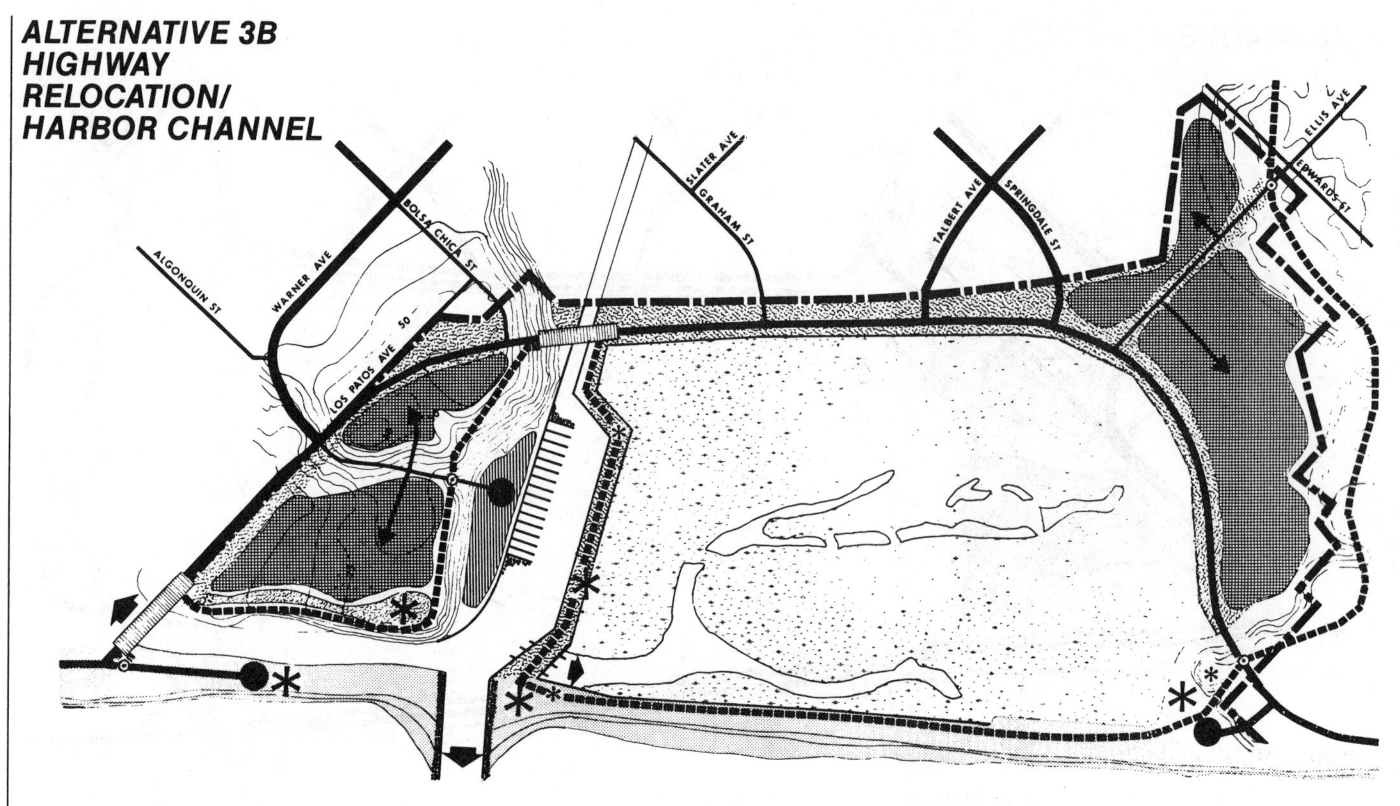
ALTERNATIVE 3B
HIGHWAY
RELOCATION/
HARBOR CHANNEL
ALGONQUIN ST
WARNER AVE
BOLSA CHICA ST
LOS PATOS AVE
50
SLATER AVE
GRAHAM ST
TALBERT AVE
SPRINGDALE ST
ELLIS AVE
EDWARDS ST

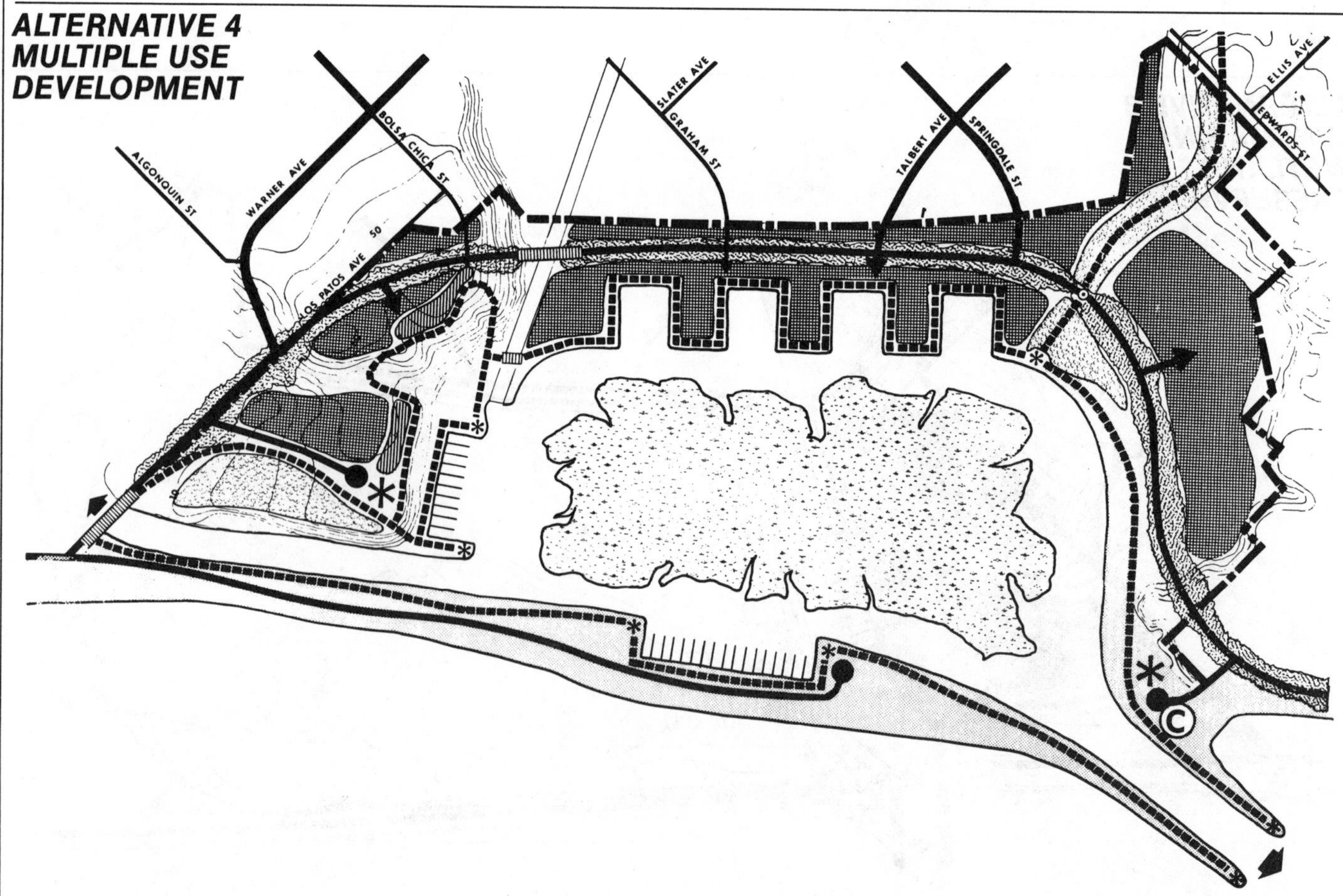
ALTERNATIVE 4
MULTIPLE USE
DEVELOPMENT
ALGONQUIN ST
WARNER AVE
BOLSA CHICA ST
LOS PATOS AVE
50
SLATER AVE
GRAHAM ST
TALBERT AVE
SPRINGDALE ST
ELLIS AVE
EDWARDS ST

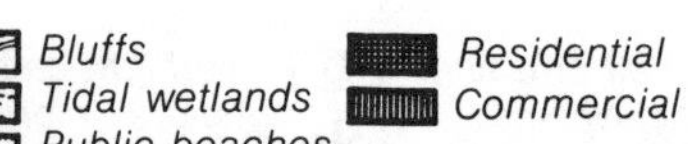

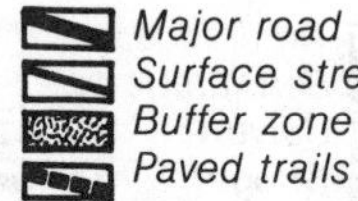
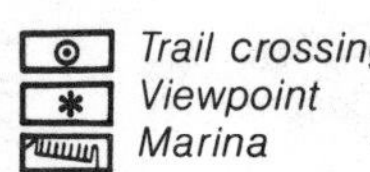
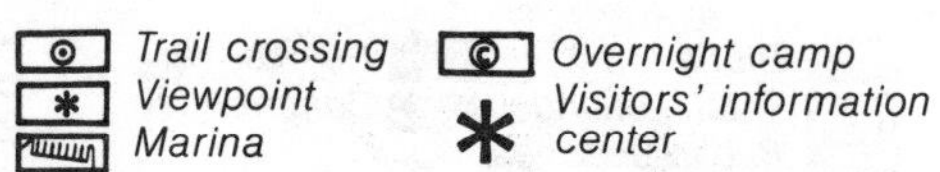
Bluffs
Tidal wetlands
Public beaches
Regional park
Residential
Commercial
Tidal access
Seaward jetty
Culvert
Bridge
Major road
Surface street
Buffer zone
Paved trails
Trail crossing
Viewpoint
Marina
Overnight camp
Visitors' information center

EVALUATION MATRIX ALTERNATIVES 1, 2, 3A, 3B, 4

Wetland enhancement: 2 Greatly benefits; 1 Moderately benefits; 0 No effect; -1 Moderately conflicts; -2 Greatly conflicts

Community enhancement: 2 Greatly benefits; 1 Moderately benefits; 0 No effect; -1 Moderately conflicts; -2 Greatly conflicts

Socio-political acceptability: 2 Strongly favors; 1 Moderately favors; 0 Indifferent; -1 Moderately opposes; -2 Strongly opposes

Public protection: 2 High protection; 1 Mod. protection; -1 Mod. exposure; -2 High exposure

		Wetland enhancement											Community enhancement														Socio-political acceptability										Public protection			
Alt.	Element	Habitat quality	Habitat size and diversity	Tidal access	Tidal fluctuation	Water quality	Salt/fresh water mix	Nutrient cycle	Min. physical intrusion	Min. noise and visual impacts	Runoff quality control	Sedimentation control	Use by low/moderate income	Housing	Commercial day/night use	Recreational facilities	Educational facilities	Visitor accommodations	Beach access	Traffic safety/convenience	Bike safety/convenience	Pedestrian safety/convenience	Visual and noise control	Air quality	Private profits	Tax revenues	Amigos de Bolsa Chica	Sierra Club	USFWS	DFG staff	State Coastal Commission	Orange County (EMA)	Huntington Beach	Army Corps of Engineers	Signal landmark	Boating interest	Flooding	Slope erosion	Salt water intrusion	Seismic disturbance
	No action alternative (existing site as is)	-1	-1	0	0	-1	0	-1	0	0	-1	2	0	0	0	0	-1	0	0	-1	0	0	0	0	1	-1	-1	-1	0	-1	0	-2	0	-1	-2	-2	-1	0	0	0
1	1,000 acre wildlife preserve	2	2	0	0	2	0	2	0	1	0	0	1	1	0	1	2	0	0	0	0	0	2	2	-2	-2	2	2	2	2	2	-1	0	1	-2	-2	1	0	0	0
1	120-acre regional park	0	0	0	0	0	0	0	0	1	1	1	2	1	0	2	1	2	1	0	1	2	2	1	-2	-2	2	2	2	2	2	1	1	1	-2	0	0	2	0	1
1	Non-navigable ocean channel	2	0	2	2	2	1	2	0	0	0	1	0	0	0	0	0	0	0	0	0	0	0	0	0	-1	2	2	2	2	2	0	1	1	-2	-2	2	0	-1	0
1	Levee-top bike/hike trail	0	0	0	0	0	0	0	-1	-1	-1	0	2	1	0	2	1	1	1	0	2	2	0	0	0	-1	1	2	1	-1	1	1	1	1	-1	0	2	0	1	0
1	Perimeter trail system	0	0	0	0	0	0	0	-1	-1	2	1	2	1	1	2	1	1	2	1	2	2	1	0	0	-1	2	2	1	1	2	1	0	1	0	2	2	1	0	0
1	2 Visitor centers (parking, etc.)	0	0	0	0	-1	0	0	-1	-1	-1	0	2	0	1	2	2	2	0	0	0	0	0	0	0	-1	2	2	1	1	2	1	1	1	0	2	0	0	0	-1
2	180-acre residential development	-1	0	0	0	-1	0	1	0	0	0	-1	1	2	1	0	0	0	0	0	0	0	-1	-1	2	0	0	0	-1	-1	0	1	0	0	2	0	0	-1	0	-1
2	Non-navigable ocean opening	2	0	2	2	2	1	2	0	0	0	1	0	0	0	0	0	0	0	0	0	0	0	0	0	-1	2	2	2	2	2	0	1	1	-1	-2	1	0	0	0
2	1000-acre wildlife preserve	2	2	0	0	2	0	2	0	2	0	0	1	0	0	1	2	0	0	0	0	0	2	0	-2	-1	2	2	2	1	1	0	0	1	-2	-2	2	0	0	0
2	Dredged "fingers" of open water	2	2	2	2	2	1	2	0	2	0	1	0	0	0	0	0	0	0	0	0	0	2	0	-2	-1	2	2	2	1	1	0	0	1	-2	-2	2	0	0	0
2	Perimeter trail system/viewpoints	0	0	0	0	0	0	0	-1	-1	2	1	2	1	1	2	1	1	2	1	2	2	2	0	0	-1	2	2	1	1	2	1	1	1	0	2	2	1	0	0
2	Visitor information center	0	0	0	0	-1	0	0	-1	-1	-1	0	2	0	1	2	2	2	0	0	0	1	0	0	0	-1	2	2	1	1	2	1	1	1	0	2	0	0	0	-1
2	Heightened levees	-1	0	0	0	0	0	0	0	-1	0	0	0	2	0	0	0	0	0	0	0	0	0	0	0	-1	0	0	0	0	0	0	0	2	0	0	2	0	1	0
3A	50-acre marina	-2	-2	0	0	-1	0	-1	-2	-2	0	0	0	1	2	2	1	2	0	0	0	1	1	-1	2	2	-2	-2	-2	-2	-1	0	-1	-2	2	2	1	0	-1	0
3A	30-acre commercial center	-1	-1	0	0	-1	0	0	-1	-2	-1	0	1	1	2	0	0	2	0	0	0	0	-1	-2	2	2	-2	-2	-2	-2	-1	0	-1	-2	2	1	0	0	0	-2
3A	25-acre earth berm	1	0	0	0	0	0	0	0	2	0	0	1	0	0	0	1	1	2	0	2	2	2	0	0	-1	2	2	2	2	0	1	-1	-2	0	0	1	1	0	0
3A	High bridge	-1	-1	0	0	0	0	0	-1	-2	-1	0	0	0	0	0	0	0	1	2	1	2	-1	-1	0	-2	1	1	1	0	-1	0	0	1	0	0	0	0	0	-2
3A	180-acre residential development	-1	0	0	0	-1	0	1	0	0	0	-1	1	2	1	0	0	0	0	0	0	0	-1	-1	2	0	0	0	-1	-1	0	1	0	0	2	0	0	-1	0	-1
3A	Blocked outer Bolsa Bay	1	1	0	0	0	0	1	0	0	0	1	1	0	0	1	1	1	0	0	0	1	0	1	-1	-1	0	0	0	0	0	0	0	0	2	2	0	0	0	0
3A	880-acre wildlife preserve	2	2	0	0	2	0	2	0	2	0	0	1	1	0	1	2	0	0	0	0	0	2	2	-2	-2	2	2	2	2	2	-1	0	1	-2	-2	1	0	0	0
3A	Visitor information centers	0	0	0	0	-1	0	0	-1	-1	-1	0	2	0	1	2	2	2	0	0	0	1	0	0	0	-1	2	2	1	1	2	1	1	1	0	2	0	0	0	-1
3A	Perimeter trail system/viewpoints	0	0	0	0	0	0	0	-1	-1	2	1	2	1	1	2	2	1	2	1	2	2	2	0	0	-1	2	2	1	1	2	1	1	1	0	2	2	1	0	0
3B	50-acre marina	-2	2	0	0	-1	0	-1	-2	-2	0	0	0	1	2	2	1	2	0	0	0	1	1	-1	2	2	-2	-2	-2	-2	-1	0	-1	-2	2	2	1	0	-1	0
3B	30-acre commercial center	-1	-1	0	0	-1	0	0	-1	-2	-1	0	1	1	2	0	0	2	0	0	0	0	-1	-2	2	2	-2	-2	-2	-2	-1	0	-1	-2	2	1	0	0	0	-2
3B	25-acre earth berm	1	0	0	0	0	0	0	2	2	0	0	1	0	0	0	1	1	2	0	2	2	2	0	0	-1	2	2	2	2	0	0	0	0	0	0	1	0	0	0
3B	Boat channel outer Bolsa Bay	-1	-2	1	1	-1	1	1	-2	-2	0	1	0	0	0	1	0	0	0	0	0	0	0	0	0	-1	-2	-2	-2	-2	-2	1	0	0	0	2	1	0	-1	0
3B	PCH relocated/buffered	-1	-2	0	0	-1	0	0	-2	-2	-2	0	0	-1	0	0	0	0	0	1	0	0	-1	0	0	-2	-2	-1	-2	-2	-2	1	0	-1		2	2	0	1	0
3B	140 acres additional housing	-1	0	0	0	-1	0	1	0	0	0	-1	1	2	1	0	0	0	0	0	0	0	-1	-1	2	0	-2	-2	-2	-2	-2	1	0	1	2	2	0	0	0	-1
3B	615-acre wildlife preserve	2	0	0	0	2	0	2	0	2	0	0	1	1	0	1	2	0	0	0	0	0	1	2	2	2	0	0	0	0	0	0	0	0		1	1	0	0	0
3B	Public beach & parking on old PCH	0	0	0	0	0	0	0	0	-1	0	0	2	0	0	2	0	2	2	2	1	1	0	0	0	-1	0	0	0	1	0	0	0	0	0	0	0	0	0	0
3B	Perimeter trail system/viewpoints	0	0	0	0	0	0	0	-1	-1	2	1	2	1	1	2	2	1	2	1	2	2	2	1	0	-1	2	-2	1	1	1	2	2	0	0	2	0	2	0	0
3B	Visitor information centers	0	0	0	0	-1	0	0	-1	-1	-1	0	2	0	1	2	2	2	0	0	0	1	0	0	0	-1	2	2	1	1	1	1	1	1	0	2	0	0	0	-1
4	250-acre beach peninsula	0	0	0	0	-1	0	0	0	-1	1	0	2	1	0	2	0	1	2	1	1	1	1	0	0	-2	-2	-2	-2	-2	-2	0	-1	-2	0	2	0	0	0	0
4	60 acres inland beaches	0	0	0	0	-1	0	0	-2	-1	1	0	2	1	0	2	0	1	2	0	0	0	0	0	0	-2	-2	-2	-2	-2	-2	0	-1	-2	0	2	1	0	0	0
4	320 acres isolated wetlands	2	0	0	0	2	0	2	0	2	0	0	1	1	0	1	2	0	0	0	0	0	1	2	-1	-1	-2	-2	-2	-2	2-	0	-1	-2	0	2	1	0	0	0
4	400 acres waterfront & mesa homes	-1	-2	0	0	-1	0	0	-2	-1	-1	-1	0	2	1	1	0	0	0	0	0	0	-1	-1	2	2	-2	2	2	2	2	0	-1	-2	2	2	0	0	0	-1
4	Small individual marinas	-1	-2	0	0	-1	0	-1	-2	-2	0	0	0	1	0	2	1	2	0	0	0	0	1	-1	2	2	-2	-2	-2	-2	-2	0	-1	2	2	2	-1	0	-1	0
4	Boat channel to Huntington Harbor	-1	-2	2	2	-1	1	1	-2	-2	0	1	0	0	0	1	0	0	0	0	0	0	0	0	0	-1	-2	-2	-2	2-	-2	1	0	0	0	2	1	0	-1	0
4	15-acre mesa commercial center	-1	0	0	0	-1	0	0	0	0	0	-1	1	1	1	0	0	2	0	0	0	0	-1	-1	2	2	0	0	0	0	0	1	1	0	2	2	0	0	0	-1
4	40-acre mesa park	0	0	0	0	0	0	1	0	1	1	1	2	1	0	2	1	2	1	0	1	2	2	1	2	-1	1	1	1	1	1	1	1	1	0	1	1	1	0	1
4	Linear park trails/viewpoints	0	0	0	0	0	0	0	-1	-1	2	1	2	1	1	2	2	1	2	1	2	2	2	1	0	-1	2	2	1	1	1	2	2	1	0	2	2	2	0	0
4	Visitor information centers	0	0	0	0	-1	0	0	-1	-1	-1	0	2	0	1	2	2	2	0	0	0	1	0	0	0	-1	2	2	1	1	2	1	1	1	0	2	0	0	0	-1
4	PCH relocated/buffered	-1	2	0	0	-1	0	0	-2	-2	-2	0	0	-1	0	0	0	0	0	1	0	0	-1	0	0	-2	-2	-1	-2	-2	-2	1	0	0	0	2	2	0	1	0
4	Overnight camping	0	0	0	0	0	0	0	0	0	0	0	1	0	1	2	1	2	1	0	0	0	0	0	0	-1	-2	-2	1	0	1	1	2	1		-2	0	0	0	0

at the beginning of the design effort? This provides a conceptually neat closing of the circle that can be quite satisfying to the goal designer, but what about the larger picture? What about all of the effects and implications that go beyond the goals, even the issues defined for the project? Environmental impact reports are required to cover these matters, but it often happens that the implications exceed the bounds of such reports. In addition to measuring their effectiveness in achieving stated goals or in resolving issues, it is important to compare alternatives with respect to their predicted effects on people and the larger environment.

Several examples of comparative analyses appear in the case studies. The evaluation matrix for the Bolsa Chica alternatives on page 179 takes a particularly broad ideal-planner's view. The alternatives are compared on the basis of their wetland enhancement—which really means ecological factors here because the wetlands are the central ecological concern—community enhancement, socio-political acceptability, and public protection. Together, these four categories summarize Bolsa Chica's influence on its environment. Within each, several specific factors comprise the particular subjects of concern. Among the wetland enhancement factors, for example, were habitat quality, habitat size and diversity, and tidal access and tidal fluctuation. For each of these factors, the major land-use component of each alternative was evaluated as to its beneficial or conflicting contributions. A brief study makes it clear that, while Alternative 4 is most desirable in terms of community enhancement and public protection, Alternative 2 is the most beneficial for the wetlands and probably the most politically acceptable.

The San Dieguito case study on pages 185–90 shows another process of comparative evaluation. Here the comparison is based on predictions of impacts on human and natural processes in both positive and negative terms.

One feature that is important in these examples is their consideration of both positive and negative factors. They recognize that human actions can benefit natural processes and populations as well as people themselves.

Another important feature is their summarization of measures of comparison in broad, relative terms, using symbols rather than quantities. The impact analyses for the projects contained quantified predictions, of course, some of them the product of intricate calculations, but the availability of quantified data varied greatly from one factor to another. It should be noted that quantifiable factors are likely to be weighted more heavily in a comparative analysis simply because more relative measures seem vague and thus less compelling by comparison. By stating all comparisons in relative terms, this bias is eliminated.

It is possible, of course, to carry the use of numbers, even in these relative terms, much further. Numerical values can be assigned to beneficial and damaging effects, and these can then be combined to give an overall indication of effectiveness. In fact, such an exercise was carried out with the Bolsa Chica matrix as an experiment. It is not clear, however, that numbers, aside from any question of their validity, serve a useful purpose. In my experience at least, choices among alternatives are rarely made in such a mechanical way. Rather, the decision makers use all available information to make their intuitive summations and judgments.

This is not to say that quantitative measures are not useful. Calculations of specific quantities are essential in many areas of impact prediction, especially those where legal decision-making processes and numerical standards have been established. Runoff volume and water quality parameters may serve as examples. Measures of economic costs in specific dollar terms are obviously of considerable importance, but they represent only one way of dealing with economic concerns.

In the Whitewater Wash example, economic factors were considered with respect to the ability of each activity to pay for itself. From the point of view of the public agencies that own the land and that will implement the plan, this is the basic question. Activities that can pay for themselves can be supported with ease, whereas those that cannot will need strong justification.

Beyond such simple comparative measures as these, the question of economic costs in relation to land use becomes extremely complex. What costs can be measured in dollars, and how? Who pays the costs and in what terms? In one 606 Studio project that involved hillside development in the City of Los Angeles, we found that a major cost would be borne involuntarily and unknowingly by owners of houses with views of the hillside in question. The development would measurably reduce the values of their properties, and the cumulative loss would be far more than the developer's profit. But how does this relate to property rights? How does one place a value on resources that cannot be fairly, much less accurately, priced? Should one even try?

These questions are all dealt with, though rarely answered in any convincing way, in the growing literature on resource and environmental economics. When detailed cost analyses are needed beyond the general comparative measures suggested both here and in the case studies, we move into a realm that is beyond the scope of this book. It is a realm that needs considerable investigation.

Another kind of quantitative measure that is very much to the point here, and that we have already dealt with to some extent, is the material and energy budget. It has become very clear through recent experience that economic costs do not adequately deal with the problem of probable future scarcities of resources, energy and water being prime examples. Although the prices of energy and water are rising to levels that may more truly reflect their real values, these increases depend on a great many factors, including government policies. So it is probably much more accurate, and perhaps more useful, to compare quantities of resources consumed rather than their monetary costs. The Aliso Creek case study takes this approach when it predicts water consumption patterns for different patterns of land use. Water being the key resource in this particular landscape, this comparative budget should probably be the major basis for deciding among alternatives. As resource scarcities become more critical in the future, budgets of this type may very well become a necessary part of all environmental impact reports. The techniques for material and energy budgeting will be discussed in the next chapter.

CASE STUDY XV

North Claremont

In an effort to maintain a "rural atmosphere," this 400-acre area in the foothills of the San Gabriel Mountains has been zoned for one-acre lots. Most of the orange groves that once covered the landscape have been removed and those that remain have aged beyond economic productivity. Citizens' groups have asked several questions. What kind of landscape will the one-acre zoning produce? What kinds of alternatives might exist? Can the agricultural use be preserved?

This project poses answers to these questions in the form of scenarios: patterns of development that might occur in response to four assumed future conditions, ranging from continuation of present trends to a serious food shortage.

Prepared by the 606 Studio, Department of Landscape Architecture, California State Polytechnic University, Pomona. Anne Nelson, with advisors John T. Lyle, Jeffrey K. Olson, and Arthur Jokela.

LAND USE PATTERN: 1947

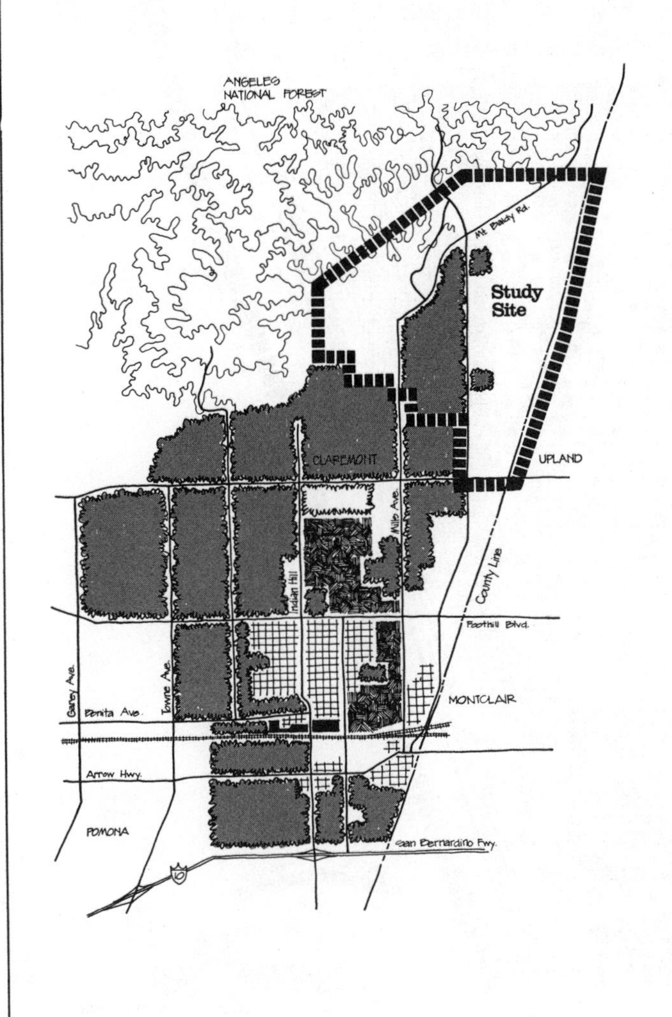

LAND USE PATTERN: 1977

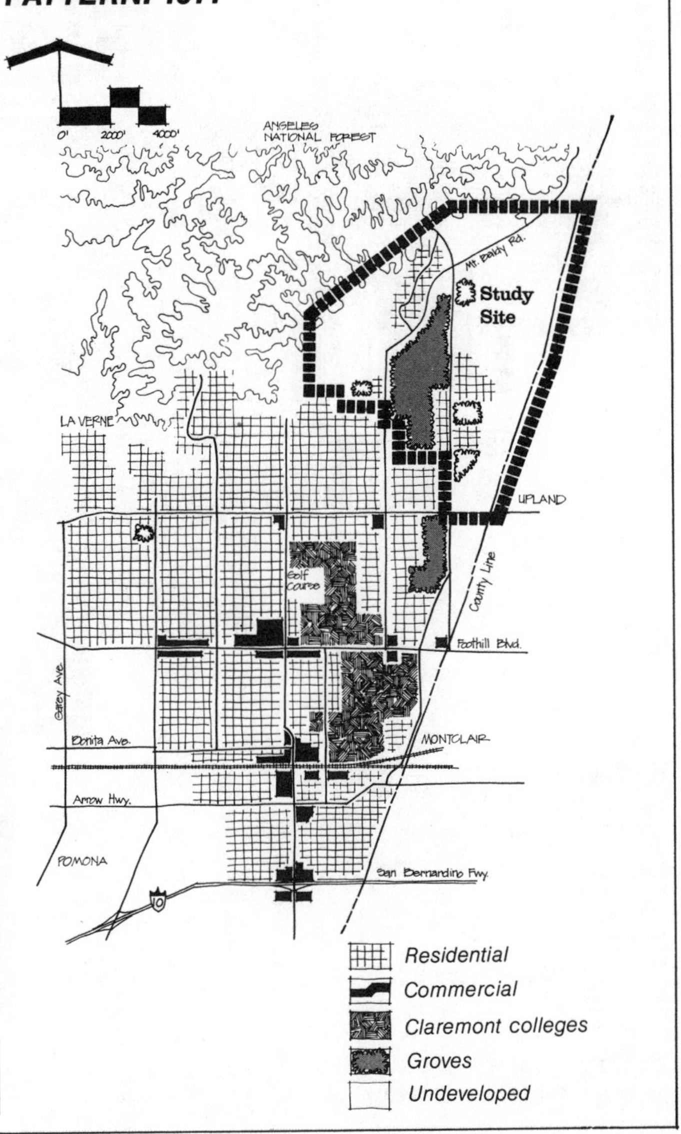

SCENARIO B FOOD EMERGENCY

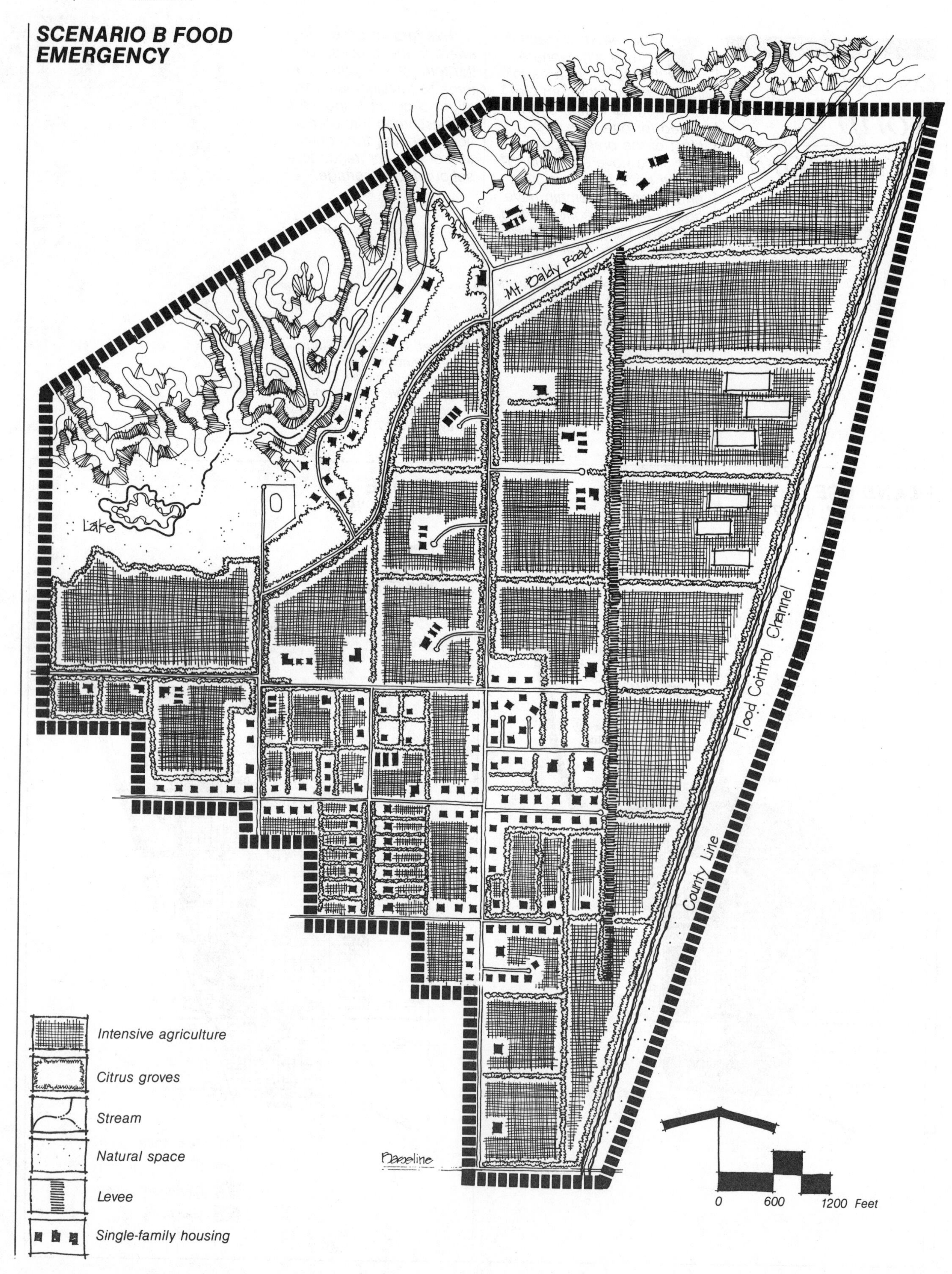

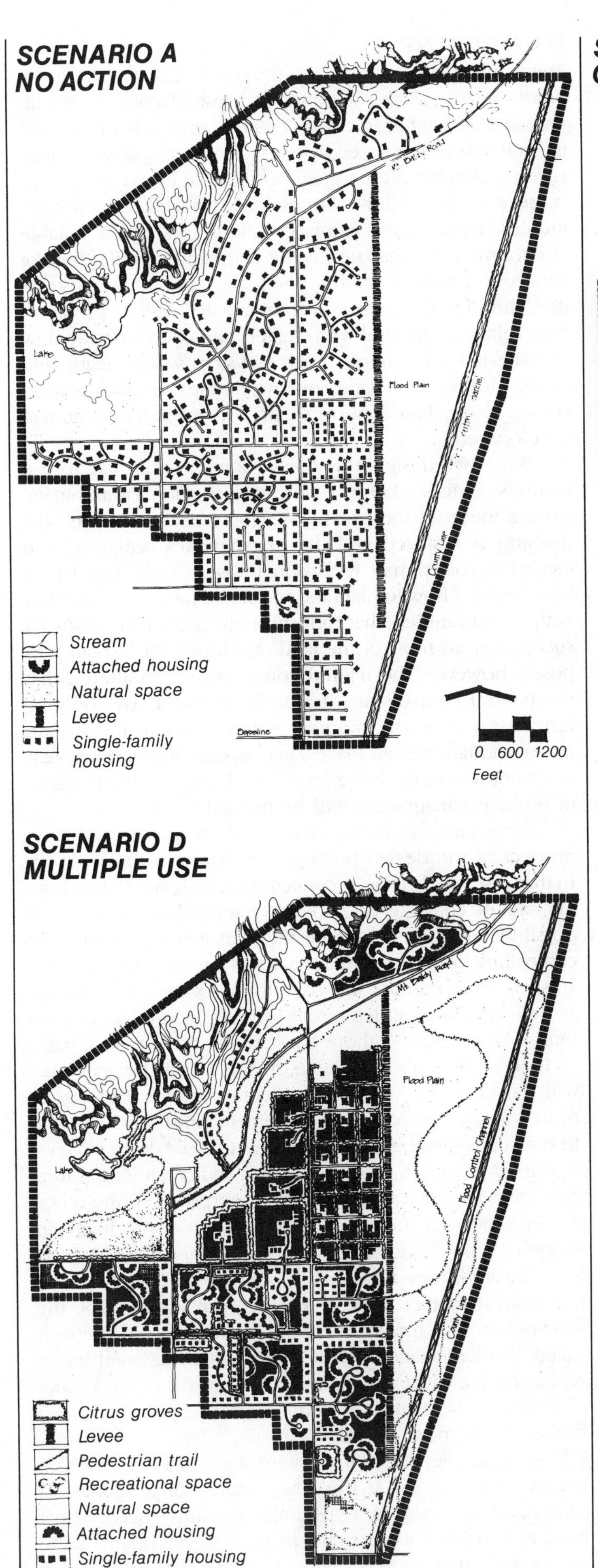
SCENARIO A
NO ACTION
Mt. Baldy Road
Lake
Flood Plain
County Line
Baseline
Stream
Attached housing
Natural space
Levee
Single-family housing
0 600 1200
Feet
SCENARIO D
MULTIPLE USE
Mt. Baldy Road
Flood Plain
Lake
Flood Control Channel
County Line
Citrus groves
Levee
Pedestrian trail
Recreational space
Natural space
Attached housing
Single-family housing

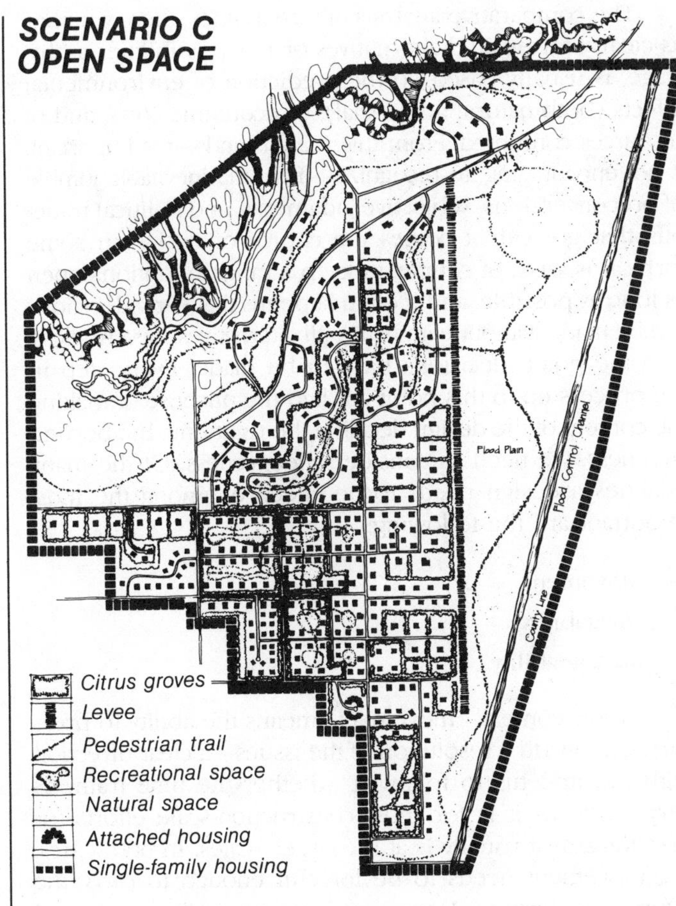
SCENARIO C
OPEN SPACE
Mt. Baldy Road
Lake
Flood Plain
Flood Control Channel
County Line
Citrus groves
Levee
Pedestrian trail
Recreational space
Natural space
Attached housing
Single-family housing

The comparative analysis of possibilities, then, whether as clearly formulated alternatives or raw possibilities at this stage, is usually based on the prediction of environmental effects (both natural and social), of economic costs, and of resources consumed. From this analysis and—it is important, if not encouraging, to recognize—from the inevitable jumble of preconceptions, subjective judgments, and political trade-offs that go with it comes the choice of design in some form. It is wise, of course, to keep a range of options open as long as possible, and it often happens that early decisions can be made without eliminating all alternatives. Nevertheless, at some point, decisions have to be made. Every step in the process up to this point has had the purpose of making the content of the design decision the right one. But beyond rightness, we need to consider effectiveness. Of the many qualities a design needs to be effective, among the most important are the following:

Movement
Imagability
Manageability

In this context, "movement" means the ability to press forward toward a resolution of the issues—a clear direction with the impetus to follow it, whether the time frame is very short, as it is for most construction-scale efforts, or very long, as it usually is at the larger scales. In both cases, the movement needs to be forceful enough to carry the effort to realization. Movement is primarily the product of steadfast purpose, a solid information base, and compelling ideas.

That said, the efforts of a great many people are needed to put most designs into operation, and these people need a clear image of what is to be accomplished if they are to work effectively. All the better if this image has the power to stir the soul. Imagability, though usually a neglected quality, is a most important one. At the site and construction scales where the plans provide a glimpse of the final result, the image comes easily. It is much more difficult, though no less important, to envision at the larger scales, where it can never be a true picture but is necessarily a conceptual sketch, an abstract pattern, or a set of ideas. Images can be graphic or verbal, of course, but graphic images are more likely to be effective. A list of programs rarely reaches the inner eye.

When Daniel Burnham made his famous exhortation to "make no little plans, they have no power to stir men's souls," he was probably assuming that only plans on a grand scale convey compelling images. If so, he was wrong. Not only can small plans have immense importance, they can project compelling images as well. Make *plans with visions*—that would be better advice. Such plans always have power to stir souls.

Finally, and perhaps most important, a design needs to be manageable, that is, it needs to provide workable guidance for the sequence of actions that will not only carry it to realization but keep it effective thereafter. This brings us to the next theme in the process. Now, the present begins to merge into the future.

MANAGEMENT

Environmental management—the continuation of design by other means—is the ongoing, purposeful control and direction of change in an environment. As such, it is a necessary, integral part of human ecosystems. In natural systems, change is controlled by internal mechanisms that evolve over time. It more or less follows successional principles, each community of plants and animals working to shape a suitable environment for the succeeding community until a state of climax is reached. At this point, the community sustains itself, limiting fluctuations with innate devices like predator-prey relationships and with triggers and checks on the flow of materials and energy. Such a system works neatly and automatically until some overwhelming agent of change comes along, like a flood or fire or volcano—or human development.

When we design a new environment, it is wise to use nature's devices wherever we can to provide for automatic management. In the San Elijo Lagoon example, once tidal flushing is restored and the marshgrasses replanted, the estuarine community will soon reassert itself. The Dutch have learned how to plant the reclaimed land of the polders with communities that go through phases of *induced* succession. In most of the land developed for human purposes, however, the natural community is so altered that the natural controls are drastically changed. The only alternative is to establish a new system of control. If we are clever enough, we can make some aspects of the new system work automatically, but almost invariably, a certain degree of human management will be needed.

Sometimes, as in the Huetar Atlantica case study, the product of the design process is a management program. In that case, the human ecosystem was designed to replicate the salient characteristics of the natural one as much as possible. However important the management problem of controlling human activity so as not to impede natural mechanisms, some control will have to be exerted over nature as well. Wherever agriculture is involved, natural succession must be impeded. Without constant efforts to hold it back, as long as its soils are viable, a rain forest, for example, will quickly move back into agricultural lands. Here, as in most human landscapes, management must work on two fronts: controlling human activity so as to avoid doing violence to natural processes on the one hand, and controlling natural processes so that they serve human purposes on the other.

In other situations, the design process produces an image of a new landscape—the places where activities are to be located, how it will work, what it will look like. The first task of management is to take over and reshape the landscape to that image. Traditionally, this process has been called "implementation." By implication, the designed image represents the climax state, that is, it reflects the characteristics of a climax, or mature, system. We will explore this idea further in the next chapters.

At the smaller scales, the mature stage might be achieved almost immediately. At the larger scales, implementation takes place in a series of stages similar to nature's successional progression. This successional phase is developmental and exploratory, that is, it is directed toward a goal—the design

CASE STUDY XVI

San Dieguito Lagoon

As evidenced by measurements of water quality, the ecological condition of San Dieguito Lagoon is slightly worse than that of San Elijo, which lies just a few miles away. Urban development has crept much closer to its banks; the famous Del Mar racetrack occupies the same flood plain only a few hundred feet away; and the San Diego County Fairgrounds are equally near.

Seven quite different thematic alternatives were seriously considered, three of them illustrated here. Each alternative is accompanied by a chart summarizing predicted environmental effects in general, qualitative form. Both alternatives and impact predictions were presented to concerned residents of the area in a public workshop. After a long discussion period, everyone attending filled out a questionnaire to express his preferences. These were tabulated immediately and the results announced to stimulate further discussion. Finally, everyone filled out a second form, again stating preferences.

The results strongly influenced the preferred alternative, which is shown here as a series of developmental stages. The considerable investment required to achieve a healthy and beautiful lagoon is not likely to occur overnight, and even if it did, a period of trial, error, and feedback would be needed. Since the dynamics of the lagoon are not entirely understood, developmental actions may well have unexpected effects. The staging of development allows time for experimentation and correction. The four phases—improvement, restoration, revitalization, and refinement—give definite management goals without demanding too much at once.

Prepared for the City of Del Mar by the 606 Studio, Department of Landscape Architecture, California State Polytechnic University, Pomona. Gregory Michael, Thomas Schurch, Ronald Tippetts, and Gregory Vail, with advisors John T. Lyle, Jeffrey K. Olson, and Arthur Jokela.

DIVERSIFIED RESERVE

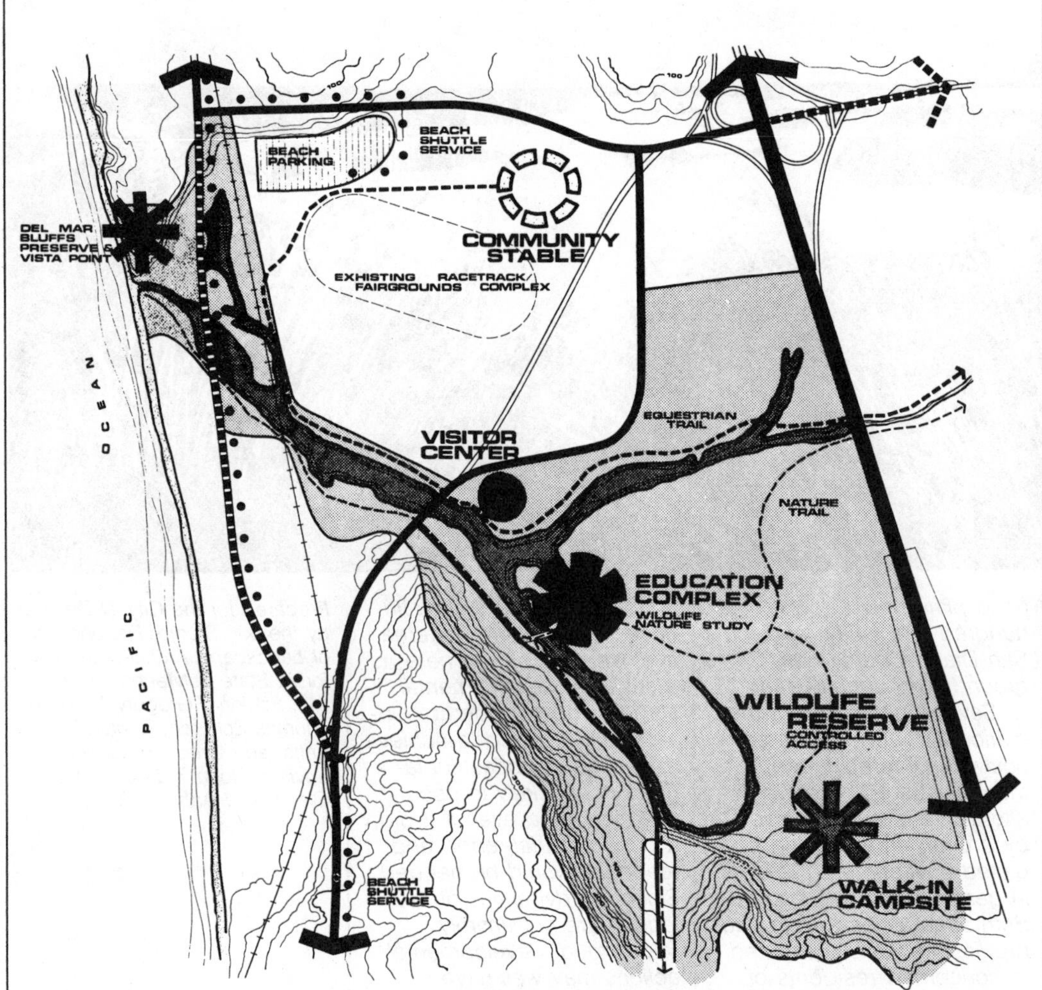

MEASUREMENT OF ENVIRONMENTAL EFFECTS

Alternative	Developmental actions	Natural processes impacted	Effects: Increase	Effects: Decrease	Human processes impacted	Effects: Increase	Effects: Decrease
Diversified preserve	Lagoon opening	Wildlife	▬		Beach access		▬
		Plantlife	▬		Maintenance	▬	
		Tidal flushing	▬				
		Beach sand replenishment	▬				
		Salt water intrusion		▬			
	Wildlife preserve	Wildlife	▬		Accessibililty		▬
		Plantlife	▬				
	Beach parking	Insignificant impact			Beach access	▬	
					Traffic congestion		▬
	Beach boardwalk	Cliff erosion		▬	Beach access	▬	
	Turf road major connector	Insignificant impact			Downtown congestion	▬	
	Visitor center	Insignificant impact			Education	▬	
					Recreation	▬	
	Education center	Insignificant impact			Education	▬	
	Beach shuttle	Energy	▬		Accessibility	▬	
		Atmospheric pollution		▬	Noise		▬
	Nature trail	Insignificant impact			Education	▬	
					Accessibility	▬	
					Recreation		
	Walk-in camp	Insignificant impact			Recreation	▬	
					Accessibility	▬	
					Maintenance	▬	
	Community stables	Ground pollution	▬		Recreation	▬	
					Revenue	▬	

REGIONAL PARK

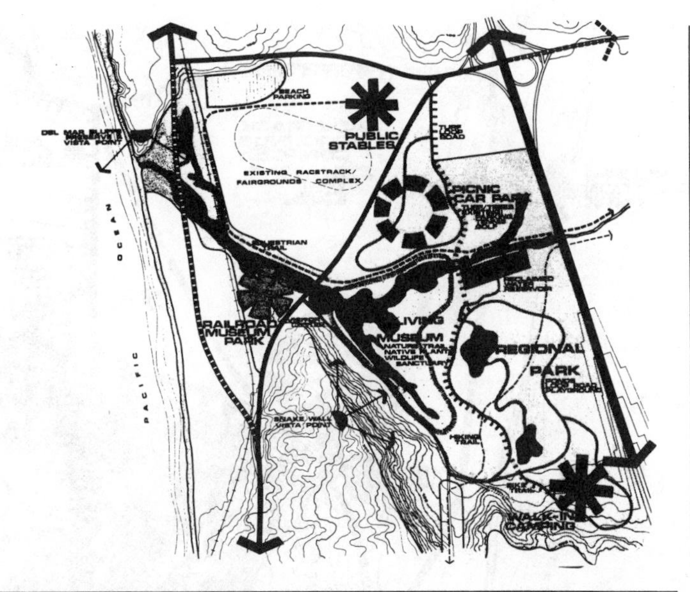

MEASUREMENT OF ENVIRONMENTAL EFFECTS

Alternative	Developmental actions	Natural processes impacted	Effects: Increase	Effects: Decrease	Human processes impacted	Effects: Increase	Effects: Decrease
Regional park	**Lagoon opening**	Wildlife Plantlife Tidal flushing Beach sand replenishment			Beach access Maintenance		
	Regional park	Water usage Native vegetation Wildlife			Maintenance Recreation Accessibility		
	Wildlife preserve	Wildlife Plantlife			Accessibility		
	Sewage treatment plant	Organic matter decay			Water recycling Maintenance		
	Walk-in camp	Insignificant impact			Recreation Accessibility Maintenance		
	Picnic park	Wildlife Water usage			Recreation Accessibility Maintenance		
	Ponds	Wildlife			Recreation Maintenance		
	Restrooms	Insignificant impact			Maintenance		
	Beach boardwalk	Cliff erosion Sand dune protection			Accessibility		
	Museum park	Insignificant impact			Recreation Maintenance		
	Public stables	Ground pollution			Recreation Revenue		
	Visitor center	Insignificant impact			Education Recreation		
	Turf road loop	Insignificant impact			Traffic congestion Accessibility		
	Equestrian trail	Insignificant impact			Recreation		
	Hiking trail	Insignificant impact			Accessibility		
	Bicycle trail	Insignificant impact			Recreation		
	Vista points	Cliff erosion			Recreation		
	Beach parking	Insignificant impact			Beach access Traffic congestion		
	Nature trail	Insignificant impact			Education Accessibility		

RESIDENTIAL COMPLEX

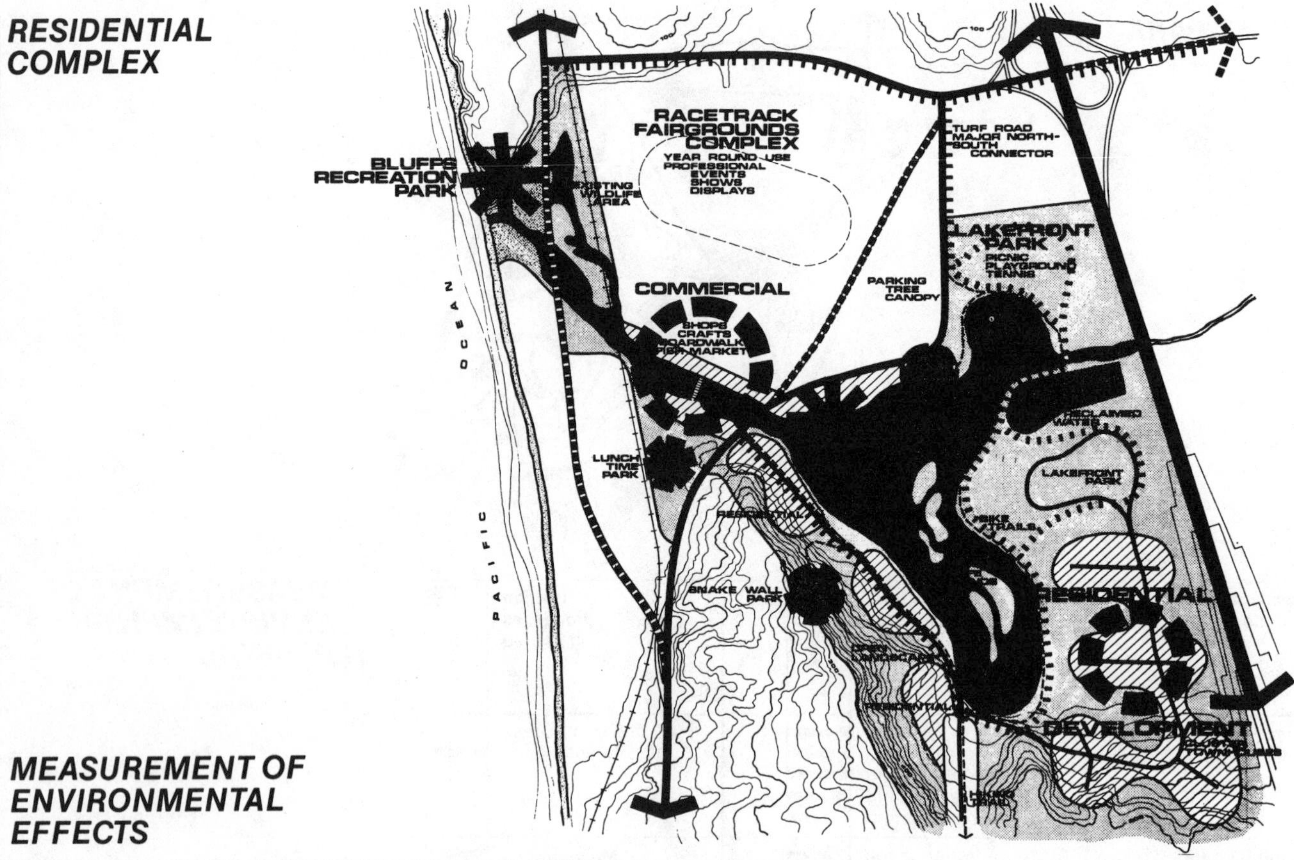

MEASUREMENT OF ENVIRONMENTAL EFFECTS

Alternative: Residential complex					
Developmental actions	**Natural processes impacted**	Effects: Increase	Effects: Decrease	**Human processes impacted**	Effects: Increase	Effects: Decrease
Lagoon opening	Wildlife Plantlife Tidal flushing Beach sand replenishment Salt water intrusion			Beach access		
Townhouse/cluster development	Water runoff Energy consumption Percolation Wildlife Native vegetation Atmospheric pollution Sedimentation			Accessibility Public services Noise Traffic congestion		
Expanded lagoon	Salt water intrusion Aquatic fauna Aquatic flora Terrestrial fauna Terrestrial flora			Maintenance		
Sewage treatment plant	Organic matter decay			Water recycling Maintenance		
Expanded race track/ fairgrounds use	Atmospheric pollution Energy use			Noise Recreation Maintenance Productivity Accessibility Labor resource		
Commercial areas	Flood plain hazard			Productivity Labor resource		
Planted race track parking	Percolation Runoff			Visual character Maintenance		
Freeway planting	Vegetation Wildlife Water runoff Sedimentation Percolation			Visual character Noise		
Lagoon area picnicking	Wildlife Water use			Recreation Accessibility Maintenance		
Turf road connector	Insignificant impact			Downtown congestion		
Bluff's Park	Cliff erosion			Recreation Accessibility		
Pocket Park	Insignificant impact			Recreation Maintenance		
Wildlife Islands	Insignificant impact			Insignificant impact		
Boating	Wildlife			Recreation Accessibility		
Hiking trails	Insignificant impact			Accessibility Recreation		
Bicycle trails	Insignificant impact			Recreation		

Tests show poor water quality, especially in the backwaters of the lagoon. The first diagram shows the existing water flow system, including sources of water and the pollutants contributed by each. The second diagram shows the redesigned flow system and the devices proposed to increase the inflow of fresh water and reduce the inflow of pollutants.

EXISTING WATER FLOW

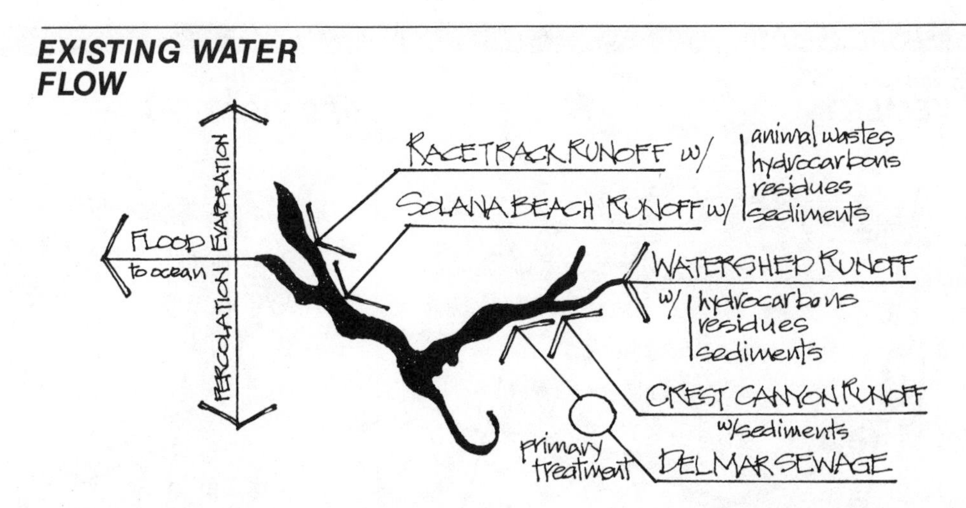

PROPOSED WATER FLOW

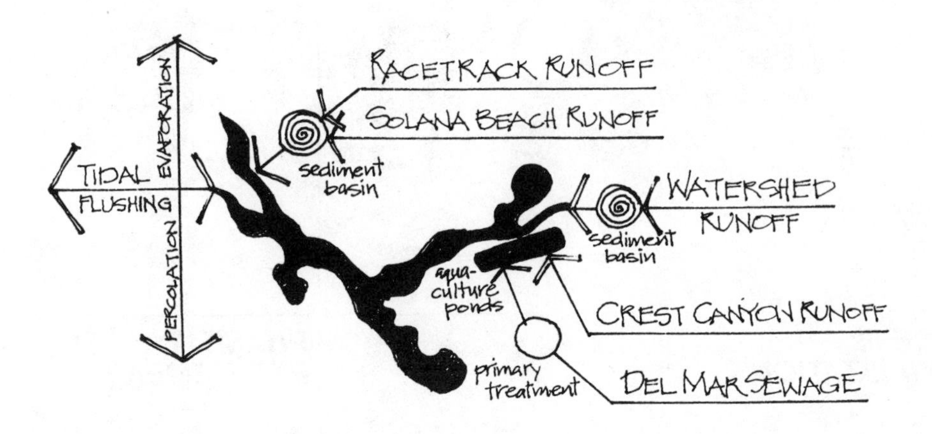

WATER QUALITY

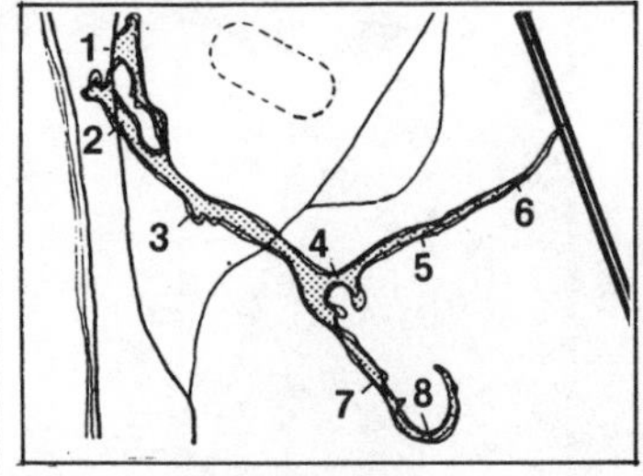

Test station locations

Station number	*	1	2	3	4	5	6	7	8
Date: 1977	NA		2/16	2/17	2/18	2/18	2/16	2/18	2/18
Time	NA		1500	930	930	100	1530	1030	1100
Air temperature °C	NA		24.5	14.0	13.5		24.5		
Water temperature °C	NS			16.5	14.5	16.5		16.0	16.5
Secchi depth feet	NA			1.5			1.25		
PH factor	7-8.5	8.70	8.75	8.75	8.73	8.77	8.35	9.05	8.98
Salinity 0/00	NA	8	5	5	6	8	4	6	6
Dissolved O_2 (PPM)	>5.0	6.60	6.29	7.34	9.38	4.90	6.05	10.55	9.44
Nitrate level MG/L	NA	1.76	2.64	3.08		5.72	0.00	2.20	1.76
Phosphate level MG/L	NA	.099	.310			.480	.118	.195	1.55
Heavy metals (lead) MG/L	NA	.04		.05					
H_2S level MG/L	.01-1.0	0.07		0.07	0.085	0.040			

*California State Water Quality Control Board standards

Tests performed by the Environmental Studies Laboratory, University of San Diego

STAGES OF REALIZATION

PHASE I: IMPROVEMENT

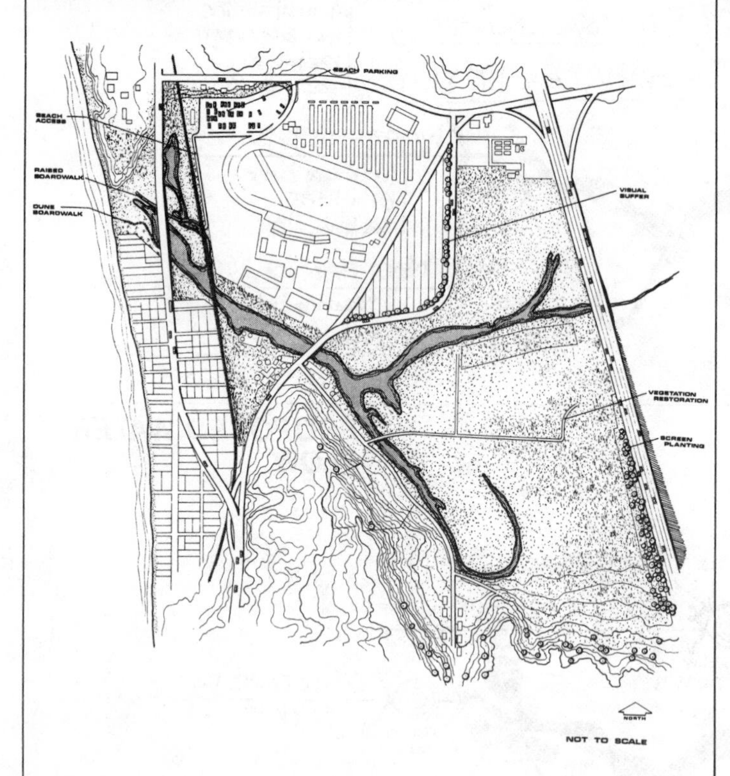

PHASE II: RESTORATION

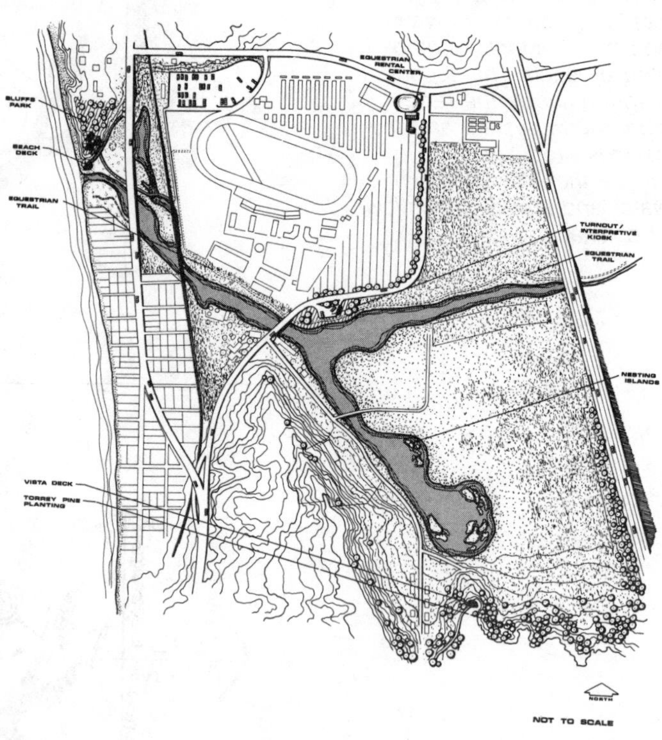

PHASE III: REVITALIZATION

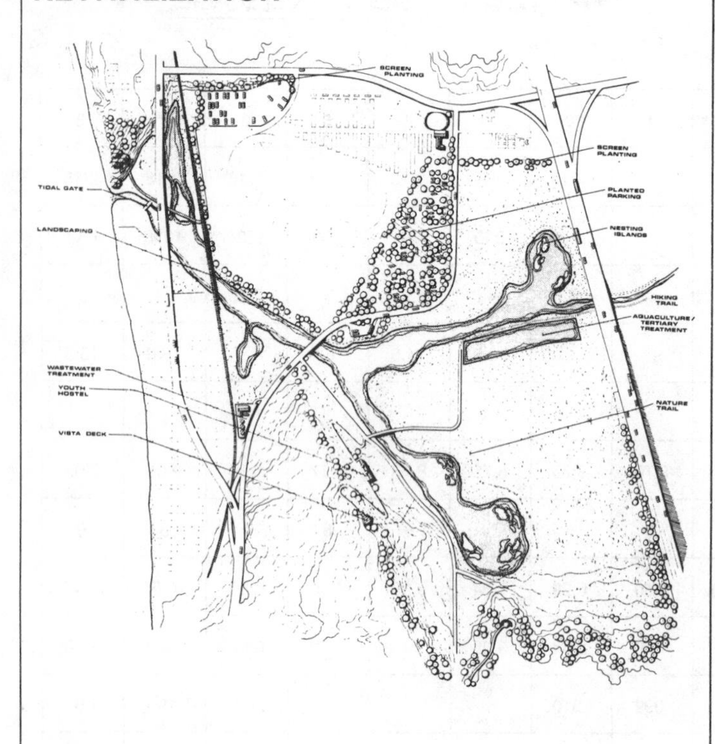

PHASE IV: REFINEMENT

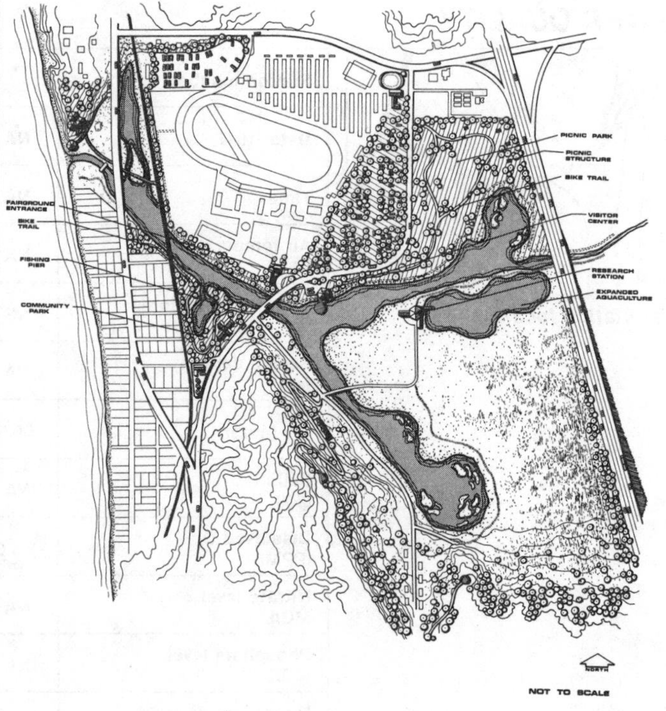

image—but it allows for experiments along the way. As in the colonization of any landscape, experiments that prove functionally viable, economically feasible, and environmentally compatible will be retained, while those that fail any of these tests will be dropped. Some seeds sprout and grow; others do not. Those that do are those that adapt to local conditions, develop mutually beneficial relationships with other species, and eventually become members of a community.

The San Dieguito Case Study

The steps for achieving such a goal are best described as part of the design. Ideally, then, a realization strategy should be considered as an integral design component. Such was the case in the design for San Dieguito Lagoon, illustrated on pages 185–90. San Dieguito is a small coastal lagoon just a few miles south of San Elijo, and like San Elijo, it is badly degraded. The issues are quite similar in both cases, as are the proposals for resolving them, and the political and economic environments are as complex as the physical ones. Achieving the design ideal will not be quick or easy. The case study plots it in a sequence of four steps, the logic of which is fairly simple. Before anything else can happen, certain lands have to be acquired and legal obstacles removed. Once this first step is accomplished, reactivation of the basic lagoon processes, including, most importantly, some degree of tidal flushing, can begin. This step is called "restoration." In the third step, wildlife habitats and some recreational facilities are established, and experiments with waste water reclamation systems in relation to aquaculture are initiated. Finally, the last step expands the range of recreational activities, completes the extensive planting program, and, based on the results of the initial experiments, increases the scope of the aquaculture activity.

At this point, the lagoon attains its climax state and is capable of sustaining itself with some degree of management control. The diagram on page 189 shows, in very general terms, how it will function—the inputs, outputs, and internal flows. Management control will be exerted by the City Planning Department and the Lagoon Preservation Committee, a particularly active citizens' group. It will receive three kinds of information to guide its decisions: goals from a larger interested public, research data from a variety of scientists who conduct the experiments in the lagoon, and data from ongoing monitoring of lagoon processes. On the basis of this information, the two management groups will make decisions that will then be passed on to two other agencies, the city's maintenance department and the State Department of Fish and Game, for implementation.

The Critical Role of Monitoring

Of the three sources of information, the most important by far is that provided by monitoring. Periodic structured observations, in fact, are the most fundamental activity in environmental management, the equivalent of taking temperature readings or blood samples to measure the health of a human body. Without some knowledge of the vitality of the system, meaningful decisions are obviously impossible. Since we cannot monitor everything, it is equally important to know the variables that furnish clear indications of health. In the case of the lagoon, these are water quantity and quality, because they effectively control everything else. Most important to know are the precise quantities of water flowing from the sources shown in the diagram and their load of sediments and nutrients. This information alone would probably be adequate for basic management decisions, although monitoring of other variables would undoubtedly be useful. For example, periodic depth soundings, the extent of marsh vegetation (which could be easily determined from aerial photographs), and bird counts might contribute to decisions. The problems of information waste and overload that apply to the original information base are relevant here as well, but it is important to limit the factors monitored to those that are truly significant.

In fact, we can view monitoring as an extension of the original information base, especially since the latter ideally acts as a management as well as a design tool. As noted earlier, feedback from later to earlier stages is a basic part of every design process, Since management usually involves periodic redesign, the feedback loop can be the critical, permanent linkage between management and design.

The interlocking, feeding-back, ongoing relationship between design and management suggested here is quite

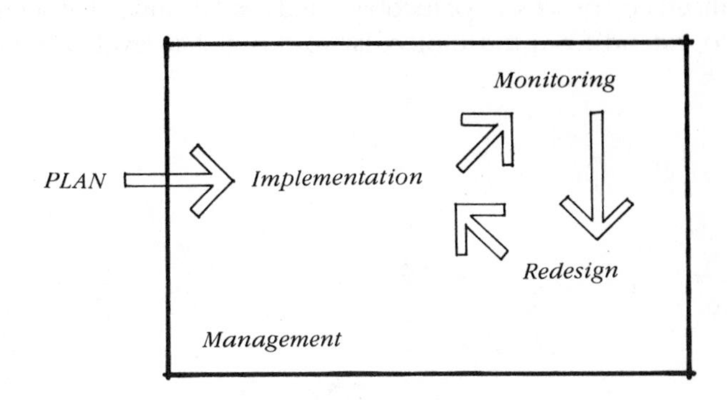

different from the usual view of both activities until very recent times, that is, that they are quite separate enterprises. Oddly enough, the literature on each subject habitually ignores the other. Without management taken into account, however, design is divorced from reality, and without design, management can have no vision of a possible future. To be effective over the long run, design has to provide the spadework for management and recognize that the vision it presents is subject to change. The illusion that we can rely only on something called "maintenance" to preserve a designer's vision for all time has faded. One need only reflect on the tangled overgrowth at Villa Lante or the enormous transfusions of energy that are needed to keep Versailles as Le Nôtre envisioned it.

This, I believe, is one of the major conceptual divides that separates us from the designers of the past: the awareness of change and uncertainty. For all the intricacy of our analytical methods, we know that none of them tells us anything for certain. To develop a model is to become aware of its crudity, of the many subtle factors with which it cannot cope. Our predictions are always couched in terms of probability, and so are our designs. We rarely have even enough understanding to estimate the odds in favor of our predictions actually occurring; we can only say that they are likely.

This uncertainty is partly due to lack of sufficient knowledge and partly to the limitations of the human intellect, but it is also largely due to the way natural processes work. So many complex factors enter into any event in nature that the outcome is always in doubt. Rarely can we establish a clear link between one cause and one effect; rather, a multiplicity of events seems to lead to a multiplicity of other events. We can find consistent patterns in sequences (among them the concepts and principles discussed in this book), but the patterns are never absolutely consistent. Thus, we are always, to some degree, uncertain.

Chance is usually an important factor in natural events as well. A rabbit lives the possibility that a coyote may perchance come his way at feeding time, and a giant sequoia lives under the risk of a lightning stroke. Whole ecosystems are perennially subject to hazards of fire, hurricane, or blight. Evolution apparently depends on chance mutation. Even if we could perfect our understanding of natural processes, therefore, our predictions would still be limited by factors of chance.

Nature deals with the continuing flux of uncertainty mostly by processes of feedback and readjustment that seem to extend from the molecular to the global level. We are beginning to understand, as a result of the work of people like Ilya Prigogine,[3] that natural systems are always in the process of self-organization and reorganization, increasing in complexity, definition, and information content, and that chance always plays a major role. Chance events are abridged or amplified by negative or positive feedback to reduce or accelerate change.

We have seen that design processes work like natural ones in some ways, and even more obviously, so does management. It is in the replication of the developmental processes of nature that the power and the great promise of landscape management lie, as well as its inevitable integration with design. Management is different from maintenance in that it is a process not of preservation but of continuing redesign. Consequently, the landscape manager must play a creative role, responding to all of the variations, even the vagaries, of behavior on the part of both human beings and natural processes—the sort that are bound to occur but cannot be predicted by even the most imaginative and meticulous of designers. In doing so, he will guide change and shape it, rather than standing squarely against it. The designer of a human ecosystem who returns to his landscape a few years later will be sadly disappointed if he sees precisely the same form that took shape on his drawing board.

[3] See Prigogine, 1976, 1978.

PART III:

Modes of Ecological Order

CHAPTER 10

Foundations of Ecosystem Design

With so many different kinds of information involved in shaping a landscape, the most difficult methodological problem is putting the information together in a sufficiently coherent form to represent reality. Analytically, we use models for this purpose, but the need for a unified vision transcends analysis. This is the very core of what we have been calling ecosystematic design. Only if we can comprehend and envision the entity we are trying to shape as a dynamic whole can we have any hope of dealing with it creatively. To do so requires strong concepts of an underlying order that holds the diverse pieces and all their hidden relations together. In architecture, this integrating role is played by the laws of physics, which provide the means for first understanding in a general way, then shaping and identifying, and finally calculating the stresses that occur in every member and joint. The whole system is envisioned as a network of members that interact through connections which depend on all other connections. The complexity of the whole is enormous, but reduced to these terms, it is not difficult to grasp. In some buildings, human creativity has managed to transform this purely physical system into poetry. Witness, for example, Chartres Cathedral, the Eiffel Tower, or a geodesic dome. In the case of Chartres Cathedral, since the understanding of physical laws at the time was entirely intuitive and the available materials were limited, the structure is a bit heavy, though nonetheless beautiful for it. By the time of Eiffel, however, the physical laws were generally known, and its forms reflect this knowledge as well as the availability of a revolutionary structural material, cast iron; members and connections are clearly expressed here as an interacting whole. It was left to the geodesic dome, however, to develop a totally different view of the parts to the whole. Buckminster Fuller articulated the concept of synergy and then expressed it by creating the geodesic dome. Synergy being a strength of the whole that is greater than the sum of the strengths of individual parts, the secret of this architectural miracle lies in its integrative form.

In ecosystem design, although the underlying concepts of order are drawn not from physics but from ecology, ecology in turn looks to physics for some of its underpinnings. Recall that the lowest, most fundamental levels of integration are at the physical levels where the building blocks of matter are put together. But at the levels that concern us here, these building blocks have already been aggregated and subsumed into ecosystems. If the additional integrative law that we proposed earlier—that predictability decreases with ascending levels—still holds, then ecosystem design will always be less predictable, both in the information it uses and in its results, than architectural design.

The ecological concepts that we will discuss and apply in this last section of the book are more or less analogous to the laws of mechanics in that they provide us with organizing principles for shaping ecosystems much as architects shape buildings, even if, in keeping with the integrative law just mentioned, they are less predictable and less precise. Our understanding of them is considerably less developed as well. On the road to reliable ecosystem design, we are probably somewhere between Chartres and the Eiffel Tower, with the geodesic dome only a vague form on the horizon.

Incomplete though they may be, structure and function are undoubtedly among the most useful of ecological concepts, at least for purposes of design. Having defined both terms at some length in the first chapter, I will provide only a brief reminder here that structure refers to the composition of the plant and animal species and the abiotic elements of a landscape, whereas function refers to the flow of energy and materials there. We might see these as the two fundamental means of making the order of ecosystems accessible to the intellect, on the one hand by showing how the pieces fit together, and on the other by revealing the distribution patterns of the essential substances that maintain and motivate them—the statics and the dynamics of the system. Their importance to design is that they give us access to the inner workings of ecosystems. They can be visualized, represented in three dimensions, analyzed, measured, manipulated, shaped, and reshaped.

There is, however, one serious difficulty with this two-part division of ecosystem order into structure and function, and this arises from the role played by spatial organization. In a human landscape, such distribution looms as an extremely important consideration. Although locational factors play a minor role in most ecological research, they have assumed a major one in most attempts to apply ecological principles to design, especially at the larger scales. In part, this is a consequence of the emphasis on location in traditional—that is, economically based—land-use planning. It also reflects the many cultural values and distinctions attached to specific locations—that is, to the importance of place—and to the plethora of abstract lines that we enforce on the land: lines defining political boundaries, jurisdictions, ownerships, zones, and other forms of control. This is why I have chosen to separate the consideration of locational distribution from the minor structural role usually assigned to it in the literature of ecology and to elevate it to the position of a third mode of order, one that is on a more or less equal footing with structure and function.

These, then, are the fundamental modes of order by which we can understand and shape ecosystems: structure, function, and location. The distinctions between them are somewhat artificial, of course, and not so much an inherent characteristic of nature as a device for human convenience that allows us to deal with the dauntingly complex, ineffable whole.

STRUCTURE

Being concerned with tangible, visible rocks, soils, plants, and animals, structure is probably the easiest of the three to understand. We can break the structure of an ecosystem down in several ways. In broad terms, we can think of it as being composed of abiotic components—rocks and soils—and three essential biological components—the producers, which means primarily green plants, which store energy through photosynthesis and comprise over 99 percent of the earth's biomass; the animals, which consume plants and other animals; and the microorganisms that facilitate decomposition. We can break each of the living components down into constituent species, each species having a population, or given number of individuals, and we can consider the populations as being organized in communities. The concept of community is one of the oldest in ecology—the earliest notice of definite associations among species having been taken before the mid-nineteenth century—but nevertheless one that is often held open to question.

The major question concerning communities seems to be this: Is it correct to consider them as homogeneous units, or do they vary along a continuum in such a way as never to be the same in two different places? Both positions can be argued convincingly. Community composition does change continuously in space, especially where environmental gradients are steep, that is, where changes in environmental conditions are great over relatively short distances. On the other hand, where environmental conditions are fairly constant over large areas, community composition is consistent enough to permit the identification of more or less homogeneous groupings of species. Though careful analyses might show continuous variations in species and numbers throughout such areas,[1] community composition is usually consistent enough for design purposes. Transitional areas where variations suddenly become greater—usually because environmental gradients become steeper—and that thus form boundaries at the edges of areas considered homogeneous are called *ecotones*. Ecotones can be identified by sampling habitats along gradients, which are often lines of altitude change.

The second major question concerns interactions. Earlier, we defined ecosystem structure as a structure comprising elements *and* their interactions; indeed, this is the definition for a structure of any sort. Elements of concern might be found at any level, but as a rule in ecosystem design, it is most useful to limit them to species and soil and other environmental conditions. The question is: To what extent do the species within a community interact and to what degree do these interactions shape the community?

The more traditional view—the one held by pioneers of ecology like Clements, Tansley, and Braun-Blanquet, and still held by most ecologists—is that communities are fundamental, natural, interacting units that can be classified almost as we classify species. Some studies have suggested, however, that interactions among species are really not all that cohesive, and that communities might be considered simply as groupings of species that share the same environmental requirements.[2] The available studies show that, analyzed two by two, most of the species within a community do not interact and that relatively few interactions are obligate (necessary to one species or the other) or exclusive. A predator, for example, usually preys on a number of species and could survive in the absence of one or several of them. The predator-prey interaction is nevertheless of fundamental importance. Thus, although the composition of most communities in any given landscape is determined primarily by the environmental conditions, virtually every species has an important interaction with at least one or two other species, and most have more. The diagram below illustrates the network of interactions among species observed in a marsh in Wales.[3] Each circle represents a species. The solid lines represent significant associations at the 1-percent probability level and dash lines at the 5-percent level.

[1] See Curtis, 1959.
[2] See Curtis, 1959, and Whittaker, 1966.
[3] After Krebs, 1978.

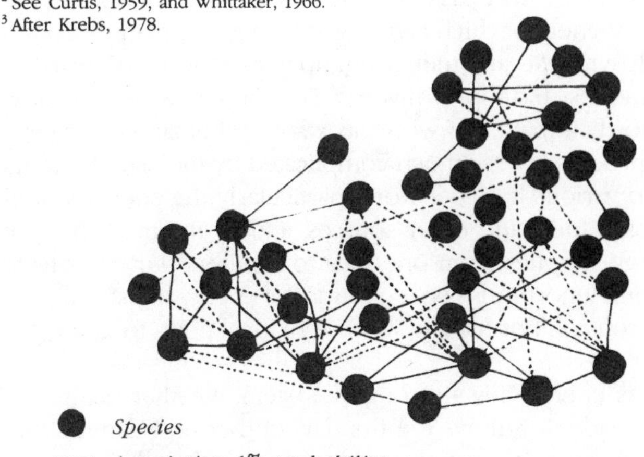

Thus we can see that while interactions are limited for any one species, the aggregated network of interactions is essential to community cohesion, and virtually every species participates in the same way. For design purposes, then, we will take the view that the network of interactions, however limited, is essential to community cohesion.

There are a number of important characteristics of community structure and several ways of classifying them. Among the important characteristics are diversity (number of species present), growth form (trees, shrubs, herbs, mosses), dominance (controlling species by virtue of size, numbers, or activities), relative abundance (relative proportions of different species), and trophic structure (who eats what). Since each of these plays an important role in defining the structural character of an ecosystem, there are a number of ways of classifying communities according to these and other characteristics. Among the more common, especially in Europe, is that of Braun-Blanquet, which bases classifications on relative measures of cover abundance, sociability, fertility, frequency, and fidelity (which measures abundance outside the community in question).[4]

Within communities are smaller groupings called "associations," which are combinations of species that consistently appear together across environmental gradients. Examples of common desert associations are given in the Whitewater Wash case study. These include the dune grass association (sand verbena, evening primrose, and panicum), the oasis association (mule fat, barrel cactus, Washington fan palm), and the riparian association (white alder, giant reed, desert willow, cotton wood, among others). They are essential to the integrity of this desert ecosystem and should be maintained in any development plans.

The concept of structure and the related concepts of community and association have practical importance because of their influence on stability and on what we have called the "sustainability" of human ecosystems. They also give us a conceptual basis for designing into human systems the conditions required by plant and animal species as well as by people.

FUNCTION

In order to understand how structural order is maintained, we need to consider functional order, the two being closely intertwined. Structural order is supported by the continuing flow of energy, which is first stored in green plants through photosynthesis and then taken up in turn by the herbivores, carnivores, and top carnivores. Thus, in nature, energy flow is mostly a matter of who eats what. In human ecosystems, the process is very much complicated by the use of energy to do various kinds of work, particularly the energy stored in fossil fuels and other sources, and by the movement of energy supplies from one area to another. Our cities have become enormous concentrations of energy transformation and control, but usually without rational goals to guide the control.

As energy flows through a system, whether natural or man-made, it follows the first law of thermodynamics, that is, it is neither created nor destroyed, although it may be

transformed from one type to another. With each use or transformation, following the second law of thermodynamics, it is degraded into a more dispersed form. These two laws make it possible to map the flow of energy through an ecosystem and thus obtain a clear understanding of the energy regime. Using the same techniques, we can design the energy flow of a human ecosystem in ways that will improve efficiency, make rational choices among energy uses possible, and permit work within limited inputs.

Because matter also can neither be created nor destroyed, we can map the movement of chemical elements as well. Water and the numerous elements that are essential to life tend to move in cycles, being stored for certain periods of time, then used by one organism, returned to the environment, used by another organism, and so on, all in repeating patterns that are more or less consistent. Human use tends to radically alter these cycles in a variety of ways, thus bringing about some of the most disastrous of the ecological impacts caused by man. In general, human use tends to speed up the cycles, sometimes to the extent that they no longer function in purely cyclic ways. In some cases, critical shortages result, but more commonly, we find materials being concentrated in places where nature cannot deal with them in her normal fashion. Examples are nutrients sufficiently concentrated in bodies of water to cause damaging levels of algae growth, the buildup of metals in animal livers, and the concentration of carbon dioxide in the upper atmosphere. The latter two problems have become such integral parts of the economic picture that changing them will be a very long-term matter.

Other problems are more immediately accessible to correction. In a great many cases, as shown in the Aliso Creek case study, the key lies in control of the hydrological cycle. Water plays a fundamentally important role in the function of ecosystems, both on its own account and as a transporter of nutrients. Through major parts of their cycles, most nutrients follow the flow of water. Water is probably

[4] See Braun-Blanquet and Furrer, 1913, and Braun-Blanquet, 1964.

the most important factor in the ecology of terrestrial organisms as well. Especially in more arid landscapes, as we have seen, it is often the limiting factor, but this is increasingly true in the human landscape also, even in damp climates. Water shortages are becoming more common, and most authorities agree that worldwide shortages are likely to occur before the end of the century.

In the light of the overwhelming dysfunctions that have been brought about by misdirected flows of materials and energy—problems that threaten to become even worse—learning to control such flows is one of our most urgent tasks. "Now in the face of man's exploding population and dwindling resource base, his very survival may depend on an accurate knowledge of ecosystem function, i.e., maintaining the continuous flow of energy and nutrients vital to the existence of ecological systems and life itself."[5]

LOCATION

As we have seen, both structure and function vary with location. In nature, the variations evolved out of the interactions of landform and climate, which produced, over millions of years, particular combinations of temperature, moisture, and surface form. These conditions, in turn, fostered the growth of particular soil regimes and communities of plants and animals. Each species has limiting conditions that depend, at a fundamental level, on temperature and moisture, but that also depend at a secondary level on the presence of other organisms with similar environmental needs. Thus, every place on earth is unique in the character of its climate and landform and in the living community that it supports, and that uniqueness is the current expression of millions of years of development.

Wherever human beings enter the picture, which is virtually everywhere, human land use enters with them, altering this ancient pattern of place and process. Man creates new places very quickly by the standards of natural evolution. Sometimes the new places fit into the patterns already evolved, work within the established limits, and create viable new regimes of structure and function. More often, however, the new places ignore these patterns, creating regimes of structure and function that have no relationship to the evolutionary past, and end up requiring continuous maintenance and repair, not to mention never-ending transfusions of materials and energy.

[5] See Likens and Bormann, 1972.

The point is that it is entirely possible for human ecosystems to fit harmoniously into evolved patterns of location, but it requires a careful mapping of the particular place and an understanding of the characteristics that combine to create it.

EXISTING AND FUTURE ORDER

Undertaking the design of any landscape, we are confronted with an existing ecosystem, the order of which can be understood only by analyzing its structure and functional and locational patterns. If it is a natural system, these elements are probably already working effectively to support the living community. If it is a man-made landscape, they are probably degraded to the point of working marginally or not at all, as at San Elijo Lagoon (page 5), Bolsa Chica (page 173), or the channel at Lake Elsinore (page 163).

If the landscape is natural, an analysis of the existing structure and functional and locational patterns will provide a working baseline. As discussed in Chap. 7, it is important to identify the key and dominant species in the structure and their roles, the factors limiting growth and distribution, the pivotal processes, and the areas where resources are particularly concentrated and why, among other things. If the landscape is man-made, it may be necessary mentally to reconstruct the order of the environment as it existed before being altered. In the case of San Elijo, such a reconstruction provided the key for design.

Once we understand the existing order, the process of reshaping it is something like designing a building. We might return here to the image of the seven-story building that we used to represent the system of scales of concern. Structural, functional, and locational patterns knit each level into a unit and also serve to join the levels together. The locational pattern—strongly influenced by the lay of the land but at the same time responsive to human needs, activities, and paths of movement—corresponds to the floor plan. The landscape structure corresponds to the skeletal framework holding all the building materials in place—a network of elements that may be inert and more malleable and predictable than the living, fluctuating structure of an ecosystem, but are nevertheless inescapably interrelated. As for function, though this is not the way the term is conventionally used in architecture, a building, like an ecosystem, has a system of flows—of water, sewage, electricity (energy), warm and cool air—that provide for life-support needs.

If a building seems too much a human artifact to serve as analogy for an ecosystem, consider the roots of the word ecology: *oikos* for "house," *logos* for "the study of." In a sense, in designing a human ecosystem, we are shaping a macro-house, and when we design a house, we are shaping a rather small ecosystem.

It is important, therefore, to understand the order of two systems: the one that exists and the one that will exist when we have redesigned it. The two, by the definition of design, are necessarily different. They may be different in only minor ways, or they may be radically different. In either case, there is yet another ecological concept that is helpful in understanding the order that underlies both, particularly with respect to their position in the changing patterns of time. The concept of succession places ecosystems in the perspective of time.

Succession

All three modes of order change over time, their state at any instant being closely related to their progress in the ongoing sequences of succession. These are interpreted in different ways by different scientists, depending on their particular viewpoints. Like other concepts of ecology, succession—the processes by which ecosystems change—is something of an unfinished sculpture, the details of which are left to be chiseled by each interpreter in whatever area suits his purposes. Eugene Odum sees succession as ecosystem development.[6] Margalef goes farther in this direction by speaking of the maturing of ecosystems. "Maturity, then, is a quality that increases with time in any undisturbed ecosystem." And, ". . . we may speak of a more complex ecosystem as a more mature ecosystem."[7] Clements, who was a pioneering researcher in the field, saw succession as ". . . the interaction of three factors, namely, habitat, life-forms, and species, in the progressive development of a formation" and as the ". . . development or life-history of the climax formation."[8]

Clements took a highly deterministic view of succession. Comparing the development of a community with that of an individual organism, he considered a single climax state as eventually inevitable. More recent researchers tend to a more flexible and probabilistic view. The dying out and immigration of species depends to a considerable degree on chance and on conditions that are largely unpredictable. Thus, the road to the climax is not a straight one. And in any case, it is sometimes argued that communities rarely reach the climax state because they are so frequently perturbed and thus set back in the sequence. The notion of a single, inevitable climax state is also widely questioned. More likely to occur are several different climaxes in a given area, depending on soil conditions, the influence of animals, and other factors that vary over small spaces.[9] Krebs argues that true equilibrium, and thus a true climax state, can never be reached because a major controlling influence, the climate, is not constant but always changing.[10] In any case, the climax concept rarely applies to human ecosystems, because of the particular demands placed on them. To replace the climax, we have dynamic management.

Although succession now looks like a less universal, predictable, deterministic process than it was once thought to be, it still seems to be clearly directional, developing, at least in the early stages, through states of increasing complexity and equilibrium. Eugene Odum has developed a set of principles describing successional change in structure, function, and locational pattern that are intriguing and most useful for purposes of design.[11] Although these remain theories, and are thus arguable, most are supported by considerable research and can thus serve as reasonable guides for practical purposes.

In brief, Odum measures successional development by changes in 24 ecosystem attributes. Of those concerned with structure, he lists total organic matter, species diversity in terms of both variety and equitability, biochemical diversity, niche specialization, and size of organisms. All of these increase with successional development. Of those concerned with energy flow, gross production and net community production of biomass both decrease, whereas the biomass supported per unit of energy flow increases. The ratio of gross production to community respiration (P/R) tends to level off at 1:1, and food chains become more weblike, less linear, and more dominated by decomposers than by grazers. Nutrient cycles become more closed—importing and exporting less to and from outside the system—and the rate of exchange of nutrients between organisms and environment slows down. As for location, patterns become finer, more diverse, more heterogeneous. In short, ecosystems become more complex and less productive with age.

In this view, the order of an ecosystem depends largely on its successional state, and if such is the case, we might see ecosystem design as a matter of mimicking certain stages of succession. Indeed, this is the theory that underlies the landscape typology proposed by Odum and described in the first chapter. Productive landscapes are ecosystems in early stages of succession, simple in structure with little diversity and simple in function with rapid flows of materials and energy and thus high biomass production. Protective landscapes are those in later stages—complex in structure, function, and locational patterns and low in productivity. Odum, equating diversity with stability, theorizes that protective landscapes will be self-maintaining and contribute to overall homeostasis by virtue of their ecological function and structure.

The relationship between diversity and stability is a matter of controversy. Although there does seem to be a general correlation between the two, researchers have found a great many exceptions, and cause-and-effect relationships are not at all clear. We will examine the questions surrounding stability and diversity at some length in the next chapter, because they point up some of the difficulties of ecological rules of thumb. Similar difficulties surround most ideas about directional trends of succession. When we examine particular situations, complications always seem to arise to confuse or cast doubt on general principles.

It is also important to understand that successional change is not linear or constant and that all changes in the variables of succession do not march to the same drummer. We may, for example, find very rapid, open mineral cycles in combination with very fine, clearly differentiated locational patterns. The directions of change, moreover, are often altered

[6] See Odum, 1969.
[7] See Margalef, 1963; p. 216.
[8] See Clements, 1916; p. 4.
[9] See Tansley, 1935, and Daubenmire, 1966.
[10] See Krebs, 1978.
[11] See Odum, 1969.

by chance events and by minor as well as major perturbations. What looks like a constant trend in the perspective of hundreds of thousands of years may look more like a series of fluctuations over a decade or two.

Altogether, then, it seems best to view succession and the related changes in ecosystem order in the probabilistic way. There are too many unpredictable factors to permit certainty in our understanding or predictions. As argued in Chap. 9, the whole process of ecosystem design is best approached in the same probabilistic spirit. Here, by understanding that we can design very little with absolute certainty, we will have to part with the analogy of the design of a building. When we design a building with a window in a certain place, we can be quite sure that the completed window will be precisely in that location. In the landscape, we might hope for that level of certainty at the construction scale but usually not even there. Much of our design is indirect; that is, since our real purpose is not directly accessible, we have to propose another action that we hope will achieve it. In the design of the settling ponds in the Los Angeles Flood Control case study, for example, islands are proposed, the purpose of which is to increase the duck population by providing ideal nesting sites. If the design is accepted, we can be reasonably sure the islands will be built in the forms shown, but we cannot be sure the ducks will use them for nesting. If they do, it is again only a matter of probability whether they will produce offspring that will grow to maturity. The farther removed from direct action a purpose is, the greater the uncertainty surrounding it. And on the other side, the larger the scale of design, the greater the distance separating the designer from direct action, and therefore the greater the uncertainty concerning results.

The probabilistic character of ecosystem design gives enormous importance to the management phase, because that is when results can be observed, probabilities reassessed, and courses of action shifted. In practice, it may turn out that techniques for control of structure and function have even greater application in management than in design. In the next few chapters, we will explore some of the practical means of working with the modes of ecosystem order and discuss some of the more complex questions concerning them.

I should not move on, however, without offering a caveat. This has been a very brief introduction to some of the more useful ecological concepts. I have presented a very complex subject in summary form, and, although the next chapters will supply considerable detail, it will be only enough to demonstrate some principles and techniques for applying these concepts. Those who intend to pursue work of this kind will need to know a great deal more and to keep up with the ever-emerging literature since ecology is a science in a state of growth and flux. The references cited will provide a beginning.

CHAPTER 11

Structure

The Roles of Plants

Plants are the creators of order in the landscape, applying their magical ability to use the sun's energy in making biomass to counterbalance, to some degree, the ever-dissipating energy of inevitable entropy. Beyond the basic level of the individual plant, order depends on cooperative life, and cooperation develops with successions. As succession progresses in the natural landscape, structure develops over time. After a disturbance—a fire, hurricane, or grading or mining operation, for example—colonizing seeds begin to germinate, usually after being imported by the wind. These are the plants we calls "weeds": small, fast-growing, widely adaptable, often beautiful, but usually scorned. Common among them are aster, goldenrod, ragweed, dandelions, and Queen Anne's lace. They tend to be highly independent individualists; they appear at random and have little to do with their fellows. In this sense, they can hardly be called a community or said to have a structure. Their interactions are minimal, but each goes about altering its own immediate environment. It provides shade and slows the wind close to the ground. Its roots stabilize the soil and after it dies, its decay enriches the soil. Thus, it prepares the way for plants that need such conditions, usually perennial grasses, which become the next stage of succession. In this more benign environment, the colonizing weeds are at a competitive disadvantage and in time die out. As the sequence advances, the number of species per unit area increases, as does the network of relationships joining one with the other and with their environment. Shrubs and trees provide various kinds of food and cover for any number of animal species as well as shade for brackens and ferns. With all these landscape elements and their interactions in a state of growth, structure, by definition, develops. At the same time, a variety of patterns, or substructures, become ever more coherent. Particularly significant among them for landscape design purposes are vertical layering, horizontal zonation, and trophic levels.

As succession advances, the two basic vertical strata—the soil and the air above it (the sites of heterotrophic decomposition and autotrophic metabolism, respectively)—subdivide into a range of levels. A well-developed soil in a deciduous forest has five distinct levels, and the vegetation it supports might be seen to have herb, shrub, and understory and overstory tree levels. Vegetation layering reaches an extreme in the tropical rain forest. Animal habitats tend to be highly stratified as well, with relatively few species occupying more than one stratum.

Horizontal zonation patterns usually reflect environmental conditions, that is, prevailing microclimates, topography, soil, and water regimes. We will deal with such patterns at length in Chaps. 14 and 15.

Trophic patterns or food chains are most easily visualized as hierarchies with at least three levels: the green plants that are the primary producers, the herbivores that eat the plants, and the carnivores that eat the herbivores. The decomposers, or detritus eaters, and the top carnivores also have important roles to play as a rule. Since trophic patterns are the pipelines of energy flow, the properties of function and structure merge in them and become virtually indistinguishable. Even purely artificial boundaries become hard to draw.

Although a great many species occupy more than one level in the trophic hierarchy (that is, many animals are omnivores), the food chain and its limitations have a major bearing on population structure. In the mature stages of succession, the trophic pattern becomes extremely complex, encompassing as it does a great many highly specialized species.

Of the 24 successional trends identified by Eugene Odum, the six measures of community structure are as follows:

1. Total organic matter (which increases)
2. Inorganic nutrients (which proceed from extrabiotic to intrabiotic)
3. Species diversity: variety component (which becomes greater)

4. Species diversity: equitability component (which becomes greater)
5. Biochemical diversity (which becomes greater)
6. Stratification and spatial heterogeneity, or pattern diversity (which proceeds from poorly organized to well-organized)

With the passage of time, then, and barring any major perturbation, the structure of a natural ecosystem becomes more complex, more diverse, more organized, more efficient, and larger in terms of its biomass. These qualities are closely associated with similar trends among functional characteristics.

Eventually, in theory, at least, the ecosystem develops to the level at which it contains the maximum volume of biomass that the environment can support. At this point, the ecosystem has achieved maturity, or its climax state, and, as far as we know at the present, it will not undergo further major changes until it is set back in the sequence of succession by a major perturbation. There remains, however, the intriguing question as to whether or not ecosystems might age as individual organisms do.

Since the available niches of a climax community are filled and any possible interactions are already at work, we might consider the structure complete, with a high level of stability. Odum's work, along with that of a great many others, has shown high correlations between the diversity associated with maturity and stability, although the cause-and-effect relationship remains unclear.

DIVERSITY AND STABILITY

Here we run head on into another ongoing controversy. Does diversity really promote stability? Since stability is a quality we fervently seek in human ecosystems, the question is one of great importance to design and is thus worth pursuing.

Part of the difficulty lies in the very general nature of the term "stability." The structure of an ecosystem fluctuates in rhythmic patterns, and stability is the tendency of the magnitude of fluctuation to remain within limits. Gordon H. Orians expands considerably on this basic meaning, defining stability as "the tendency of a system to remain near an equilibrium point or to return to it after a disturbance."[1] He goes on to explain it in terms of seven different characteristics. These are *constancy* (lack of change in some parameter of a system), *persistence* (survival time), *inertia* (ability to resist external perturbations), *elasticity* (speed with which the former state is returned to following a perturbation), *amplitude* (area over which the system is stable), *cyclical stability* (oscillation around some central point or zone), and *trajectory stability* (movement toward some end point).

Since the factors influencing these different characteristics of stability are different, any given ecosystem, Orians points out, is likely to be stable in one or more ways and unstable in others. It thus becomes difficult to generalize on the subject of ecosystem stability. Rather, he recommends that inquiries concerning ecosystem stability be focused on specific characteristics of stability and the causal processes associated with them. Interaction matrices of the sort used by ecologists to analyze population equilibrium (in which each element of the grid shows the effect of one species on another) can be useful for this purpose. The trouble is that we encounter again the common dilemma: How can we apply ecological information to design in a practical way? Such analyses require years of research for natural ecosystems and are well beyond our present reach for human ecosystems. Certainly, they present a promising direction for ecological research, but meanwhile, there are decisions that have to be made. We need more general guidelines.

As a move in this direction, we can reduce Orian's seven characteristics, which overlap somewhat, to only two: resistance and resiliency. Furthermore, we can examine some specific ecosystems whose structures and stability characteristics are well known. In earlier chapters, we dealt at some length with two such systems: the salt marsh and the tropical rain forest. The salt marsh features a simple structure that is only moderately stable with respect to resistance but highly resilient. The tropical rain forest is an extremely diverse environment that is somewhat resistant but has very little resilience. (As we pointed out, once rain forests are cleared and their soils laid bare, the vegetation is extremely slow to reestablish itself.)

Robert May explains such situations by postulating a relationship between the environment of a system and its stability.[2] In an environment subject to only minor perturbations, a diverse structure may be the more stable, whereas in an environment open to major, random perturbations, a simple structure may be the more likely to persist. Thus, in the relatively constant conditions of the tropics, where few outside forces enter to upset the natural balance, the complex rain forest is virtually imperturbable. Whenever human beings go to work with chain saws and bulldozers, introducing a truly massive new force, however, the whole structure can quickly collapse. The collapse is due largely to the tragic flaw of its soil structure, although it might be argued that a diverse system with a large number of interactions harbors more opportunities for such flaws to develop than a comparatively homogenous one. The simple salt marsh, on the other hand, persists in the midst of enormous fluctuations between land and sea, through storms, floods, dry years, and tidal waves. As we have seen, the human hand can be its undoing as well, but only after decades of abuse. When the conditions it needs are made right again, moreover, it can very quickly reestablish itself. Finally, it is worth noting that the structural collapse of a rain forest is only partial, being limited to that area where the soil has been exposed, whereas in a lagoon, failure, when it occurs, is total. Complexity tends to protect a system from total collapse.

Eugene Odum presents a somewhat different explanation for the relative stability of the salt marsh.[3] He hypothesizes that diversity is closely related to energy flow and that stability seems to be associated with low diversity in eco-

[1] See Orians, 1975; p. 141.

[2] See May, 1975.

[3] See Odum, 1975.

systems that enjoy large, high-quality energy subsidies or nutrient inputs and with high diversity in systems dependent on limited energy inputs and internal recycling of nutrients. Estuarine salt marshes, as at San Elijo Lagoon, have the luxury of energy subsidies provided by the tides. (This provides another good example, incidentally, of the intertwining relationships between structure and the flows of materials and energy.)

All of this is fundamentally important to landscape design in at least two ways. First, an understanding of the qualities that make for stability can help us maintain the integrity of a natural ecosystem whenever we find it necessary to alter it in some way. And second, such an understanding can provide guidelines for shaping stable structures for human ecosystems. Unfortunately, we can draw from the existing data no simple rule of thumb. In the broadest terms, the evidence is quite strong that complex, diverse structures are likely to be more resistant and probably more resilient than simpler ones and thus that mature systems are more stable than younger ones.[4] This agrees with what common sense tells us: That a diverse structure is more likely to retain its basic form and function after the loss of some of its members and that it will retain more interactions and more alternative pathways to survival than a simple system. In short, a diverse system simply has more glue to hold it together, but we cannot conclude from this that more diverse always means more stable.

We might find in the existing data reason to believe that both extremely diverse systems (like tropical rain forests) and extremely simple systems (like cotton fields) are less stable than those in between. Thus, we might someday be able to define optimum, moderate levels of diversity. Once again, more investigation is needed.

For design purposes at smaller scales, it is probably best to get beyond generalities as quickly as possible and base consideration of structure on the specific elements and interactions at hand. Although there is little hope of knowing every element and interaction in either a natural or a man-made ecosystem, we can identify and make provisions for the major ones, as well as determine the general conditions for development of the rest. While all ecosystems continually fluctuate, by design we can hope to keep the magnitude of fluctuations within acceptable limits.

Design for Diversity

At the larger scales, where generalities abound, it is probably best to design for the highest level of diversity that remains compatible with both the environment and human purposes. Besides the likelihood that such an ecosystem will be more stable, there are several other arguments for preferring it. Important among these are the conservation of energy and materials that result from the tight, closed flows accompanying diverse structures, the improved preservation of species obtainable, the high information content, and the likelihood of less damaging effects on the larger environment. However, since human purposes often demand simple ecosystems, such demands may outweigh the advantages of diversity. Man-made environments will probably always include systems ranging in diversity from the very simple to the very complex. If Odum's thesis is right, and it seems likely that it is, the simpler systems will require large energy inputs, and it will be most important to recognize this fact and plan for these inputs from the beginning. As energy becomes more scarce and expensive, we will probably need to design more diverse systems merely to reduce this subsidy.

If this discussion has a rather theoretical ring to it, it is not because considerations of structure are not practical or applicable, but because such considerations have seldom entered into landscape design in the past. Landscapes shaped by human beings, more often than not, are like great skyscrapers with no internal steel or concrete skeleton to hold them up.

Agricultural landscapes, in particular, have notoriously simple structures. For the most part, they are monocultures, and most of the interactions they support are undesirable ones that involve pest species. Their corresponding instability is equally well known. The resistance of most crops to insect infestations and blight is almost nonexistent. A fungal blight wiped out the cacao production of Costa Rica in less than two years. Only massive sprayings have prevented similar disasters in the United States at the hands of the boll weevil, the Mediterranean fruit fly, and myriad other pests. As for resilience, a look at any abandoned farm will confirm that the usual agricultural monoculture has none whatsoever. When the plowing, planting, weeding, and spraying stop, colonizing weeds are quick to move in and begin a new sequence of succession.

The character of the agricultural landscape clearly supports Odum's thesis that the stability of simple, nondiverse ecosystems is dependent on high-energy inputs. The amount of energy required to maintain agricultural productivity in the United States is considerable. Three to five times more energy in fuel and electrical power is expended in growing crops like broccoli, cauliflower, and lettuce than is contained in the final product.

May offers a somewhat different explanation. The instability of agricultural monocultures, he says, is due to the fact that they did not coevolve—that is, develop ongoing relationships over time—with pests and pathogens.[5]

Structure in the Agricultural Landscape

The debate on correlations and causes will probably go on indefinitely, but whatever new understanding may eventually emerge, we will continue to create and maintain agricultural landscapes, and we will continue to pay the price in energy and money to keep them productive. However, with energy prices rising, there will be strong incentives to reduce energy inputs. If we accept Odum's thesis, the way to do so is to diversify the structure of the agricultural landscape. Some historic examples and some current research suggest ways of accomplishing this goal by design.

The hedgerows that line the fields and roads of England and much of Europe, creating the distinctive mosaic that one sees from the air, provide an example of one approach. These have been planted from the Middle Ages on and by now have developed highly diverse structures of various

[4] See Odum, 1975, and Margalef, 1975.

[5] See May, 1975.

tree and shrub species, herbs, grasses, and all the wildlife, from insects to birds and small animals, that goes with their presence. While their original purpose was to define the boundaries of the fields, they also help to break the wind, thereby sheltering crops and soil, and provide habitats for spiders and other predators that prey on the insects that prey on crops. In addition, they provide food and shelter for a great many species that would otherwise not be able to exist were the environment otherwise.

The relatively complex structures of the hedgerows thus contribute to the stability of the agricultural fields, but it would be very difficult to say to what extent. Certainly, the high quality of the glaciated European soils, the long tradition of mixed farming that provided for a balance of plants and animals, and the small sizes of the farms have all played major roles in the impressive long-term record of stability and productivity in English and Northern European agriculture. Whatever their ecological role, the hedgerows are the dramatically visible expression of the whole biological and social system.

The hedgerow approach to the creation of diversity consists of the following: the interspersing of more complex structures among the simple ones needed for particular reasons and providing for beneficial interactions among them. A second approach would be to diversify the human ecosystem itself. High levels of diversity are fairly common in small-scale intensive agriculture, especially in some of the less developed countries. Otto Soemarwoto, for example, has described the complex structure of the typical home garden in a Javanese village. The highest stratum is occupied by the crowns of coconut trees, which grow to a height of over 60 feet. Below that level are the fruit trees, mango

and rambutan. Crops like corn and cassava grow between the lower branches of the fruit trees and the ground, and at ground level are surface and root crops like sweet potato. Climbers such as passion fruit wind up the trunks of the trees and into their branches. The pattern of vertical layering is thus complete. The village ecosystem, Soemarwoto notes, "resembles a tropical forest."[6] He also points out that it is a stable ecosystem. Pests are not a problem, whereas in nearby monocultures, plant pests and diseases definitely are problems. Productivity is also high. According to one report cited by Soemarwoto, people derived more income in monetary terms, and certainly in satisfaction as well, from their home gardens than from their rice fields.

Although such intensive-diverse methods have long histories in some parts of the world, they may not be really applicable to the highly mechanized, large-scale farms of the industrialized countries. Nevertheless, they might be useful on small plots in cities and their surrounds. And they might supply part of the answer to the severe food problems in less-developed countries, as, for example, on the very limited alluvial soils of Costa Rica.

We have touched several times now on the problems of agricultural development in the tropical rain forest and Costa Rica's Huetar Atlantica Region. At this point, I would like to look at Huetar Atlantica again, this time as a problem in structure. We have seen that the irreversible breakdown of the rain forest ecosystem that follows clearing stems from the interactions of the lateritic soils with water. The problem, then, is to prevent these interactions from occurring by designing structures for that purpose. The structural requirements for each of the management zones will be somewhat different, according to the conditions of the environment. In the rugged, steep upper slopes (Zone V), with their lateritic soils and very high rainfall, we are forced to recognize that little can be accomplished by design; any disturbance of the natural structure is likely to lead to irreversible damage. In the foothills, where rainfall is lighter and slopes more gentle, a less complex structure will be adequate to protect most of the soil. The proposed forest production and tree crop uses will require soil to be occasionally bared, but the smaller volume of rain and slower rate of runoff will probably allow time for regeneration before serious damage is done. In some of the flatter areas, if the need for food becomes overwhelming, clearings might be made for food crops; for these to be sustainable, however, soil amendments must be made to the extent of establishing a new soil regime in order to eliminate the problems of laterization. To do so will be expensive, and thus probably limited to small areas. Even where soils are completely rebuilt, at least a few of the canopy trees must be left in place when the clearings are made to slow the fall of rain and so that their roots can help hold the soil in place. In fact, this practice is often observed now in the region. As shown in the photograph on page 55, lanky towers of isolated rain forest trees, looking rather naked and vulnerable without their fellows clustered around to hide their skeletons, often create a strangely dramatic landscape. They create little shade because their massively buttressed trunks rise

[6] See Soemarwoto, 1975; p. 280.

the height of several men into the air before spreading out to support their foliage. For effective protection, there should be more of them.

As shown in the drawings, we can visualize the four variations of the basic rain forest structure most easily in terms of vertical stratification patterns. This provides only a vastly simplified indication of the real structural differences involved, of course. Only a completely unaltered structure is capable of retaining the full diversity of plant and animal life that characterizes the natural rain forest. Altered structures become progressively less diverse, and probably less stable in terms of resistance to change as well. In the perspective of nature, all but the natural structure might be seen as a compromise of nature's ideal. With human needs entering the picture, we might more reasonably see the alterations simply as four different ecosystems with four different sets of characteristics of structure and function adapted to four different sets of purposes.

Structure in the Urban Landscape

Most existing urban landscapes are hardly more structurally cohesive than agricultural landscapes. When settlers move into a new area, they bring with them sets of landscape images from myriad entangled sources—their childhoods, the places where they came from, history, pictures in magazines, television commercials—and more often than not, rather than seeking out the natural qualities of the new place, they try to bring to life their stored imagery. Thus, when the waves of Midwesterners arrived in Southern California in the late nineteenth century, they created a sea of emerald green lawns with islands of lush temperate forest on its semiarid soil, fashioning a kind of idealized version of the places they had come from—all with the craggy, brown, treeless slopes of the natural mountainsides looming in the background. The natural plant communities—the scrubby chaparral and coastal sage—were seldom thought hospitable. The climate was sunny and hot, and few native shade trees were to be found outside the riparian areas, which were few and far between. It was difficult even to make one's way from one place to another through the thorny thickets of hillside chaparral. One could hardly blame the settlers for wanting to plant something new.

What they soon discovered was that in this warm climate practically immune to frost, almost any plant would grow—given plenty of water and fertilizer. So they brought exotic plants from all over the world, especially brilliantly flowering plants. The effect was spectacular, and it still is, but the price paid in terms of water, fertilizer, and energy is high indeed. Recall the enormous water imports needed to support this urban region described in Chap. 2 and illustrated in the Southern California case study. Almost half of the urban water use in this region is for watering ornamental plants. Much of it ends up in the ground water supply, of course, but the high energy cost of moving it around remains.

Though probably somewhat extreme, this is not an atypical situation. The structural difficulty with urban landscapes is seldom the result of a lack of species diversity, as it is with agricultural landscape, but of a lack of interactions between the species and the transfusions needed to support them. The plants brought from all over the world and thrown indiscriminately together do not form a cohesive community. Each left its own network of relationships behind: the animals it fed and protected, the other plants it shaded or took shade from, the soil that it held in place and drew nutrients from. Some important elements are conspicuously missing, especially medium-sized and larger mammals. These have been replaced by pet dogs and cats, both of which interact only minimally with any landscape and, in fact, subsist on food brought from elsewhere. The processes of decay, not considered fitting in most idealized landscape images, are also missing, thus eliminating a whole regime of ground-level interactions and breaking nutrient cycles.

INTERACTIONS: PLANTS—ANIMALS—PEOPLE

Clearly, then, there is good reason to pay more attention to structure in the design of human ecosystems, however incomplete our understanding. We have to live with the fact that the species present in any ecosystem can be only partly known, and their interactions only partially understood. In practice, however, if we can identify the major species and the key interactions, most of the others will develop over time, and if the management stage works as it should, monitoring, feedback, and corrective design (management) will take care of most mistakes.

As shown in the Costa Rica example and others, the baseline for any new ecosystem is the preceding natural ecosystem. This source alone can provide the salient facts about a particular environment: the limitations it presents, the way its conditions vary in time and space, the characteristics of the plants and animals that have adapted to it, its key interactions. The more profound our understanding of this natural system, the more sound our plans for a man-made system.

Once we understand the baseline structure of the natural system, we can seek out ways of expanding or modifying or even reshaping it to accommodate human purposes. Too often, especially in severely disturbed natural environments, we find a strong bias in favor of restoring the natural structure without considering its potentials for contributing to human aims, much less strengthening the natural system itself. In the San Elijo case study, we saw that a designed structure can be more stable—both more resistant and more resilient—as well as richer in plants and wildlife and more hospitable to human use than the natural structure. In particular, controlled infusions of treated sewage effluent can make the fresh water input into the lagoon both regular and consistent, thereby eliminating the destabilizing fluctuations that occur in the natural state as a result of the virtual absence of fresh water flow throughout the dry summers and the often large, if sporadic, flow in the rainy winters. It must be said again that the baseline state is not necessarily the ideal.

To develop an understanding of the interactions among populations at a level useful for design purposes, we might begin by grouping populations into the three largest possible categories, that is, into human beings, plants, and animals. This is not to ignore the fact that human beings are undeniably also animals but to emphasize the very different roles they play. We humans not only live differently, but more important, occupy the controlling position, at least for the short term.

Among the most critical interactions, and certainly most subject to control by design, are those between these three groups. For design purposes, the linkage between human beings and animals is somewhat less critical than that between human beings and plants because human influence on animal populations is exerted mostly through manipulation of plants and water.

Because the terms used to classify interactions between populations can be confusing, it will be helpful to review them. *Competitive* interactions are those that adversely affect both sides. *Amensal* interactions are those in which one side is adversely affected, the other unaffected. *Parasitism* and *predation* refer to interactions in which one side benefits but the other is adversely affected, in the latter case by being eaten. *Commensal* and *mutual* interactions are favorable to both sides, the latter being obligatory, the former not. Both of these mutually beneficial types of relationships are subsumed in the term *symbiosis*.

In this sense, most interactions between humans and plants are symbiotic. The benefits human beings derive from plants can be loosely abstracted under the headings of environmental regulation, climate control, productivity, visual quality, and emotional satisfaction. The benefits plants derive from human beings include design, provision of water and nutrients, physical care such as trimming and pruning of individual plants, and at the population level, propagation, diversification, and dispersal. All of the possible benefits are not achieved in every relationship, of course. In fact, among the serious difficulties with most man-made landscapes is that the benefits provided to human beings by the plant community are far less than they might be with more thoughtful design, and the benefits provided by human beings to the plants are often far greater than they really need have been. In other words, we might obtain far more benefits with less investment by means of more thoughtful design. This is especially important because so many of the benefits must otherwise be provided by technological means at very high economic cost.

Since they are so important in shaping the structure of a landscape, a brief review of the major interactions between plants and human beings will be useful here. The first two are indirect; that is, they involve interactions between plants and other environmental elements, which in turn profoundly affect humans.

Environmental Regulation

Most obvious among the roles plants play as functional regulators is their literal creation of much of the air we breathe. The capacity of plants to utilize carbon dioxide and give off oxygen whereas animals, including ourselves, do exactly the reverse may be the neatest and most important interaction of them all. Until quite recently, it was believed that most of the oxygen needed by human beings was produced in the ocean, but now it seems likely that most of it comes from terrestrial green plants.[7] Global patterns of air circulation move oxygen quickly about the world. There is no general agreement concerning the volume of photosynthesizing plant matter needed to provide enough oxygen for each human being. Bernatzky estimates that it takes at least 150 square meters of leaf surface to supply one person's needs.[8] Reilly calculates that this means enough oxygen for 14 people can be produced by one average-sized tree.[9] Other studies have produced somewhat different figures. In any case, the oxygen production of trees can definitely dilute the concentration of pollutants in the air.

In Chap. 5, we discussed the global problem of carbon dioxide concentrations in the upper atmosphere. Plants can, to some extent, mitigate the buildup of carbon dioxide by using it in photosynthesis as well as by diluting it with their oxygen output. To do so, however, takes a great many plants. Boer estimates that about 75 trees are needed to absorb the carbon dioxide produced by the average city dweller in the course of his daily round of activity.[10] Kuehn says that three acres of woods can utilize the carbon dioxide produced by four people, or one average household.[11] Both these figures suggest that it takes an enormous number of tress to have any noticeable effect on the carbon dioxide balance.

In addition to literally creating the air we breathe, plants also cleanse it. They are probably more effective in dealing with excess airborne particulates than they are with gaseous pollution. The presence of more plants could, without doubt, reduce the density of urban dust domes. Trees, in particular, remove significant amounts of particulate matter from the air, mostly by filtration. As air moves among the branches, particles are trapped and held on the stems and leaves. Rough and pubescent leaves are especially effective collectors of small particles, most of which cling until they are washed off and carried to the ground by rain, where they are absorbed in the soil and rendered harmless or even beneficial by being devoured by microorganisms or taken up as nutrients by the plants.

A number of researchers have substantiated the particle-removing abilities of trees. A German study found that two and a half acres of beech trees had removed from the air every year about four tons of dust, which was eventually found in the soil.[12] Measurements in Frankfurt showed six to eighteen times as many particles in the air in the center of the city as in nearby Rothschild Park. Bernatzky found that particulate levels on the edges of treeless roads were three to four times those on roads lined with trees. In several research projects, Russian investigators measured air quality beyond the "sanitary clearance zones," or greenbelts, that are sometimes provided around industrial areas in that country and reported considerable reductions in pollutants attributable to the planting. One of these studies showed a 22-percent reduction in sulphur dioxide and a 27-percent reduction in nitrogen oxides beyond a planted area 500 meters wide.

The plants themselves, however, can be damaged by the matter collected on their leaves. Unless washed away

[7] See Woodwell, et al., 1978.

[8] See Bernatzky, 1969.
[9] See Reilly, 1976.
[10] See Boer, 1972.
[11] See Kuehn, 1959.
[12] See Meldan, 1959.

soon, the particles can clog breathing pores and choke the trees. This is particularly a problem for trees with sensitive leaves and for conifers that retain their needles for more than two years. On the other hand, beech, ash, gingko, plane, and elm, all deciduous trees that grow new leaves every year, are among the tough-leaved trees that seem resistant to air pollution.

The capacity of plants for cleansing the air argues for more green areas in precisely the places where they are now most rare: in densely developed city centers and industrial areas. At the very least, we should plant dense masses of trees and shrubs along the edges of every highway and major arterial street. The benefits of even small plantings of trees in such places will extend far beyond the immediate environment. Nevertheless, we have to work with care because unexpected interactions can occur, and they are not always beneficial. In the Lake Tahoe Basin in California and Nevada, researchers found that terpene emissions from the forests were combining with nitrogen oxides from automobile exhausts to form ozone, which is the area's major summer air pollutant. In this case, plants would seem to be contributing to air quality problems rather than helping solve them. Since the terpene itself has no polluting effects, the only solution appears to lie in reducing automobile emissions.

In the hydrological cycle, plants serve as both media and regulators of flow. A tree canopy catches rain as it falls and then releases it slowly to the ground. Since trees commonly drip for two or three hours after a rain, there is more time for the water to percolate into the soil and the ground water supply and consequently less runoff. At the same time, the roots of the trees, along with those of other plants, help hold the soil in place and keep erosion to a minimum.

Plants provide water cleansing services as well. In nature, they do so regularly. Rainfall, particularly over industrial regions, is full of nitric and sulfuric acids and heavy metals like lead, nickel, and copper. On the basis of their input-output studies, Bormann and Likens concluded that much of this material is removed after the water flows through a forest ecosystem, leaving it relatively clean. We cannot, of course, ignore the fact that the forest may be damaged by the materials left behind, as in the case of water bodies polluted by acid rain. But by contrast, if the same rainwater falls on a paved urban area, it will pick up even more particles of various sorts and, by the time it flows out of the city, achieve a level of quality about the same as that of secondarily treated sewage effluent.

The cleansing process is partly a matter of time. Most of the raindrops that strike leaves or branches are held there for a short period, during which their particles can settle out. Once on the ground, they flow through grasses, shrubs, and other ground covers, which filter out more particles. Some of the water soaks into the soil, part of it being taken in by the roots of plants to supply their own life processes. The water given off by photosynthesis and transpired back into atmospheric storage by leaves helps to maintain a certain level of atmospheric humidity and, at the same time, provides another means of removing particles from the air. Drops of moisture on leaves as well as in the atmosphere attract airborne particles and carry them to the ground.

For all these virtues, we should recognize that the contributions of plants are not entirely beneficial. In dry climates, evaporation and transpiration from plants can cause serious water losses, especially when plants native to humid regions are planted in dry climates. The quantities involved can be very large. An acre of turf can surrender about 2400 gallons of water to the atmosphere on a sunny summer day. Where water is scarce, needless to say, it is important to avoid losses of that magnitude if we can. In semiarid Southern California, riparian areas have sometimes been cleared of all vegetation in order to curtail evapotranspiration losses. That is a rather extreme measure, and one that is probably not often justified when we consider all the benefits that plants provide.

Control of Microclimates

Living plants help control microclimates primarily by cooling, The most familiar method is by intercepting incoming solar radiation and using its energy in photosynthesis or dispersing it by reflection (or, stated more simply, by providing shade).

Since generally less than 1 percent of the energy is used in photosynthesis, the principal means of radiation control is reflection. The amount of energy reflected depends on the form of the tree. According to the best available estimates, a dense tree in full leaf intercepts about 75 percent of the radiation that strikes it. This figure might rise to over 90 percent for the densest trees or fall to less than 60 percent for those with sparse foliage. Deciduous trees reflect about 30 to 65 percent during leafless periods.

A second way in which trees reduce heat levels is by creating a zone of calm air under their canopies. In any close grouping of trees, as in a forest, a zone of heat exchange, and thus turbulent air mixing, builds up around the edges. According to experiments conducted by Rudolph Geiger,

this turbulent zone extends only to the edge of the canopy, the latter providing an enclosing frame for the still air within.[13] Geiger found this zone to be 4°C cooler than the air at the top of the trees because of its insulated position and reduced level of solar radiation. The diurnal variation—or difference between maximum daytime and minimum nighttime temperatures—was 5.4°C less under the canopy than at the top. Thus, not only is the temperature among the trees more constant, the thermal inertia is greater. Both of these are significant differences. The last one may be particularly important in the light of Ralph Knowles' hypothesis that environmental stress can be measured by the differences between the extremes of affecting forces—in this case, temperature.[14]

The third way that plants lower ambient temperatures is by releasing water vapor from their leaf surfaces through evaporation and transpiration. This vapor cools the surrounding air, and the temperature difference can be considerable. Michael Reilly, using data developed by Pinkard, estimated that one tree, through evaporation alone, can produce cooling effects approximately equaling those of ten room-size air conditioners working 20 hours a day.[15] Though it would be no easy matter to use that cooling capacity to the full, there are obviously potential energy savings here. Studies of mobile homes have shown that shading by carefully designed plantings can reduce energy consumption for air conditioning by over 50 percent.[16] Similar savings are possible for poorly insulated conventional houses. In well-insulated buildings, the savings are less, but generally still well over 10 percent.

On the larger scale, the ameliorating influence of plants, in contrast with the tendency of most building and paving forms and materials to exaggerate climatic extremes, makes the planted areas within cities usually cooler than the built-up areas. One of the earliest and still one of the most thorough studies of the subject was carried out in St. Louis in 1952 and 1953. Winter temperatures in Forest Park turned out to be 13 degrees cooler than in the downtown area about 5 miles away. In June, the park temperature was about 9 degrees cooler. Since the warmer downtown temperatures seem desirable in winter but not in summer, heavy plantings of deciduous trees in the built-up areas should make it possible to keep the winter warming while creating a cooler summer. A number of other studies have shown similar results. See, for example, Duckworth and Sandberg,[17] Bernatzky,[18] and Bryson and Ross.[19] Estimates of the shade coverage needed to overcome the urban heat island effect range from 20 to 50 percent.

Computer simulations have been developed to measure shadow coverage of trees through time, and others estimate effects on building heat loss or gain. When their reliability is proven, such programs may be extremely useful. Nevertheless, there are many considerations that enter into the selection and location of plants. A high level of quantified precision in the assessment of one factor might easily lead to less concern for other factors that can be at least equally important.

Productivity

The development of agriculture some 12,000 years ago was probably the most decisive turning point in human history. Since then, human beings have made plants produce for their own purposes in an endless variety of ways and with an endless number of techniques. Some of these have brought destruction to the landscape, leaving its soil ultimately barren. Examples surround the Mediterranean Sea. Other regions, like most of Northern Europe, have sustained productivity for thousands of years. We thus have a long record of trial and error to look back upon for useful lessons.

The next agricultural revolution—the control of water for irrigation—came some five to six thousand years after the first and seems to have led to the formation of the earliest urban civilization.[20] And after another five or six thousand years came the third, this one putting the energy in fossil fuels to work. Through two centuries of the fossil fuel age, agriculture has become increasingly specialized, mechanized, and large-scale. The results, in terms of productivity per farm worker, have been extraordinary, but the energy costs—adhering to the relationship between ecosystem diversity and energy inputs that we have discussed—are very high.

In California, the average energy efficiency for crop production is about 0.6, which means that the calorie content of the crops is about six-tenths of the energy input. When the energy needed for transportation and processing are considered, the food on the table ends up having an energy cost several times its calorie content. As shortages of fossil fuels become more acute and their costs correspondingly rise, this problem is likely to become more and more serious. Food and fiber costs will also rise, and more land in urban and suburban areas will probably be turned over to agricultural production. This phenomenon has happened

[13] See Geiger, 1965.
[14] See Knowles, 1975.
[15] See Reilly, 1976, and Pinkard, 1970.
[16] See Hutchinson, et al., 1982.
[17] See Duckworth and Sandberg, 1954.
[18] See Bernatzkv, 1969.
[19] See Bryson and Ross, 1972.
[20] See Wittfogel, 1956.

before. John Stilgoe has shown how suburban vegetable gardens proliferated as food prices rose in the late nineteenth and early twentieth centuries.[21] Similar trends became more and more visible in the 1970s and 80s, and there are strong reasons to believe that they will continue. Richard Meier predicts that about a third of the caloric value of the average diet and over two-thirds of its economic value will eventually be produced in the city.[22]

To bring people back into touch with their sources of sustenance in this way would be an enrichment not only of the landscape, but of urban life in general. As Aldo Leopold put it a half century ago,

> ... bread and beauty grow best together. Their harmonious integration can make farming not only a business but an art; the land not only a food factory but an instrument for self-expression, on which each can play the music of his own choosing.[23]

The possibilities of biotic productivity in urban and suburban landscapes go far beyond backyard vegetable gardens. In fact, of the four basic resources needed for agriculture—land, water, nutrients, and energy—the urban landscape features an abundance of all but the first. Water and nutrients are available in great quantities in sewage effluent that normally goes unused. Energy is available in a variety of forms, a great deal of it in substances usually regarded as waste—garbage, refuse, and tree trimmings. Even the dissipated heat that creates the heat island effect can be of some benefit. According to H. E. Landsberg, it is not unusual to find the growing season lengthened by three to four weeks in urban environments.[24]

The most important resource of all, however, is high-quality human energy, and in this respect cities have an embarrassment of riches, a surplus they have never learned to put to good effect. With careful management that is informed by scientific knowledge, productivity per unit of land can be several times higher in cities than in less intensive rural settings. Biotic production is one way of putting the city's surplus management skills and labor supply to beneficial work. In the process, the shortage of land can, to some degree, be overcome.

The North Claremont and the University Village case studies (pages 181–83 and 111–13, respectively) illustrate a few of the ways in which productivity might become an integral part of the urban landscape structure. As explained in Chap. 4, the University Village design resulted from an effort to explore the full technical possibilities of urban food production for a small community, one that could satisfy its ongoing needs to a reasonably practical extent. The plan called for a carefully controlled system of energy and water flows, which in turn required a diverse community structure of plants and animals.

The series of what-if scenarios in the North Claremont study posed images of the suburban landscape that might develop under a number of future circumstances. These ranged from a one-acre-lot subdivision should energy and water continue to be abundantly available at low cost to a community of intensively managed small farms should widespread food shortages occur. The first represents a loose structure of the sort associated with urban development, whereas the second is a very closely interacting structure designed for high levels of productivity. By comparison with a natural ecosystem, the latter is a relatively simple structure, but compared to conventional agriculture, it is quite diverse.

Aside from these rather specialized environments, food production can be integrated with other roles of plants. In China, the need to support an enormous population on meager resources has encouraged some practices that might very well be useful in other parts of the world. There, the integration of production and consumption has become both an ideological goal and a matter of public policy. Urban parks have particularly followed this ideal. Galen Cranz provides a description:

> The soft growth on a park hill might be a tea plantation, while a stand of trees may be a spice or fruit orchard. Bamboo is harvested in parks for building construction. In lakes the Chinese raise edible white lotus, crabs, mother-of-pearl, and fish. Trees planted for beautification are placed close together, so as they grow some can be cut down for lumber. At the Summer Palace outside Beijing, workers harvest fish five or six times a year for market during festivals according to an economic plan. Fifteen tons of freshwater fish are collected each year from Hangzhou's West Lake—once only a tourist spot.[25]

This is a very different ecological structure from those of Western urban parks, or cities, one that has an added dimension of integration with human life. In fact, over 85 percent of the vegetables consumed by China's urban populations are grown within urban areas. Shanghai and Beijing grow all their own vegetables.[26]

Urban forestry also presents considerable possibilities. The city of Chicago has experimented with grinding dead trees into wood chips for the roofing materials market. It seems possible that all of the city's trees might be managed on a sustained-yield basis.[27] Other cities are working with firewood production. Whether or not such efforts can eventually make significant contributions to the vast energy needs of our age remains an open question, but it does seem possible.

Visual Relationships

That the human eye is most sensitive to light at a wavelength of 553 microns—which is a yellow-green approximately the average color of plant leaves—is no accident. The physiological structure of our eyes evolved during the long period when they were almost continuously exposed to a green landscape. Apart from the human face, our most essential visual imagery is probably that of the landscape.

Although the importance of green landscape to the emotional well-being of people has long been widely recognized,[28] human reactions to particular types of landscapes have not remained the same throughout history. Perceptions

[21] See Stilgoe, 1982.
[22] See Meier, 1974.
[23] See Leopold, 1933; p. 642.
[24] See Landsberg, 1956.
[25] See Cranz, 1979; pp. 4–5.
[26] Figures from Hough, 1983.
[27] See Hartman, 1973.
[28] See, for example, Lewis, 1979.

and attitudes have varied enormously. During the Middle Ages, for example, the forest was thought of as the abode of the devil—feared and avoided or cleared away to make farmland when the opportunity arose. The love of scenery, as reflected in landscape painting, did not appear in Europe until the seventeenth century. By then, of course, there were few natural landscapes left; they had become domesticated and there was little left to fear. The Chinese and Japanese traditions in landscape painting go back to much earlier periods, though it is sometimes argued that these represent a philosophical, rather than a truly sensual, relationship with nature. The tiny figures—priests, perhaps, or poets—wandering among vast, misty mountains in early oriental landscape paintings seem to depict a search for unity with nature that we can well understand.

Without a doubt, there are profoundly philosophical elements in our contemporary regard for green landscapes. As Rene Dubos points out, many of those who write most eloquently in defense of nature rarely experience her vagaries. Nevertheless, the regard is real. Some of the most heated environmental battles have been over the potential loss of scenic values.

Although we usually accept the value of beautiful scenery as self-evident, it is important to recognize that our views have other stimuli than purely esthetic ones. Sometimes, in the arguments for preserving scenic beauty, one detects a clear note of escapism. Natural beauty gives surcease from the ugly city. In studies of landscape preference, man-made elements are almost always scored as being far less desirable than natural ones.

Presumably the Parthenon or the Temple of Poseidon at Sounion would rate in most such studies as a less satisfying visual experience than the badlands of North Dakota.

Aside from questionable esthetic standards, there is an important principle at stake here. Among the crucial roles of the visual landscape is the imparting of information. Not so many eons ago, an individual's very survival often depended on the information he could draw from his inspection of the landscape. Jay Appleton has described the importance of the imagery of "prospect" and "refuge" in our visual impressions of landscapes—images seemingly deeply imbedded during a distant past when attaining a prospect or a refuge at the right moment could mean living a little longer.[29] It is entirely possible that the information we draw from a landscape can still have survival value. Ancient symbols aside, a landscape may still tell us much about our sources of support. To some degree, this may still be true for us collectively as a society.

To illustrate the point, let us reconsider the Whitewater Wash case study (pages 89–100). Recall that the objections to the wind generators arose from the desire of the people of Palm Springs for unobstructed views of the mountains. Considering that this is a resort community to which people retreat for varying lengths of time from the urban trials of metropolitan Los Angeles and other cities, the desire is entirely understandable. On the other hand, one can plausibly argue that the wind generators, at least to some degree, should be visible. We not only should be able to see where our energy is coming from but should actually have it imposed on our consciousness. It is important for us to have some knowledge of our life-support systems, and much of this knowledge can be acquired by seeing them working in the landscape. This is not to say we should allow such objects to overwhelm the landscape or make it unsightly. Wind generators can, in fact, be beautiful. The landscape of Holland once had a distinctive, kinetic beauty that was the gift of its windmills. They not only provided graceful forms, they also communicated something about the complex hydrological system that kept the country literally above water.

The same argument applies to other man-made features. such as powerlines. A great deal of thought has gone into hiding power transmission lines that might better have been spent on designing them as expressive complements to the landscape. The basic point is that these man-made objects are essential elements of the structure of human ecosystems, and if we are to make sound decisions concerning the management of such systems, they should become a part of our consciousness. Although there are obviously some beautiful landscapes where they should not be allowed to intrude, to hide them all is escapism and bad esthetics.

Although not for these reasons, the importance of scenic values is now being increasingly recognized in land-use decisions. "Visual quality objectives" have been institutionalized and made matters of public policy in the U.S. Forest Service and Bureau of Land Management's Visual Resource Management Systems (VRMSs). The latter attempt to classify and evaluate the visual quality of landscapes according to sets of universal, though very simple, analytical criteria. Although there are a great many philosophical concerns about the arrogance of giving grades to nature and trying to measure objectively what are, after all, subjective, emotional, and ever-changing relationships, these techniques represent the only useful ones we have to deal with visual resources.

Unfortunately, the VRMSs have focused on the traditional elements of visual composition—line, form, color, texture, and so on—with the implication that these are ends in themselves, rather than signs pointing to deeper qualities. Thus, in leaving out of the equation the ability of the landscape to convey information and in applying uniform criteria to widely varied settings, there is a very real danger of losing the unique qualities of a place.

Every spot on earth has a unique set of qualities, produced by a combination of natural evolution and human development, that sets it apart as different from all others, and that is perceived as unique by those who live there. This human relationship with place has long been celebrated in literature and painting and seems, though it can hardly be proven, to be important to our sense of connectedness with the world. "The catalyst that converts an environment into a place," according to Rene Dubos, "is the process of experiencing it deeply—not as a thing but as a living organism."[30]

[29] See Appleton, 1975.

[30] See Dubos, 1980.

Two other important qualities of the visual landscape are its ability to impart order and to evoke symbolic meaning. Perhaps the oldest example of using plants to give visible form to an underlying order is the hedgerow pattern already mentioned, which marks roads and property lines in England and Northern Europe. Other example abound, especially in our cities. Rows of trees marking major boulevards, mats of flowers lending color to important intersections, clusters of shrubs at entranceways, all help to create a sense of order and continuity in surroundings that are otherwise often discontinuous, haphazard, or even chaotic.

The use of plants as symbols goes back much further. In primitive cultures, a particular tree or group of trees often took on profound spiritual significance: the abode of a god or perhaps a god in itself. The Greeks had their sacred groves, and the Jews had the Tree of the Fruit of Good and Evil. Gardens became complex symbolic expressions. The Hanging Gardens of Babylon, if we can believe the legend, were built by Nebuchadnezzar for his wife to evoke the forested mountains of her homeland, which she desperately missed. The Western garden traditions that still strongly influence our landscapes gave form to complex ideals of the relationship between human beings and nature. At Versailles, André Le Nôtre managed to fuse order and symbol. He made visible the passion of his age, articulated by Descartes, for absolute clarity of understanding of nature, derived from mathematics and the scientific method. Perceived in terms of its underlying order, nature was seen as a matrix. The English landscape garden presented an idealized pastoral landscape that looked more natural than, but was in reality at least as artificial as, the French. It was just as simplified in structural terms, just as lacking in interactions among its parts. The Japanese garden, with its myriad intimations and nuances, is probably the most highly developed of symbolic landscapes. A few rocks can evoke a mountain range, and a grouping of plants and water can suggest evolving flights of the human spirit.

Our own views of nature are in some ways like all of these but in other fundamental ways very different. Thanks to science, we perceive nature in different terms, but we rarely express our views beautifully or clarify them beyond the superficial. It is important for us to recognize that the man-made landscape is fundamentally different from the natural landscape because it is shaped for human purposes, and among those purposes is the need for cultural meaning expressed in abstract form.

Visual quality, then, in our present context, is a goal to be achieved by making the underlying structure of the landscape both visible and meaningful, not merely pretty.

Emotional Satisfaction

Although most of us spend most of our time in man-made environments, our species evolved in the natural environment. It is often suggested that through those eons of evolution, our ancestors may have developed certain emotional attachments to green plants along with the visual ones already discussed, attachments that are still a fundamental part of us. This would go far in explaining the feelings that most people seem to have for plants. Charles Lewis wrote that "plants take away some of the anxiety and tension of the immediate Now by showing us that there are long, enduring patterns of life."[31]

Although research on the subject is limited, actual contact with plants through gardening seems to produce the greatest satisfaction. Several government-sponsored programs to promote neighborhood gardening in low-income areas have consistently resulted in greater community pride, increased social interactions, and less vandalism. Although behavior changes have been documented, gardeners usually describe the greatest benefit of gardening as a sense of peacefulness and tranquility.[32] Even without the actual involvement of gardening, just being in the presence of plants seems to produce the same kinds of feelings.

DESIGNING INTERACTIONS

We can summarize the interactions between plants and people as follows:

Plants Contribute to People:	People Contribute to Plants:
Regulation of ecological functions:	
By absorbing carbon dioxide	Design
By giving off oxygen	Water
By filtering air	Nutrients
By controlling water flow	Propagation
By filtering water	Diversification
By absorbing and reflecting solar radiation	Dispersal
By cooling air	
By creating pockets of still air	
Productivity	
Food	
Fiber	
Energy	
Visual Quality	
Structure	
Scenery	
Place	
Symbolic expression	
Emotional satisfaction	

In the process of urbanization, then, as the native plant community ceases to perform the roles listed in the first column, they must be either performed artificially (flood and microclimatic control), or relegated to faraway landscapes (food, fiber, and oxygen production and scenery), or not performed at all (water filtering and some small portion of carbon dioxide absorption). In each case, there are energy costs, and difficulties are created both in the subject landscape and the source landscape.

Clearly, given the high concentrations of human activity and the vastly accelerated flows of materials and energy, a community of plants cannot perform these roles as successfully in human ecosystems as in natural ones. Nevertheless, as demonstrated by the instances already mentioned,

[31] See Lewis, 1979; p. 334.
[32] Ibid., p. 335.

they can make a significant contribution. As energy and water supplies run short, we will rely on them increasingly to do so—by expanding, strengthening, and enriching the structures of human ecosystems. This has already begun to happen, as evidenced by the growth of urban agriculture, reforestation projects around the world, efforts to slow urban runoff (reflected in a number of "zero runoff" laws, especially in the West), research in passive solar technology, and Visual Resource Management. But unfortunately, most efforts in these directions are limited, and perhaps doomed, by the same conceptual flaw that created the conditions they sought to overcome—the one problem/one solution habit of thought, which results in single-purpose communities with few interactions and correspondingly weak structures.

Criteria for Selection

The fact that plants can play such important roles in helping to maintain a healthy environment clearly argues for planting more plants, but there is more to it than that. Plants can also have undesirable effects like the terpene emissions at Lake Tahoe and like the evapotranspiration losses they create in dry climates. And they incur costs, such as those for maintenance, fertilizers, and cleaning up litter in urban areas. Much more important than the expense of growing a tree is the time and space involved. It takes at least 15 years for most trees to grow large enough to produce any of the desired effects to any significant degree, and trees preempt a considerable amount of space where space is scarce, especially in the urban environment where they are most needed. Selecting trees that will perform as expected thus becomes an important matter.

Although the need to select plants to fit particular conditions of soil and macro- and micro-climate has long been recognized, selection based on their ability to contribute to environmental control is far less common. Plants have different capacities for performing these services, depending on physical characteristics such as overall size, growth habits, and the size and texture of leaf.

Another consideration is the ability of a plant to play more than one ecological role. There is no reason why a single tree cannot help filter air and water, lower surround-temperature, produce fruit, and provide a visual focal point. Grouping into cohesive communities a variety of plants that fit local conditions and interact in full measure with human beings and other species is one of the best ways to obtain more sustainable ecological structures for human communities.

Too often, however, plants are selected for more limited or single-minded reasons. When the volcano, Irazu, erupted on the high central plateau of Costa Rica in 1962, lava rolled down over its slopes and volcanic ash floated over most of the country, obscuring the sun for several days. Square kilometers of rain forest and farmland were covered over. The Costa Rican government decided to experiment by replacing a major portion of the native rain forest with production forest. With the help of international forestry consultants, they planted over twenty stands of various species. The species were probably chosen on the bases of fast growth, marketability of their timber, and availability of seedlings. All were complete strangers to the landscape. (For the most part, they were species of pine and eucalyptus.) None of the twenty was even a member of the rain forest community or had any relation to the other plants or animals of the natural community.

This is one small example of an issue that will probably take on global significance. Major reforestation projects will probably become much more common over the next few decades. The most recent global forest survey showed the earth's land surface to be only about 29 percent forested. This means that the earth's forests have diminished by half since human beings came on the scene, and most, if not all, of the reduction is attributable to human activity.[33] Clearing is still going on at a rapid pace in many less developed countries. The Global 2000 Report projects a net deforestation rate of 18 to 20 million hectares (about 45 to 50 million acres) per year, leading to a reduction in global coverage from 20 to less than 17 percent by the year 2000. At the same time, however, a number of countries deforested in past centuries are now reforesting, some of them very rapidly. The reforestation programs undertaken in China since 1945 have probably been the most ambitious. In 1982, the National People's Congress of China began a massive tree-planting campaign that will involve every able-bodied Chinese citizen in the planting of a total of three billion trees every year. If the goals are reached, this campaign will increase land areas covered by forest from 12.7 to 20 percent in just three years.

In Western Europe, most of which had been deforested by the seventeenth century, reforestation has been progressing for well over a hundred years. Rene Dubos describes the planting of pines in the vast marshes of Les Landes during the Napoleonic era.[34] The French Ministry of Agriculture is presently beginning work on reforesting the Mont D'Aree region, involving several hundred thousand hectares. For economic reasons, about 60 percent of the trees planted will be Sitka spruce and the rest will be other conifers. The natural community, long since removed, was an oak and beech forest, which was far richer in both plant and animal species, especially birds, than its replacement. Local citizens and environmental groups have protested for both ecological and esthetic reasons but seemingly to little effect.[35]

This is not an uncommon situation. The temptation to reforest with fast-growing stands of marketable timber is not hard to understand, but it can negate the promise of reforestation for achieving a variety of interactions and for enriching the global ecosystem.

Predicting Interactions

The alternative to the usual practice of designing to maximize a single benefit is to provide for a broad range of interactions. In each situation, of course, some will be more important than others, and priorities will resultingly vary. At the same time, there are costs to consider. These will vary also, according to the kind of environment and the adaptability of the plants in question.

[33] See Eckholm, 1976.
[34] See Dubos, 1980.
[35] See Saurin, 1980.

For more detailed design, we need to consider the roles of all the dominant plants in a community as well as their interactions. The latter can present difficulties, since in many cases so little is known about them. Here again, however, we can summarize their complex patterns in a way that will be useful for design purposes. The chart on page 122 summarizes the interactions in the plant community designed for the Simon residence. On this sloping site, soil retention and the control of runoff were important considerations, and because of the semi-desert environment, microclimate control was also a matter of concern. The owners are sophisticated gardeners who want to raise a significant portion of their own fruits and vegetables on the property and who have a serious interest in its esthetic quality as well.

The site is located on the suburban edge of the City of Riverside in an area that was once covered with orange groves. Across the street is a large remnant of a major grove. Along the ridgetop at the site's southern boundary is a contrasting stand of the native chaparral community.

The problem, then, is one of developing a plant community with the range of interactions described above, but one that will also fit into an established natural/historical context. The plan on page 121 shows the design response to these requirements. A large, producing avocado grove occupies the northern part of the site, imparting a sense of continuity with the grove across the street. To the south, on the slope that leads up to the ridge, a planting of chaparral natives provides a connection with the natural landscape. The avocado grove makes a gradual transition into a grove of varied citrus and other fruit trees; inside it is a fairly large planting of asparagus, vegetables, and herb gardens, as well as a gentle slope covered in strawberries. The steeper slopes are stabilized by tougher, soil-holding ground covers, primarily coyote bush. Deciduous trees spread over all the south walls of the house, and dense shrubs cover most of the west wall. Concentrated about terraces and in areas viewed through the glass walls are plantings of azaleas, rhododendrons, ferns, and other plants arranged purely for esthetic pleasure. These areas will need continuing inputs of water and energy to keep them going, but they are very limited in size. Except for this last group and the food producers, all plants were chosen for their low water requirements. (Water is the most limiting factor in this very dry climate.)

The basic structure of this community is conveyed by the chart in concise form. On this one acre, then, we have areas of protective, productive, and "compromise" landscapes that blend together into a structural unit. Using devices like the chart helps us maintain our awareness of the range of possible interactions and thus design for them. The same principle applies for much larger projects as well.

CHAPTER 12

Structure

Design for Animals

As a rule, human relationships with wildlife are far more distant then those with plants. This sad situation leaves a void in our lives, because there can be something profoundly satisfying about the presence of wild animals. Seeing them, merely being near them, even knowing that they are there gives most of us a sense of connection with nature, a timeless, primordial thrill that nothing else, not even plants, can provide. Coming upon a deer in the forest or a coyote or a fox in a city park is somehow deeply rewarding. "If all the beasts were gone, man would die of great loneliness of spirit," Chief Sealth (Seattle) of the Duwamish tribe wrote to President Pierce in 1855.

Unfortunately, when interactions do occur between human beings and wild animals, they are often uneasy, even destructive. The fact that a wild animal is beyond our control can be frightening. Having no sense of property or civilized order, animals can wreak havoc without intending to. Bears in campgrounds destroy food and tents; raccoons on the edges of cities rip up roofs and destroy fences; coyotes devour pet cats.

Most of our attitudes toward wildlife, then, are ambivalent. We value their presence, and on the larger scale we know we need them as vital elements of the biosphere. Intellectually, we know it is important to preserve them. Nevertheless, when they come too close, there is often nervousness and fear as well as excitement.

It may be that pets act as controllable stand-ins for wild animals in the affections of human beings, but pets interact only minimally with ecosystem structures and thus play no ecologically significant role, except possibly as parasites.

Given both the emotional and biological roles of wildlife, and the fact that the environments in which they live are largely controlled by man, the design of animal habitats is an important matter. On the smaller scale, there are ways of easing the tensions between people and animals by shaping the proper interface. On the larger scale, we can significantly contribute to the preservation of species, which means the preservation of the biosphere.

ANIMALS IN THE HUMAN ENVIRONMENT

Although we are often unaware of them, wild animals are always with us. A number of studies confirm this fact. The Finnish zoologist Nuorteva, for example, compared bird populations in central Helsinki to those in a rural agricultural area and an uninhabited forest. He found that there were about three times as many birds per square kilometer in the city as in the rural area and almost four times as many as in the forested wilderness. More surprisingly, he also found that the total bird biomass in kilograms per kilometer in the city was roughly seven times that in the country and almost ten times that in the forest. In other words, there were far more birds in the city than in either the country or the forest, and their average size was over twice as large. On the other hand, there were fewer species in the city—only 21 as compared to 80 in the country and 54 in the forest. What is most surprising here is that the greatest diversity is found in the rural area. This is probably the result of the great variety of food created by agricultural activity. In any case, man's impact on the bird populations, in this case at least, has not been entirely damaging. Other studies have consistently yielded similar results.

This does not mean, however, that all is well with wildlife in human environments. The lack of species diversity reflects a number of problems with serious implications both for the larger web of wildlife that extends from the city to the remotest wilderness and for the human population.

Human land use improves the habitats of some species, degrades or eliminates those of others, and makes room for a number of exotic domesticated species. As a rule, development makes life especially difficult for the top predators, especially the bigger, more specialized ones, while creating inviting conditions for smaller omnivores, especially those that are less specialized and can adapt to varied food and cover situations.

Wolves and mountain lions, for example, need enormous foraging areas and are entirely unable to tolerate the presence of people. As a result, they have been driven away from much of their natural habitat. Some bears, on the other hand, are willing to live near people, although they need the cover provided by large forested areas. They seldom venture near cities, but they often congregate near campgrounds or vacation cabins in national forests where they become scavengers of human waste. It seems unlikely that urban areas will ever be attractive to animals like wolves, mountain lions, or bears. Coyotes, by contrast, are probably the most adaptive of all the predators. They live in considerable numbers around the edges of cities in the western United States and sometimes, even in Los Angeles, they are seen on busy downtown streets.[1]

Some predator birds, for example, eagles, are rarely seen in cities, whereas others, like the red-tailed hawk and the white-tailed kite, manage to survive in much reduced numbers. Sometimes they find certain urban conditions that, however different from those in the wild, nevertheless manage to meet their needs. In California, the freeways have provided ideal hunting areas for the kites. One can

[1] See Gill and Bonnett, 1973.

often see them soaring over the streams of traffic and occasionally pouncing on small rodents that abound in the grassy edges. Owls, being nocturnal anyway, often thrive in the quieter parts of cities where they can do their hunting when there are no people around to disturb them.

These, however, are exceptions. By and large, there are far fewer predators in urban areas, and the populations of prey species thus tend to burgeon, often with no natural controls at all. For smaller birds and mice, there is an abundant food supply of human origin and extremely varied cover in the form of building structures and ornamental plantings. Such small creatures have so often been carried by people, inadvertently or on purpose, from one city to another that the more adaptable species proliferate in cities all over the world. Among the more common of these almost universal species are the starling, the English sparrow, the black rat, the Norway rat, and the house mouse. It is not coincidental that these are also among the most troublesome of animals, often multiplying in prodigious numbers at the expense of other, less adaptive species and sometimes eliminating the natives entirely. The starling, for example, is often blamed for driving away a great number of the less aggressive hole-nesters, including bluebirds, titmice, nuthatches, swallows, wrens, and woodpeckers.[2]

Among the medium-sized mammals, those between the top predators and the smallest rodents, the man-made environment is benign to some and forbidding to others. The difference seems to be a matter of their behavior. The raccoon is extremely adaptive, having learned how to ferret out food in the unlikeliest corners of the human domain and even to use human tools and artifacts to his own advantage. By contrast, his smaller look-alike, the ring-tailed cat, is extremely shy and rarely survives urbanization. Among the other species that learn to take advantage of the human presence are the opossum and the striped skunk.

THE NEEDS OF WILDLIFE

The global imperative to preserve wildlife diversity demands not just a few protective measures but the ability to design habitats and whole environments. A species cannot survive without the network of interactions that provide for its daily needs. Without careful attention to habitat design, human land use will almost always lead to takeover by generalist species. Although designing for diverse, balanced, interacting animal populations is more complex and less certain than designing for plant communities, habitat design is inseparable from plant community design.

Because of the lack of direct interactions, we can usually design for wildlife only indirectly, that is, by providing or enhancing—or, negatively, by eliminating—the environmental conditions that they need. These needs include food, cover, water, and territory, and the requirements of each species are at least slightly different from those of any other. Since the requirements for food and cover are provided primarily by plants, control over wildlife populations is exerted largely through control of vegetation, that is, by what species are planted and in what form.

Food Needs

Since there is a high energy loss with each step up the food chain, there are usually many more herbivores (plant eaters) than omnivores (plant and meat eaters) and more omnivores than carnivores (meat eaters) in any given environment. The vegetarian mule deer, for example, requires roughly two square miles of territory, whereas the smaller carnivorous coyote needs at least twenty square miles. If an animal has less space than it needs, it is likely to cause problems. Crowded deer will strip all vegetation within their range, leaving a denuded landscape, and then, should they be in an urban or suburban situation, eat their way through any available gardens. Whenever shortages of space become severe, populations decline, and it is therefore important to strike a balance between population and feeding area. In natural areas, the combination of predation and limitations on food supply keeps the population of each species under control. In human ecosystems, natural controls are altered by human activities, and it is usually necessary to exert some management control.

Cover Needs

In general, cover means protection from the extremes of weather, from the eyes of predators, and from anything else that may seem to pose a threat. For most species, cover is vegetation, although some of the smaller animals are able to find cover in an endless array of small spaces. Starlings and pigeons, for example, are known for their ingenuity in finding cover in attic vents, on covered ledges, and other parts of buildings. Before they were all but eliminated by DDT, peregrine falcons used to find adequate cover for their nests on the narrow ledges of New York skycrapers.

It is for their nests that most animals demand the greatest privacy and the best protection from predators, and therefore the most secluded cover. Some birds that are entirely willing to feed in plain view refuse to nest where there is any sign of human movement or habitation. The need to protect their young is paramount.

Most species require cover for a variety of activities. Aldo Leopold lists cover functions as including shelter, escape, refuge, loafing, nesting, roosting, shade, and sun.[3] The California Department of Fish and Game lists different types of cover needed for five separate activities of quail—nesting, loafing, escape, roosting, and feeding—each with distinctly different requirements. Most of these can be met by low-growing, shrubby vegetation as can the needs of most other game birds and small mammals. If a variety of leaf litter, dead and rotting logs, and rocks is available, it will harbor rich populations of even smaller creatures as well, including insects, amphibians, and reptiles.

Low or sparse stands of vegetation do not, of course, provide adequate cover for larger animals like deer, elk, or moose. These rely on a combination of their own mobility and fairly dense forest cover for protection. One solution to the different requirements is to provide clearings in wooded areas that permit sunlight to reach the ground and thus encourage the growth of low, shrubby cover. Especially

[2] See Martin, et al., 1951.

[3] See Leopold, 1936.

around the edges of clearings, then, rich communities of small animals are likely to thrive, while larger ones inhabit the denser woods. We can expect the borders between any two different environments to be heavily populated with diverse wildlife species.

Birds of prey usually prefer to nest in high places, in tall trees near clearings, for example, where they can spot their prey on the ground. The nighthawk has adapted such habits to urban conditions by nesting on the flat gravel roofs of multistory buildings and roosting in nearby trees. A study carried out in Detroit found a close relationship between the number of flat roofs in certain areas of the city and the nighthawk populations.[4]

Water Needs

The third of the major needs—water—varies in importance from species to species. Most browsers, such as deer and elk and some of the smaller herbivores like rodents and rabbits, get all the water they need from the leaves of plants. Others need sources of drinking water, and still others require standing water. Wetlands, as we have seen, are essential to the survival of waterfowl, which are mostly migratory.

In arid and semiarid cities, wildlife diversity is almost directly proportional to the amount of water available. In other areas, water is less critical but still important. Planning for wildlife, therefore, goes hand in hand with planning the water regime. Spreading water over the ground and slowing its flow makes it available for longer periods, even in wetter areas. In Columbia, Maryland, ducks thrive in the sediment ponds originally provided to lessen soil losses during construction. In the very different environment of western coal mining regions, the same phenomenon has been noticed in sediment ponds dug for mining runoff. On the other hand, concrete-lined channels and ponds and lakes with hard banks have little value for supporting wildlife.

Cities in dry regions usually have more water standing free in the landscape than do the surrounding natural areas. As a result, the wildlife in such cities can be richer and more diverse than it is in the countryside. The limiting factor has been removed, or at least made less limiting. On the other hand, in the humid east-coast megalopolis, urbanization usually means less water in the landscape. One authority believes that the disappearance of several salamander species from suburban areas on Long Island is due to the elimination or disruption of breeding ponds and streams.[5]

Bodies of water of almost any size are attractive to wildlife. Ducks will congregate in ponds hardly more than a few feet wide. Urban runoff that is not pure enough for human use can be a boon to wildlife. As shown in the Los Angeles flood control case study (pages 69–74), retention basins and spreading grounds, in which water drained from urban areas is held temporarily until it can percolate to the ground supply, can double as wildlife habitats. Wherever we can incorporate features that are particularly attractive to wildlife, we can be reasonably sure that sizable populations will be drawn to them. Small islands make safe places for roosting. Artificial nesting structures, particularly for birds of prey, considerably increase the chances of their reproducing. Rafts surrounded by water will be heavily used for roosting and nesting.

Edge conditions are also important. The greater the edge area, the more extensive the habitat. Such areas are particularly important for frogs, toads, turtles, and snakes. Designs that maximize them, particularly those that provide heavy vegetation including canes, marshgrasses, and riparian shrubs, are likely to be the most successful.

Territorial Needs

Of the major wildlife needs, the requirement for territorial space is the least understood. We know that many animals need territories to call their own, and that the need generally has to do with mating and hunting. Each species has its own means of marking territorial boundaries, usually by employing a scent. Sizes of the territories vary, and their boundaries seem somewhat elastic. One result of territorial behavior is the fairly even distribution of a population over the available space. Another is the weeding out of the weaker individuals who are unable to claim and hold onto a space of their own.

As a result of the apparent rigidity of territorial requirements and the variable amounts of space available in cities, the more territorial species seem to be at a distinct disadvantage in environments dominated by human beings. Schlauch points out that the nonterritorial Fowler's toad has survived in suburban Long Island, whereas the territorial spadefoot toad has not.[6]

So far, we know very little about the exact spatial areas needed by individuals of various species. And we do not know how much, if any, these requirements might be able to contract or be replaced by other behavioral adaptations when population densities rise, as seems to be the case with human beings. Nor do we understand the relationships between human territories and those of other species. Territoriality usually has to do with relationships among individuals of a single species, but a number of animals—like ground squirrels, for example—express considerable hostility whenever human beings enter their territories.

When more is known about territorial behavior, it may turn out to be the most limiting of urban wildlife requirements. No matter how many ways we find to increase supplies of food, cover, and water, space will always be at a premium.

HABITAT TYPES

A habitat is a place where the full range of the requirements of a particular species is provided. We can distinguish six fundamentally different habitat types, all differing in size, form, population potentials, and management practice. These are wild areas, wild patches, wild enclaves, corridors, exotic greens, and wildlife parks. Although the first type probably plays the key role in preserving the global ecosystem structure, the gene pool mentioned in Chap. 5, the other types play important roles as well. In general, the smaller the area, the more intensive the management needed.

[4] See Armstrong, 1965.

[5] See Schlauch, 1976.

[6] Ibid.

Wild Areas

Wild areas are stretches of land not radically altered by human use and large enough to provide a habitat for the full range of species native to that location. In the United States, they are usually public lands, including national parks, national forests, and lands administered by the Bureau of Land Management. Considering their size and importance, wild areas are usually planned at the subcontinental or global scale.

Management practices in natural areas are primarily directed to keeping human uses under control, usually by restricting the numbers of users, areas of use, and types of use. Areas that are logged and mined are regulated to keep the impacts of these activities on wildlife to a minimum. In the United States, wilderness areas and wildlife refuges, along with several other categories, are special types of wild areas in which human use is even further restricted in order to maintain their primordial features.

How large should a wild area be? If they are to serve as habitats for the full range of wildlife species, including the top carnivores, quite large areas of land would seem to be implied. Minimum sizes probably vary according to the range needed for the top carnivore in each particular habitat. Sullivan and Shaffer have estimated that the minimum size for the usual full-range wildlife reserve should be between 600 and 760 square kilometers.[7] This is based on providing enough space for at least a few individuals of the top carnivore species. Each grizzly bear has been estimated to need about 75, each wolf about 60, and each mountain lion about 95 square kilometers. The needs of other top predators are probably within the same range. Since eight, or even twelve, individuals is a very small number, however, and probably not enough to maintain a population indefinitely, this has to be considered a very risky minimum. In practice, few wild areas, especially in the densely settled European countries, are this large.

Given the state of the world's development, virtually all wildlife habitats, even wild areas in this size range, are surrounded by human activity; they are all islands. Wild areas, however, are more like continents in that they appear to be boundless, at least from within. In the sense that the number of species they can support is not limited by their extent, moreover, they are indeed boundless. The other types of habitats, being seriously limited by their extent, are clearly islands.

This island character is important in considering the wildlife population potential of any landscape. MacArthur and Wilson's theory of island biogeography provides some guidance.[8] These authors show that an island develops a community over time that includes a more or less constant number of species. Some species die out while others migrate in, the number of extinctions equaling the number of immigrations. The rate of extinction correlates with island size; the smaller the island, the larger the number of extinctions per unit area. The immigration rate correlates with the distance to the origins of the migrating species; the closer the island is to other islands, the more new species are likely to arrive. Thus, the larger the island, the larger the number of species it is likely to have.

As always, however, there are exceptions. Higgs and Usher found that among the chalk quarries of England, two reserves featured more species than a larger one the size of the two combined.[9] The reason probably lies in the fact that the two were sufficiently different from one another to have more diverse environments. Gilpin and Diamond similarly found several instances in which two islands had 5 to 10 percent more species than one island with the area of both. They hypothesize that the subdivision of reserves favors species with small area requirements, whereas larger reserves favor those needing larger areas, which are usually top carnivores.[10]

In general, however, in the light of subsequent research, the MacArthur-Wilson theory seems to hold true, which argues that, other factors being equal, wild areas should be as large and contiguous and as close together as possible. One large preserve is better than several smaller ones of equal total area, except where particular species requirements or habitat qualities create exceptions. It often happens that areas with extraordinary habitat environments are in locations severely limited in their extent by surrounding uses. In such situations, the need for wild patches, enclaves, and corridors becomes essential.

Wild Patches

Wild patches are large enough to be self-maintaining systems with working feedback loops to control their populations, as ecologically complete as any area can be without top predators. Usually, they are set aside because of environmental characteristics particularly attractive to wildlife. Their size is ordinarily measured in square miles, and usually, they require considerably more intensive management than do wild areas.

Among the best known and most dramatic wild patches in the United States is Jamaica Bay, a thirty-square-mile marsh less than ten miles, as the egret flies, from Times Square. Wild patches are usually managed for high levels of species diversity and for large populations. The intensity of management varies, and Jamaica Bay is one of the more intensively managed. Herbert Johnson, who was its superintendent from the time it was named a wildlife refuge in 1953 until the early seventies, made it a model of the use of urban resources to benefit nonhuman populations.

In the early 1950s, the old railroad trestle over the bay burned, and the Transit Authority proposed to build in its place a new embankment on which to run an extension of the subway. The Parks Department agreed, on the condition that, while dredging, they would build two circular dikes that would be filled with fresh water, thereby creating a new type of environment and considerably expanding the number of species attracted to the area. Once the dikes were in place, they were stabilized with plantings of marshgrasses and other plant species favored by the birds. At the same time, sewage sludge was pumped into the lagoon to cover an island called Canarsie Pol. Marshgrasses quickly

[7] See Sullivan and Shaffer, 1975.
[8] See MacArthur and Wilson, 1967.
[9] See Higgs and Usher, 1980.
[10] See Gilpin and Diamond, 1980.

invaded this rich new land and made an ideal nesting environment for shorebirds.

Over the years, Johnson planted a variety of tree species that were particularly favored by certain bird species. Among them were the autumn olive, Russian olive, chokeberry, holly, and Japanese black pine. There was no intention to provide a purely native landscape. Rather, it is one designed specifically to accommodate the greatest possible number of bird species. The results are spectacular. Over three hundred bird species are seen in the bay in a typical year, far more than would have appeared there in its natural state.

Wild patches this close to urban areas usually lead precarious lives. In the late 1960s, Kennedy Airport, which lies adjacent, proposed to expand its runways in a manner that would have seriously reduced wildlife populations. After a public furor and an exhaustive impact analysis by an interdisciplinary group, the plan was abandoned. In 1972, Jamaica Bay become part of the newly created Gateway National Recreation Area and thus was gathered under the wing of the National Park Service. Since then, the number of urban wilds under Park Service control has grown considerably, and the results have been promising. Nevertheless, urbanization continues to make encroachments. It now appears that the entire edge of the bay will become a ring of subsidized housing within a few years. The vitality of the refuge will probably depend on the design of this critical edge.

Wild Enclaves

Wild enclaves are smaller than patches, not large enough to be self-sustaining systems but still able to support significant wildlife populations and important for that reason. Sometimes rich habitats manage to survive all the threats and encroachments of surrounding human use and become small spots of wildlife diversity in the midst of cities or farms. Almost invariably, these areas have some exceptional feature that makes them especially attractive to animals, and more often than not, that feature is water. Ponds, streams, marshes, even drainage ditches, usually support rich populations. Among the richest are marsh enclaves, which are invariably alive with waterfowl and other species that thrive around water. Too often, they are continually threatened with obliteration.

In Southern California, the 11-acre Madrona Marsh, discussed in Chap. 5, is such an area. This fresh-water marsh supports a rich population of microscopic and macroscopic plants and invertebrates, amphibians and reptiles, and, in particular abundance, migratory waterfowl and resident birds. The small size and the character of the surrounding uses required a complex design process and will require an equally complex management process.

Systems that identify critical zones provide a strong basis for resolving conflicts and for future control. Considering the precarious existence of most enclaves in the midst of high land prices and urban exigencies, strong conflicting purposes, and volatile emotions, processes of this kind are much needed.

Most enclaves have survived as a result of historical accidents, and we can expect that sooner or later pressures for "higher and better uses" will be brought to bear. The Madrona Marsh continues to exist as an inadvertent byproduct of oil extraction. Some enclaves are the remains of man-shaped environments that have been reclaimed by natural succession.

In areas whose natural vegetation is dense forest, we sometimes find small stands of native woodland surviving in the cities. Though too small to support complete communities of animals, these often feature heavy concentrations of smaller birds and mammals. Typical of this sort of enclave is Wimbledon Common in a suburban area near London. (According to landscape architect Sylvia Crowe, the cost of maintaining Wimbledon Common is less than 3 percent of the cost of maintaining Kensington Gardens, which is a highly developed London park.)

Wild Corridors

In general, the longer the available routes of movement in an area, the greater will be the diversity of wildlife. In a large natural area, the number of routes can be almost infinite. But between smaller ones that are not too far apart, and certainly between patches and enclaves, corridors of natural landscape can expand the possibilities of movement in relatively little space. Interconnection by corridors provides an increased array of possibilities for wildlife movement and habitat choice and is particularly useful for expanding the potentials of the smallest areas and reducing the island effect. Corridor systems are best considered at the regional level.

Urban and rural landscapes are characteristically laced with networks of various kinds, some of which can be fairly easily adapted to attract wildlife. Waterflow networks are perhaps the best, because they provide not only water but the abundant vegetation that usually goes along with it. If the edges of streams, gullies, washes, flood plains, and even flood control channels can be kept open and unimpeded, they can serve as a network of corridors interweaving the wilds with the man-made landscape.

Among the man-made corridors are various types of rights-of-way, including railroads (especially abandoned ones) and powerlines. With some care in design, these can be made to attract animals. Mostly, it is a matter of opening up movement routes that will be beneficial to wildlife and closing those that are likely to bring animals into conflict with human activity. For example, along abandoned railroad rights-of-way, the California Department of Fish and Game recommends using one side for bicycling and hiking and reserving the other side for wild animals.

Wild corridors are especially desirable in areas where it is important to provide habitats for large predators because of human alterations of the natural landscape. In Costa Rica, for example, government policy calls for the creation of new farming settlements in which small tracts of arable land, usually about eight hectares, are made available to landless families. Since there are few alternatives, these settlements are often carved out of the rain forest, with all of the attendant difficulties already discussed. Because of the national concern for their preservation, some undisturbed forest is usually left within the bounds of each settlement, more or less in the pattern shown in the first diagram on page 63. The size of most of these forest remnants ranges from that of enclaves to that of patches—far short of the territory needed by the top predators. The magnificent, elusive jaguar is especially threatened. Little is known about its habits or the extent of its range, except that the latter is immense, probably covering hundreds of square kilometers.

The best solution would probably be to maintain a continuous matrix of rain forest, with the settlements as islands within, rather than the reverse, as shown conceptually in the second sketch on page 63, but this approach would be expensive and would tend to spread settlements far apart, resulting in communication and transportation difficulties. It would also require more land, thus driving settlement deeper into natural areas.

Another possibility would be to group the preserves in larger units spaced closer together, as shown in the third diagram. Island theory tells us that these would be far richer habitats, even if they cover no more land than scattered ones.

Finally, if we connect the preserves with corridors, as shown in the fourth diagram, they would be richer still. The corridors might follow streams and rivers and would have to be wide enough to keep animals like the jaguar well away from human settlements. Although they would not effectively replace the larger natural areas of rain forest, they would at least provide a minimally complete structure in the midst of human activity.

Exotic Greens

Far removed from rain forests and jaguars, exotic greens are man-made landscapes, usually in cities, that are not designed or managed with wildlife as a major concern but nevertheless feature surprisingly large animal populations. It is important to remember that urban landscapes shaped and maintained for specific human uses are far from being in a natural state. Among them are school grounds, urban parks, cemeteries, college campuses, and areas of residential developments, even back and front yards. Usually, they feature exotic species of vegetation that are chosen more for visual effect than for the benefit of wildlife or other ecological reasons. Extensive areas of lawn are common, usually punctuated with a few stands of trees. Often, these places are teeming with wildlife, but diversity is normally limited to a few species of small birds and mammals since their structures are usually simple and incomplete.

The vegetation of an exotic green is designed to serve the particular functions and image of that place, and these are necessarily different from those of other greens. Thus, each attracts a wildlife population related to its own unique community of plants. Melvin Hathaway has found, for example, that a reason for the high squirrel population in a cemetery he studied was its particular variety of plants, which produced foods at different times throughout the year, assuring a continuous feed supply. He points out that "man's efforts to maintain the beauty of the cemetery had stabilized the habitat by stopping ecological succession at just the stage preferred by these animals."[11] The closed hours of the cemetery—5 P.M. to 8 A.M.—also seemed to benefit the squirrels, since they had the landscape to themselves during this period. On the other hand, two factors presented greater dangers than they would find in the wilds. One was the automobile traffic, which, although light and slow-moving, killed a large number of them, and the other was disease, which seemed to spread faster than in the wilds. Predation probably took a toll about equal to that in the wilds, since a few red foxes also lived in the cemetery. All in all, Hathaway concludes that the squirrels lived much as they would have in a rural or wilderness area, with a population that was stable, if relatively high.

In the eastern United States, the red fox is one of the few predators able to adapt to the environment of the exotic green. They have been seen fairly frequently in the larger urban parks of most eastern cities. Their role is an important one because without larger predators, squirrel populations would be likely to grow beyond the capacity of the environment to support them. When that happens, they usually cause severe damage to trees.

Exotic greens are the most intensively managed (for human, not wildlife, purposes) and heavily used of urban habitats. Consequently, they usually attract the more sociable species, those most tolerant of the human presence.

Wildlife Parks

The types of this category of wilds are relatively few and highly specialized. They are areas of some size that are specifically designed and managed to accommodate specific exotic wildlife populations. The zoological garden is the most extreme example, although zoos are usually so artificial as to deserve a category of their own. Somewhat less artificial are the several types of wild animal parks where varied species of animals are brought together and allowed to roam freely in a setting that provides them some space and privacy, although it might be quite different in its physiography and vegetation from their native environment. Such parks have been fairly common on royal estates for centuries. A number of estates in England, most notably Woburn Abbey, maintain them. In the United States, wildlife parks usually have a public orientation, functioning essentially as theme parks. Wild Animal Park near the City of San Diego is perhaps the best known example.

The Wild Animal Park Case Study

Like several others of the type, San Diego's Wild Animal Park has increasingly devoted its energies to the preservation of endangered species. This, in fact, is its main purpose, the other two being recreation and public education. It has established a very active and successful breeding program. Some of its offspring have been sent back to their natural environments in an effort to bolster the native species. In 1978 and 1979, eight specimens of the Arabian oryx, bred in the Park, were released in a wild area of Jordan where the species is now extinct. Other species that no longer exist in the wilds have been preserved and bred here as well.

[11] See Hathaway, 1973.

CASE STUDY XVII

Wild Animal Park

The San Diego Wild Animal Park occupies an 1800-acre site—of which only about 600 acres are not in use—in rolling, dry hills about thirty miles north of downtown San Diego. Although its main purpose is breeding wild animals, the park has become a popular weekend attraction with a sizable village of shops and restaurants at one end. These activities generate revenues to support the facilities and programs. Assuring the visibility of the animals to the public requires relatively dense populations, and this has resulted in rather serious damage to the landscape, especially in the form of erosion.

This project develops four alternative plans for future land use and management in the park that will sustain the quality and capability of the landscape. Each alternative achieves this goal in a somewhat different way, with emphasis on different priorities and at different levels of investment. All four call for more intensive management. To support large numbers of animals without landscape deterioration will require denser, faster growing vegetation—ideally a replication of the plant communities with which the animals coevolved. In this naturally dry landscape, most of these exotic communities will require irrigation. The suitability analysis defines areas where different biomes might be created.

Prepared for the San Diego Zoo Association by the 606 Studio, Department of Landscape Architecture, California State Polytechnic University, Pomona. Cecile Holloway and Janet Yarbrough, with advisors John T. Lyle, Jeffrey K. Olson, Arthur Jokela, and Robert Perry.

LAND RESOURCE INVENTORY

LAND USE

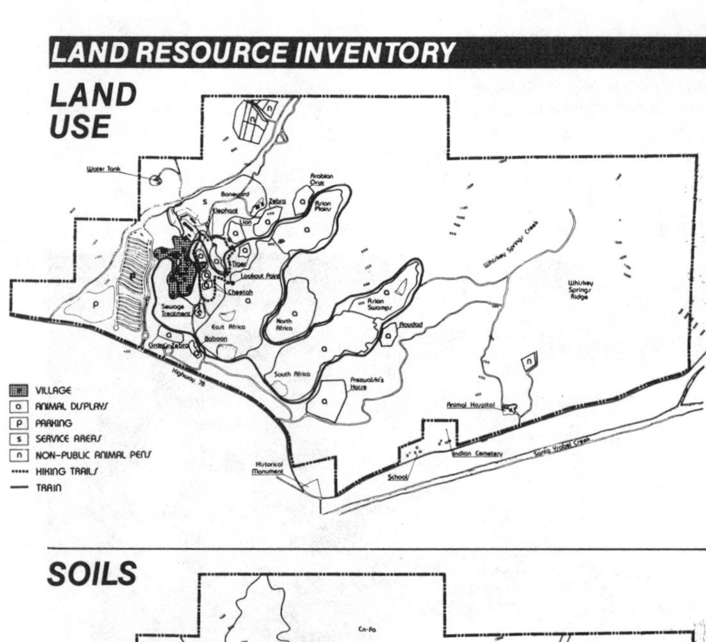

GEOLOGY/HYDROLOGY

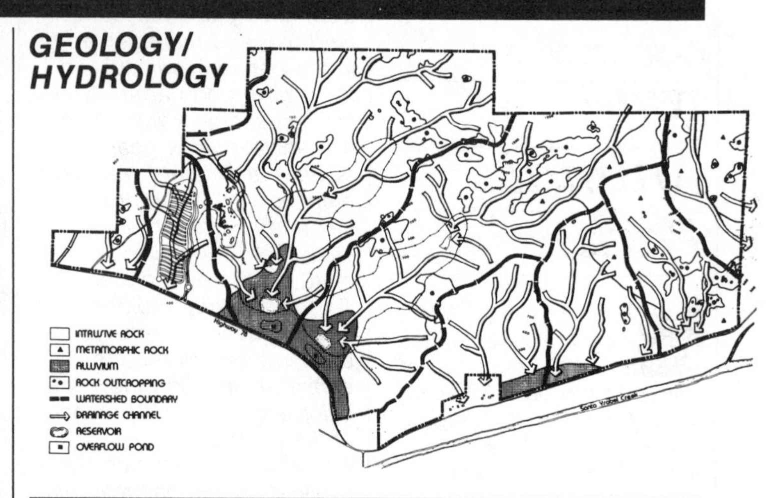

SOILS

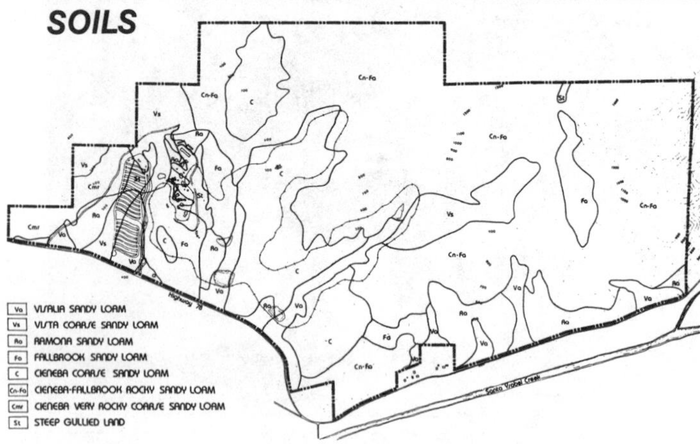

SLOPE

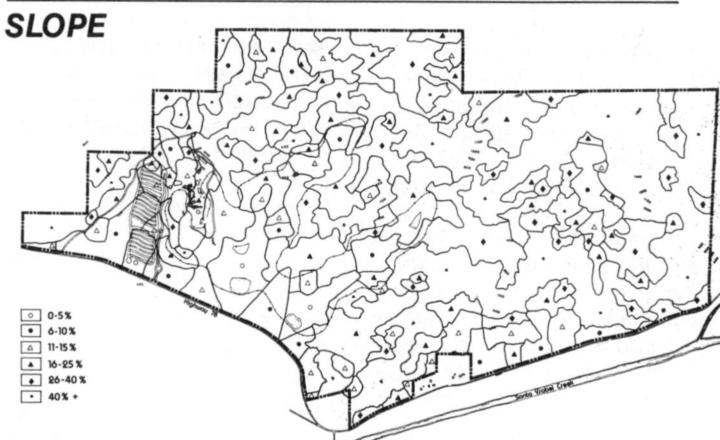

○ Slight
◍ Moderate
● Severe

	Soil properties						Soil capability ratings							
	Erosion hazard	Runoff	Fertility	Shrink/swell	Depth	Instability	Building	Campsites	Picnic	Paths & trails	Topsoil	Range	Roads	Water retention
Va	○	◍	○	○	○	●	◍	○	◍	○	○	○	◍	◍
Vs	◍	◍	◍	○	◍	○	○	◍	◍	○	◍	○	◍	◍
Ra	◍	◍	◍	◍	○	○	○	◍	◍	○	◍	○	●	◍
Fa	◍	◍	◍	◍	◍	○	○	◍	◍	○	◍	○	●	◍
C	●	◍	●	○	●	○	◍	◍	◍	◍	●	◍	●	●
Cmr	●	◍	●	○	●	○	◍	◍	◍	◍	●	◍	●	●
Cn-Fa	●	●	●	○	●	○	◍	◍	◍	◍	●	◍	●	●
St	●	●	●	◍	●	○	◍	◍	◍	◍	●	●	●	●

VISUAL FORM

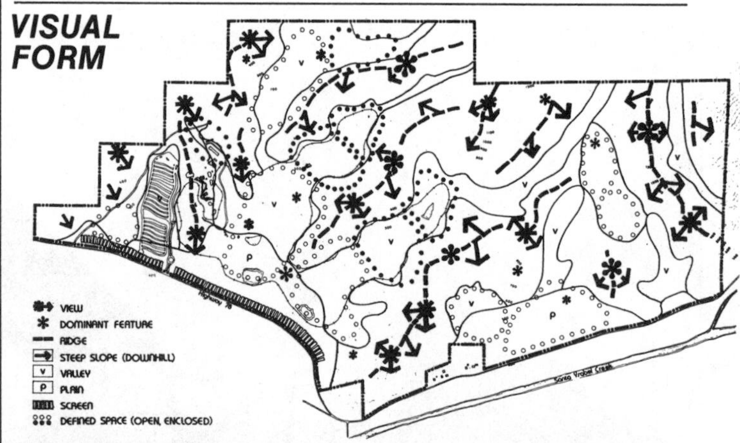

VEGETATION

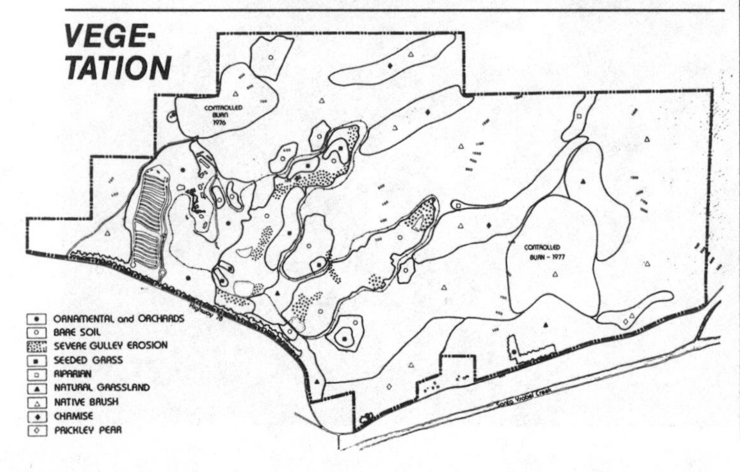

MICROCLIMATE

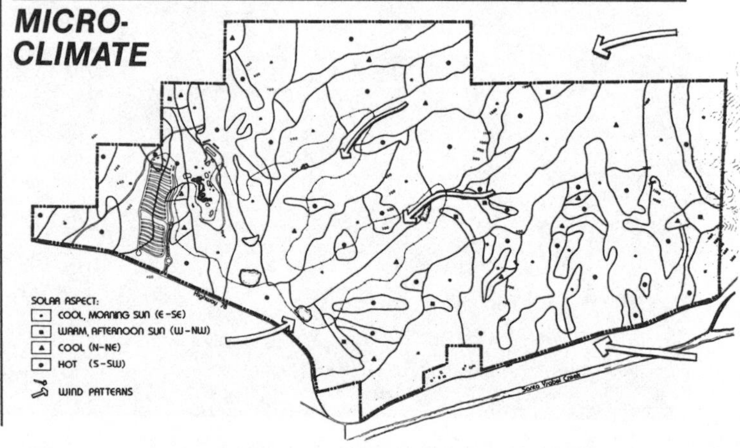

SUITABILITY MODELS

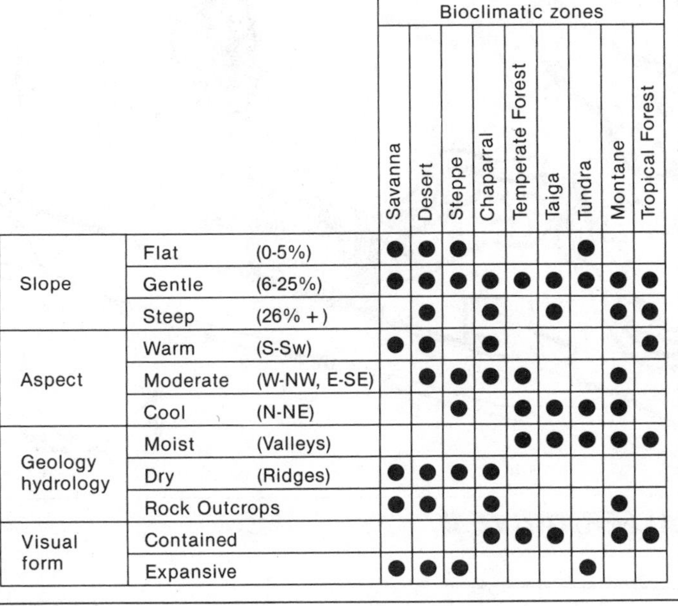

			Bioclimatic zones								
			Savanna	Desert	Steppe	Chaparral	Temperate Forest	Taiga	Tundra	Montane	Tropical Forest
Slope	Flat	(0-5%)	●	●	●				●		
	Gentle	(6-25%)	●	●	●	●	●	●	●	●	●
	Steep	(26% +)		●		●		●		●	●
Aspect	Warm	(S-Sw)	●	●		●					●
	Moderate	(W-NW, E-SE)		●	●	●	●			●	
	Cool	(N-NE)			●		●	●	●	●	
Geology hydrology	Moist	(Valleys)					●	●	●	●	●
	Dry	(Ridges)	●	●	●	●					
	Rock Outcrops		●	●		●				●	
Visual form	Contained					●	●	●		●	●
	Expansive		●	●	●				●		

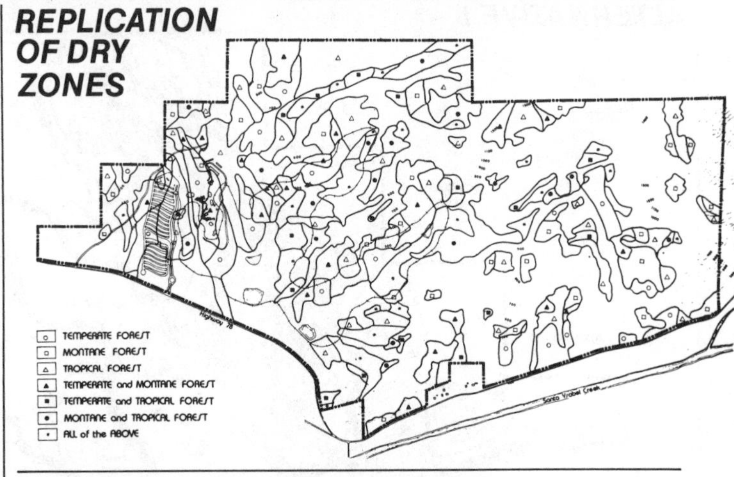

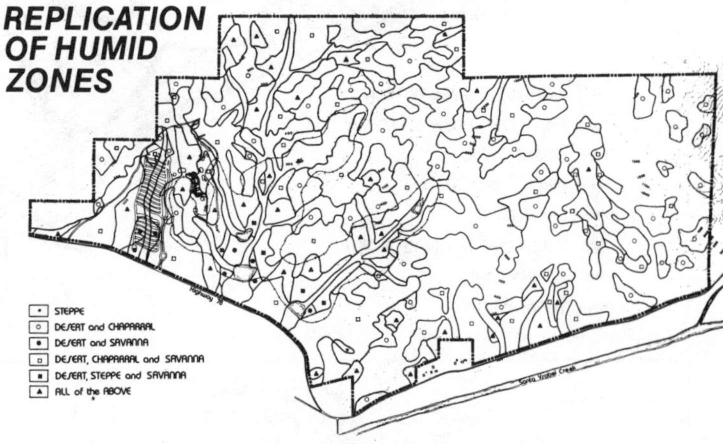

Severe constraints
- *Steep slopes (over 40%) or*
- *Moderately steep slopes (25-40%) and erodible soils (Cl, Cn-Fa, Cmr) or*
- *Severely eroded soils (St) or*
- *Riparian vegetation*

Moderate constraints
- *Moderately steep slopes (25-40%) or*
- *Erodible soils*

Minor constraints
- *Moderate slopes (16-25%) or*
- *Fair soils (Fa, Ra) or*
- *Hot aspect (south, southwest)*

No constraints
- *Good soils (Va, Vs) or*
- *Gentle slopes (0-16%) or*
- *Rock outcroppings*

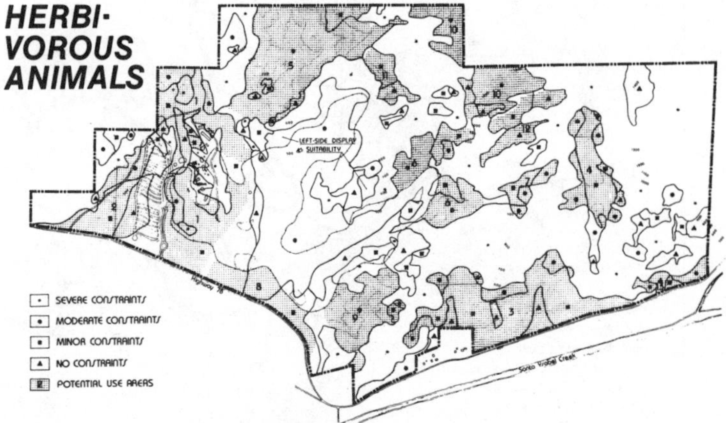

Severe constraints
- *Steep slopes (over 40%) or*
- *Moderately steep slopes (25-40%) and unstable or erodible soils (Va, Cn-Fa, Cl, Cmr, St) or*
- *Moderately steep slopes and unstable or erodible soils and major drainage channels or*
- *Riparian vegetation or*
- *Rock outcroppings*

Moderate constraints
- *Moderately steep slopes (25-40%) or*
- *Moderate slopes (16-25%) and unstable or erodible soils (Va, Cn-Fa, Cl, Cmr, St) or*
- *Moderate slopes and unstable or erodible soils and major drainage channels or alluvial rock*

Minor constraints
- *Unstable or erodible soils (Va, Cn-Fa, Cl, Cmr, St) or*
- *Moderate slopes (16-25%) and fair soils (Fa, Ra) or*
- *Major drainage channels or alluvial rock*

No constraints
- *Good soils (Vs) and gentle to moderate slopes (0-25%) or*
- *Fair soils (Fa, Ra) and gentle slopes (0-15%)*

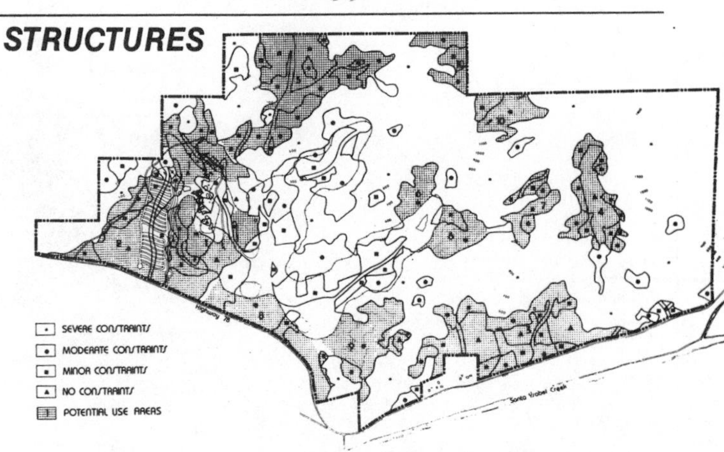

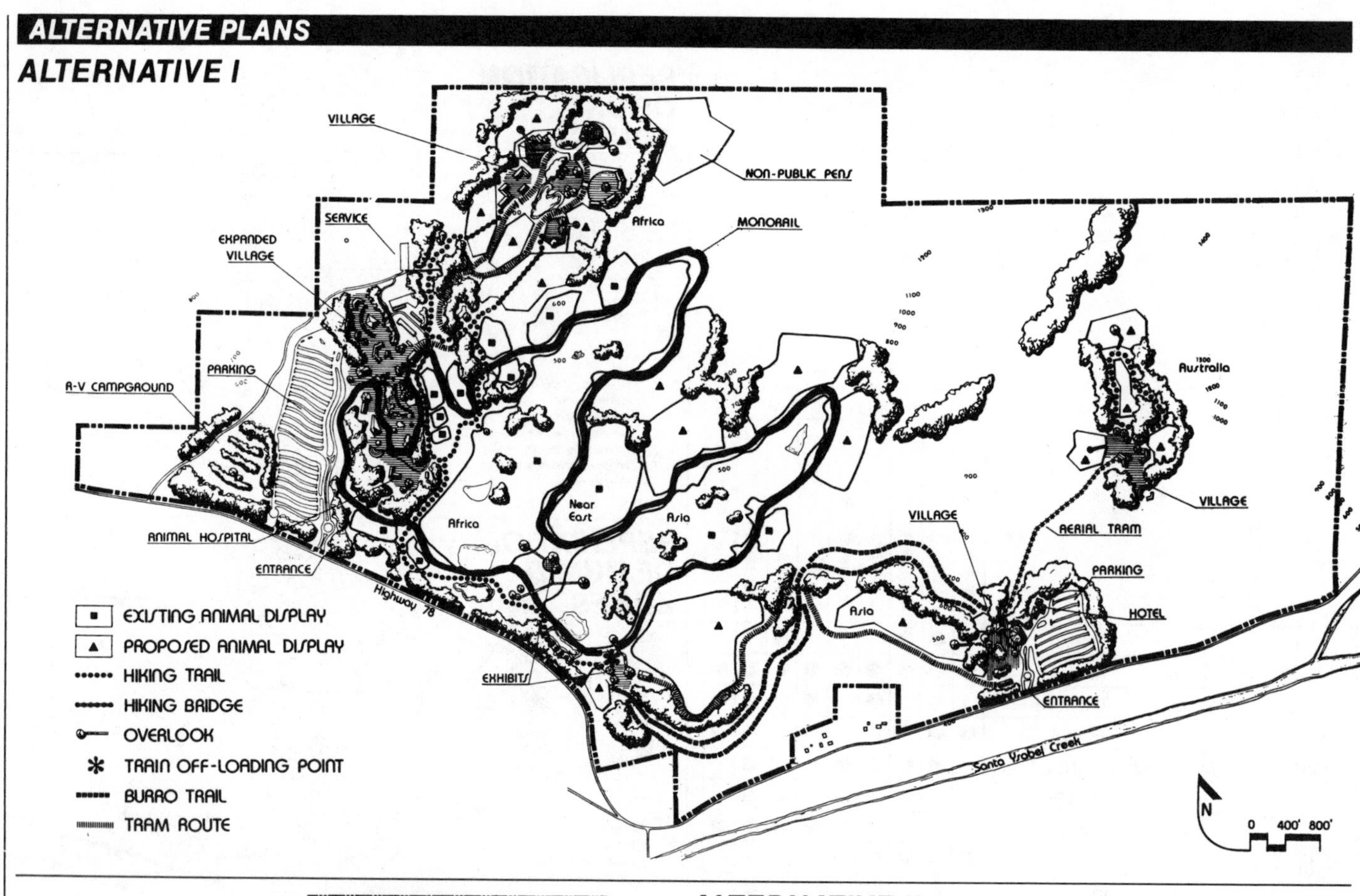
ALTERNATIVE PLANS
ALTERNATIVE I
VILLAGE
NON-PUBLIC PENS
SERVICE
EXPANDED VILLAGE
Africa
MONORAIL
PARKING
R-V CAMPGROUND
Australia
Near East
Asia
Africa
VILLAGE
ANIMAL HOSPITAL
ENTRANCE
VILLAGE
AERIAL TRAM
PARKING
HOTEL
Highway 78
Asia
EXHIBITS
ENTRANCE
Santa Ysabel Creek
EXISTING ANIMAL DISPLAY
PROPOSED ANIMAL DISPLAY
HIKING TRAIL
HIKING BRIDGE
OVERLOOK
TRAIN OFF-LOADING POINT
BURRO TRAIL
TRAM ROUTE
N
0 400' 800'

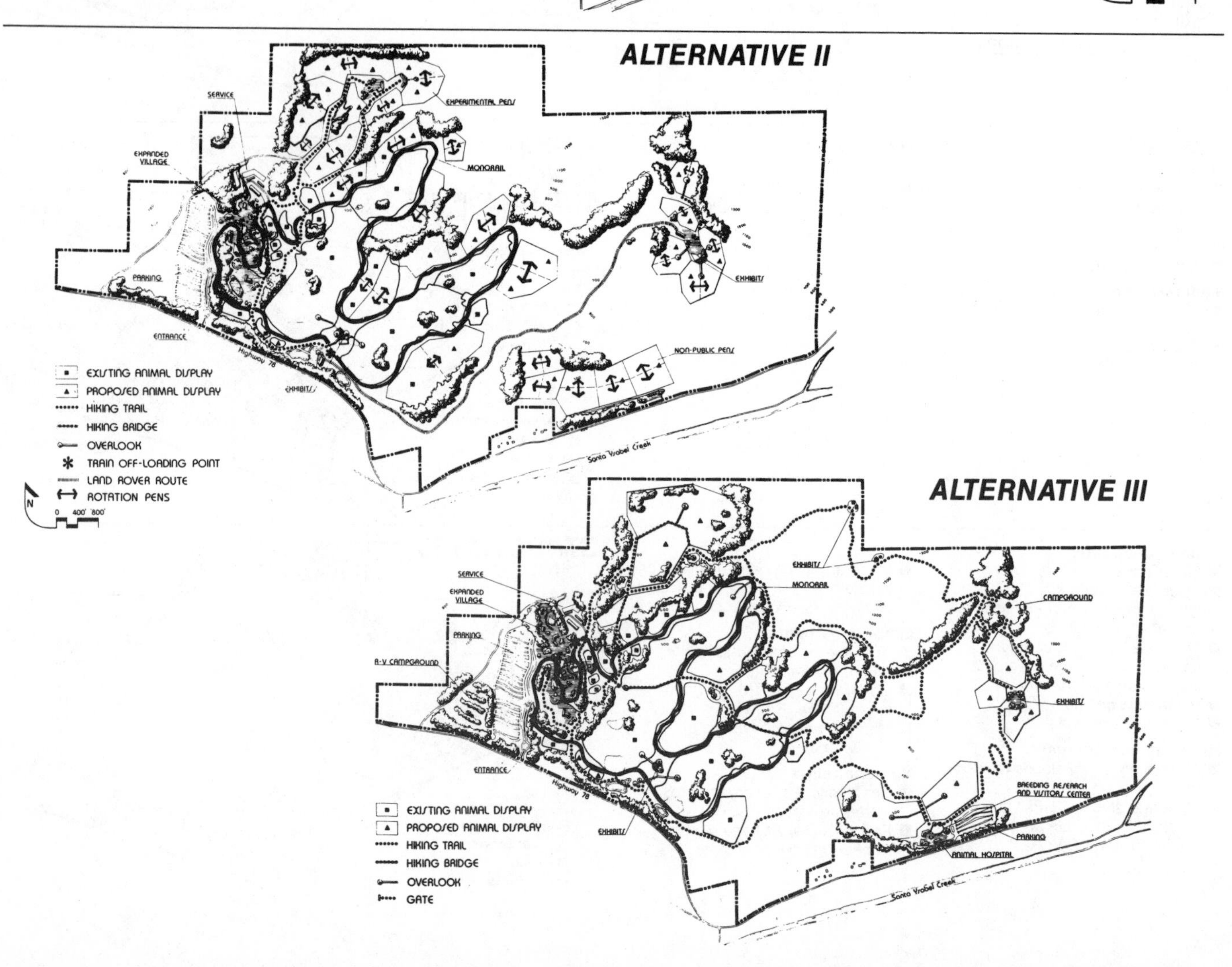
ALTERNATIVE II
SERVICE
EXPERIMENTAL PENS
EXPANDED VILLAGE
MONORAIL
PARKING
EXHIBITS
ENTRANCE
Highway 78
NON-PUBLIC PENS
EXHIBITS
Santa Ysabel Creek
EXISTING ANIMAL DISPLAY
PROPOSED ANIMAL DISPLAY
HIKING TRAIL
HIKING BRIDGE
OVERLOOK
TRAIN OFF-LOADING POINT
LAND ROVER ROUTE
ROTATION PENS
N
0 400' 800'
ALTERNATIVE III
SERVICE
EXPANDED VILLAGE
EXHIBITS
MONORAIL
CAMPGROUND
PARKING
R-V CAMPGROUND
EXHIBITS
ENTRANCE
Highway 78
BREEDING RESEARCH AND VISITORS CENTER
EXHIBITS
PARKING
ANIMAL HOSPITAL
Santa Ysabel Creek
EXISTING ANIMAL DISPLAY
PROPOSED ANIMAL DISPLAY
HIKING TRAIL
HIKING BRIDGE
OVERLOOK
GATE

ALTERNATIVE IV

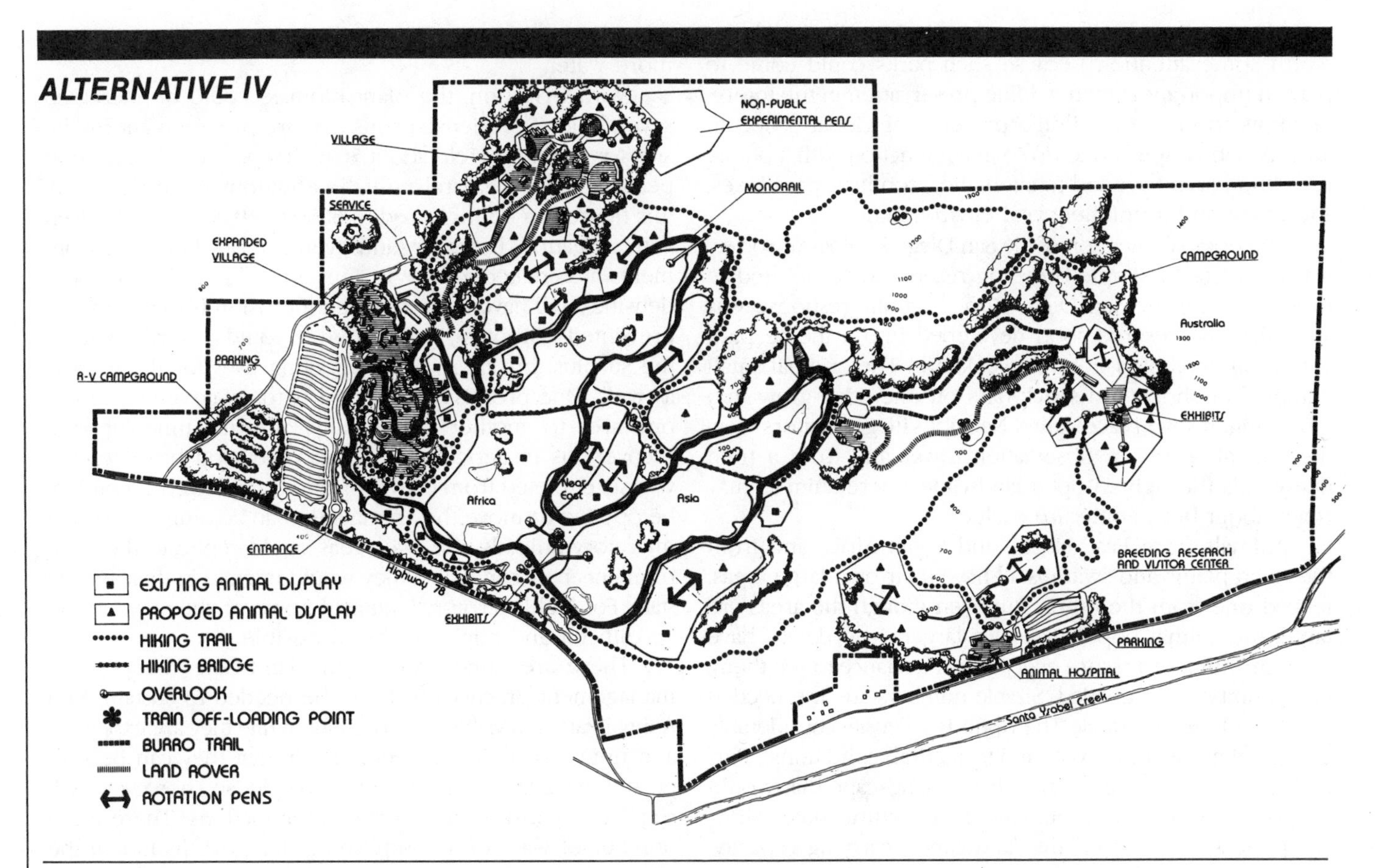

	Systems			Density			Management							Transportation							Safari		
	Exotic	Convergent	Dryland	High density	Low density	Mixed densities	Population control	Rotation	Fencing gulleys	Irrigation	Terracing	Keyline	Game ranching	Train off-loading	Alternate system	Aerial tram	Hiking trail	Burro trail	Extend train	Second entry	Expand displays	Slight expansion	Expand and rotate
Recreation	●			●					●	●	●	●		●	●	●		●	●	●			
Breeding		●				●			●	●	●	●	●				●				●	●	
Stable landscape			●		●		●	●				●			●				●				●

	Village			Themes									Geographic zones							Bioclimatic zones								
	Second village	Village expansion	Camping	Paleontology	Archaeology	Natural history	Human culture	Animal behavior	Botanical	Conservation history	Media	Game ranching	Asia	Africa	Australia	Europe	South America	North America	California	Desert	Chaparral	Savanna	Grassland	Tropical forest	Temperate forest	Taiga	Tundra	Montane
Recreation	●				●	●	●			●	●		●	●	●	●	●	●	●	●	●	●	●	●	●	●	●	●
Breeding		●	●			●		●				●	●	●	●	●	●	●	●	●	●	●	●	●	●	●	●	●
Stable landscape		●		●		●			●				●	●	●	●	●	●	●	●	●	●	●					

The history and development of Wild Animal Park are worth some attention because such parks could come to play an important role in wildlife preservation in the future. Attempts to create a wildlife preserve of global scope in an alien landscape and to do so in conjunction with a public recreation area have led to anomalies, conflicts, paradoxes, successes, and a uniquely vital environment.

The original intention of the San Diego Zoological Society was to create "a simple breeding reserve to accommodate overflow from the San Diego Zoo."[12] Public response was such that the reserve soon developed into a theme park with thousands of visitors on weekends. The original small complex of shops and snack bars on one edge grew into a bustling, festive place called Nairobi Village. Visitors view the animals from an observation tower and from a train that winds through an open landscape where animals may roam about but people are excluded.

Animals from East, North, and South Africa and from the Asian plains and swamps all have their own large areas, fenced one from the other. But even though the areas are large, the animal populations are larger. In order to have enough animals to assure every visitor a chance to see them, large numbers are needed. Sizable numbers are also needed to ensure breeding stock. The result is a density considerably greater than would exist under natural conditions. Furthermore, this is a dry, rather brittle landscape that would support very few large animals in its natural state. Since the populations exceed the landscape's carrying capacity, the resulting overgrazing and overtrampling have led to severe denudation and erosion. These are common conditions in landscapes (whether natural or man-made) where populations are larger than the land can support.

In 1978, the Cal Poly Graduate Studio group undertook to develop a plan to deal with these problems and to propose future uses for several hundred acres that were still undeveloped. The results, as presented by Cecile Holloway and Janet Yarbrough, are summarized on pages 219–23. The group concluded that placing such diverse species of exotic animals in a landscape to which they are not naturally adapted will probably always present serious problems. Nevertheless, with intensive management efforts and at some expense, an ecosystem might be created that could support them.

The first need is for a varied and sophisticated irrigation system to provide for faster growth and more varied plant communities. Under present conditions, the sparse, slow-growing native vegetation is not able to maintain ground coverage or provide for the animals' needs. The limiting factor for plant growth is water. Given appropriate amounts of water, the topographic variety of the site could make it possible to replicate the native environments of most of the animal populations now in the park and some others as well. Thus a fit, albeit a man-made one, could be achieved by planting the requisite vegetation communities. The chart on page 221 shows the range of possible environments and the landscape attributes needed to replicate them. The dryland and wetland suitability map shows the areas with the conditions needed to replicate various environments if given more water.

With irrigation, the plant biomass could be increased severalfold. The Cienaba soils that are predominant on the site, for example, yield about 50 to 300 pounds of vegetation per year under natural rainfall conditions. With irrigation, this figure can be increased to 6000 to 10,000 pounds. Even with irrigation, however, and even if more fitting environments are created for the various animal populations, their densities will have to be reduced if erosion is to be entirely prevented. The now unused lands could absorb some of the surplus, but there are competing uses for these areas as well. One promising proposal is to rotate animals from one area to another periodically, allowing time for land recovery as in agriculture. When vegetation becomes obviously depleted from browsing and trampling, a gate might be opened to move the animals into an adjoining area. For migratory animals, the two areas could replicate the environments between which they would migrate in their natural state. For highly territorial animals, however, the move would be difficult and might prove impossible.

These are only a few of the landscape changes and management practices that will be needed to sustain Wild Animal Park as a viable environment. While they are expensive and time-consuming, it seems likely that they can be supported by admission revenues. To do so, however, will require expansion of recreational attractions. There are a number of ways of accomplishing this goal, including the addition of new villages, restaurants, and even a hotel. The first alternative is based on maximizing economic return in just this way. The second alternative provides for only a minor increase in recreational activity while increasing breeding activities and thus the amount of space and facilities devoted to them. The third alternative also provides for a limited expansion of recreational activity while restricting animal populations to the carrying capacity of the land. This would mean densities ranging from 1 to 2.39 animals per acre, depending upon the type of landscape. If the landscape is planted in vegetation patterns similar to those of the animals' native environments, it should be possible to establish a balance at these densities. That is, the landscape will probably be able to provide for all the animals' needs except food, which is brought from the outside in any case. After these changes are made in the landscape and its use, the results would be monitored with a view to reducing future populations should deterioration continue. Thus, a steady-state condition could be reached within a few years.

The fourth alternative calls for a major expansion of the park and a major investment. It would implement the key ideas from the first three alternatives.

The choice among these alternatives clearly represents a policy decision that will be based on goals concerning future roles and priorities. Although the fourth alternative is in many ways desirable, it represents a definite commitment to a highly visible, intensively managed enterprise with a scope and aim quite different from Wild Animal Park's original purpose.

These, then, are the kinds of questions and issues likely to confront any wildlife park. It cannot be denied that such

[12] See Holloway and Yarbrough, 1978; p. 9.

a park offers a vital, stimulating, highly educational environment, one that is doing important work in species preservation. Nevertheless, it is also a very artificial landscape that will probably have to become even more artificial in the sense that it will have to be more intensively controlled and managed if it is to survive. Its various goals provide for fascinating symbiotic interactions of the sort that probably will be much more common in the landscape of the future. Unfortunately, they also provide for vexing conflicts. Wild Animal Park does not fit most people's notions of an appropriate environment for the preservation of wildlife, but for some, possibly a great many, species it may be the only way.

HABITAT DESIGN

The Wild Animal Park example illustrates several important aspects of design for wild animal environments, including carrying capacity and the importance of suitable plant communities. Although these principles are equally important under more natural conditions, wherever animals live with both their predators and prey and wherever they live entirely by their own devices, especially when their environments are in close proximity with those of human beings, a number of other factors enter into habitat design.

In each type of wild environment except those set aside as pristine wilderness, there are definite advantages in encouraging the largest, most diverse communities that the environment can sustain. The notion that this objective can be achieved, that humans can systematically design landscapes to accommodate and nourish wildlife populations, originated only a few decades ago in the work of Aldo Leopold.[13] Leopold listed environmental design as the fifth of five means of wildlife management, the others being restrictions on hunting, predator control, reservation of game lands, and artificial replenishment. Since his time, habitat design has become considerably more important and sophisticated.

The two general goals of wildlife management are usually summed up as species richness and featured species enhancement.[14] The former involves general management practices, while the latter requires specifically focused practices. It is important to keep in mind that these two goals can conflict. By designing for diversity, we might eliminate a particular species, and by designing for one species, we can limit diversity.

Design for Featured Species

Earlier we mentioned the different habitual activities of the quail: nesting, loafing, escape, roosting, and feeding. Based on an analysis of these activities, the California Department of Fish and Game has made some specific recommendations for the design of all small game bird habitats. For the basic nesting and loafing cover, they recommend that clusters or hedgerows of shrubs or small evergreen trees, twenty-five feet or so across, be provided, preferably alongside gullies or stream beds. Thickets and brushpiles located nearby and twenty to thirty feet apart will furnish the necessary escape routes. Where ground level vegetation is heavy, especially in riparian areas, clearing of wide strips will help provide suitable feeding areas. The clearing is best done just before the rainy season. Room for these three activities can be provided even in small spaces. The varied cover conditions should be attractive to most upland birds—including the partridge, grouse, pheasant, and turkey—as well as to quail and a wide variety of other birds and small mammals. The major purpose in this case, however, is to provide optimal conditions for the premier game bird, the quail. This is an example of design for a featured species.

With these rather modest habitat requirements, we can contrast those of one of the larger browsers, the elk. Rodiek defines the elk's cover needs as hiding cover, thermal cover, travel lanes, and calving areas.[15] Food needs include those for forage and browsing. Hiding cover is defined as vegetation capable of hiding 90 percent of an elk from view at a distance of 200 feet or less (the sight distance), which means a combination of tall timber and lower shrubbery. Thermal cover means combinations of timber-shaded areas and sunny slopes within which the animals can seek out optimal conditions. Openings are also needed to serve as forage areas. The optimum forage-to-cover-area ratio is given as about 3:2, and the ideal area of forage openings is 4 to 16 hectares.

The quail and the elk happen to be two species whose habits have been extensively studied. Information at this level of detail is not available for most species, although the pace of research has quickened in recent years. In the future, it will probably be possible to design specific habitats for increasing numbers of species.

Design for Species Richness

For most design situations, providing for species richness is probably more useful and more readily open to implementation. Here, we are dealing with a set of general principles with wide-ranging applications. Usually, the areas of greatest concern are riparian zones and ecotones, or transition zones, between two different types of environment.

Riparian zones normally feature the most diverse range of species to be found in any landscape, and the drier the climate, the more consistently true this phenomenon is. The reason stems not only from the presence of water, which all plants and animals need, but also from the varied environmental conditions associated with this presence. Varying levels of wetness within a riparian zone provide somewhat different sets of conditions, which in turn provide habitats for different living communities. This is important at every scale of concern. We have discussed the roles of river corridors at the regional scale, and those of streams, ponds, and drainage ditches—even debris basins—which can be equally important at smaller scales.

The tendency toward greater species diversity and population density in an ecotone is called the "edge effect." It results from the fact that the ecotone usually supports some species that are not present on either side in addition to most of those that are. Thus, the zones between forests and meadows, or between two forest types, are unusually rich in species.

[13] See Leopold, 1933.
[14] See Rodiek, 1982.
[15] Ibid.

This is a principle with wide-ranging implications for landscape design. Since man-made environments usually feature large numbers of different environments within relatively small areas, they contain a great many edges. Since the length of the edge of an area of given size varies with its form, and since transition zones can vary greatly in depth and composition, there are endless possibilities for design.

Of all the ecotones in the man-made landscape, the most critical and challenging ones are those that link it with natural landscapes, that is, with wild areas, enclaves, and corridors. Edge conditions are extremely important to the integrity of habitats on both sides. Ideally, they foster interactions that are beneficial to both wildlife and people. This might mean an area where people can observe wildlife without being seen, as at Madrona Marsh and San Elijo Lagoon, or it might mean an area of restrained motion and activity on both sides. The latter tends to attract wildlife that are most adaptable to people as well as people who are fond of wildlife, while excluding the shyer and more hostile types on both sides. The studies of elk habitat mentioned above found that elk would not venture near populated areas and recommended buffer zones to separate elk forage areas by 800 meters from pedestrian trails and by 400 meters from roads.

Nevertheless, confrontations are bound to occur. The fact that wildlife corridors share long edges with human populations presents considerable problems. Some animals, like deer, eat garden plants and otherwise disrupt residential communities. Raccoons ransack garbage cans and sometimes damage buildings in trying to build their nests. Parks and other large exotic greens located along the corridors can provide ideal buffers. Whenever a corridor must butt against an area populated by humans, as will inevitably happen, other kinds of barriers will be needed. Carefully chosen plantings at critical spots can sometimes do the job. Plants disliked by deer can be cultivated in rows and masses to discourage their crossing. According to some reports, deer will also not go near small bags of slaughterhouse tankage strung at intervals along their paths, but that solution will not appeal to everyone. Extensive stretches of lawn are unattractive to most animals. As a last resort, when all else fails, there are fences.

The most complex buffering problems usually occur at the edges of wild enclaves. Because they are small, intensively used by wildlife, and, by definition, very different in ecological character from their surroundings, these areas are especially vulnerable to urban encroachment. The design of buffers in this situation can be a complex, subtle process.

In the Madrona Marsh case study, for example, it was proposed (1) to replant an area several feet wide around the edges of the wetlands with thick marshgrasses, which are almost impenetrable by humans; (2) to construct a boardwalk across this buffer zone as a means of reaching a well-screened observation blind; (3) to plant an earthen berm in shrubs to provide further screening on the street sides; and (4) to locate a passive-use area between the habitat and more active uses on the park side.

PLANTING FOR WILDLIFE

Because vegetation provides both food and cover for most species, wildlife diversity is usually directly proportional to plant diversity. Plant species differ greatly, however, in their attractiveness to wildlife. The choice of plantings, therefore, can be an effective means of encouraging (or sometimes, discouraging) the presence of wild animals. In natural areas, where native vegetation must predominate, the choice is necessarily limited. In exotic greens, which usually feature greater varieties of plants as well as exotics, the opportunities for creating wildlife diversity are greater.

Although virtually every part of every plant can be eaten by one or another species, some parts are far more edible than others. Since fruits, nuts, berries, and seeds are particularly favored by a great many animals, plants that produce them in large numbers are especially attractive. Ideally, they should be available the year round. Holly and toyon have particular value because they are among the few plants that bear fruit in the winter. Nuts are also valuable because they are available over most of the year and keep for long periods. Acorns are the most commonly available nuts for wildlife and among their most important foods.

Seeds are the major or sole source of food for a number of birds and small mammals. Unfortunately, many seed-bearing annuals are considered to be weeds and thus excluded from managed landscapes. Among those important to wildlife are ragweed, crabgrass, bristle grass, goosefoot, and tarweeds. Eliminating them creates severe problems for seed-eaters.

Thick, low-growing shrubs and ground covers provide food or cover or both for a number of animals. Since these types of plants are not particularly popular, animals that need them often have trouble surviving in cities. Urban parks usually follow the English landscape style, being planted with sweeping lawns, which have little value for wildlife, and informal masses of trees. Shrubs, especially low, dense ones, create policing problems since they provide cover for rapists, muggers, and other violent types as well as for birds and small mammals. As a result, shrub masses in some city parks have often been removed at the insistence of police, who want unobstructed views of landscapes considered dangerous. This problem becomes more serious as urban crime rates rise, and there is probably no solution to it until we can find ways of making cities safe again.

Across the United States as a whole, oaks and pines are by far the most valuable of the woody plants for wildlife. Next in importance are blackberry, wild cherry, and dogwood, followed by thirty others considered to be of major significance. Among the upland weeds and herbs, bristle grass, ragweed, pigweed, and panic grass rank highest. By taking advantage of such information, we can be reasonably sure of attracting considerable populations of wildlife, although the composition of communities will not be entirely predictable.

Building design can affect wildlife habits as well. In a particular area of Columbia, Maryland, studies have shown that the varied nooks and crannies offered by the style of the houses were more responsible for the high population of starlings and sparrows than any other factor.[16]

Some light has also been shed on the habitat preferences of a number of species by a study of two very different residential areas bordering Rock Creek Park in Washington, D.C., by Robert D. Williamson. According to Williamson, the native species, including cardinals, robins, and mockingbirds, preferred a wooded park area and an affluent residential district to the west of it that featured richly planted yards. Introduced species, like pigeons, starlings, and house sparrows, thrived in higher density residential areas to the east of the park, which had little vegetation. Starling populations, in particular, increased with higher human densities. House sparrow populations correlated with the number of architectural niches available for nesting sites and increased in number with distance from the park.[17]

In rural and more natural landscapes, the possibilities of accommodating more diverse populations than just birds and small mammals are far greater. One situation in which ecological structure has been of particular concern is the reclamation of mined landscapes. Most mining states now have laws requiring that such lands be restored to approximately the state that existed before mining and call for the use of native species. Some wildlife managers believe that far more might be done to improve their wildlife habitats, making their populations richer than they were in the natural state:

> . . . the establishment of riparian areas, wetlands, and specialized habitats for threatened or endangered species or species of high value within a particular region may be products of mining activity. Effective planning can thus create opportunities for wildlife that would be difficult to accomplish in the absence of mining activities.[18]

Moore, et al., recommend the establishment of communities of a wide variety of successional stages to promote species diversity.[19] In some cases, they point out, introduced species of grasses, forbs, and shrubs can further this goal because they are better adapted to disturbed environments than native species. Why should this be so? The answer is not clearly known, but some researchers believe that, for some reason, no highly competitive annual weed species ever evolved in the western United States, in contrast to other parts of the world where an intensive selection for such species was stimulated long ago by agricultural activity. In any case, human intervention can help fill the gap.

Along the same lines, Streeter, et al., recommend a number of practices to increase the diversity of wildlife.[20] Most depart significantly from favoring the natural state, either by creating new forms or conditions or maintaining those created by mining. The list of their suggestions is instructive:

1. Increase diversity of seed mix.
2. Leave small plots for natural seral development.
3. Sod native vegetation.
4. Grade to develop topographic variation.
5. Reestablish drainage patterns.
6. Leave islands free of grasses for shrub and tree planting.
7. Mulch with native range hay cut after seed maturity.
8. Use seral seed mixes for varying sites, instead of a universal mix.
9. Leave water-collecting depressions; these will initially be small potholes or marshes and ultimately mesic meadows or forest pockets.
10. Plant hedgerows through and shrubby borders around mines in areas of adequate rainfall.
11. Salvage and spread topsoil as quickly as possible after seed maturation of most species.
12. Transplant various age classes of trees and shrubs.
13. Leave high walls and spires for raptor nesting sites.

Some of these suggestions will be useful in situations other than mine reclamation, whereas others are rather specialized. In any case, they help to make the fundamentally important point that by shaping a landscape purposefully and by imaginatively taking advantage of conditions that resulted from an entirely different purpose, we can create human ecosystems capable of supporting an extraordinary diversity of wildlife populations.

[16] See Geis, 1974.
[17] See Williamson, 1973; p. 55.
[18] See Streeter, et al., 1979; p. 61.
[19] See Moore, et al., 1977.
[20] See Streeter, et al., 1979; p. 47.

CHAPTER 13

Function

Controlling Material and Energy Flows

In human as well as natural ecosystems, flows of energy and materials are the engines that drive everything else. As a rule, they are taken for granted. We tend to forget the very fundamental fact that life, for human beings as well as for other organisms, is dependent on a continuing supply of energy, water, and the chemical elements necessary to its dynamics and that the most essential function of all ecosystems, both human and natural, is to make the needed energy and materials available to their inhabitants.

Man-made systems, in particular, experience enormous difficulties in regulating flows. We are continually faced with difficult problems of too much or too little energy, water, or nutrients, and these account for a great many of the dysfunctions in the human environment.

Eutropication, drawn-down ground water levels, urban heat islands, fuel shortages, even acid rain—all these are symptoms of dysfunctional flows. Other symptoms, such as the massive movements of energy and water from one system to another, have affected the very core of our economy. Increasingly, these dysfunctions take their place among the most critical issues of our times.

Despite the fact that maintaining energy and material flows is largely a matter of how we use the land, these flows are rarely considered very seriously in landscape design. Too often, flow issues are defined as isolated problems with single-purpose solutions. The early development of the Los Angeles flood control channels described earlier is a case in point. The logic there went like this: If flooding is a problem, then the solution is to direct the rainwater to the ocean as quickly as possible. As we have seen, however, although this solved the flooding problem, it altered the flows in a way that caused numerous difficulties in other parts of the hydrologic system. We should always remind ourselves of one of the major lessons of the systems approach: Consider the environment of the problem at hand—the *whole system*.

Here, then, we shall discuss flows of materials and energy as functions of whole systems and show how they can be dealt with in the design process—first identified as issues (as described in the information phase), then analyzed through the use of models, and finally redesigned as possibilities for the future. We will begin with a brief review of the basic features of natural flows.

Water and nutrients circulate through natural ecosystems in established patterns that can be predicted with some accuracy. They are taken up by certain organisms and used and then put back into the environment to be used by other organisms in ongoing cycles. To illustrate the workings of typical cycles, the diagram on page 229 shows the flow of nitrogen (one of the three most important nutrients) through the environment. Although the mechanics of the nitrogen cycle are well known, a brief description might be useful to establish the relevance of all such processes to the human context.

The major storage reservoir of nitrogen is the air, of which it composes 80 percent. Atmospheric nitrogen is fixed in the soil by certain bacteria and algae that have the unusual ability to accomplish this feat. Once in the soil, nitrogen is taken up by plants. Through plants, it is passed on to human beings and other animals.

Some nitrogen is lost in the flow of sediments to the ocean depths, and an approximately equal amount is gained from igneous rock through volcanic action. These quantities constitute the output and the input of the nitrogen cycle, but they are very small. By and large, each organism finds the nitrogen that it needs already available within the system and later makes it available for use by others.

Energy, like nitrogen and other essential materials, is continually transferred from one form of organism to another. The sequence is fundamentally different from that of matter, however, in that it does not cycle. The second law of thermodynamics holds equally true in natural and man-made systems. Some energy is dispersed, or converted into heat, at each transfer. Since living organisms are particularly inefficient users of energy, the losses involved are great, usually between 70 and 95 percent at each transfer. For this reason, all living systems need large, continuing inputs of energy.

The continuing process of degradation of energy into more dispersed form is called *entropy,* whereas the use of energy to build more complex forms of matter is called *syntropy.* In the global biosphere, syntropy is initiated by the photosynthetic construction of complex living cells. In the dual process of building and degradation, both degraded and stored energy are continously replaced by solar radiation. Succession and evolution, both of which develop toward higher levels of organization, are long-term syntropic processes. Combustion, on the other hand, is an entropic process. Human design can be—should be—a syntropic process.

In nature, syntropic order develops slowly—one step at a time. Jacob Bronowski calls the process "stratified stability," which begins with the forming of molecules from atoms, and bases from molecules, and goes on to higher levels, simple animals from cells, and so on:

> The stable units that compose one level or stratum are the raw material for random encounters which produce higher configurations, some of which will chance to be stable. So long as there remains a potential of stability which has not become actual, there is no other way for chance to go. Evolution is the climbing of a

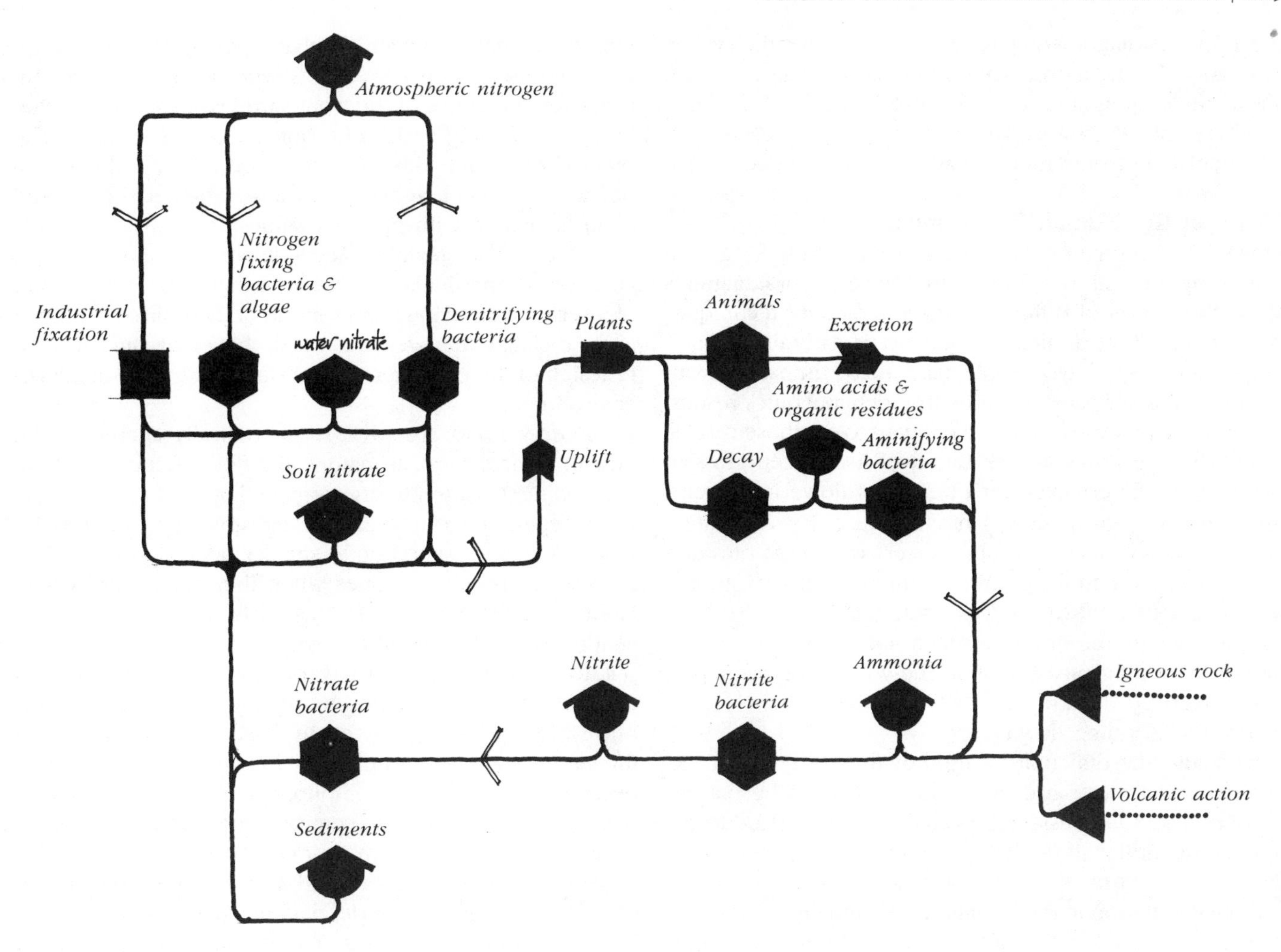

THE NITROGEN CYCLE

ladder from simple to complex by steps, each of which is stable in itself.[1]

We can reasonably hope that ecosystematic design might contibute to the climbing of the ladder to some degree greater than chance encounters alone might provide. It may be too much to hope that every design will result in a step up the ladder, but with thoughtful organization of the flows, we can create "stable units," or islands of order, with a reasonable rate of success. These we might see as contributions to the stream of ecosystem development that further the cause of increasing order and complexity. At the very least, we should succeed in not contributing to entropy.

SOME PRINCIPLES OF FLOW

In general terms, then, we say that materials flow in patterns that repeatedly recycle through the biosphere, whereas energy flows follow dual paths of increasing complexity and degradation, syntropy, and entropy. Since the survival of a system depends on the provision of energy and materials to its members, selection in nature favors systems that most effectively keep the flows going. Cities, as we have seen, have serious problems in distributing energy and materials in the right quantities—that is, in quantities that will satisfy the needs of the urban populations without concentrating too much of anything in places where it might cause trouble as well as in quantities that can be sustained.

Difficulties in regulating flows tend to be much the same in all cities, however different their economies. In nature, since flows of energy and materials are universal, following the same basic rules everywhere on earth, they exhibit common characteristics even in very different environments. Under natural conditions, they tend to be localized. Each environment has its own complete flows and cycles of energy and materials that local organisms can draw upon. Human systems have introduced circulation patterns of such a scope as to create a tight network of global interdependence, with all the consequences discussed in Chap. 5. It was argued there that we need to learn to use the processing capacities of local landscapes more fully in order to reduce pressures on the global ecosystem.

Even as an ecosystem can be considered as a unit of any size, energy and material flows can be considered at any scale. Ecosystems are interconnected by flows moving across their boundaries, and these interconnections may be thought of as their inputs and outputs since anything taken in by a system must have been put out by another system, and vice versa. This viewpoint allows us to cast ecosystem function in a large perspective. Compare, for

[1] See Bronowski, 1973; p. 349.

example, the diagram on page 229, which shows the cycling of nitrogen at the global scale, with that on page 10, which shows the cycling of nitrogen in San Elijo Lagoon. The basic processes are the same, but at the small scale, each compartment is not only much smaller but more specific.

Flows in the Human Environment

Flows in the human environment are generally faster and less complete than those in nature. The San Elijo diagrams show the effects of different sewage treatment techniques on nitrogen flow. Similar changes occur naturally in landscapes all around us. People tend to be impatient with nature's leisurely pace, however; they demand quick results. Technology provides us with the means to get those results. A suburbanite wants a thick carpet of lush green grass in his front yard, perhaps even a border of flowering shrubs, the very same year he moves into his house. Thus he waters heavily and applies chemical fertilizers with large nitrogen contents, usually in the form of ammonia, urea, or nitrate. The water runs off, carrying the nutrients in the fertilizers with it, through the drainage system into a river or stream. There, by adding measurably to the volume of water previously carried, it may contribute to faster stream cutting or even to flooding downstream.

Usually, the difficulties caused by the nutrients carried in this runoff water are even more serious. The natural nitrogen cycle is considerably altered by the introduction of large quantities of nitrates contained in fertilizers. These disrupt the natural state by fostering the growth of algae and phytoplankton in the stream, and eventually in the river, lake, or estuary into which it flows. Algae blooms—the large surface concentrations of algae commonly seen on the surfaces of urban water bodies—inevitably appear. The decomposing bacteria go to work on this feast, and their expanded activity uses the oxygen dissolved in the water at a faster than normal rate. If enough of the oxygen supply is used up, then fish begin to die for lack of it. This problem is more commonly caused by sewage effluent, as in the San Elijo example.

Whenever excessive nitrates reach surface waters, the nitrogen cycle is drastically altered. The excess accelerates the growth of one set of organisms at the expense of other organisms, resulting in a drastic imbalance that reduces the natural diversity and creates unsightly bodies of water that contribute less and less to the quality of human life.

Still more damaging are the effects of nitrogen compounds leached through the soil into the ground water. Once such materials are in the ground supply, they are virtually impossible to remove. Nitrates dissolved in drinking water are toxic and have already become a problem in several parts of the United States.[2]

None of this unhappy state of affairs is the result of anybody's conscious intentions. The homeowner simply wanted an attractive front yard that would conform to his image, and that of his neighbors, of what a yard in a good residential district should be like. The land planner who laid out the lots was responding to the dictates of the residential market, which say that most people want to live in single houses on private lots with deep front yards. In fact, even the depth of the front yard was probably dictated by a local zoning code requiring minimum setbacks. The officials enforcing this law were concerned with the appearance of the neighborhood and, whether they would admit it or not, with property values.

[2] See Johnson and Hester, 1981.

Thus, nobody really wanted to impose an altered nitrogen cycle on this particular landscape. It came about as a result of a series of individual aims, none of them directly related to the cycling of nitrogen, and the upshot of this unintentional redesign of the ecosystem of the watershed was a degraded environment.

Energy flows are even more radically altered in the human environment. In nature, the flow of energy follows the trophic structure. Energy is gained by organisms at every level depending on what they eat, put to work through respiration, and passed on to smaller organisms when they defecate and to larger ones when they are in turn eaten. Human beings complicated this fairly simple order by inventing work. Energy (from wind, water, and steam) was put to work to do a variety of jobs that human beings considered desirable, thereby greatly expanding its role. And once the energy stored in fossil fuels was added to the supplies already available to do work, the energy flows became so enormously complex that we no longer understand them well enough to control them or to know whether or not we can sustain them. Our cities have become gigantic transformers, concentrators, and directors of energy but are often almost completely dependent on enormous energy inputs from outside sources that will be hard to sustain in the future. To make matters worse, (as discussed in Chap. 5) energy sinks—the final resting places of dissipated energy and its byproducts—are presenting global problems of increasing magnitude and may prove to be even more limiting to highly mechanized systems than limits of supply.

The same principle applies to all energy flows and material cycles. All have limited capacities. The limits may lie in the incapacity of source landscapes to provide the necessary inputs, as is true of the Los Angeles water supply, or they may lie in the incapacity of the sinks to process the outputs (which we are accustomed to calling "wastes") in an acceptable way. Herman Koenig has described the present situation in these terms:

> . . . as regional densities of industries and human population increase, a point is reached (as is now the case) where flow rates of wastes exceed the capacity of the natural environment to dilute and assimilate them. This limit represents a point of ecological breakdown beyond which the natural environment ceases to function in providing the required process functions essential to human life.[3]

[3] See Koenig, et al., 1971; p. 9.

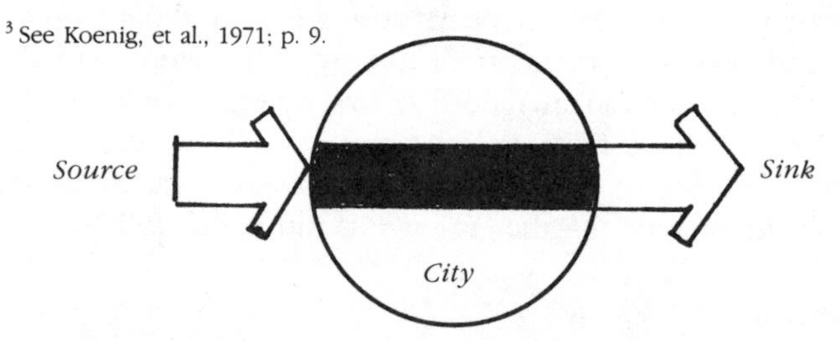

It should be noted that the limits on either the input or the output side depend to some degree on system design.

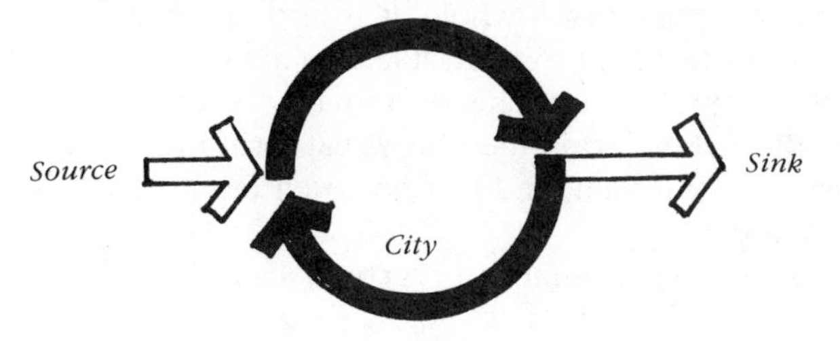

The Need for Conscious Control

Human activity, especially when supported by technology, inevitably changes natural processes. It seems likely that flows in the human environment will always seem hyperactive when compared with those of the more patient wilderness. Once we understand the consequences of our actions, however, we can control and stabilize them to a considerable degree. Although we usually know what the remedies are, it is not enough simply to recognize and relieve the problems that occur. Most planners are well aware nowadays of the effects of runoff nutrients on water bodies. The Federal Water Pollution Control Act Amendments of 1972 made a national goal of eliminating "pollutants" in all navigable waters by 1985. In some cases, rigid local interpretations of this goal have created yet other problems by making it illegal to introduce water containing materials that may actually be highly beneficial even though technically classed as "pollutants." Control of this sort may turn out to be almost as bad as no control at all, partly because it is very expensive and partly because it fails to realize the contributions that such materials might make. Here, as in the flood control examples, a problem-solving approach has led to a solution that will create other problems elsewhere.

Only if we deal with flows in their entireties as unified systems can we see the benefits as well as the drawbacks that may result from reshaping them to accommodate an urban situation. The aquaculture system proposed for San Elijo Lagoon, for example, consciously uses the large concentration of nutrients produced by the human settlement to increase the lagoon's productivity. Instead of unintentionally encouraging the growth of enormous quantities of algae that can only die off and cause problems, the nitrates are intentionally used to foster the growth of algae that will be consumed by fish, and the fish, in turn, will provide food for human beings and other carnivores. Ideally, once the water has been treated by biological means to the standards of secondary treatment processes, it could be discharged into the lagoon, where it would not only enrich the entire food chain but help stabilize the salinity level.

No doubt, the design of a system like this is a tricky business. The period of experimentation proposed for San Elijo is not an academic nicety, but a necessity—a period during which we will have to learn by observation exactly how the proposed arrangement works. We will probably encounter difficulties with water quality in the beginning, but these can almost certainly be solved. They can only be solved, however, by practical experimentation. We will need considerable experience with human ecosystems before we can make them function at optimum levels. What we can do now is to establish some guidelines for the design of flows, using as our starting point the very characteristics that promote stability and survival in natural systems.

The Characteristics of Stability

According to Eugene Odum, the material and energy flow characteristics of stability—and we will here take the term "stability" to comprehend both resistance and resilience—are those associated with mature natural systems.[4] For design purposes, these can be summarized as follows:

1. *Slow and Relatively Complete Internal Cycling:* Materials are reused within the system in closed loops. Comparatively small quantities are taken in from other systems or exported to them. This operation implies a high degree of local self-sufficiency.
2. *Varied Pathways of Flow:* Material and energy move in complex and diverse patterns, making it probable that if a flow is broken at some point by some unusual event, the overall movement continues through parallel, if not identical, routes.
3. *Filled Niches:* Each link in the chain is occupied by a species adapted to that particular role. Niches tend to be narrow in mature systems, and the species occupying them are comparatively specialized.
4. *High Volume of Life (or Biomass) per Unit of Energy Flow:* Energy is used sparingly and efficiently to support life.
5. *Low Net Production:* Material produced by the system is mostly used within the system. There is little accumulation and little export.
6. *High Information Content:* This goes with complexity. In natural systems, information is stored primarily in the genes.

It is unlikely that any man-made environment can achieve these qualities in full measure. Indeed, there are some fundamental conflicts. Man is himself a notably unspecialized creature in the biological sense, for example, filling as he does an extremely broad niche. It is thus improbable that urban systems can ever feature the tight networks of niche specialization that characterize the most stable natural systems. Moreover, the desire for high levels of productivity that provide surpluses for export and consequent balances of payment is a basic drive in the economies of cities. Indeed, there are a number of conflicts between long-term ecological needs and the relatively short-term economic compulsions to which cities are so heavily committed. In establishing guidelines for flows of material, it seems likely that we will always have to balance the ecological goal of stability with other human needs and desires. There is no ideal or perfect state, no easy answer to aim for. Nevertheless, with careful design, it should be possible to achieve high levels of stability in combination with reasonable degrees of productivity and human amenity.

[4] See Odum, 1969.

PREDICTING FLOWS FOR NEW DEVELOPMENT

By predicting the flows and cycles of human ecosystems in systematic ways and making these predictions integral parts of the design process, we can assess the degree to which the new system will incorporate the necessary conditions of stability. We can also estimate the quantities of energy and critical materials that will have to be imported to support the system on an ongoing basis. Different materials will prove to be critical in different situations. Energy, water, soil, and the major nutrients will normally be important considerations, even though the degree of importance will vary in ways that are hard to foresee. The natural system always constitutes the baseline on which man-made systems are founded. It is absolutely vital that we understand the natural situation first.

Given the experiences of the 1970s, it has become fairly obvious that the supply of energy or any of the essential materials can be a limiting factor in future urban development. Not so obvious is the fact that controlling the sinks may be at least as important as assuring the sources. The world's nitrogen supply seems to be reliably adequate, for example. Difficulties with the nitrogen cycle usually arise only when too much of it collects in places where a surplus can do a great deal of damage. Examples of the damage that sinks can do to the urban environment include the heat islands produced by dissipated energy, smoggy air, atmospheric carbon, and sewage outfalls—not to mention the horrendous problems caused by nuclear wastes.

Predictions of material and energy flow in the urban environment have four major purposes:

1. To determine the general quantities of energy and critical materials that will be needed to support a system in the future
2. To ensure that proposed changes in the landscape will not seriously impair natural flows
3. To define the pathways of movement through a system as an indication of its potential stability and flexibility
4. To analyze what happens to materials and energy after they have been used as a means of seeking out possible reuse and avoiding harmful sinks.

The progress researchers have made in measuring and predicting flows since mid-century has been dramatic enough to reshape our way of thinking about ecosystems. Technical improvements in tracers, automatic monitoring devices, and computer technology promise an even higher degree of sophistication for the future. Nevertheless, the information provided by research is usually excessively detailed and fragmentary. Researchers, bound as they are by the need for precise controls and by disciplinary boundaries, rarely take whole systems into consideration. Consequently, when attempting to assemble information on flows, we often find ourselves faced with a mass of precise data on certain parts of a system and little or nothing on others. As often as not, the data we need are hidden in obscure charts and tables or buried helter-skelter in massive research reports.

Not only do we have the task of assembling information in a form that makes the critical flows understandable as parts of a continuous whole, it is also important to make them visible, in the abstract, at least, if not in reality. Material and energy flows are elusive primarily because we cannot see them. Rendering them in visual form makes them far easier to deal with, and perhaps even more important, to communicate.

There are a number of techniques for doing so. The nitrogen flow diagram provides a simple example of a qualitative representation. Pictorial representations like those accompanying the Aliso Creek and Lake Elsinore case studies (pages 33 and 163) help clarify understanding the relationships between flows and the landscape. Both of these studies deal with water concerns, but in rather different ways. For example, the illustrations in the Aliso Creek example show how the water flow system of the landscape after development will compare with that of the natural landscape.

The Lake Elsinore Case Study: Predicting Water Flows

The pictorial representations shown with the Lake Elsinore case study (page 163), on the other hand, are concerned with the quantities of water involved. The arrows are vectors whose widths give approximations of the amount of water moving through the system at critical points. As one can see, Lake Elsinore is a very complex hydrological system, its flows integrally connected with the larger region. In such situations, it is especially important to consider the flows in terms of the whole unit. The lake level fluctuates enormously, the difference in level from a dry to a wet year being on the order of 30 feet. Although the lake is a sink with no regular outflow, a small overflow channel does traverse the center of the small town of Lake Elsinore on the lake's edge. The inadequacy of this channel caused severe flooding in the town when the lake overflowed during the heavy rains of 1980. Flooding was therefore a major issue addressed by this project. It was critical to represent water quantities because redesigning the channel was an important part of the project.

Another problem was the deteriorating downtown shopping area along Main Street, which runs parallel to the channel and just a block away. As the map on page 164 shows, it is also located quite close to the lake, but visually separated from it. That is, from downtown one cannot see the lake and, indeed, has no sense at all of its being there.

The critical issues, then, are lake-level stabilization, flood control, and downtown revitalization, and the channel is the key to all three. With careful regulation of the flow of water through it, the channel can carry overflow water away during flood periods and can also carry water into the lake during dry periods. The flow will be far from enough to stabilize the lake level, but it will ensure the presence of sufficient water for recreational and esthetic purposes in the ponds near the center of town. This reversible flow design was made possible by the flatness of the land over which the channel passes. The sections on page 166 show how it will work. The channel bottom will slope very slightly toward the lake from a high point (1255 feet above sea level) a few hundred yards from the lake's edge. Beyond this high point, the channel will slope slightly away from the lake. Thus, when the lake level rises above 1255 feet during heavy rains, the channel will carry away the overflow.

During dry periods, when the lake level is dropping, the process is reversed. From an area just inside the high point, water is pumped from underground into the channel, which feeds it through a series of ponds into the lake. In this way, the ponds are always kept full and the lake level more nearly constant. The ground water level is drawn down slightly, but the water removed is simply stored in the lake rather than in the aquifer. Because the lake surface is larger when it is higher, there will be some increase in evaporation loss, but the amount will be negligible. Moreover, water will continue to seep back into the ground through the lake bottom to complete the cycle.

The channel and ponds will have natural edges and bottoms, with riparian vegetation growing along them to support wildlife populations. Some of the ponds will also be designed for recreational use—fishing, paddle boating, strolling, picnicking. This riparian corridor will be linked with Main Street by a series of tree-lined alleys, providing an environmental amenity that the downtown now lacks.

Since the driving force in this environment is clearly the flow of water, redesigning that flow was the basic step in shaping a healthier, more useful environment.

In both this and the Aliso Creek examples, a descriptive model was used to represent the existing flow in the information phase. This baseline tells us what the important factors and events in the flow are and what the basic functional pattern is. We then go on to redesign the flows using predictive models to determine how they are likely to function. Predictive models can be formulated in purely mathematical terms, of course, but in environmental design where we are dealing with physical form, however, it is often more useful to work with more visual techniques. The pure abstractions of mathematics can separate us too completely from the physical reality that is our subject matter and that must ultimately be our overriding concern.

The University Village Case Study

The University Village case study takes us to another level of abstraction, but the function of the system is still represented in visual terms, including energy, water, and nutrient flows. Recall from its description in Chap. 5 that the project was to design an experimental student community that would be able to support itself with on-site resources to the greatest practical and reasonable extent. The design is illustrated on page 111. Dwellings clustered on a south-facing hillside use passive solar devices for heating and photovoltaic cells for generating electricity. Food is to be grown on the surrounding land using intensive agricultural techniques. Sewage is to be treated in biological treatment ponds at the bottom of the hill and pumped back up by windmills for use as irrigation water. A number of other soft technologies are to be used to minimize consumption and maximize production, but these are enough to explain the basic concept. The two flow diagrams show inputs, outputs, and patterns of internal use of three basic resources: food, energy, and water. In the beginning, as shown in the initial phase model, considerable resource imports from outside will be needed—about a fifth of the food needs, over a third of the energy needs, and over half of water needs. On the export (right-hand) side, considerable energy will be dissipated as heat, of course, and some water will filter into the ground or be lost by evaporation. Virtually everything else, however, will be recycled. The technical means of accomplishing this plan are shown in the diagram.

Later, as shown in the model for the progressive phase, imports will probably be considerably reduced by fine-tuning the system and getting the various recycling devices working at peak efficiency. This model also traces the import of raw sewage from other parts of the campus for treatment and recycling. A similar vector technique was used for showing the flow of water into and out of the Southern California Metropolitan Region.

A Flow Language

Ultimately, the development of diagramming techniques may be as important to the design of human ecosystems as the development of orthographic projection was to the design of buildings. Considerable progress has already been made. The diagram used to represent the general nitrogen cycle was based on ecologist Howard Odum's technique for representing energy flows. This is an especially interesting technique because it provides a visual language to express the changes and events along the pathways. Fourteen symbols represent the basic types of operations that use energy as it flows through a system. These symbols are then connected with lines to represent the "pathways of power flow and action of forces."[5]

Odum's technique can be applied in a broad range of situations to describe the flow of materials as well as energy. Potentially, it is a powerful tool for the design of man-made ecosystems. To improve readability and thus their utility for our own purposes, the Laboratory for Experimental Design abridged Odum's 14 symbols to six very fundamental ones, as follows:

INPUT OR OUTPUT
Importation or exportation of material or energy to or from a given system.

STORAGE
Temporary retention of material or energy in a certain level of a given system; such as the storage of water in a reservoir.

WORKGATE
A material flow acted upon by the energy of some outside force; such as evaporation of water.

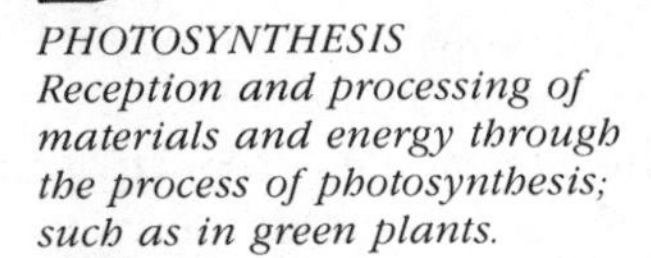

PHOTOSYNTHESIS
Reception and processing of materials and energy through the process of photosynthesis; such as in green plants.

RESPIRATION
Reception and processing of materials and energy through the process of respiration; such as in herbivores and carnivores in a grazing food chain.

HUMAN ENTERPRISE
A systematic or purposeful human activity, transforming material and energy; such as agriculture or manufacturing.

[5] See Odum, 1971; p. 37. The adaptation of these symbols for design purposes was accomplished by Mark von Wodtke and his students.

These six symbols encompass most of the operations in man-made or natural systems that use or transform energy or materials.

The theoretical underpinning of this technique is the first law of thermodynamics, which tells us that material and energy can be transformed from one state to another but never created or destroyed. Thus, in theory at least, all materials and energy can be accounted for. For the energy flow diagrams, the second law, which tells us that some energy will be converted into entropy or unusable form with each use, is also applicable.

The model on page 9 shows how the basic soil flow diagram can be used for making predictions. The first step in predicting the effects of upstream development on a coastal lagoon is to assess the transformation brought about by the grading and baring of slopes. The model is a concise visual forecast that the increased volume of sediments will leave less room for the storage and movement of water, thereby decreasing the force of tidal action, thereby trapping salt in the lagoon and cutting off the distribution of nutrients to some of its parts and leaving them stranded in others, thereby inhibiting the growth of marshgrasses and creating an unhealthy environment for fish and shellfish, with the final result of diminishing their populations and reducing the food supply for birds. If such a chain of events were allowed to continue, the lagoons would soon be filled with sediments and the estuarine life would disappear.

In this particular case, the environmental impacts of human activity are all damaging, or at least adverse. Flow diagrams can also be used, however, to explore the possibilities of beneficial impacts. For example, the last of the San Elijo nitrogen flow diagrams illustrates the redesign of the nutrient cycle. The symbols show how sewage is fed back into control tanks adjacent to the lagoon instead of being pumped out of the system through the ocean outfall. There it is used to feed algae, which provide a large and rapidly growing food source for the fish and shellfish in the lagoon. The recycled water can then be used for irrigation and eventually fed back into the lagoon to stabilize the inflow of fresh water (assuming that local water quality regulations can be modified for this purpose). The internal cycling apparatus of the lagoon can be much improved by this plan and its productivity increased as well. By enriching natural systems and serving human needs at the same time, this kind of flow exemplifies the exciting possibilities of human ecosystem design.

Quantifying Flows

Not only do qualitative models allow us to visualize flows in an ecological system as a whole and to analyze the changes that human activities induce, they are also useful for synthesizing concepts needed for adapting ecosystems and for predicting and later evaluating the results. They can also help environmental management identify key components that require monitoring and show how these components interrelate.

Qualitative understanding precedes quantitative understanding, and often it suffices. Nevertheless, it is sometimes useful to know how much. In the soil flow example, the watershed can obviously be subjected to some grading without seriously altering the lagoon system. On the other hand, if all the hillsides in the watershed were graded, the resulting erosion would overwhelm the lagoon. Somewhere between these extremes is a certain amount of grading, or more likely a range of amounts, that will permit the continued functioning of the lagoon while making it possible to satisfy at least a part of the need for housing in the area. The precise quantity would be difficult, if not impossible, to determine. It would vary according to the amount and intensity of the rainfall, to degrees of slope and soil type, and to grading techniques and mitigation measures. Given this difficulty, it will probably be necessary to subject the entire watershed to erosion control measures, such as those shown on the Control by Design charts, and to monitor the results.

The diagram on page 235 shows the flow language applied to the flow of energy in a human ecosystem. The subject is the Integral Urban House in Berkeley, a very special environment in which both house and landscape have been designed to make use of every feasible energy transformation and pathway. Incoming solar energy is utilized in four ways: to grow food crops, ornamental plants, and algae and to heat water for domestic use. A certain amount is reserved for space heating, but the exact quantity is not known because the heating is accomplished by a passive system. Other inputs include food and various types of fuel. Outputs are the heat and carbon dioxide produced in the processes of respiration and combustion.

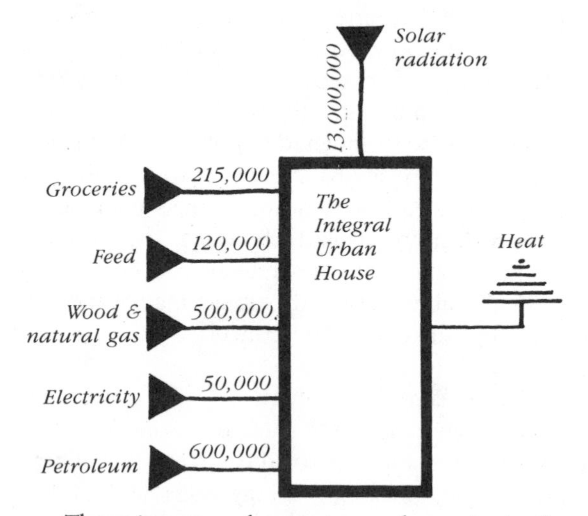

These inputs and outputs are shown in the first diagram, the internal flows being lumped together as a black box. This is a useful technique whenever we are concerned only with inputs and outputs, when what we need to know is the amounts of energy and materials that must be obtained from source systems on one side and the amounts that are put out for absorption by outside sinks on the other. In this case, the substances imported and exported are those normally taken in and put out by any household, but the quantities are considerably smaller because of the complex internal system of reusing energy in various forms at it is degraded. This system is explained by the larger diagram. Thus, solar energy is used both for heating and growing crops, which are eventually fed to humans, animals, and fish. Wastes are composted or digested and returned to the

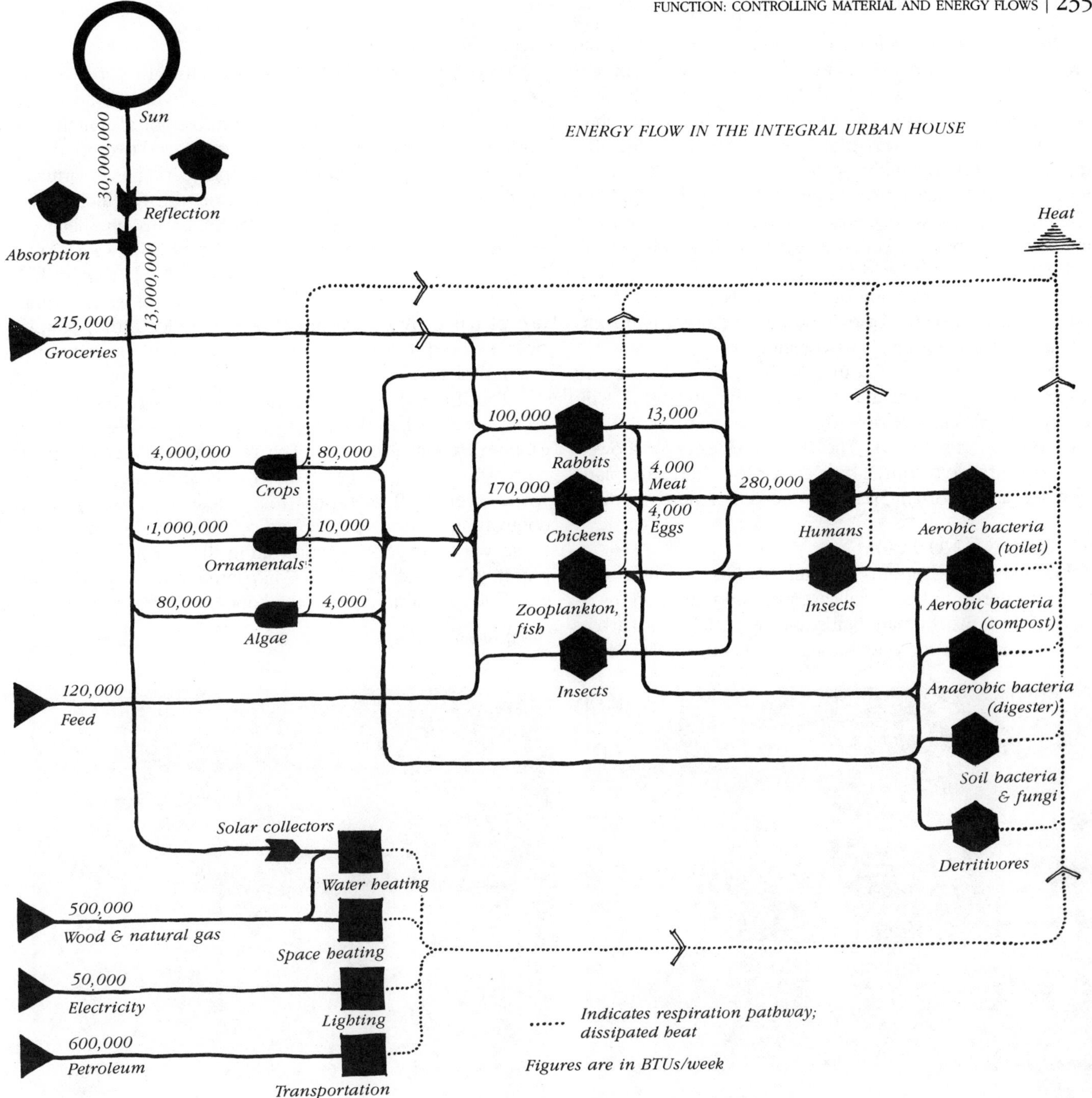

soil, thus making use of the detritus pathways that human ecosystems seldom employ.

The basic flow is quite similar to that in the University Village example, although the latter features a greater diversity of species and energy uses, and therefore of flows, because of its larger size, and hence more abundant opportunities for energy transformations. There is room for a number of different species of animals and fish and a variety of field and tree crops. The topography provides opportunities for gravity flow, sites for windmills, and a range of solar orientations. Although the Integral Urban House is an impressive example of resource-conserving design on a small urban lot, the comparison with diagrams for University Village suggests that a consideration of energy flow in detail at the site scale, or perhaps even at the project scale, probably makes possible a greater diversity of pathways and thus a more effective design.

Material and Energy Budgets

Since budgets remind us of our limits and conjure up images of accounting sheets, we usually think of them as an unhappy subject. But they are as necessary for the use of resources as for family finances, simply because resources are limited. We have seen that flows can be diagrammed at every scale and that patterns of use and nonuse can be described in simple and understandable terms. We have also seen that we can usually predict the quantities of materials or energy involved in various operations along the pathways. As with money, unless we have specific knowledge of where resources are going, they tend to get out of control. When we analyze the distribution of energy or materials in an environment, we usually find that the way we use resources is not at all the way we would want to use them. It came as a surprise to most people, for example, to discover, when the energy shortage was first becoming a problem, that heating and cooling buildings accounted for over 30 percent of our national energy consumption. Even more surprising, when the need for water conservation in the arid Southwest first became obvious, was the discovery that between 40 and 50 percent of the domestic water use in Southern California was for irrigating ornamental plants, mostly lawns. In both cases, these discoveries pointed to fairly easy means of quickly reducing levels of consumption.

It is much easier, of course, to budget in advance and design an environment that can live within its means. University Village showed how this can be done for the full range of support resources. The Aliso Creek example showed how comparative resource budgets for five alternative plans provided an important basis for choosing among them.

FUNCTION AND FORM

Embedded in all diagrams and numbers used to represent flows of energy and materials are profound implications of the form of the human landscape, especially the urban landscape. Most of these implications reveal themselves most clearly in the flow of water.

The Chinese seem to have been first to perceive the relationships joining the flow of water with the shape of the land and with the social and philosophical milieu. According to Joseph Needham, China produced two opposing schools of thought in hydrological engineering as in virtually every other area of human endeavor: the Confucian and the Taoist.[6] The Confucians were disciplinarians who believed in strict rules and strong measures of control. They advocated "high and mighty dykes, set nearer together." They called for rigid and precise control over the flow of water by the use of narrowly contracted channels. Indeed, they simply overpowered it. The result was a very simple system with few interactions.

The Taoists, or expansionists, were more inclined to let water take its own course as far as possible, giving it plenty of room to spread. The result was a very complex network of flow. An early Taoist engineer, one Chia Jang, wrote over 3000 years ago that "those who are good at controlling water give it the best opportunities to flow away; those who are good at controlling the people give them plenty of chance to talk."[7] Confronted with a situation in which towns located along the banks of a river were being

[6] See Needham, 1954.
[7] Ibid., p. 235.

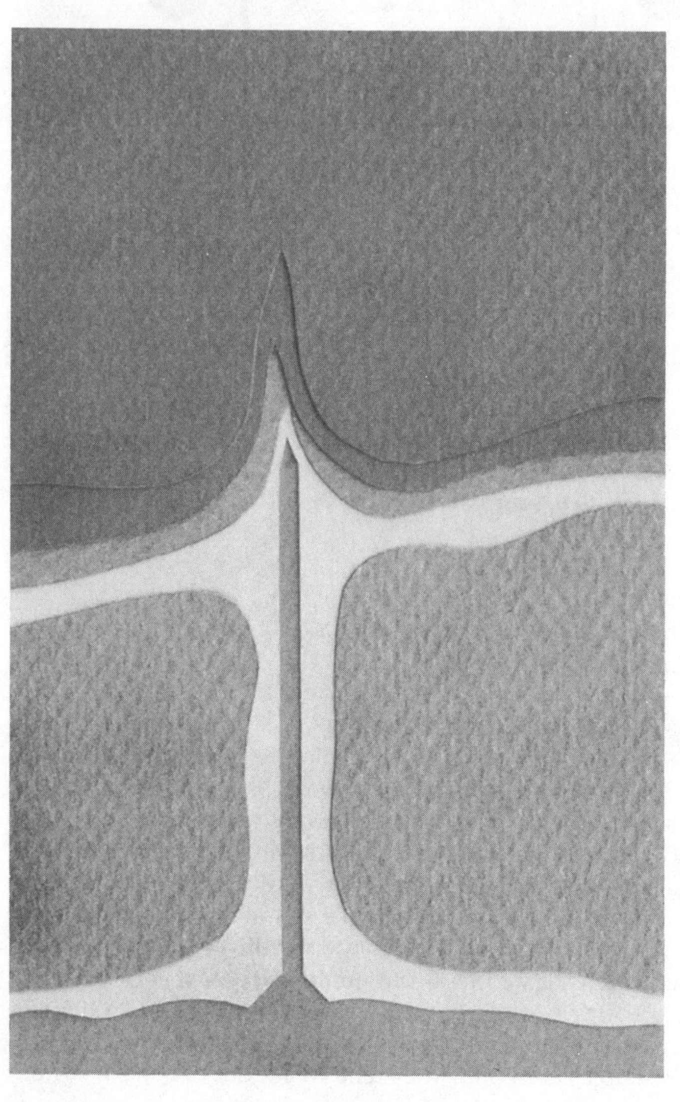

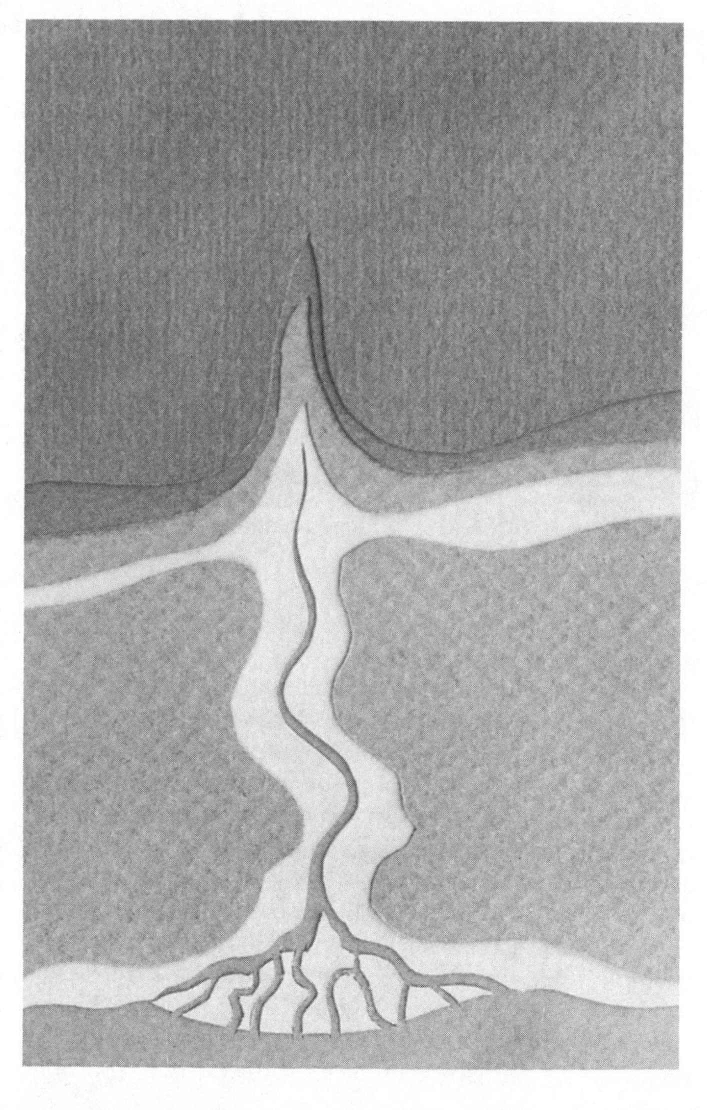

regularly flooded, Chia Jang recommended resettlement of their populations outside the area of flooding.

As one might expect, that was not a politically popular recommendation. In fact, political acceptability seems always to have plagued the Taoists. Partly for that reason, the Confucians appear to have won most of the arguments. A narrow, constricted water course leaves the greatest amount of land for human use, an especially desirable aim if people are already living on it.

In practice, however, what seems to have evolved over time in China were combinations of the two approaches. In places where streams had been narrowly channelized, large retention basins were often provided along the way at a later date to allow for spreading the waters of occasional major floods. And often where the Taoists had succeeded in building widely separated dykes along a stream, lower parallel dykes were subsequently built closer to the river bank to contain minor floods and permit the growing of crops on the land between the two sets of barriers.

In the United States, the control of water flow has usually taken the Confucian approach; witness the large-scale schemes to control the nation's rivers developed in the first half of the twentieth century. The Army Corps of Engineers channelization of the Missouri River is a particularly striking example. And in most American cities, one can find networks of narrow, concrete-lined channels, whose only purpose is to carry runoff water out of the city as rapidly as possible. The attitude is one of strict control, and the assumption is that man can, indeed, conquer nature.

Perhaps, in a sense, for a while he can, but the cost comes high. Narrow channels, in conjunction with the urban development that tends to crowd right up to their edges just as it did in ancient China, speed up the runoff and outflow of water enormously. Little time or space is allowed for it to percolate into underground storage. Ground supplies become depleted as more water is taken from them than flows in. This is a serious problem because underground water-bearing rock strata, or aquifers, are vitally important to a healthy water regime.

For a number of reasons, aquifers provide the best possible place for storing water. First of all, water is purified by being strained through soil. Second, it is protected in the aquifer from most kinds of contamination. Third, there is no evaporation loss. Since the ground water supply is an essential resource, a primary objective for any urban water flow system is to maintain it at stable levels and acceptable quality. Furthermore, water that flows through narrow channels is not only lost to the ground water supply, it is often lost to all other uses as well. In coastal cities, it usually flows rapidly into the ocean, thus depriving not only the human population of its use, but also all the animals that normally live along the banks of streams.

Flowing out with the water are the silts that, in natural situations, would wash down from the upper watersheds and be spread out by flood waters along the banks of streams, thereby replenishing valley soils. Not only are these silts lost to productive use, they usually end up in places where they can cause serious problems. Silts tend to collect, for example, in the bottoms of downstream lakes and estuaries, causing them, as in the case of San Elijo Lagoon, to fill up much faster than natural rates.

The Confucian approach, at least in its modern application, thus seems more often than not to be just another case of simple, single-purpose solutions for complex, multifaceted problems. In this instance, the purpose is to move water as quickly as possible out of an urban system to a place where it can do no harm, thereby ignoring the many opportunities for achieving a variety of other benefits as well. The Taoists seem never to have argued their case in just those terms, but we can surmise that they must have had some feeling for the complexity of the situation.

The Taoist Approach in Recent Practice

During recent years, a more Taoist approach has appeared in the designs of a number of water-control networks. The system proposed by Wallace, McHarg, Roberts, and Todd for the new town of Woodlands in southern Texas is a case in point, and particularly interesting for its economic implications.

The site covers a flat, forested area of 18,000 acres. The soils, for the most part, are heavy and impermeable. A conventional Confucian drainage system would have meant grids of underground drainage tiles, a network of storm sewer pipes, and perhaps concrete-lined flood control channels. These would have required extensive deforestation and the elimination of the habitats of various wildlife populations. They would also have radically altered the ground water supply, and not for the better.

The planners therefore proposed a system that the Taoist engineers would have liked. In their own words, their objective was "to develop a method utilizing the attributes of the existing drainage system . . ."[8] Natural drainage patterns were augmented by a system of man-made streams and swales. Rather than moving the water as quickly as possible off the land and into the ocean, as conventional piping would have done, this system slows the flow in a controlled manner. Water is impounded for brief periods along the way by using berms and check dams. Trapped in settling ponds located over the more permeable soils, the water has time to percolate down into the ground water supply. Development is primarily concentrated on the more impermeable soils. Ian McHarg wrote that "it became evident that the natural balance of the hydrologic regime was the key to successful environmental planning and an organizing concept for development."[9]

There was little doubt that this approach could effectively achieve the stated goals of maximizing the ground water table, diminishing runoff, retarding erosion and siltation, increasing the base flow of streams, and protecting natural vegetation and wildlife habitats. What was perhaps most decisive in convincing both the developers and the engineers of its superiority, however, was the cost analysis. When comparative studies of costs were made, it turned out that the Taoist system would be less than one quarter as expensive as a conventional piped system. As McHarg pointed out, "there is no better union than virtue and profit." However,

[8] See McHarg and Sutton, 1975; p. 81.
[9] Ibid., p. 78.

there is still no denying that, as the ancient Chinese discovered, the Taoist approach takes more land.

The contrast between Confucian and Taoist approaches is equally dramatic on the output side of the water flow picture—in the treatment of sewage. The conventional treatment systems—involving transportation in pipes over long distances and then a highly concentrated sequence of straining, settling, and removal of dissolved organic materials—are fast, direct, energy-intensive, and thus Confucian in nature. Where sewage water is to be recycled, treatment by such advanced techniques as reverse osmosis is even more intensive and usually prohibitively expensive.

A Taoist alternative, mentioned in connection with several of the case studies in this book, is biological, or aquacultural, sewage treatment. The processes by which it works provide a good example of the Taoist way, which is in turn a good example of a sustainable approach to the design of human ecosystems.

The Aliso Creek case study, for example, proposed that sewage water be screened and given primary settling treatment so as to remove most of the solids before it is fed into biological treatment ponds. There, through the work of plants such as bullrushes and water hyacinths, the water is purified to a quality somewhat better than that achieved by conventional secondary treatment. Some of it is then fed into a series of community lakes, where it is further treated by more plants. Here it also helps support the growth of freshwater fish for sport fishing and as a food source. After flowing through three more lakes, the water is of a high enough quality to be used for swimming and boating, thus becoming a valuable recreational amenity. From the last lake in the series, it flows into Aliso Creek, where it provides a stable stream flow throughout the year.

This system is based on the use of several small local treatment plants that process wastes near their source. The site plan on page 38 shows the integration of lakes and ponds into the residential environment. Dwellings are clustered together around the bodies of water and the irrigated recreational areas.

This is a simple, economical, and quite practical process. Although the nutrients dissolved in domestic sewage water, especially nitrogen and phosphorus, are difficult and expensive to remove by conventional treatment methods, aquatic plants, such as water hyacinths, duckweed, cattails, reeds, and rushes, are designed by evolution to accomplish the task with ease. They grow rapidly in the process and can be regularly harvested. Although such schemes seem radically different from the more conventional and familiar mechanical ones, the principles underlying them are not really new. Basically, they represent little more than an intensification of nature's way of achieving the same purpose. Chinese farmers have been converting these processes to their own ends for at least 4000 years. Farmhouses in China are often built either over small ponds or on their edges. Household wastes are thrown into the ponds, where they are either eaten by the ducks and fish living there or consumed by fast growing algae, which are in turn eaten by carp, some of which grow very large on this diet. The ducks and the carp then provide food for the farmers.

Several treatment plants in the United states are using this principle on a regular basis. The town of Orange Grove, Mississippi, uses a combination of a three-acre lagoon and a one-acre pond of water hyacinths to treat the sewage for its population of about 1500 people. Solids settle out in the lagoon; from there the water flows into the hyacinth pond. After two weeks among the hyacinths, the quality of the water is high enough to permit its discharge into local streams without impairing their water quality.

Besides their extensive root systems, which provide for rapid nutrient uptake, and their fast growth habits, water hyacinths have two distinct advantages as usable crops. Their exceptionally high protein content (about 20 percent) makes them excellent cattle feed, and they can be fermented in sealed chambers to produce a biogas that is a useful fuel.

A larger, more elaborate system at Michigan State University provides a similar kind of treatment, while at the same time making its ponds available for recreational purposes. The system consists of a series of four man-made lakes that cover about 12 acres. These are filled by a flow of about two million gallons per day of primarily treated sewage for the East Lansing treatment plant. Here, the biological treatment is accomplished by another fast growing aquatic plant, Elodea, which, like the hyacinth, is useful for cattle feed. It is periodically harvested by a small barge with a giant sickle mounted on its bow.

The bottom layer of water from this lake, laden with nutrients, is pumped into nearby fields for irrigation. The top layers flow into a second lake, where treatment is continued by still more aquatic plants, and from there to a third lake and a fourth. This last lake in the series maintains a large fish population and is heavily used for recreation. Its water, now sufficiently high in quality to meet stream discharge standards, is released into the Red Cedar River.

In Solana Beach, California, a biological treatment process developed by microbiologist Alice Tang Jokela and called the Managed Ecological Wastewater Treatment System (MEWTS) has undergone tests in a pilot plant with a capacity of 1500 gallons per day. After primary sedimentation, which lasts about 30 minutes, the water is fed through a series of three tanks. The first two are aerated with diffused air, utilizing the sanitary engineering technology of aerated lagoons. Water hyacinths are grown on the surfaces. Retention time for each of these first two tanks is one day, and for the third, nonaerated tank, two days. The third tank is designed for growing fish as well as water hyacinths.

The quality of effluent produced by this system has been carefully monitored and proven to be extremely high. Yields of Biological Oxygen Demand and suspended solids were under 10 parts per million, and nitrogen removal was over 80 percent.[10] These figures can be compared with EPA standards for secondary effluent that require a BOD and SS of under 30 ppm and require virtually no nitrogen removal.

Both capital and operating costs of the MEWTS system, like those of other biological treatment processes, are exceptionally low.[11] A cost estimate was prepared for the nearby city of Del Mar, which is proposing to use the system

[10] See Jokela and Jokela, 1978.
[11] See Robinson, et al., 1976.

instead of conventional secondary treatment. According to the Jokelas' figures, the total net cost of MEWTS during its first year of operation would be $244,000 as compared to $260,000 for the conventional systems. By 1984, the costs would be $284,000 and $417,000, respectively.[12] These amounts do not take into consideration the value of the hyacinths and the fish produced. Other studies of the economics of biological treatment have shown similarly favorable cost comparisions.

Terrestrial plants can also be used to take up nutrients in the biological treatment process, but comparisons developed by George Woodwell indicate that aquatic and marsh environments work best.[13] According to Woodwell, the presence of sewage increases the growth of all plants to the extent that harvesting is a necessary adjunct of biological treatment processes. Without harvesting, nutrients are likely to build up to levels greater than the environment can absorb. For this reason, the plants employed should have some economic uses.

We can see, then, that biological sewage treatment has a great many advantages that will probably make it an important and widely used technology in the future. For all that, it is no panacea. Like other Taoist approaches, it needs a great deal of land.

The sequence of treatment usually follows these steps:

USE → SETTLING → POND → IRRIGATION
→ GROUND WATER SUPPLY

The two essential central stages, the ponds and irrigation, both monopolize considerable areas. As much as one acre of pond is required for every 200 people, and about one acre of sprayed fields per 50 to 100 people. Herein, in our own day even more than in Chia Jang's, lies a major difficulty. When land values are so high and weigh so heavily in land-use decisions, such large parcels of land are hard to find and even harder to pay for. This means that economic feasibility is closely related to the proximity of irrigable recreational and agricultural lands. Thus, we see that diverse Taoist technologies tend to encompass multiplicites of issues and varied uses of the land.

Also arrayed against Taoist technologies are the heavy investments that most industrial countries have committed to Confucian technology. Neither the miles of flood control channels nor the conventional sewage treatment systems already in existence are likely to be abandoned. Nor is the rapid pace at which they work likely to be willingly sacrificed.

What we will probably see developing over the next few decades, then, are varied combinations of Confucian flows ameliorated by Taoist augmentations, and Taoist systems modified by infusions of Confucian efficiency. The overall evolution, in fact, is likely to follow a pattern something like the one Needham described as having occurred in ancient China. We can see an example of such an evolutionary process in the Los Angeles flood control case study. There, the Confucian channels are being gradually moderated—Taoized—by the needs for ground water replenishment, recreation, and wildlife habitat.

[12] Ibid., p. 8.

[13] See Woodwell, 1977.

Energy Paths

There are Confucian and Taoist approaches to the design of energy flow as well. The Confucian way is essentially the intensive, fossil-fuel-powered industrial civilization that we know so well. The Taoist way is much the same as the soft energy path mentioned in Chap. 5, featuring renewable flows, diversity, flexibility, and the matching of scale and energy quality to end-use needs. As with water, the Taoist soft energy path preempts more land. Solar, wind, and biomass energy are by nature less concentrated and therefore require larger areas for collection. A study carried out in California for the U.S. Department of Energy estimated that about 86 percent of that state's energy needs could be met by renewable sources by 2025, but that large areas of urban land would be needed to accomplish it. In the extreme case, if all solar-heating and electricity-generating facilities were located at the sites of the industries they serve, they would cover about 25 percent of all urban areas. These figures suggest that the soft path will require a considerably lower overall population density.

At this juncture, we come to the intriguing question of the relationship between energy flow and population density. Is there such a thing as a most energy-efficient density? The question has been approached by a number of researchers in a variety of ways.

Ecologist Howard Odum argues for very low development densities, believing that "in times of declining energy quantity and quality, the economies of scale shift to smaller dispersed units, and energies are needed for lower quality uses."[14] He bases this conclusion on the need to make use of the low-quality energy that can be harvested from the landscape. His calculations suggest that densities greater than one person per acre will probably not be sustainable in the future because they require too much high-quality (primarily fossil-fuel) energy.

A number of architectural theorists, on the other hand, argue in favor of larger units and higher densities. Ralph Knowles, following the theory that Schrodinger developed for biological systems, hypothesizes that susceptibility to environmental stress varies with the surface-to-volume ratio of a building.[15] Since energy must be used to overcome stress, this formula seems to provide a measure of energy efficiency for architectural forms. Because a larger form of the same relative shape has a smaller surface-to-volume ratio than a smaller one, it suggests that that the larger the form, the greater its energy effectiveness. Knowles goes on to develop this notion with an eloquent series of models showing variations of form in response to different conditions of solar exposure. The theory, however, is geometrically pure at the expense of ignoring the internal arrangements of a building, the flexibility of response that smaller structures can provide, and the potential energy contributions of the landscape. Thus, whereas Odum assumes a high energy contribution from the landscape, Knowles assumes none.

[14] See Odum, 1976; p. 269.

[15] See Knowles, 1975.

Some studies have attempted to determine an ideal urban density, based on a specific concern or set of concerns. The New York Regional Plan Association, for example, studied the relationship between density and energy consumption in the New York metropolitan area. The results indicated that per capita energy consumption decreased with increasing density up to an average of about 25,000 people per square mile, which is roughly 39 people or 13 dwelling units per acre.[16] Above that level, consumption increased with increasing density. This information is useful because it gives us a general indication that some sort of optimum density seems to exist for a given pattern of urban development and energy flow (i.e., that of New York City). It says nothing about other patterns of settlement and energy flow, however, such as those suggested in the University Village and North Claremont case studies.

A number of researchers have concluded that higher levels of urban concentration than now exist will probably be brought about by energy shortfalls in the future.[17] These are likely to be considerably less dense, however, than the ones envisioned in the megastructure concepts of architects like Paolo Soleri, who propose densities greater than that found in present-day Manhattan.

In sum, it seems that each case for a particular theoretical energy-efficient density is based on a limited range of variables and that each is debatable. The arguments for higher density center about the theoretical energy efficiency of large building mass (as in Knowles) and about the need to minimize circulation distances, but in view of our changing means of communication, movement from place to place may be replaced to a large extent by electronic connections. It is fortunate if not purely fortuitous (perhaps there is some deep-seated explanation for the fact) that the exploding development of electronic communications has coincided with the depletion of fossil fuels. Beyond the industrial city very likely lies the electronic city, which promises some relief from energy intensiveness but brings about some problems of its own, problems that are at least as perplexing.

On the plus side, electronic communications can replace face-to-face contact in a great many situations and make moving about from place to place unnecessary. The telephone with satellite links has already begun to bring this about. Computer networks equipped with central data banks and remote terminals and closed-circuit video systems offer even greater potentials. The electronic conferences held in recent years in which people on several continents participate via satellite give some notion of the possibilities. As such technologies become more common, more and more people will be able to work in their homes or in dispersed locations, and mass commuting may be considerably reduced.

The implications of all this for the forms of future cities are at least as profound as those of the development of internal-combustion engines were a century ago. Freed from the old-fashioned umbilical cord to the workplace, a sizable number of people will have a vastly greater range of choice of dwelling place. This maneuverability will bring new population pressures to bear on areas of natural beauty, desirable climate, or other amenities that have been sparsely populated until now for lack of employment opportunities. Photovoltaics, wind generators, solar heating, and biological sewage treatment will certainly make the electronic city possible, and it is likely to be far more dispersed than the industrial city. It may not even be visible as a city at all. Land-use patterns will probably be more related to recreation and esthetic values and less to the realm of work than those of today.

At the same time, the work and living patterns of a considerable portion of the population will no doubt continue to require a great deal of moving about. In view of the need to use less fossil fuels, these people will probably be able to live more effectively and comfortably in higher-density communities. Thus, we can see existing trends that suggest future patterns of development that are both lower and higher in density than those of today.

One conclusion that we can safely draw from all of the studies and opinions on the subject is that density in itself does not ensure energy efficiency. Each admixture of human needs and desires and local landscape character will have its own most fitting pattern of energy flow. Thus, the design of energy flow becomes an important part of environmental design processes. In every case, the character of the landscape is an important factor in determining the appropriate location and density of settlement and its pattern of energy flow. We will consider locational factors in the next two chapters.

[16] See New York Regional Plan Association, et al., 1974.
[17] See, for example, Van Til, 1979.

CHAPTER 14

Location

Patterns and Landscape Suitability

In natural landscapes, locational patterns develop in response to various combinations of evolving conditions. Beginning with the interactions between the rocks pushed up to the earth's surface from its core and the movement of air and water above driven by differing radiation levels, an incredible diversity of environments has taken shape over the face of the globe. Each combination of rock and climate produced, over time, a community of plants suited to the specific conditions provided there. Hot, dry climates, for example, have tended to foster plants with narrow, rough-textured leaves and deep roots—leaves that transpire little water and roots that can probe far into the earth to tap ground water supplies. Hardwood deciduous forests develop in cool, wet climates, conifers usually in mountains.

The plants in their turn help to reshape the environments in which they grow, providing shade to the ground, more moisture to the air, and certain nutrients to the soil. Animals then find their own places among the varied conditions created by rock and climate as ameliorated and brought to life by vegetation. Every animal species fits into a particular set of physical circumstances limited to a certain area or areas of the earth's surface.

Change in the natural landscape comes about through trial and error. Seeds are blown on random winds to new places. If they grow, they may join the native community; if they do not, that experiment is written off as a failure—to be tried again, however, when chance dictates. And animals wander. What was Hemingway's leopard doing high on the slopes of Mount Kilimanjaro?

For a very long time, human beings distributed themselves in very much the same way. Over hundreds of thousands of years after he first appeared in the grasslands of eastern Africa, man stayed put there. Slowly, he learned to use tools, to keep himself warm with skins and eventually with fire, and to expand his habitat. Not particularly well suited to any one environment, he proved himself adaptable to a broad range of them and began to push his adaptive skills to the limit. And then, only about 12 thousand years ago, upon learning how to grow plants, he turned the whole sequence of adaptation and fit upside down. He could—at least to some degree, in a few places—force the environment to adapt to his own requirements. Six thousand years later, after he had learned to control the flow of water, he could at last create wholly new ecosystems.

The place of human beings in the global pattern of organisms and habitats has been in a state of confusion ever since. We have never understood exactly how far we can go in reshaping the landscape to suit ourselves or, on the other hand, to what degree we must fit our activities into the evolved patterns.

Knowledgeably or not, our ancestors used their new power over nature to build settlements in places of their own choice. Usually, the choices, whether through intuitive good judgment or trial and error, seem to have been quite judicious. Recall, for example, that in Huetar Atlantica the areas where settlements and farms had long been established conformed closely with those analytically deemed suitable for such uses. Although such cases are not unusual, the reasons for the choices often had more to do with trade and the availability of raw materials than they did with the natural processes that had dictated locational patterns to man as well as other species during the pre-agricultural period. We know from the records that mistakes were sometimes made. A great many towns were unknowingly built on lands that were hazardous for one reason or another. Then as now, the problems often had to do with soils. The original town of Winchelsea in England, for example, somehow developed and thrived for a while on a steep hillside overlooking a beautiful expanse of ocean. Unfortunately, some of its buildings slid into that beautiful expanse when the unstable soils underneath them began to give way. When this happened, the residents simply moved out and rebuilt in a safer location nearby. Not far from Winchelsea, the town of Sarum also had to be relocated and rebuilt in the thirteenth century. Its population could no longer tolerate the harsh winds that battered them almost continuously on the exposed hillside where the town was originally located. And to top things off, the town's well had run dry.

Perhaps more bothersome, but still not entirely disastrous, a number of seaport towns were built on harbors that proved subject to rapid siltation. The Roman port of Ostia, where hundreds of ships from all over the empire were docked on any given day in the time of the Caesars, is now several miles from the ocean. And the exquisite Flemish town of Bruges, which was one of Europe's major seaports throughout the Middle Ages, is now in a similar situation.

These, however, were isolated incidents. Not many towns seem to have suffered greatly from natural hazards. And the towns in turn had little direct effect on natural systems. Their major impacts on the environment were indirect. Concentrated urban populations required large quantities of food and water, which severely overtaxed the capacities of rural areas in some times and places. The city of Rome wore out the soils of vast regions in its efforts to feed its urban hordes. The North African fields that once provided wheat for much of the Roman Empire have been barren for nearly two thousand years. By and large, until the nineteenth century, the great human alterations of the natural landscape were brought about by the agrarian activities of farming and grazing, which required clearing forests. Except for a few occasional comments like the famous strictures of Plato on the disasters of soil erosion, people seem to have been largely unaware of the damage they were doing to the land.

Then, after the industrial era began, rural populations began flocking into the cities to find jobs and diversions. As mechanized transport made it increasingly possible for people to live farther and farther from where they worked, urbanization covered more and more of the landscape and began to have increasingly profound and more strikingly obvious impacts on natural systems. Productive soils were surfaced over. In the United States, the eastern hardwood forests began to disappear, and with them, wildlife populations. Rivers became polluted with urban wastes. Even parts of urban areas more often fell victim to old-fashioned hazards. On the east coast, houses built on unstable sand dunes at the ocean's edge are periodically torn apart by hurricanes, for example. On the west coast, those built on slippery clays periodically slide into the sea like those in Old Winchelsea.

GESTALT DESIGN

Over time, some awareness of cause and effect develops. During the earlier phases of human development, this understanding was passed on from one generation to the next, probably improving slightly with each generation as more trials were made and more errors remembered. Often we find the villages of primitive people located near rivers just beyond or above the areas likely to be flooded. There is no way of knowing how many earlier villages were washed away before a more fitting location was found—close enough to the river for easy access to food and water but far enough away to avoid floods. The trial-and-error process may have taken thousands of years, but now it is part of the collected wisdom—the communal information system, if you will—of that culture.

Not having entirely lost our collective wisdom as yet, we can sometimes select fitting locations in the same way. We do so by using the gestalt method referred to earlier in the context of information. (Recall that a *gestalt* is an integral unit that is not considered as the sum of its parts.) The gestalt designer does not see a landscape in terms of its compositional elements, its geology, soils, and vegetation. Instead, he examines the landscape as a whole with great care, perhaps walking over it again and again at different times of day, perhaps camping out on it for a period of time, perhaps even sitting upon it in one spot and meditating for hours as Zen garden designers do. When all of the impressions have been sifted through his mind, he decides how human activities should be distributed over the landscape.

Generally, this approach works provided two conditions are met. The designer must have a considerable fund of experience to draw upon, preferably with that same kind of landscape, and the site must be small enough to be comprehended as a whole. These conditions are often met at the construction and site design scales. The locational analyses illustrated with the Simon Residence (page 120) and Madrona Marsh (page 115) case studies provide two examples. At larger scales, however, and in the very complex situations—far-reaching issues and broad client participation—that we so often encounter nowadays, the gestalt approach usually proves to be inadequate. As a result, we find it necessary to substitute more analytical approaches in dealing with land-use locational issues. For an intimately evolved knowledge of the land, we substitute the land resource inventory, and for a hands-on understanding of its behavior, we substitute analytical models. Thus, we seek to provide ourselves with bases for locational decisions that are founded in some comprehension of the likely consequences.

There will always be consequences. When we superimpose human uses on landscapes whose living communities have coevolved with the rocks and the climate over millions of years, changes are bound to occur. Depending on the particular location and the particular use, the changes will range from the minuscule to the positively overwhelming, and from very detrimental to very beneficial for human and natural populations and the integrity of the ecosystem. As mentioned in the first chapter, some authorities dispute the possibility of human uses ever benefiting nonhuman species or the integrity of ecosystems. By now, however, I believe I have presented enough examples to state with confidence that beneficial effects are indeed possible, although admittedly difficult to achieve.

LOCATIONAL CONCERNS

In general terms, the concerns of ecosystem design can be grouped under the headings of hazards, resources, processes, and consumption.

History, as mentioned earlier, has left us with a great many examples of *hazards*. Unstable soils like those that were the undoing of Winchelsea and Old Sarum, floods, hurricanes, earthquakes, sandstorms, tornadoes, volcanic eruptions, and forest and brush fires are all examples of dangers that are more likely to occur in some locations than in others. Thus, it is wise to consider future uses in such locations in view of the risks involved. In some cases, the likelihood of a particular disaster occurring on a particular piece of land can be computed with considerable accuracy. The best-known example probably concerns flood plains, which are classified by the U.S. Army Corps of Engineers as being likely to flood every 5, 10, 50, or 100 years. A 5-year flood plain has a 20 percent chance of being flooded in any given year, a 10-year flood plain a 10 percent chance, a 50-year flood plain a 2 percent chance, and a 100-year flood plain a 1 percent chance. This information provides a clear basis for deciding whether or not to build in a particular flood plain, whether or not to provide flood control structures, and how much to budget for the replacement or repair of facilities likely to be periodically destroyed by flooding. For most other hazards, the risks have not been so carefully calculated, and decisions cannot be so finely tuned.

Resources include all substances or qualities that are likely, at some time, to be useful. The commonly accepted distinction between renewable or nonrenewable resources does not always hold literally true. Consider the tropical rain forest, which, being almost impossible to regenerate once cleared and converted to other uses, is not really renewable but which nevertheless falls into the renewable class because it is a type of forest. Thus, although the distinction is useful, the terms are somewhat misleading. For planning purposes, it is worthwhile to consider the distinction in somewhat more functional terms.

Returning for a moment to the discussion of function and the flows of energy and materials, we might pose the questions about the compatibility of land uses and resources in this way: How does the proposed use fit into the network of flows on the land in question? We can define the ideal fit as a relationship in which human use benefits from, but does not disrupt or destroy, natural resources. While considering resources in terms of function, it will be useful to think of them in two categories:

1. *Stored resources* (sometimes called *capital resources*), which do not, for the time being, play an active part in ecosystem functioning. Usually, these are nonrenewable in terms of human time spans. Examples are sand and gravel, coal, oil, and other minerals.
2. *Flow resources* (sometimes called *income resources*), which participate in flows of energy and materials and are thus living, changing, being created and broken down. Usually, they are considered renewable. Examples are plants and animals, solar radiation, and wind.

The ideal fit described above is just that, of course—an ideal. In reality, any human land use that occurs in the same location as any of these resources will change them, at least to some degree, but in what ways and to what degree? Among the questions that determine the potential fit between human use and resources are the following:

1. Are the land resources utilized?
2. Are they made unavailable for the future (as when a parking lot is built over a gravel deposit)?
3. Are they damaged or destroyed?
4. Will they be depleted?

Processes, the third category of locational considerations, are resources too, of course, in the largest sense. They need separate consideration, however, because they occur over time and because they extend far beyond the boundaries of the landscape being considered.

Stored and flow resources are physically, measurably, mappably present at any given instant, whereas that instant represents no more than a minutely incremental part of a process. To consider the process as a whole, we have to consider it over greater spans of time and space. At different points along the span, it is linked to different localities in different ways. If we examine the water cycle, illustrated for the Aliso Creek watershed on page 33, for example, we note that the upper slopes are watershed lands. If these are cleared of vegetation, the water will not soak in but rather run off all the more rapidly, carrying ever greater loads of soil along with it. This process has implications that extend far beyond the specific land in question. There, it causes increased siltation in the coastal lagoon. In other places, it might lead to severe flooding. Other critical resources specifically locatable on the land are the aquifer recharge zones that admit water into underground storage, lakes and marshes that store it and absorb surges, and streams and rivers. As shown in the Aliso Creek case, human use can sometimes augment the flow of water, thus intensifying its functions. In other situations, it can disrupt the flow, as when an aquifer is paved over, a marsh filled in, or a stream dammed. All of the material flows discussed in Chap. 13 are linked with specific places in the landscape, although most, as mentioned, follow the flow of water.

Ambient energy levels vary with climate, topography, and latitude. The fact that solar radiation gain is greatest on south-facing slopes is an important consideration for urban development.

Whitewater Wash is an example of a piece of land having a unique situation with respect to energy flow. The high-velocity winds that blow through San Gorgonio pass into the desert valley create one of the best sites for wind generators on the West Coast. Since this is also an area of exceptionally high insolation, the potentials for power production to serve the entire Southern California urban region are enormous.

The fourth class of locational considerations is *consumption*. Although levels of potential consumption are not as closely linked with specific locations as the first three concerns, there are enough relationships to warrant some attention.

Among the more important factors affecting levels of resource consumption are distance and climate. In general, the greater the distance of any human use from large population concentrations, the more energy that must be consumed in transportation. As for climate, the more extreme it is, the more energy will be consumed to maintain comfort. This principle applies to the location of buildings and activities within a macroclimate at the larger scales as well as to those within the microclimate of a particular site.

Consideration of resource consumption also reinforces the principle stated earlier that uses are best located wherever they can benefit from the resources already available in the landscape. The more locally available required resources are, the less the need to bring them in from outside and the lower the attendant energy costs.

THE SUITABILITY MODEL

The major design tools for dealing with these four basic locational concerns are capability and suitability models, which are simply maps showing the varying degrees of ability of land to support a given use or set of uses as they are distributed over a given area in accordance with its distribution of physical attributes. Capability is the inherent capacity of the land to support a given use on the basis of a given resource (usually soil). It is generally either capable or incapable. Suitability is more relative, as it depends on the degree of change considered desirable or acceptable, and it is usually represented in ranges from most to least. Commonly, a capability analysis is applied first to eliminate land areas that need not be considered further.

The maps in the Land Resource Inventory provide the data base from which these models can be developed. Since assessments of suitability are made according to the combination of attributes that exists in a given location, a number of different maps must be combined to determine these aggregate conditions. This may be done by manually overlaying maps that are all at the same scale (as described in Chap. 8), or it may be done mathematically by combining numbers, which is the way of the computer. In general, manual overlays are the usual technique adopted in simple situations involving only a few variables or requiring only a single or several models, whereas the computer is the common choice in more complex and longer-term situations. Examples of manual overlay models are shown in the High Meadow, Whitewater Wash, and Wild Animal Park case studies, and computer-generated examples in the San Elijo, Alaska, Nigeria, and Boston case studies.

The first thing that one must understand about suitability models is that they represent not absolute facts but relative judgments. Presumably, these are the best-informed, most objective judgments that could be made under the circumstances, but they are merely judgments nevertheless. Since models are abstractions, or simplifications, of reality, they require highly selective choices of the factors to include from among the infinite number that affect even a small piece of land. They also require judgments about the relative importance of the four basic concerns. These derive largely from the issues as defined early in the effort, but there are still judgments to be dealt with. Is it more important to protect the wildlife population on a site or to make good use of the fertile soil? Is it more important to keep a high-quality gravel deposit available for future use, or would we rather have a new shopping center in a convenient location? Models can be developed in such a way that they pose questions like these along with the arguments on both sides, but if no judgments are made along the way, then we can end up with no model at all, only questions.

Models can vary greatly in their degrees of restrictiveness. For example, we might consider only the most fertile soils (Type I on the Storie Index) suitable for agriculture, or we might include all those on which crops are at all possible. Often the difference will depend on the need for agricultural land and the amount of different types of soil available.

Suitability models also require assumptions, the most difficult of which concern the controllability of the expected consequences. For example, it is fairly common to consider flood plains unsuitable for urban development, but this is a very gross assumption. Drainage channels can be built to prevent flooding, or in certain situations, buildings can be constructed on stilts or mounds above the high watermark. There will be extra costs involved, but as urban land becomes more scarce, such costs become increasingly justified by land values. In practice, most hazardous conditions can be rendered reasonably safe if enough money is available, and a great many impacts on resources and processes can be controlled or mitigated by design. Note the charts with the San Elijo case study that show different construction methods and estimate their relative degrees of impact.

For this reason, when considering potential damage to resources and processes, it is useful to distinguish between mitigable and nonmitigable impacts and to define the means of mitigation. Often, at later stages, these are translated into guidelines for smaller-scale planning and design, another instance of the means being determined at the next smaller level of integration. Such an approach is shown in the High Meadow case study.

Types of Suitability Models

A great many different types of suitability models have been devised, and no doubt a great many more will appear in the future. They vary in the terms they use to define suitability and the means by which they combine variables.

One of the earlier approaches, called "sieve mapping," was developed in England and widely used in the planning of the British new towns after World War II. This approach involves the use of a series of "sieves," or maps delineating areas that present particular difficulties. The land under question is "passed through" the "sieves" one at a time in a definite sequence, and with each passing through, areas revealing the difficulties defined by that sieve are eliminated from consideration. Once the land has been through all the sieves, all difficulties have been eliminated, and the area remaining is considered suitable for the use contemplated.

Sieve mapping is a simple, reliable method that is still employed on occasion. For example, Deitholm and Bressler used it in laying out new ski runs for Mount Bachelor in Oregon.[1] Their sieves, which are typical, defined four types of areas:

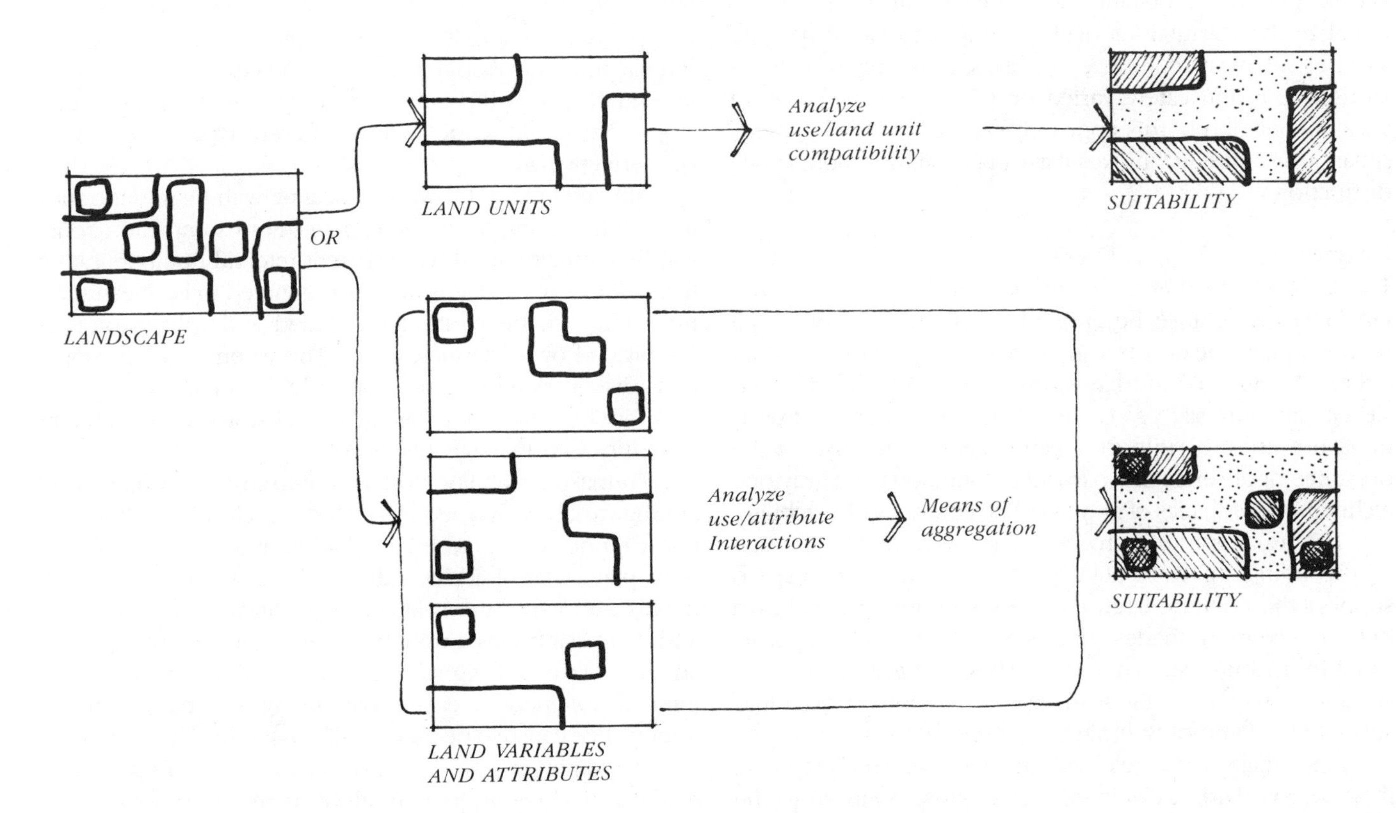

1. *Hazard areas,* those subject to rock slides, snow avalanches, subsidence, high winds, floods, fires, and blowholes
1. *Problem areas,* those that were difficult to develop or where care was needed to avoid environmental damage or economic loss
3. *Resource areas,* those with economic, social, or environmental values to be considered for protection or conservation
4. *Preempted areas,* those that were not available.

The sieves, one can see, are all negative; that is, they define factors that argue against locating uses in particular locations but do not deal with factors that make a location desirable or attractive. To define the latter, the designers developed "attractiveness" models for the various proposed uses. These models were then passed through the four sieves to yield locations that were both attractive and physically suitable for each use.

This dialectic, the balancing of a set of characteristics representing inherent suitability with another set representing economic, social, or other concerns that create a public "pull" for certain uses to a certain piece of land—here called "attractiveness"—is a device that has often proven useful.[2] Among other examples are the Honey Hill project[3] and a number of analyses carried out by Carl Steinitz and his Harvard group. The use of sieves, however, does not depend on the dialectical balance. They can simply be applied to any land area as a whole.

Landscape Units

A quite different, almost opposite approach—which we might call the "landscape unit approach"—is one that begins by classifying land according to a set of physical characteristics. Rather than beginning with problems and potentials, it begins with the natural character of the land. Once individual landscape units have been classified, each sharing a common set of attributes, each is analyzed to define its limitations and potentials, and it is from these assessments that suitable uses are derived. An early proponent of the landscape unit approach, G. Angus Hills, based his classification of landscape units on microclimate, geological composition, and landforms.[4]

The landscape unit approach was used in developing the management zones for Huetar Atlantica (pages 54–64). In this case, the extreme importance of the critical soil-water interaction determined the basis for classification. The pivotal variables, in order of importance, were soil type, rainfall, and general land slope. Since these three variables are closely interrelated, their boundaries coincided fairly closely, and in the course of analysis it became clear that vegetation communities and temperature ranges also correlated closely with the three basic variables. Using this yardstick, there turned out to be five distinct zones in the entire region, all fairly contiguous.

The landscape unit approach is simple, direct, and highly workable whenever the landscape is not enormously varied and only a limited number of uses are being considered.

[1] See Deitholm and Bressler, 1982.
[2] See Marušič, 1979.
[3] See Steinitz, 1978.

[4] See Hills, 1966.

People readily understand the results because it is easy to visualize the distinctions between one landscape type and another. In more complex situations, however, where the landscape is intricately varied or where a wide range of uses is considered, this approach becomes unwieldy. Such situations require techniques that can encompass more subtle distinctions.

Graytones

The graytone technique, which is usually associated with Ian McHarg because he applied it on several pioneering regional planning efforts, can express subtler variations than either the sieve or the landscape unit approach.[5] Whereas sieves can only say yes or no (that is, whether an area is or is not able to support a particular use because of the presence or absence of a particular condition), the graytone technique uses shades of gray to delineate shades of suitability.

The usual graytone process begins with the classification of the ability of each attribute in each variable map to support the contemplated use. Levels of ability are shown on each map in shades of gray, normally with the least suitable attributes shown in the darkest shade, and the more suitable ones in the lightest shade. For the sake of readability, shades are commonly limited to three or four.

Once maps for all relevant variables have been prepared, they are overlaid. Areas in which the grays build up to the darkest levels are the least suitable; those to the lightest levels, the most suitable. On McHarg's composite map for the Richmond Parkway, for example, the ideal highway route would be directed through the lighter tones.

Though simple, straightforward, and easily understood, the graytone method has several serious shortcomings that are typical of the technical pitfalls into which suitability modelers can easily tumble. The first of these concerns the relative importance of the variables selected. Although all are given equal influence on the results, we know that in fact some are more important than others. In a simple analysis involving a limited number of variables and possible uses, this might not seem to be important. In more complex situations, however, the relative importance of variables can be a matter of serious concern.

Technically, it is a cumbersome and time-consuming operation to reconstruct the land resource inventory—by definition, an assembly of facts—in terms of graytones, which represent values. If numerous uses are under consideration, numerous sets of graytone maps will be required. There are also difficulties in combining the variables. Lewis Hopkins, who analyzed the mathematical implications of the graytone process at some length, concluded that its logic is fallacious for two reasons.[6] The graytones, he argued, are actually equivalent to an ordinal number system, that is, to a one-two-three ordering of attributes, where each number represents a place in a sequence rather than a quantity. The overlaying of grays is thus an additive process that amounts to the same thing as adding ordinal numbers, which is a mathematically invalid operation.

The second fallacy stems from the handling of the variables. Overlaying or adding them implies that they are independent, whereas many of them are, to some degree, interdependent. Hopkins gives the example of the sliding potential of a well-drained soil over clay on a 25-percent slope. The same slope with a different type of soil might be perfectly safe.

Although there are ways of dealing with these difficulties, they tend to complicate the process. For example, the tones can be explicitly used to represent interval numbers rather than ordinal ones, that is, they are assumed to be measurable units that can be legitimately added and whose intervals are divided by equal increments. This assumption, of course, is a vast oversimplification of reality, even in the context of a model whose purpose it is to abstract the complexity of reality to a manageable level.

To make suitability models more sensitive to differences among variables, we need to introduce a level of complexity that cannot be expressed in shades of gray. One solution is to play what Hopkins calls the "weighting game." This means estimating the relative importance of each attribute and expressing it as a numerical factor or multiplier, called an "importance weight," and then multiplying the basic interval assigned to each attribute by its weight. Relative importance can also be assessed by assigning a range, rather than a weight, to each variable, using any of a number of mathematical techniques. In all of them, scores of one sort or another are assigned to each attribute and then added to arrive at a measure of relative suitability. One of the ways to assign scores is shown with the models in the Nigeria Federal Territory case study.

In the scoring technique used by the Environmental Systems Research Institute for its ecological sensitivity model for the Federal Capital Territory in Nigeria (pages 248–56), positive scores between 1 to 10 were assigned to four of the five variables: vegetation types, distance from streams, ecological edges, and slope gradients. Negative scores were assigned to the fifth: distance from human disturbances. Given a maximum score of 10, vegetation type was the most important variable. Stream proximity, with a maximum of 5, was second.

Combining Variables

Scoring, although flexible and convenient, does present a number of difficulties. Besides converting the process from a basically graphic to a basically mathematical one, it often requires estimates of the relative importance of variables for which we have insufficient information.

If we can consider professional judgments as a reliable source of scores, weighting does provide a mathematically valid technique, but it is still plagued by the problem of interaction. All combinations of attributes must be examined to identify any that might interact to affect the results. These must be dealt with separately, adding still another level of analysis.

In general, it is best to use the simplest technique that can be justified as conceptually valid and that yields a model reliable enough for the purpose at hand. For simple situations involving only a few variables, we can make the tones in the graytone technique represent equal intervals and deal

[5] See McHarg, 1969.

[6] See Hopkins, 1977; p. 390.

with possible interactions separately.

It is possible to avoid mathematical difficulties altogether by expressing the combining of variables verbally rather than in terms of shades or numbers. For example, we might state that all land with sandy soils is to be considered suitable for residential development except that in which the slope exceeds 25 percent or that which is a flood plain. After similar statements have been made for the other variables, the maps are overlaid to determine where each combination of variables exists. Since this can also be a laborious process, it is important to limit the variables to those one considers truly significant.

In the Whitewater Wash case study (pages 89–100), a suitability matrix was used to analyze the relationships between land attributes and each of the three major uses proposed for the area: energy generation, preservation, and recreation. A matrix of this sort is a useful tool for summarizing large quantities of information in a small space. Levels of suitability for each use were defined by the sets of rules based on information in the matrix. Suitability for preservation, for example, was determined by number of preservation attributes assigned to it by the matrix in combination with a visual sensitivity rating. Thus, the areas most suitable for preservation include those with four or more preservation attributes and a "most sensitive" visual rating. The second most suitable were areas with three preservation attributes and a "most" or "highly sensitive" visual rating. In like manner, the fourth most suitable featured only one preservation attribute and a less sensitive rating, and the fifth featured no preservation attributes, only a visual sensitivity rating.

The strong interest in visual quality repeatedly expressed by the residents of the area is reflected in the fact that visual sensitivity was considered at every level. Although one might argue with the rules adopted, they are clear and explicit and agreed upon by those immediately concerned.

In more complex situations, rules can often become so numerous that it is difficult to understand the significance of each in shaping the model. In such cases, it is useful to proceed in a definite sequence of steps that can either be linear or take a hierarchical form. In the Nigeria Federal Territory case study, for example, the first step in the modeling process was the development of nine interpretive models, each of which, like the sensitivity model shown, involved only a limited number of variables. These first-round models were then combined in various ways to shape the suitability models.

The habitability model shows how the interpretive models were combined to generate a suitability model. In its case, a system of rules of aggregation was used rather than numerical scoring. Each category within each interpretive model was rated on its suitability for habitation. The ratings used were high, medium, low, and incapable (H, M, L, and I). These were then combined in accordance with the rules shown.

The first-round models reveal the limitations and opportunities that each of the concerns imposes on the land. Had the suitability models been developed without this intermediate interpretive phase, these patterns of limitation and opportunity would have been subsumed in the aggregated patterns of suitability and their individual significance lost. Whatever the importance of understanding the implications of the various concerns, the hierarchical sequence provides a measure of flexibility and control that would be otherwise lacking in complex modeling processes.

Sensitivity Models

Sensitivity models are somewhat simpler in concept. They are based on the premise that sensitivity is a condition that characterizes certain attributes of the landscape and that inherently sensitive land is likely to be heavily and detrimentally altered by any form of human development. The model in the Nigeria case study, described above, is one example.

The sensitivity model used in the High Meadow design identified sensitivities in terms of magnitude—high and low—and possibility of mitigation—whether impacts can or cannot be mitigated. The sensitivities were based on five key variables: slope, soil type, geology, hydrology, and vegetation. The model was developed by sequential combination. Each attribute was placed in one of the four sensitivity categories, and each area was then typed according to its most sensitive attribute. Limiting the variables made it possible to keep a record for each land area. The model thus provides a specific basis for evaluating a plan in terms of potential impacts and for requiring mitigation measures in smaller-scale design.

At the regional scale, Cooper and Zedler have proposed using the inherent sensitivity principle for general assessment of the suitability of land to support any type of development.[7] They determine sensitivity on the basis of three qualities: significance, rarity, and resilience. These are applied only to plant and animal communities, thus severely limiting the range of concerns. Nevertheless, it is worthwhile examining them further, since they are typical of the criteria often used in estimating the quality of living communities.

The ecological significance of an ecosystem or species is a particularly difficult quality to measure. Cooper and Zedler admit that significance is a "subjective value judgment of biological importance"[8] but nevertheless list the characteristics that should be considered in determining it as follows: the importance of the role of the system or species in regional ecosystem function; its uniqueness; its actual and potential esthetic, scientific, and economic value; its relative size or rarity; and its prospects for continued persistence. Since each of these, in turn, is difficult to measure, determining significance remains elusive. It is, however, a useful list to remember since the question of ecological significance regularly arises, whatever modeling approach we may apply.

Rarity, or abundance, by contrast, is fairly easy to measure. It is a quality often studied by ecologists, readily subject to quantification, and accessible through aerial photographs and other techniques.

Resiliency, like significance, is difficult to measure, but agreement on rankings is easier to reach. Cooper and Zedler

[7] See Cooper and Zedler, 1980.
[8] Ibid., p. 288.

CASE STUDY XVIII

Federal Capital Territory of Nigeria

The Federal Capital Territory (FCT) was established in 1976 as the site for a new Federal Capital of Nigeria. The creation of the new capital was part of a national effort to relieve urban and administrative pressures being experienced in Lagos, the current capital, and to provide a symbolic focus for the growth and development of Nigeria. Master planning for the new capital and surrounding region has been ongoing since establishment of the FCT.

The FCT covers approximately 8000 square kilometers in central Nigeria lying just above the hot and humid lowlands of the Niger-Benue trough and river systems. The Territory consists of broad, undulating plains separated by rugged hills, many of which are composed of exposed granite bedrock. Nearly the entire FCT is drained into the Niger River by tributaries of the Gurara River. Savanna-type vegetation predominates, although remnants of earlier rain forest and woodland vegetation communities are found on steeper slopes. Human habitation of the FCT has been limited to small but numerous settlements; their economic activity centers around subsistence and cash crop farming and traditional rangeland grazing.

The automated Geographic Information System (GIS) illustrated here was developed for the Territory and surrounding lands. It was applied to a systematic assessment of environmental opportunities and constraints in the region and to the evaluation of land capability/suitability for various broad land-use types. The data bank was created at a scale of 1:100,000. All data were rectified and registered to a set of four 1:100,000 base-map modules aggregated from 1:50,000 topographic quadrangles of the region.

Included in the data base were the following variables:

Manuscript #1: *Surface Hydrology*
- Streams
- Drainage basins

Manuscript #2: *Roads/Infrastructure/Settlements*
- Roads
- Trails
- Pipelines, pumping stations, and storage facilities
- Transmission lines
- Airports
- Dams
- Settlements

Manuscript #3: *Administrative Boundaries*
- Local development areas
- Census units
- Forest reserves

Manuscript #4: *Climate*
- Elevation
- Precipitation
- Temperature
- Pan evaporation
- Relative humidity
- Potential evapotranspiration
- Temperature-humidity index

Manuscript #5: *Integrated Terrain Units*
- Bedrock geology
- Surficial geology
- Physiographic provinces
- Landforms
- Slope gradient
- Land resource units (dominant soils)
- Vegetation and land use
- Surface configuration
- Laterite inclusions

Manuscript #6: *Committed Land Uses*
- Existing and proposed settlements
- Proposed and committed land uses

Manuscript #7: *Special Features*
- Faults
- Shear zones
- Mines

Prepared by Environmental Systems Research Institute, Redlands, California, as consultants for environmental inventory and analysis to Doxiades Associates International Ltd., planners for the Federal Capital Territory of Nigeria.

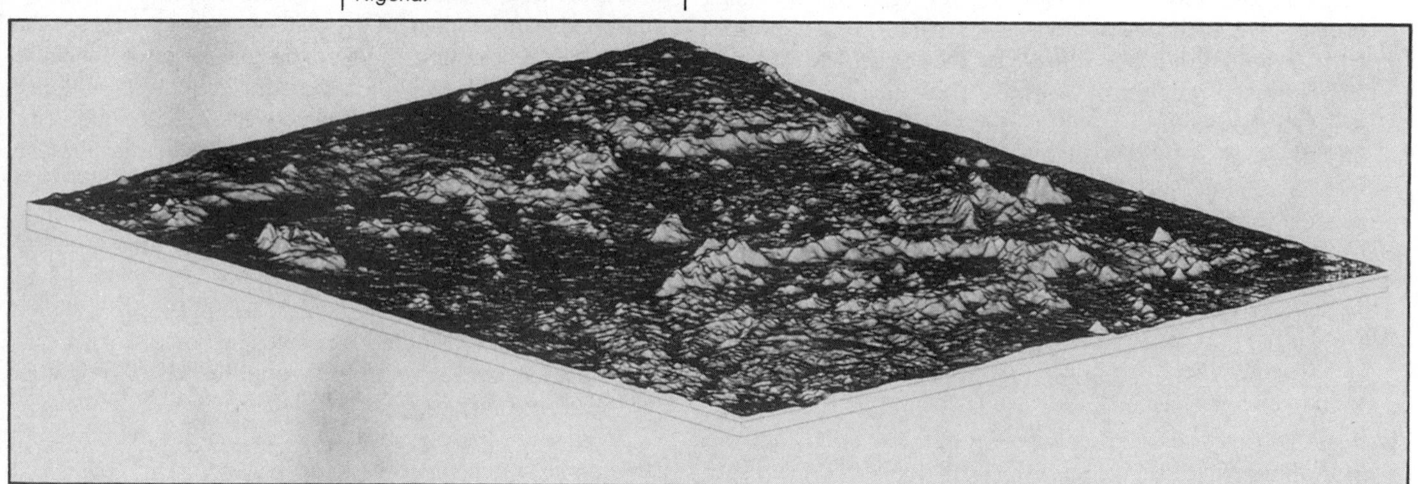

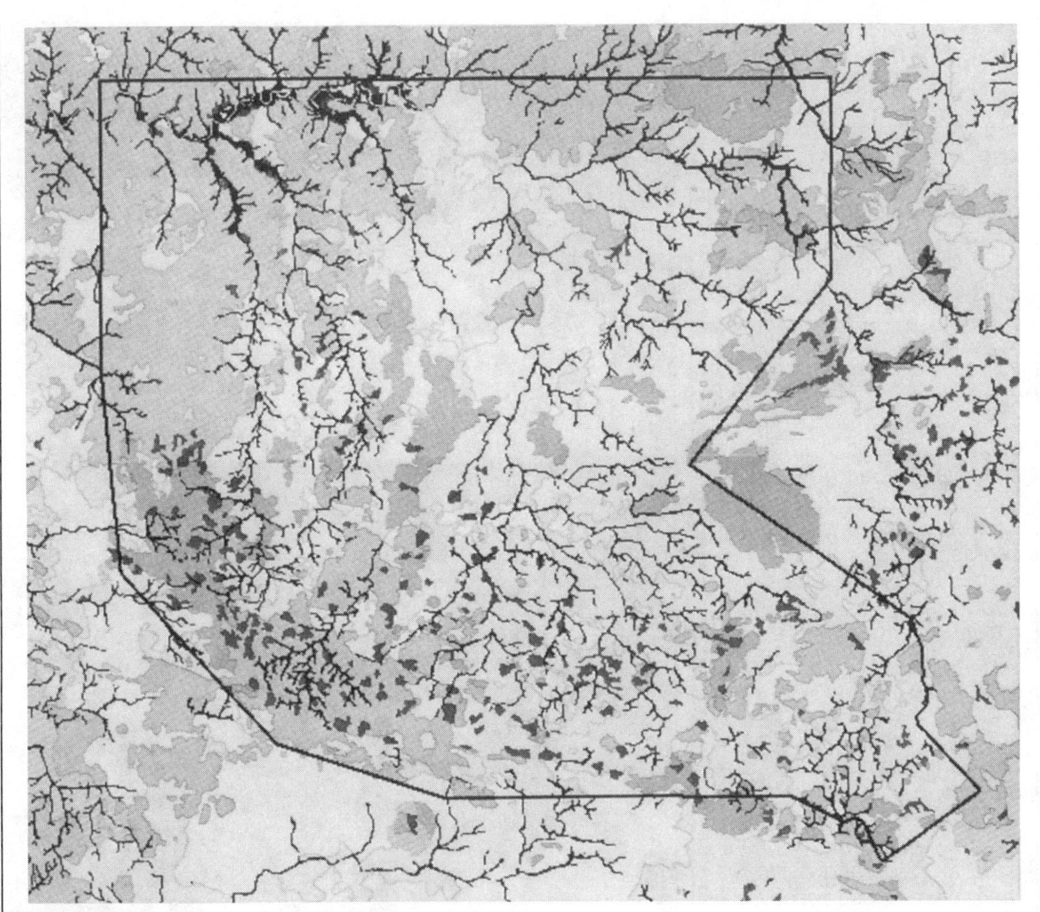

The mapped data displayed on the manuscripts were automated by a process of x,y coordinate digitizing. The computerized data files—composed of polygons, line segments, and points—were used to create a number of plotter-drawn maps of the area as well as to create a parallel set of data files in a grid format. A uniform four-hectare grid was laid atop each of the original data files in the computer, and the data values were transferred into it and recorded by individual grid cell. This grid-cell data bank, ultimately formatted as a grid multivariable file, was used to produce a grid map atlas of the region and to display the mapped results of the environmental analyses that were conducted. Maps describing vegetation cover and soil types are given here as examples.

VEGETATION

GRASSLAND AND WOODLAND
- AQUATIC
- GRASSLAND AND WOODED GRASSLAND
- SHRUBLAND AND THICKET
- WOODED GRASSLAND/WOODLAND TRANSITION
- WOODLAND

FOREST
- MATURE DISTURBED
- RIPARIAN

DISTURBED VEGETATION
- FARMLAND AND SETTLEMENT MOSAIC
- CONCENTRATED SETTLEMENT
- WATER

SOIL TYPE

SHALLOW SOILS (LT 50 CM)
- SAND TO SANDY LOAM
- SANDY LOAM TO SANDY CLAY LOAM

MODERATELY DEEP SOILS (50-100 CM)
- SAND TO SANDY LOAM
- SANDY LOAM TO SANDY CLAY LOAM

DEEP SOILS (GT 100 CM)
- SAND TO SANDY LOAM
- SAND TO SANDY CLAY LOAM
- SAND TO LOAM
- LOAMY SAND TO SANDY CLAY LOAM
- LOAMY SAND TO CLAY LOAM
- SANDY LOAM TO SANDY CLAY LOAM
- SANDY LOAM TO CLAY LOAM
- SANDY LOAM TO CLAY

The Ecological Sensitivity Model

The ecological sensitivity model identifies areas of increasingly greater ecological value because of the productivity and diversity of their flora and fauna. Since modification of areas of higher ecological value could potentially result in a significant loss of productivity and diversity, such areas were considered to be more sensitive. Major assumptions that served to refine the concepts of productivity and diversity for the study area included the following:

- *The forested areas represent remnants of the most diverse climax communities in the area.*
- *Woodlands and aquatic grasslands are remnants of less diverse climax communities.*
- *Other natural vegetation communities are ecologically more diverse and productive than the simplified plant communities of agricultural and urban settings.*
- *Water or proximity to water enhances the ecological productivity of an area.*
- *Ecological edges or ecotones are the most productive portion of natural plant communities because of enhanced diversity resulting from the proximity of other communities.*
- *Proximity to roads and settlements tends to reduce an area's ecological value because of the increased likelihood of manmade disturbances.*

The major factors considered included vegetation type, stream hierarchy, slope, existing roads, and settlements. Vegetation type was the basic factor because it was most closely related to ecological diversity and productivity. Specific vegetation types were assigned initial weighted values that reflected their relative diversity/productivity value. The other factors were used to modify these initial values through a process of addition or subtraction. Streams and proximity to larger streams added value because they represented a water source. Larger perennial streams added more value than the smaller intermittent streams. Steeper slopes were assigned additional value because they tended to discourage human intrusions. A separate submodel was also used to identify and add value for ecological edge conditions. The factors related to roads and settlement were subsequently modeled throughout the study area and the resulting values subtracted because of the increased possibility of human disturbance.

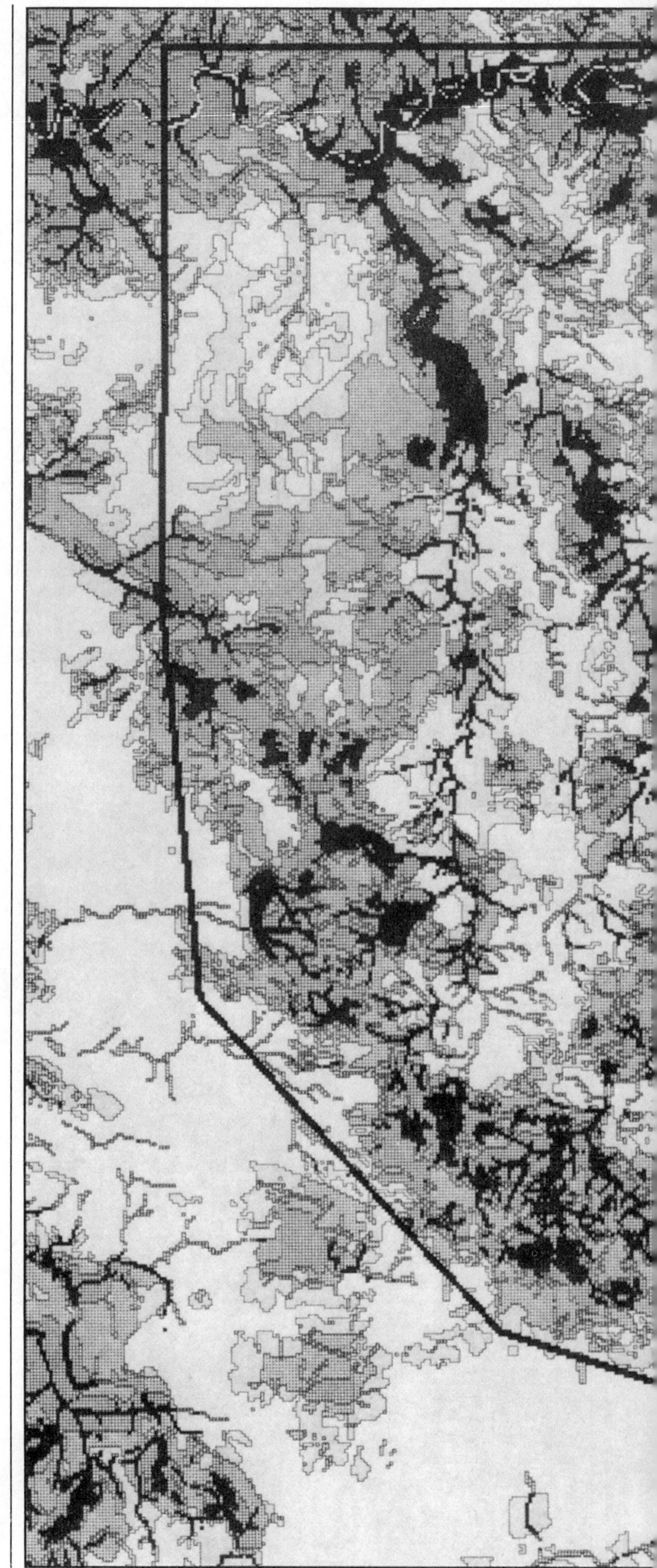

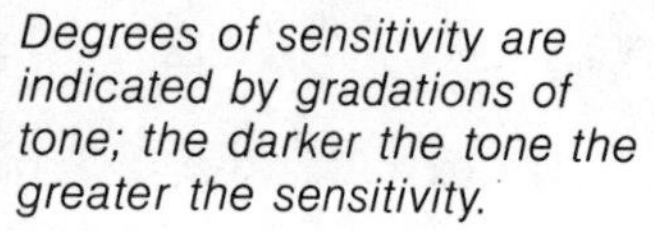

Degrees of sensitivity are indicated by gradations of tone; the darker the tone the greater the sensitivity.

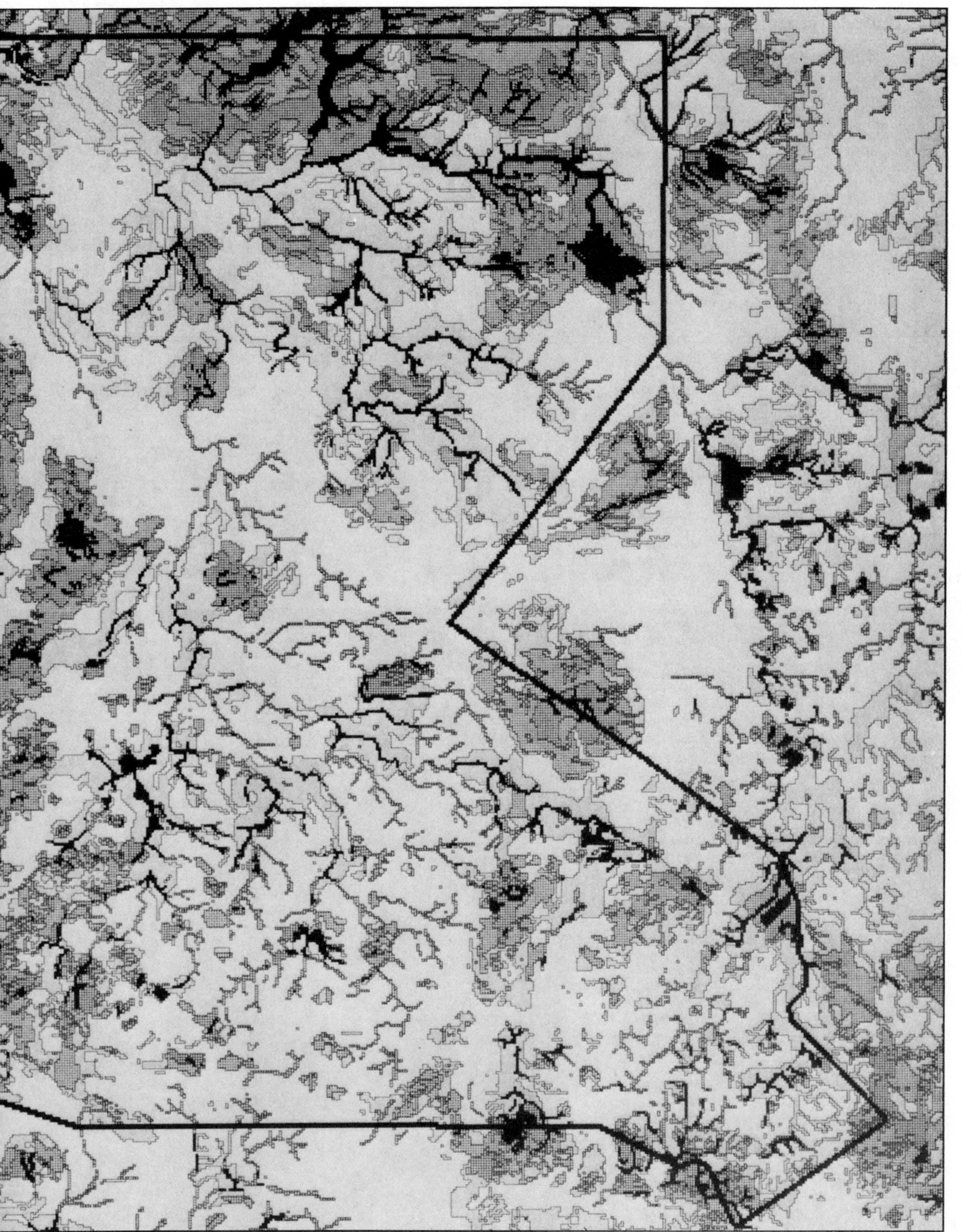

Wintersburg Channel

HABITAT CONCEPTUAL MODEL OUTLINE

MODEL SUMMATION RULES

Ratings are scanned within each general category encompassing more than one factor, and the most severely constraining rating is used to provide the overall rating for the category. In effect, each general consideration—landform, soils, water availability, etc.—has a single rating when summation begins. The following summation procedures are used:

Rating	***Value***
Low capability	*(GT 4M or EQ 1,2,3 or 4L and Not EQ I)*
Low amenity	*GE 1L*
Moderate amenity	*GE 1M and Not EQ L*
High amenity	*GE 1H and Not EQ M or L*
Moderate capability	*(EQ 1,2,3 or 4M and Not EQ L or I)*
Low amenity	*GE 1L*
Moderate amenity	*GE 1M and Not EQ L*
High amenity	*GE 1H and Not EQ M or L*
High capability	*(GE 1H and Not EQ M, L or I)*
Low amenity	*GE 1L*
Moderate amenity	*GE 1M and Not EQ L*
High amenity	*GE 1H and Not EQ M or L*
Incapable	*(GT 4L or GE 1I)*

CONSIDERATION	*SPECIFIC DATA CLASS*	*VALUE (Incidence)*	*VALUE (Proximity)*
HAZARD/CONSTRAINT SUBMODEL			
Flood/drainage hazards (Flood/Drainage Hazard Model)	*Hazard rating*		
	No or low flood potential		
	Well drained soils	*H*	
	Imperfectly drained soils	*H*	
	Poorly drained soils	*M*	
	Moderate flood potential		
	Well drained soils	*M*	
	Imperfectly drained soils	*M*	
	Poorly drained soils	*L*	
	High flood potential		
	Well drained soils	*L*	
	Imperfectly drained soils	*I*	
	Poorly drained soils	*I*	
	Very high flood potential		
	Well or imperfectly drained soils	*I*	
	Poorly drained soils	*I*	
Geologic/topographic hazards (Geologic/Topographic Hazard Model)	*Hazard rating*		
	Very low	*H*	
	Low	*H*	
	Moderate	*M*	
	High	*L*	
	Very high	*I*	
Soil engineering constraints (Soil Engineering Constraints Model)	*Constraint rating*		
	Very low	*H*	
	Low	*H*	
	Moderate	*M*	
	High	*M*	
	Very high	*L*	
Soil erosion constraint (Soil Erosion Model)	*Constraint rating*		
	Very low	*H*	
	Low	*H*	
	Moderate	*H*	
	High	*M*	
	Very high	*L*	
Ecological sensitivity (Ecological Sensitivity Model)	*Sensitivity rating*		
	Very low	*H*	
	Low	*H*	
	Moderate	*M*	
	High	*L*	
	Very high	*L*	
Water pollution sensitivity (Water Pollution Sensitivity Model)	*Sensitivity Rating*		
	Very low	*H*	
	Low	*H*	
	Moderate	*H*	
	High	*M*	
	Very high	*L*	

Continued

CONSIDERATION	SPECIFIC DATA CLASS	VALUE (Incidence)	VALUE (Proximity)
Water availability (Stream Proximity and Ground Water Potential Model)	*Stream hierarchy*		
	Seventh order or greater	H	
	Less than ½km distance		H
	½–1km distance		H
	1–2km distance		M
	2–5km distance		M
	Greater than 5km distance		L
	Fifth to sixth order	H	
	Less than ½km distance		H
	½–1km distance		M
	1–2km distance		M
	Greater than 2km distance		L
	Third to fourth order	M	
	Less than ½km distance		M
	½–1km distance		M
	Greater than 1km distance		L
	Second order	M	
	Less than ½km distance		M
	Greater than ½km distance		/
	Ground water opportunity rating		
	Very low	L	
	Low	L	
	Moderate	M	
	Less than ½km distance		M
	½–1km distance		M
	Greater than 1km distance		L
	High	H	
	Less than ½km distance		H
	½–1km distance		M
	1–2km distance		M
	Greater than 2km distance		L
	Very high	H	
	Less than ½km distance		H
	½–1km distance		H
	1–2km distance		M
	2–5km distance		M
	Greater than 5km distance		L
AMENITY SUBMODEL			
Proximity to high visual quality (Visual Quality Model)	*Visual quality rating*		
	High and very high	L	
	Less than 1km distance		H
	1–3km distance		H
	Greater than 3km distance		M
Climatic comfort (Temperature-Humidity Index Calculation)	*Mean Annual THI Rating*		
	Less than 21 (0% Uncomfortable)	H	
	21–21.5 (0–7%)	H	
	21.5–22 (7–15%)	H	
	22–22.5 (15–25%)	H	
	22.5–23 (25–35%)	H	
	23–23.5 (35–42%)	M	
	23.5–24 (42–50%)	M	
	24–24.5 (50–60%)	M	
	24.5–25 (60–70%)	L	
	Greater than 25 (GT 70% Uncomfortable)	L	
Unsuitable land uses or designations	*Forest reserves*		
	Confirmed forest reserves	OFF	
	Existing land use		
	Extraction areas	OFF	
	Committed land uses		
	Committed reservoirs	OFF	
Settlements and development areas	*Existing settlements*		
	Concentrated settlements	OFF	
	Rural settlements (points)	OFF	
	Committed development areas		
	Planned settlements	OFF	

ECOLOGICAL SENSITIVITY OUTLINE

CONSIDERATION	SPECIFIC DATA CLASS	VALUE (Incidence)	VALUE (Proximity)
General context	Water	SKIP	
Vegetation diversity and productivity*	Vegetation type		
	Mature disturbed forest	10	
	Mature disturbed with rock outcrop	10	
	Riparian forest	10	
	Aquatic grassland	8	
	Woodland	7	
	Woodland and rock outcrop	7	
	Wooded shrub grassland	6	
	Wooded shrub grassland and rock outcrop	6	
	Wooded grassland	4	
	Grassland	4	
	All other natural vegetation	2	
	Plantations	2	
	Farmland mosaic	1	
	Farmland	0	
	Settlement or extraction	0	
Proximity to streams	Stream order		
	Seventh order and greater	5	
	Less than 1km distance		5
	1–2km distance		3
	2–5km distance		1
	Fifth or sixth order	5	
	Less than 1km distance		5
	1–2km distance		3
	Third or fourth order	3	
	Less than 1km distance		3
	First or second order	3	
Ecological Edge	Number of natural vegetation types		
	With distance of ½km		
	Two types	1	
	Three types	2	
	More than three types	3	
Proximity to human disturbance	Existing road class		
	Primary or secondary roads	−3	
	Less than 1km distance		−3
	1–2km distance		−2
	2–5km distance		−1
	Tertiary roads	−2	
	Less than 1km distance		−2
	1–2km distance		−1
	Rural collectors and trails	−1	
	Less than 1km distance		−1
	Existing land use		
	Settlement (area)		
	Less than 1km distance		−3
	1–2km distance		−2
	2–5km distance		−1
	Farmland and farmland settlement		
	Less than 1km distance		−2
	1–2km distance		−1
	Farmland mosaic		
	Less than 1km distance		−1
	Plantation		
	Less than 1km distance		−1
	Extraction areas		
	Less than 1km distance		−1
	Settlement points (only when not settlement area)		
	Settlement point	−2	
	Less than 1km distance		−2
	1–2km distance		−1
General Accessibility	Average slope gradient		
	30–50%	1	
	Greater than 50%	3	

MODEL SUMMATION RULES

Rating	Value
Very low	0 or less
Low	1–3
Moderate	4–6
High	7–9
Very high	10 or greater

* Vegetation categories employed in the analyses are: Aquatic grassland (110); grassland (120); shrub, wooded shrub, and wooded grassland (130–160); shrub and thicket (210–220); wooded grassland/woodland transition (310–320); woodland (410–420); mature disturbed forest (510–530); riparian forest (520).

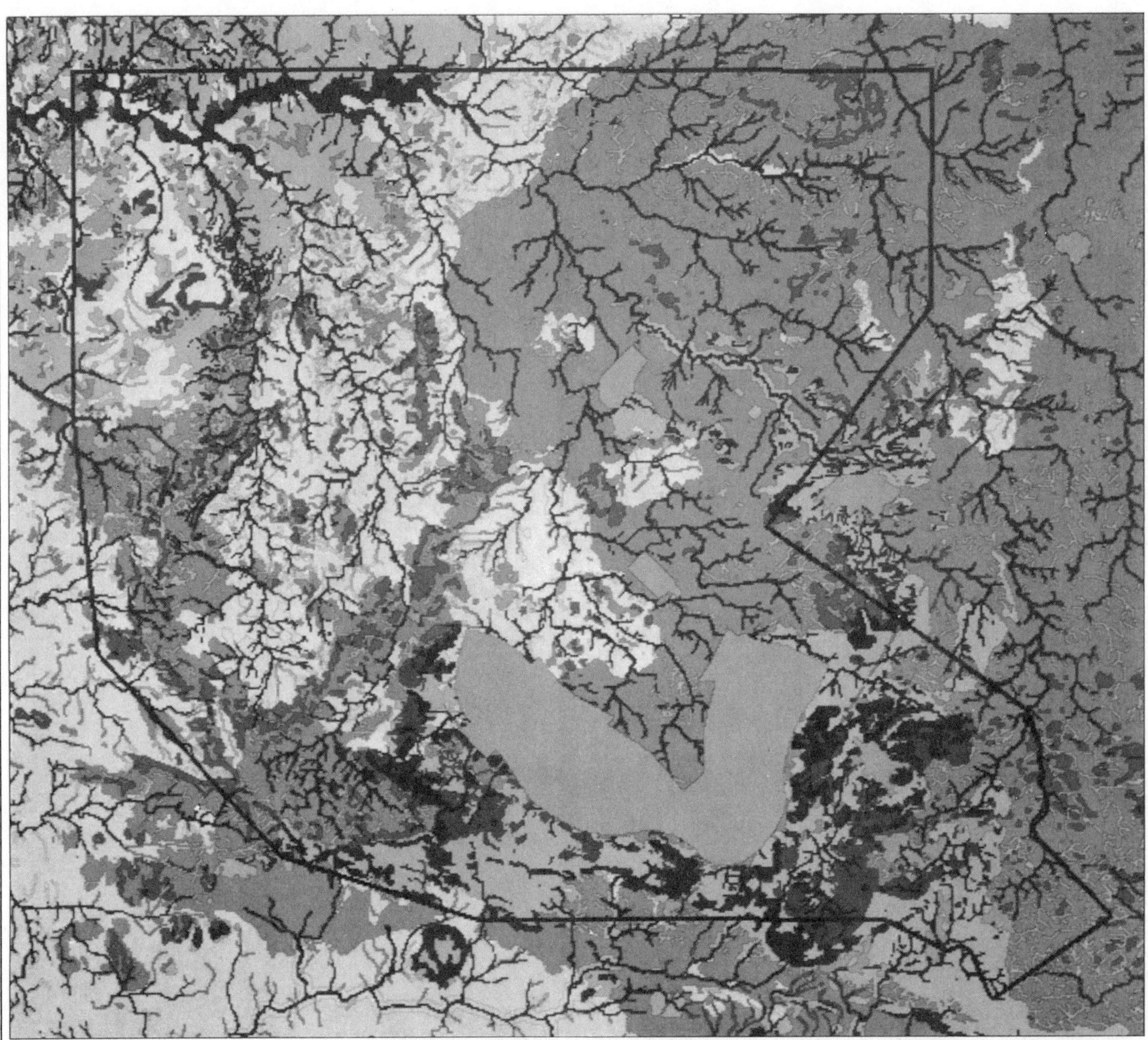

The Habitability/Suitability model evaluates the FCT and its surrounding area for their capability and suitability to support concentrated settlement. Capability for concentrated settlement was based on environmental factors representing hazards to or constraints on urban development or providing amenities for it. Suitability was defined in terms of the conflict between urban settlement and existing land uses.

The constraints, hazards, and amenities considered in the model included: flood/drainage hazard, geologic/topographic hazard, soil engineering constraint, soil erosion constraint, ecological sensitivity, water availability, visual quality, and climatic comfort. With the exception of climatic comfort and availability of surface water, these considerations were previously defined and analyzed by the opportunity/constraint models. The Habitability/Suitability model analyzed and evaluated the composite or interrelated influence upon habitability of these factors. In performing this analysis, various assumptions were made, including the following:

- *Capability ratings are influenced primarily by the level of constraint to site development.*
- *Areas are considered incapable of habitation only when engineering, planning, or design solutions for mitigating hazards or constraints are considered practically or financially infeasible.*
- *Both surface water and ground water will be used as sources of potable water.*
- *Proximity to areas of very high or high visual quality is preferable to development within these areas in order to avoid degradation of the visual resources.*
- *Forest reserves, extraction areas, and reservoirs represent long-term commitments to nonurban uses.*
- *Existing and planned settlements represent areas of concentrated settlement and do not require evaluation for future habitability.*

The Habitability/Suitability model was structured as two submodels—one to evaluate constraints and hazards and the other to evaluate amenities. The hazard and constraint submodel served as the most important factor in the evaluation and the amenity submodel as a subordinate consideration.

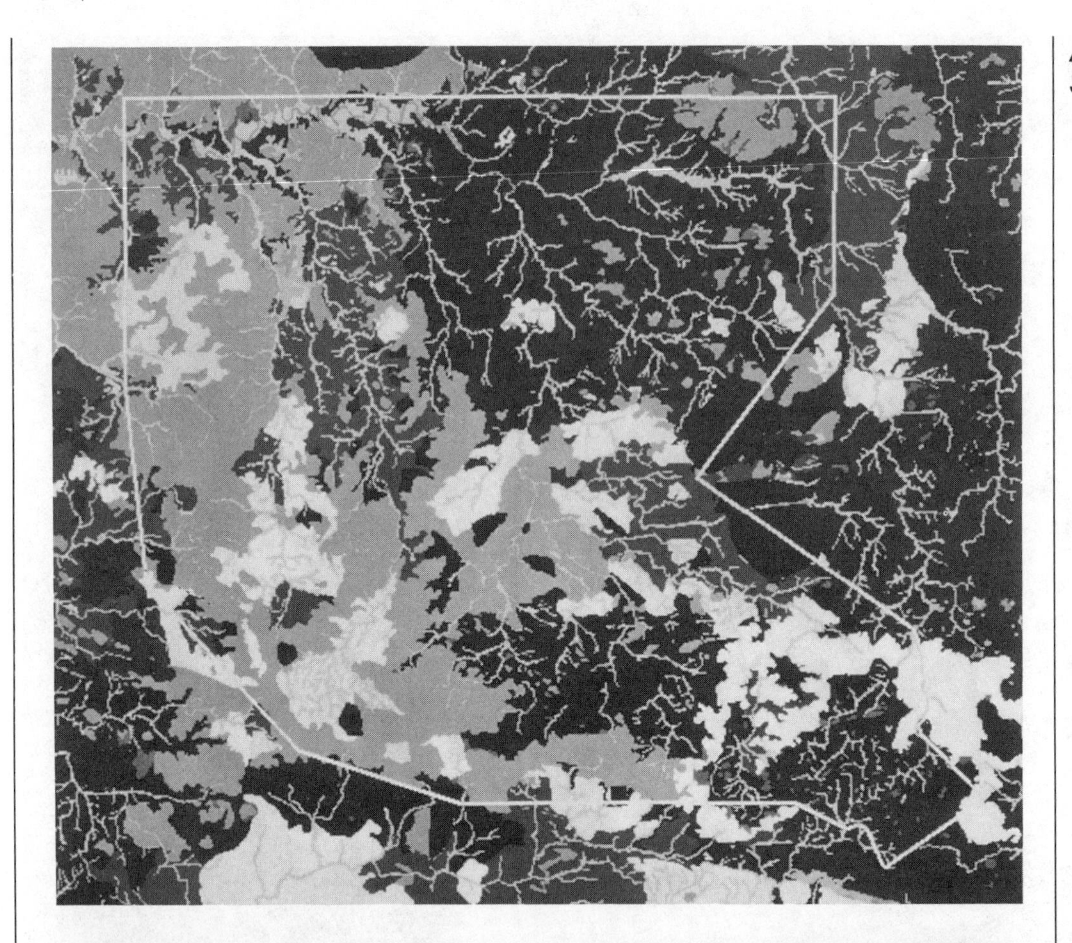

AGRICULTURE SUITABILITY MODEL

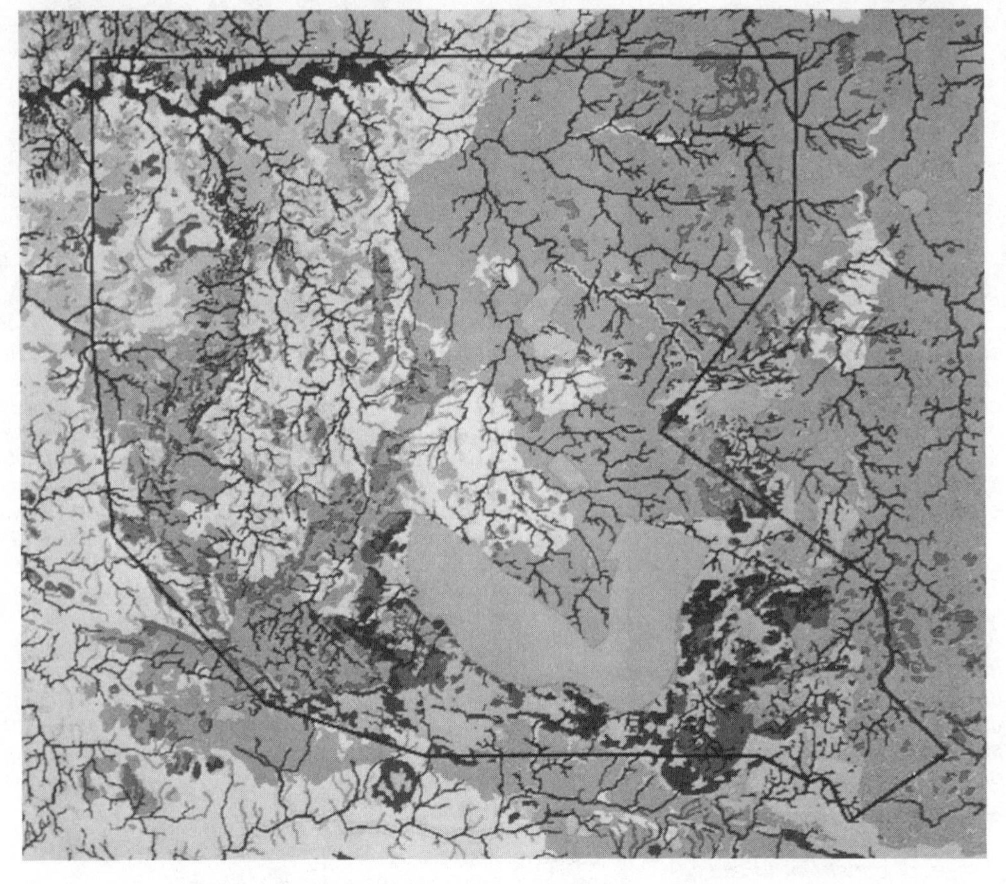

CONSERVATION SUITABILITY MODEL

Lighter tones indicate higher degrees of suitability.

list six responses to environmental stress to be considered in evaluating resiliency: mortality; changes in birth rates; displacement; changes in coverage, growth, or vitality; changes in behavior; and disruption of interactions.

Thus we see that determining levels of sensitivity, while seeming simple at first glance, quickly becomes a complex task involving a great many judgments. The major advantage of this particular approach is the larger perspective that it affords. Too often, suitability models have no larger frame of reference than the boundaries of the land areas they are dealing with. A model may tell us that residential development will wipe out several hundred acres of a certain plant community, but we have no way of knowing how undesirable this may be unless we have some understanding of its value and rarity.

Summary of Approaches

At this point, we can step aside to take an overall look at the range of suitability modeling techniques discussed so far with a view toward typing them by conceptual approach. All work on the basic assumption—the cornerstone of all suitability modeling—that the landscape varies in its ability to support human use according to physical variables that are distributed over its geographic area. The first clear distinction that we can draw is that between the *landscape unit* approach, which begins by identifying landscape units independently of potential uses, and those approaches that begin by analyzing expected interactions between land attributes and potential uses. The landscape unit approach is typified here by Angus Hills' work, by the Huetar Atlantica project, and by landscape sensitivity analyses such as those advocated by Cooper and Zedler.

Approaches that begin with interactions can be further categorized as those that work by a process of elimination and those that work by aggregation of compatibility. The sieve and graytone techniques eliminate unsuitable land step by step until only that deemed suitable is left for consideration. Aggregations of compatibility are accomplished by examining the relationship between each land attribute and each potential use, as in the Whitewater Wash and Wild Animal Park case studies. The judgment of compatibility might be based on any or all four of the basic locational concerns discussed earlier: hazards, resources, processes, and consumption. Once compatibilities have been determined, they are aggregated—by weighting and scoring, by sequential combination, or by some other means—and applied to the land inventory to produce a map that summarizes the compatibilities in graphic form.

MEANS OF AGGREGATION

The means used for aggregating relationships is critically important to the technical validity of the resulting model. We have already discussed graytones, weighting and scoring, explicit rule statements, and sequential combination as methods with particular advantages and shortcomings. There are a great many other ones possible, most of them probably as yet undiscovered. Although the discovery of a perfect all-purpose means of aggregation seems unlikely, the development of more varied means will provide more choices and thus probably more workable models. Two other less commonly used methods warrant mention here: cluster analysis and fuzzy sets.

Cluster analysis identifies land areas that, however widely separated in space, have similar combinations of attributes across a range of variables and that can therefore be assumed to perform similarly in similar use situations. Once these areas, which are called *clusters,* have been identified, the attributes can be analyzed for their ability to support different uses. The suitability map then displays clusters as homogeneous areas and defines the suitability of each for various uses. The major advantage of cluster analysis lies in its avoidance of weighting or comparative judgments, which are always open to question in arriving at the basic locational pattern. Its major disadvantage lies in the complexity of the computations required, an effort that does not seem to be entirely justified by the results. In one planning effort for a new town and airport, the analysis produced 25 clusters among 1350 grid cells, an unwieldy number from which to develop suitability profiles.[9]

Fuzzy set evaluation was proposed by MacDougall.[10] As the name implies, it accepts the intuitive and rather imprecise nature of qualitative professional judgments and provides an orderly, rational way of converting them into quantitative terms. First, as in the other scoring systems, each place or attribute is rated for the degree to which it satisfies each criterion. In the second step, the criteria are compared two at a time, deciding for each pair which is more important and how much more. Both of these first steps can be performed by an individual or by a group. Then, by a technique called *eigen-analysis,* the judgments are combined to assign weights of relative importance to the criteria. These have an average value of 1, with the more important factors being rated higher than 1 and less important ones lower than one. These weights are applied by exponentiation to the ratings developed in the first step, with the result that higher ratings are reduced and lower ones raised. Finally, the suitability of any place is determined not by an average or total score, but by its lowest rating. That is, the factor for which it scores as being least suitable determines its level of suitability for the use in question, however high its other scores might be. That criterion is, in effect, its limiting factor.

These two examples illustrate two fundamentally different ways of dealing with the central dilemma of suitability modeling. Because of its holistic nature, such modeling deals with a great variety of imprecisely known factors in qualitative ways. One way—the way of cluster analysis—is to find means of analytically subdividing all operations into discrete parts and isolating qualitative judgments in limited groups at later stages. Another way is to accept the fuzziness and put it in orderly terms from the very beginning. Most techniques of aggregation follow one way or the other.

Computer-Aided and Manual Overlay Techniques

All the suitability modeling approaches considered here can be applied with either computer-aided or manual overlay techniques. The logical principles are the same for both,

[9] See Sharpe and Williams, 1972.
[10] See MacDougall, 1981.

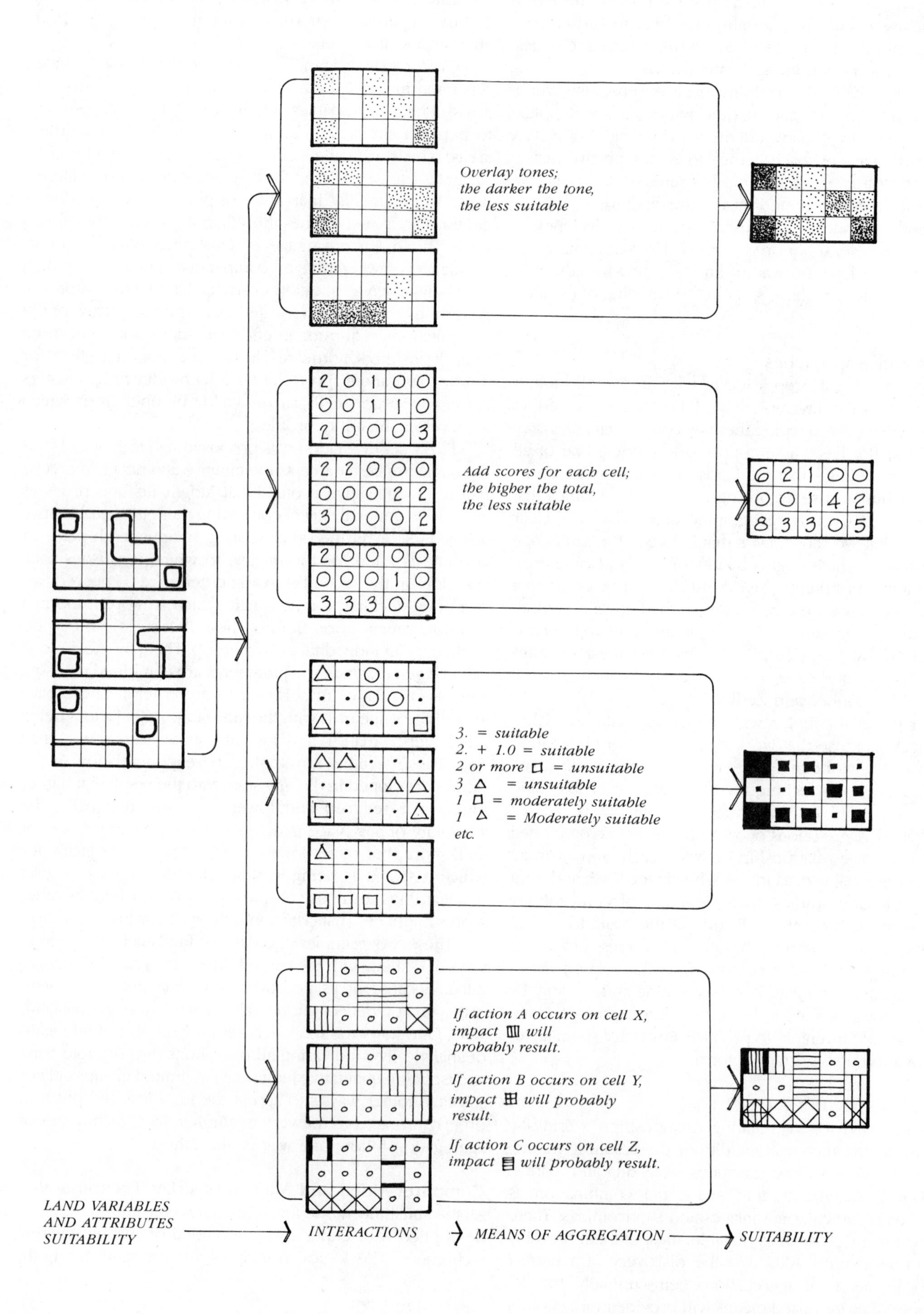

Overlay tones;
the darker the tone,
the less suitable
Add scores for each cell;
the higher the total,
the less suitable
3. = suitable
2. + 1.0 = suitable
2 or more □ = unsuitable
3 △ = unsuitable
1 □ = moderately suitable
1 △ = Moderately suitable
etc.
If action A occurs on cell X,
impact will
probably result.
If action B occurs on cell Y,
impact will probably
result.
If action C occurs on cell Z,
impact will probably result.
LAND VARIABLES
AND ATTRIBUTES
SUITABILITY
INTERACTIONS
MEANS OF AGGREGATION
SUITABILITY

but the computer has strong advantages in certain kinds of situations. In Chap. 8, I contrasted the one-time or short-term resource inventory with the long-term inventory or information system, the latter being assembled for a variety of applications over a long period of time. For the information system, computer use is virtually necessary. Once an inventory has been coded and stored on magnetic tape or disc, models can be generated very quickly. This is especially true whenever a modeling logic has already been established and programmed so that all that is required for each new model is to establish a new set of values and apply them to the relevant variables.

The computer's ability to carry out complex mathematical operations quickly is another advantage that becomes important with the more complex methods of aggregation. Both scoring and sequential combination become extremely time-consuming if large, complex land areas with a number of variables are involved. Computations for techniques like cluster and fuzzy set analysis are usually not feasible without the use of a computer, even in many one-time situations.

Manual-overlay techniques are most useful in one-time situations requiring relatively simple methods of aggregation. The manual resource inventory might identify land attributes by colors, by shades of gray, or by symbols such as letters or numbers. Maps drawn on mylar, acetate, or other types of transparent or translucent plastic are the easiest to use. Usually, two maps are overlaid at a time to define areas sharing suitable or unsuitable attributes. One variable after another is then added until the picture is complete. It is important to develop the guiding logic and list the scores or rules before beginning to overlay maps. If the overlay process is then carried out step-by-step, confusion can be kept to a minimum. Each step should follow definite written rules. When one comes to a step for which there is no rule, a rule has to be devised in keeping with the guiding logic and written down. If large numbers of variables are involved, introducing an intermediate step of interpretive maps can be most useful. A major advantage of manual techniques is the opportunity they provide to make judgments, change course, discuss, and readjust along the way.

A difficulty that plagues both computer-aided and manual-overlay techniques is unreliable accuracy. The levels of inaccuracy discussed in Chap. 8 are increased when maps are overlaid. The possible horizontal error in a model generated by overlaying a given number of maps is the total of the allowable errors of all the maps. MacDougall points out that in an overlay of six maps, all conforming to the U.S. National Map Accuracy Standard allowable horizontal error of 0.5 mm and a purity of 0.80 (typical for a good soil map), the possible horizontal error in boundary locations will be 3 mm and the purity might be 0.80^6, or 0.21. The limited utility of such a map is obvious. It could even be dangerously misleading if one were unaware of its inaccuracy.

In practice, such a level of inaccuracy is unlikely, partly because inaccuracies tend to diverge much less than that and partly because it is possible to correct for inaccuracies to some degree in the overlay process. Given the inevitable levels of error, and the high potential for multiplying inaccuracies, it is obviously important to exercise great care in assembling overlays. Maps displaying attributes separated by transition zones rather than by actual linear boundaries, such as vegetation and slope maps, are particularly subject to error. Considering that the accuracy of an overlay can be no greater than that of the least accurate of the variable maps from which it was produced, such maps present particular difficulties.

It often happens in assembling overlays that certain physiographic features—a ridgeline, a valley, even a fence—account for boundary conditions for several variables but are not present on every map. As overlays are assembled, the missing boundaries can be correlated to these given lines, thus maintaining or even improving the accuracy. When overlaying two variables, it is often a good idea to lay both over a reference base map with all physiographic features accurately defined.

Oddly enough, the computer, for all its precision in computation, has particular difficulties in dealing with problems of map accuracy. It is completely unable to make minor corrections and adjustments like these. When overlaying polygons, one is likely to end up with myriad tiny, meaningless slivers. When using the grid cell system, one finds errors easily multiplied. A line incorrectly positioned by half a grid cell can cause a shift in the output to an adjacent cell, causing an error of a full grid cell's width. Thus, to assure accuracy for an automated data base, it is usually necessary to redraw all data maps at a common scale, overlay them, and adjust for incongruencies before coding or digitizing. The Integrated Terrain Unit method for accomplishing this correcting process is illustrated in the Alaska case study.[11]

These difficulties point out a larger problem in using the computer for suitability modeling. The heavy reliance of the process on human judgment is somewhat at odds with the blind precision of the computer. Human judgment is exercised most visibly in the establishing of values and adjusting of parameters, and these are easily combined to produce models by computations that use the information stored in the computer. As in any design process, however, less significant, sometimes unconscious, often invisible judgments are also rendered all along the way, an example being the adjustment of lines on maps just discussed. In the aggregate, these minor judgments can be important to the results, but the computer is incapable of making them, at least in the present state of its development, and we would be reluctant to allow it to do so if it could. In some situations at least, the myriad small judgments lost could make a world of difference.

In any case, both manual and computer-generated models are inaccurate to some degree, a degree that we have as yet limited ability to measure, and it is important to remember this fact. Since accuracy is closely related to scale, maps at the appropriate scale for the level of concern should always be used. When using suitability models, one should also double-check critical points and lines by referring to maps of known accuracy or by field checking.

[11] See Dangermond, 1982.

CHAPTER 15

Location

Impact Prediction

Although all suitability models make predictions, one type makes specific predictions of environmental impact, thus not only joining suitability modeling and impact analysis into a single process, but also integrating the design process with the environmental impact report.

The requirement for environmental impact reports for all Federal projects, which came into effect with the National Environmental Policy Act of 1970 (NEPA), has far-reaching implications. Some of the state legislation that followed—for example, the environmental quality laws in California and Massachusetts—is even broader in its coverage. Some states now require environmental impact reports for all private development projects of significant size.

We might even see NEPA as symbolizing a new era in the shaping of the human environment. We can act responsibly only if we give thought to the consequences of our actions. It hardly seems believable now that predicting the effects of a reshaped landscape should have become common practice only so recently. As we have seen, some designers, planners, and concerned citizens have urged attention to what we now call environmental impacts for over a century. We can date scientific understanding of the far-reaching landscape alterations wrought by man back to George Perkins Marsh's *Man and Nature,* published in 1864. Only a century later did such concern finally become institutionalized in NEPA.

What NEPA does not make sufficiently clear is that simply to design something and then predict its impacts is not enough. Whenever this approach is taken, impact reports tend to become voluminous lists, superficial analyses, and justifications that do little for environmental quality. To be effective, impact prediction has to be built into design processes from the very beginning. The very term "impact" tends to discourage this approach because it has such a negative sound, suggesting as it does a mailed fist hammering away at the environment while failing to import any possibility of beneficial effects. Since it is so commonly accepted, I will have to use it anyway, recognizing that if we include beneficial impacts as well, we will have a powerful direction for design. Most of this book has been devoted to describing approaches to predictive design, and we can see by now that these approaches involve methods, techniques, and even ways of thinking that played no part in design practice until very recent times. Uncertain though they may be, predictions are an important prerequisite to responsible action.

Suitability models are predictive devices that can be applied to specific impact prediction. The sensitivity model developed for the High Meadow project, for example, was transformed into an impact analysis by laying the final plan over the model and making the chart that summarizes the impacts for each homogeneous land area. With some detail added, it became the heart of the environmental impact report.

IMPACT-PREDICTING SUITABILITY MODELS

It is also possible to generate suitability models using the maximizing or minimizing of specific impacts as modeling criteria. To understand this approach, we might begin by envisioning a specific piece of land with its varying attributes, the developmental actions that are brought to bear when it is converted to some human use, and the resulting environmental changes as three interacting sets of factors.

Given the proposed developmental actions and the location, we can make predictions of environmental impact. Given a set of potential environmental impacts to be controlled, minimized, or maximized and a set of developmental actions, we can determine the most suitable locations. Following this line of thought, we can integrate the landscape planning and impact-prediction processes in a way that permits us to plan directly on the basis of expected consequences.

Such predictions are neither precise nor entirely reliable, of course. Our ability to predict environmental impacts remains very general and crude, and although it is likely to improve in the future, progress will probably be slow. Because of the role of chance in the natural world, we know that our predictions can never be exact. There will always be the unexpected hurricane or infestation or meteorite. Impact prediction, then, is a matter of probability rather than certainty; as we learn more about the relationships between developmental actions and environmental consequences, we will very likely begin to predict impacts with odds of the kind gamblers use in betting on horse races. Until ecology achieves that level of sophistication, however, such precise statements of probability will remain beyond our ability. (We are, of course, dealing with far more complex combinations of possibilities than obtained at the track.)

Another difficulty lies in the obscurity of cause-and-effect relationships. We find a great many events in nature that are clearly related, that occur in conjunction or in sequence, but for which we can show no definite cause-and-effect connection. An example is the relationship between diversity and stability discussed at length in Chap. 11. In fact, the whole notion of cause and effect is increasingly called into question the more it becomes clear that few events have single causes, that chance plays a much greater role in nature than anyone used to think, and that the importance of feedback is fundamental.

This does not mean that impact prediction is not possible or feasible, but it does mean that it is naive to think in terms of one action bringing about one impact. Rather, we need to think in terms of changing the environment in a way that will increase the likelihood of certain combinations of events. Some impacts are certain, of course. If we pave over prime agricultural soil, it will be lost to productive use, but if we pave over the habitat of a particular wildlife species, it may simply move elsewhere or it may be eliminated entirely. If it is a rare, endangered, or especially important species, it is probably not wise to take the risk.

The greater risk, however, is that events will combine in ways that we did not, or could not, predict—ways that produce seriously damaging impacts. Since this is entirely possible and even likely in some situations, some environmentalists habitually argue against all land development. Our powers of prediction, they say, are simply not good enough to be sure of preventing disaster.[1] As we argued in Chap. 9, this is the very reason why the management phase is so important. Especially in valuable and sensitive environments, the uncertainty of our predictions makes feedback loops and mechanisms for correcting mistakes absolutely essential.

With the understanding, then, that impact prediction deals with fields of change rather than simple cause and effect and with what we believe to be the most likely events rather than with certainty, we will discuss several types of suitability models based on impact predictions.

The first of these was prepared for the first case study in this book: San Elijo Lagoon. The criteria for the computer-generated suitability models shown on page 6 were the minimization of certain impacts on the ecosystem in conjunction with the feasibility of the use in question. That is, the most suitable sites for the activity were considered to be those that offered the attributes needed for it and at the same time provided for minimum levels of the major impacts. Since the two overriding issues in this planning effort were filling the wetlands for residential development and preserving wildlife habitats, these became the most important criteria for the model. Specific predictions of the impacts of wetland filling on the wildlife populations were needed. The impact matrix on page 8 shows how these impacts are likely to occur in a long chain of events. An advantage of the matrix format is that it can show how events combine to bring about other events in a network of impacts. The sum-total of the causes and effects here amounts to the destruction of the lagoon as a functional ecosystem and the disruption of all its wildlife habitats. The only way to avoid the disruption is to avoid disturbing the tidal marsh areas. The other major impact to be avoided is siltation of the lagoon. Avoiding siltation means minimizing erosion on the slopes. It also means controlling the inflow of sediments from the watershed (a concern outside the boundaries of this model).

In this way, then, much of the understanding of the function and structure of the estuarine ecosystem that we have discussed was brought to bear in the suitability models. It is important to recognize, however, that suitable locations for human uses in this situation are not enough to assure a healthy ecosystem. As in most other situations, suitable locational distribution provides only a framework, a framework that has to be augmented by design and management control over structure and function.

[1] See Ehrenfeld, 1972.

Quantified Models

For the purposes of such processes, more precise predictions of consequences are often useful. If hazardous land is developed, how much damage is likely to be done? What will the cost be? How much will mitigation measures cost? Sometimes we can answer questions like these with mathematical simulation models in combination with the land resource inventory. A relatively simple example is the formula for estimating stormwater runoff, as follows:

$$Q = CIA$$

where:

Q	=	peak runoff rate
C	=	coefficient of runoff
I	=	rainfall intensity in depth per hour for a peak rainstorm
A	=	area of the watershed

The coefficient of runoff, being the percentage of rainfall that runs off the land, varies according to soil infiltration rates, vegetation cover, and extent of paved areas. Usually, it is estimated, or guessed at, for large areas. Given a computerized land resource inventory, however, fairly precise calculations of C can be made on the basis of the conditions actually present in each grid cell. These can be aggregated for areas featuring similar conditions to produce a more exact value for each variable in the formula. The same process can also be used to predict the changes in runoff rates that will result from changes in land use. The runoff resulting from one pattern of use can then be compared with that from others, as well as with the pattern of the natural situation.

This approach can, of course, be carried to much higher levels of sophistication. The model for flood plain management developed by Hopkins, et al., for example, predicts levels of flooding and extent of flood damage to be expected for alternative land-use patterns and then compares the net costs of various flood control policies.[2] The principal use of this model is for comparing the costs of runoff controls, the latter usually involving detention structures and flood plain controls, which in turn involve prohibitions on building in flood plains.

To give some idea of the workings of models of this type, I will briefly summarize the development of this one. The first step was to divide the watershed into sub-basins, each of which drains into a particular reach of the stream network. The next was to define possible land uses in terms of hydrologic characteristics (those affecting runoff, such as amount of impervious surface) and economic value. The

[2] See Hopkins, et al., 1980.

third was to calculate, using a hydrograph, the runoff at particular storm levels for each land use in each sub-basin. Flooding levels were then calculated for each reach of the stream by combining its runoff with flows from upstream reaches. Given these flooding levels, the extent of damage to each land use can be predicted by applying a damage function, which relates damage to depth of flooding. The economic value of a particular land use can then be calculated as the sum of land rents for that use minus the annual cost of the predicted damages.

Mathematical models of this type are fundamentally different from the types of suitability models previously discussed in two ways. First, they deal with single factors—runoff levels, soil loss, flood plain management—rather than with holistic combinations of factors. Given this limited range, they are able to consider their subjects in greater detail. The limited range also makes possible the second fundamental difference: quantification. Mathematical models deal with measurable quantities (cubic feet per second, dollars) rather than relative ones (high, medium, low), that is, with ratio scales rather than with intervals.

The detail and quantification of mathematical models offer a number of clear advantages. They can be used to establish enforceable limits of tolerance like water quality or emission standards. Being empirically verifiable—at least in theory—they provide data that can often be translated into economic terms. Perhaps most important, they have a feel of objective solidity, a foundation in reality that relative measures often lack.

On the other hand, mathematical models have some serious shortcomings. Even the simplest ones, like the runoff and soil loss models discussed, require data that are rarely available. The more complex ones require sophisticated programming as well as immense quantities of data, and are thus very expensive. Furthermore, the results are usually difficult to interpret. The information they provide may be too complex to be easily assimilated into the decision-making apparatus. Moreover, even the most sophisticated models require radical simplifications of reality. Especially for the more complex ecological issues, the simplification may so distort reality as to make the results meaningless or misleading. This charge has been convincingly leveled at the famous Limits to Growth models of the Club of Rome.[3] The same is true of the more inclusive suitability models, of course, but their relative values and tonal or color scales do not make any claims to high levels of precision.

Moreover, there are strong advantages in the holistic models that seek to deal with ecosystems as wholes by combining diverse variables, however relatively and crudely. Since they reflect the interrelated unity of nature, they force those working with them to keep in mind that their subject is, after all, indivisible. When we separate certain parts out of the whole for analysis, we risk losing their connections with the larger context, thus risking a repetition of one of the great fallacies of industrial-age engineering. Moreover, we tend to become preoccupied with the parts to the exclusion of the whole and particularly with the quantifiable to the exclusion of the ineffable.

[3] See Meadows, et al., 1972.

Still, despite these difficulties, and if their limitations are borne in mind, mathematical models have considerable utility and enormous promise for making predictions about land suitability.

The Metropolitan Boston Case Study

The Metropolitan Boston information system attempts to be both quantitative and holistic. It uses 28 different mathematical simulations for predicting the consequences of alternative patterns of suburban growth. It thus combines the standard economic modeling approach used in urban planning with similar techniques that deal with ecological concerns at the regional scale. The Boston information system was developed by Carl Steinitz and an interdisciplinary group at Harvard under a grant from the Research Applied to National Needs (RANN) Program of the National Science Foundation.[4] The subject area is the rapidly suburbanizing southeastern sector of the Boston metropolitan area, which includes eight towns.

The models are of two basic types: *allocation models,* which assign specific land uses to particular locations on the basis of certain rules, and *evaluation models,* which then predict the environmental, fiscal, and demographic impacts of these allocations.

Allocation models are included for six land uses: housing, industry, commerce, public institutions (schools), conservation, and recreation. A public expenditure model, which predicts budget size and investment priorities for each town, and a solid waste model, which predicts amounts of solid waste produced by alternative patterns of population and land use, are also included in this category, even though they are not land uses.

The rules on which allocations are based are mostly economic. For example, the housing model assigns residential development to the most profitable locations, which are determined by estimating costs and selling prices at the available sites. The commerce model allocates commercial uses to areas with high potential sales volume and a lack of existing facilities. The conservation model considers locations already identified as areas of critical environmental concern in existing legislation; these include coastal and inland wetlands, beaches, dunes, estuaries, flood plains, unstable soils, ecosystems deemed rare or valuable, watershed protection areas, and designated scenic and historic areas. These locations are ranked and allocated so as to protect the maximum number of resources on the basis of costs and user preferences.

Evaluation models are used to predict impacts on soils, water quality, water quantity, vegetation, wildlife habitats, critical resources (those included in the conservation model), visual quality, transportation, air quality, noise quality, historic resources, land values, public fiscal accounting, and demographic patterns. A legal implementation model contains a list of existing land-use laws, which is used to regulate model operations and to warn of violations.

The inventory, or data base, uses a grid-cell storage and mapping system with 1-hectare (2.47-acre) cells. It includes four types of data: physiographic and natural systems,

[4] See Steinitz, et al., 1976, 1981.

man-made land use and land cover, political jurisdictions, and functional zones and data that is created by combining the first three types or derived from the first type.

This combination of allocative and evaluative models with the geographic data base, along with the program routines for linking all three, provides great flexibility for a variety of operations. The authors list nine different operational modes. This list is highly indicative of the capabilities of information systems of this type:

1. The Pre-analysis Mode provides information from the data base about existing conditions and combinations of conditions in the study area.
2. The Single Model Analysis uses the allocative models to locate particular activities on the basis of the criteria built into each model.
3. The Project Evaluation Mode uses the evaluation models to predict impacts of particular land-use allocations within the range of the model's capabilities.
4. The Plan Evaluation Mode uses the evaluation models to predict impacts of proposed land-use plans.
5. The Sensitivity Analysis determines the sensitivity of the study area to given value changes by changing a single assumption, value, or parameter and repeating the operation for each.
6. The Gaming Mode predicts results of assumed planning strategies and values.
7. The Optimizing Mode first determines a maximum or minimum impact level within one area of concern and then runs other models so as to maintain that impact at a constant level.
8. The Legal Testing Mode predicts the results of enforcing existing laws or new ones.
9. The Alternative Strategy Simulation allocates land uses according to a given planning strategy and evaluates the resulting pattern using all of the evaluation models.

The Boston case study includes illustrations of several allocation maps produced by this system and summaries of several predictive evaluations.

Simulation and Suitability: Some Questions

The Metropolitan Boston System shows the flexibility and the sophistication that can be achieved through the linkage of mathematical simulation models with a land-use inventory. Although this is a powerful and versatile tool, the question of the breadth of its potential applications remains open. In fact, there are a number of questions concerning the applications of geographic information systems that will probably be answered by repeated use over the long term. Among the more important questions are these:

Are the predictions provided by mathematical simulations accurate enough to provide reliable bases for decisions? As mentioned earlier, they represent vast simplifications of reality, perhaps to the point of not being accurate enough to be useful.

Is enough sound data available to give meaningful results? For the geographic variables, data are rapidly becoming available to the point of saturation, but other information remains hard to obtain. Crop management factors and conservation practice factors are difficult to measure with sufficient accuracy for use in the soil loss formula, for example. In the testing of the flood plain management model, it was necessary to use very roughly estimated bid price data because of the difficulty of obtaining real data.[5]

Do they answer the right questions? The Metropolitan Boston System is extremely flexible in its ability to answer a range of question, but only putting the answers into practice will tell whether these were the questions that needed to be answered and whether the answers were couched in truly useful terms. It is important to remember that planning decisions are fundamentally political. The answers may be of no help here.

Will planning agency staffs be able to adapt to highly complex information systems? The answer so far has often been no. Major reorientations will be needed to accommodate what amounts to a whole new foundation for planning. If the advantages seem great enough, the accommodation can be made.

All these questions can be, and have been, debated at great length. There is no reason to join the debate here. Only time will give significant answers. Very likely, information systems will be extensively used by some agencies for some types of planning and very little or not at all used by others. In any case, uses will remain for a broad range of suitability modeling techniques, from the very simple to the very sophisticated. More complex is not necessarily better. In fact, as we have repeatedly argued, the simplest technique and the least information capable of providing the desired results are usually the best alternative in any given situation. One should also not forget the problem of communication. Sophisticated models are difficult to understand and can thus limit participation in the design process.

In actuality, the situation at hand should determine the choice of model type and its complexity. Although the range of types discussed here is already considerable, given the variety of planning situations, different approaches will probably continue to emerge. It is important that landscape designers not feel restricted by formulas or typologies. Rather, they should feel free to explore any ways that might better deal with the issues at hand, remembering always the need for conceptual validity and the need to be no more obscure, convoluted, perplexing, or perplexed than is absolutely necessary.

[5] See Hopkins, et al., 1978.

CHAPTER 16

A New Era of Design?

The material in this book represents our progress to date in learning how to design sustainable human ecosystems. None of these concepts or techniques is perfect; all can bear improvement. Nevertheless, I do believe a fruitful direction has been established. We are beginning to learn to design the landscape in a responsible way. To act responsibly, we need to have some understanding of the consequences of our actions, that is, to make predictions. Ecosystematic design is based on predictions of the structural and functional forms that will take shape in the landscape following our actions.

In the perspective of history, we might even view this predictive approach as a new phase in the shaping of the physical environment. Looking back, we can see at least three earlier phases that brought us to where we are now.

First, there is the instinctive phase, which is still the way of all nonhuman species. Building is done without plan or forethought, by following programs somehow implanted in the genes. Some incredibly complex, technically sophisticated creations have come about in this way. Witness the spider web, the termite hill, and the beehive.

If human beings once had such instincts, they are considerably less effective now than those of the spider, the termite, or the honeybee. The earliest living environments made by man, and those still being built by primitive peoples today, reflect the next phase, which is shaped by tradition, or the application of accumulated knowledge. Each generation passes on its store of technical wisdom to the next, which then builds in essentially the same way, usually with minor improvements. Thus, the traditional forms are both conserved and continuously bettered. Over time, they achieve an environmental fit that serves their purpose well.

Next is the form-making phase, in which the completed form is envisioned and described on paper before it is built. Human capacities for reason and insight are brought to bear, and a high level of invention becomes possible. This phase reached its highest development between the beginning of the Renaissance and our own day. At their best, the results are as formally logical and symbolically meaningful as Villa Lante or Versailles, but they seldom achieve the environmental fit.

Finally, the emerging era of predictive adaptation uses the skills developed in earlier phases and adds to them our abilities for storing knowledge and for technological manipulation, for reason, insight, and invention, for predicting the behavior of an imagined form in relation to its environment, and for shaping it on that basis. It is unlikely that prediction in design, occurring as it does in the infinitely variable world of reality, will ever be entirely deterministic. We have a great many tools and techniques, considerable volumes of information, many data—all useful, even indispensable—to help us in our predicting, but there is no *method,* no formula for success. The intangible and the ineffable always enter in. Even though the possibilities are infinite, we need to be willing to explore them. The exploring involves both halves of the brain: imagining future landscapes and analyzing their behavior. I believe it is one of the promising achievements of our times that so many people are taking part in the imagining and predicting. Genuinely participatory processes are beginning to take place, in design if not in government, but these do not simplify or clarify design processes. Rather they add layer upon layer of varied values and perceptions.

It is important to recognize that, however sophisticated our tools and techniques, the unexpected can always occur. With prediction goes uncertainty. Recognizing the inevitability of uncertainty and having done all we can to shape a system that will behave as predicted, we need to minimize the possibility of catastrophe when it does not. We need to be as sure as humanly possible that when the unexpected does occur, it does not bring about a disaster. This means checks, balances, feedback, complexity, sometimes a conservative stance. In ecosystem design, the whole stake should never ride on a roll of the dice.

The second quality we must ensure is a capacity for adaptive change. Creative change should always be possible as a designed system interacts with its environment. Herein lies the importance of the linkage between design and management. Forms should be pliable, not rigid. The Taoist way is really the way of human ecosystems, although we also have to adapt to an economic world that operates in a generally Confucian way. All in all, we have not learned to do it very well yet, but I believe we have at least reached the point where we know what we are trying to do.

REFERENCES

Alden, Jeremy, and Morgan, Robert. 1974. *Regional Planning: A Comprehensive View.* New York: John Wiley and Sons.

Alexander, Christopher. 1964. *Notes on the Synthesis of Form.* Cambridge: Harvard University Press.

Allen, Robert. 1980. *How to Save the World.* Totowa, New Jersey: Barnes and Noble.

Appleton, Jay. 1975. *The Experience of Landscape.* New York: John Wiley and Sons.

Armstrong, J. T. "Breeding Home Range in the Nighthawk and Other Birds: Its Evolutionary and Ecological Significance." In *Ecology* 46, 1965.

Assagioli, Roberto. 1971. *Psychosynthesis.* New York: The Viking Press.

Barney, Gerald, ed. 1980. *The Global 2000 Report to the President of the U.S.: Entering the 21st Century.* Vols. I and II. New York: Pergamon Press.

Bastian, Robert K. "Natural Treatment Systems in Wastewater Treatment and Sludge Management." In *Civil Engineering*, May, 1982.

Bennett, John. 1976. *The Ecological Transition: Cultural Anthropology and Human Adaptation.* New York: Pergamon Press.

Bente, Paul F., Jr. *The Food People Problem: Can the Land's Capacity to Produce Be Sustained?* Washington, D.C.: Council of Environmental Quality, 1977.

Bernatzky, A. "The Performance and Value of Trees." In *Anthos*. No. 1, 1969.

Berry, Wendell. 1977. *The Unsettling of America: Culture and Agriculture.* New York: Avon Books.

Boer, K. "Tree Planting Reconsidered." In *Landscape Architecture*. Vol. 62, No. 2, 1972.

Bormann, F. H., and Likens, G. E. "The Watershed-Ecosystem Concept and Studies of Nutrient Cycles." In *The Ecosystem Concept in Natural Resource Management*, ed. George M. van Dyne. New York: Academic Press, 1969.

——— "The Fresh Air–Clean Water Exchange." In *Natural History*. Vol. 86, No. 9, 1977.

Braun-Blanquet, Josias. 1964. *Pflanzensoziologie, Grundzugie de Vegetationskunde*, 3rd ed. Vienna: Springer.

Braun-Blanquet, Josias, and Furrer, Ernst. "Remarques sur l'etude des groupements de plants." *Bulletin Societe Languedocienne de Geographie* 36, 1913.

Bronowski, J. 1973. *The Ascent of Man.* Boston: Little, Brown, and Company.

Brown, Lester. 1978. *The Twenty-Ninth Day.* New York: W.W. Norton Co.

Brown, Lester R., with Eckholm, Erik P. 1974. *By Bread Alone.* New York: Praeger Publishers.

Brubaker, Charles William, and Sturgis, Robert. "Urban Design and National Policy for Urban Growth." In *American Institute of Architects Journal*, October, 1969.

Bruner, Jerome. 1965. *On Knowing: Essays for the Left Hand.* New York: Atheneum.

Bryson, Reid A., and Ross, John E. "The Climate of the City." In *Urbanization and Environment*, ed. Thomas R. Detweiler and Melvin G. Marens. Belmont, California: Duxbury Press, 1972.

Bureau of Land Management. *The California Desert Conservation Area Plan Alternatives and Environmental Impact Statement.* Draft. Riverside, California, 1980.

Churchman, C. West. 1979. *The Systems Approach and Its Enemies.* New York: Basic Books.

Clements, Frederic E. "Plant Succession, An Analysis of the Development of Vegetation." *Carnegie Institution of Washington Publications* 242, 1916.

Cooper, Charles F., and Zedler, Paul H. "Ecological Assessment for Regional Development." In *Journal of Environmental Management* 10, 1980.

Cranz, Galen. "The Useful and the Beautiful: Urban Parks in China." In *Landscape*. Vol. 23, No. 2, 1979.

Curtis, J. T. 1959. *The Vegetation of Wisconsin.* Madison: The University of Wisconsin Press.

Dangermond, Jack; Derrenberger, Bill; and Harnden, Eric. "Description of Techniques for Automation of Regional Natural Resource Inventories." Redlands, California: Environmental Systems Research Institute, Inc., 1982.

Daubenmire, R. F. "Vegetation: Identification of Typal Communities." In *Science* 151, 1966.

Deithelm and Bressler, Inc. 1980. *Mt. Bachelor Recreation Area: Proposed Master Plan.* Eugene, Oregon.

Department of Conservation, State of Illinois. *Illinois Streams Information System.* Urbana, 1982.

Diesing, Paul. 1962. *Reason in Society.* Urbana: The University of Illinois Press.

Dubos, Rene. 1980. *The Wooing of Earth.* New York: Charles Scribner's Sons.

Duckworth, F., and Sandberg, J. "The Effect of Cities Upon Horizontal and Vertical Temperature Gradients." *Bulletin of the American Meteorological Society* 35, 1954.

Earle, Eliza. *Conservation of Natural Areas in the Context of Urban Development.* Unpublished M.L.A. thesis. Pomona, California: State Polytechnic University, 1975.

Eberhard, John P. "A Humanist Case for the Systems Approach." In *American Institute of Architects Journal*, July, 1968.

Eckholm, Eric. 1976. *Losing Ground.* New York: Norton.

Ehrenfeld, David. 1972. *Conserving Life on Earth.* New York: Oxford University Press.

Ehrlich, Paul, and Ehrlich, Anne. 1981. *Extinction*. New York: Random House.

Emmelin, Lars, and Wiman, Bo. 1978. *The Environmental Problems of Energy Production*. Stockholm: Secretariat for Futures Studies.

Environmental Protection Agency. "Aquaculture Systems for Wastewater Treatment/Seminar Proceedings." EPA 430/9-80-006, MCD-68. September, 1979.

——— "Aquaculture Systems for Wastewater Treatment—An Engineering Assessment." EPA 430/9-80-007, MCD-68. June, 1980.

Evans, Francis C. "Ecosystem as the Basic Unit in Ecology." In *Science* 123, 1956.

Fabos, Julius G. 1979. *Planning the Total Landscape: A Guide to Intelligent Land Use*. Boulder, Colorado: Westview Press.

Fabos, Julius G., and Caswell, Stephanie J. *Composite Landscape Assessment*. Research Bulletin No. 637. Amherst: Massachusetts Agricultural Experiment Station, Amherst, 1977.

Holling, C. S., and Clark, William C. "Notes toward a Science of Ecological Management." In *Unifying Concepts in Ecology*, eds. van Dobben, W. H., and Lowe-McConnell, R. H. The Hague: Dr. W. Junk B.V. Publishers, 1975.

Holloway, Cecile, and Yarbrough, Janet. *San Diego Wild Animal Park Land Use Study*. Pomona: Department of Landscape Architecture, California State Polytechnic University, 1978.

Hopkins, Lewis. "Methods for Generating Land Suitability Maps." In *Journal of the American Institute of Planners*. Vol. 43, No. 4, 1977.

Hopkins, Lewis D.; Brill, E. Downey, Jr.; Liebman, Jon C.; and Wenzel, Harry G., Jr. "Land Use Allocation Model for Flood Control." In *Journal of Water Resources Planning and Management Div*. ASCE. Vol. 104, No. WRI, 1978.

Hopkins, Lewis D.; Goulter, Ian C.; Kurtz, Kenneth B.; Wenzel, Harry G., Jr.; and Brill, E. Downey, Jr. *A Model for Floodplain Management in Urbanizing Areas*. Final Report. Water Resources Center, University of Illinois, Urbana, 1980.

Hough, Michael. "Metro Homestead." In *Landscape Architecture*. Vol. 73, No. 1, 1983.

Hutchinson, Boyd A.; Taylor, Fred G.; and the Critical Review Panel. "Energy Conservation Mechanisms and Potentials of Landscape Design to Ameliorate Building Climates." In *Landscape Journal*. Vol. 2, No. 1, 1983.

Hutchinson, Boyd A.; Taylor, Fred G.; and Wendt, Robert L. 1982. *Use of Vegetation to Ameliorate Microclimates: An Assessment of Energy Conservation Potentials*. Oak Ridge National Laboratory Environmental Sciences Division.

Hyams, Edward. 1976. *Soil and Civilization*. New York: Harper Colophon Books.

Jantsch, Erich. 1975. *Design for Evolution*. New York: George Braziller.

Janzen, Daniel. "Tropical Agroecosystems." In *Science* 182, 1973.

Johansson, Thomas B., and Steen, Peter. 1978. *Solar Sweden: An Outline to a Renewable Energy System*. Stockholm: Secretariat for Futures Studies.

Johnson, Arthur H., and Hester, James S. "Nitrate as a Consideration in Planning Future Land Use." In *Landscape Planning* 8, 1981.

Jokela, Alice Tang, and Jokela, Arthur W. "Water Reclamation, Aquaculture, and Wetland Management." Paper delivered at Coastal Zone '78, Symposium on Technical, Environmental and Regulatory Aspects of Coastal Zone Planning and Management. San Francisco, California. March 14–16, 1978.

Jokela, Arthur. "Institutional Ecology: Another Agenda for the Landscape Planner." In *Landscape Research*. Vol. 5, No. 1, 1979.

Kilpack, Charles. "Computer Mapping, Spatial Analysis, and Landscape Architecture." In *Landscape Journal*. Vol. 1, No. 1, 1982.

——— 1982. *Innovating Landscape Architecture*. Washington, D.C.: Architecture Technical Information Series, American Society of Landscape Architects.

Knowles, Ralph. 1975. *Energy and Form*. Cambridge: Massachusetts Institute of Technology Press.

Koenig, Herman E. "Engineering for Ecological, Sociological, and Economic Compatibility." In *IEEE Transactions on Systems, Man and Cybernetics*. Vol. SMC 2, No. 1, 1972.

Koenig, Herman E.; Cooper, William; and Falvey, James M. "Industrialized Ecosystem Design and Management." Unpublished working paper. East Lansing: Michigan State University, 1971.

Fabos, Julius G., and Joyner, Spencer A., Jr. "Landscape Plan Formulation and Evaluation." In *Landscape Planning* 7, pp. 95-119, 1980.

Falini, Paola E.; Grifoni, Cristina; and Lomoro, Annarita. "Conservation Planning for the Countryside: A Preliminary Report of an Experimental Study of the Terni Basin (Italy)." In *Landscape Planning* 7, 1980.

Federer, C. A. "Trees Modify the Urban Microclimate." In *Journal of Arboriculture* 2. 1976.

Feibleman, James K. "Theory of Integrative Levels." In *British Journal of the Philosophy of Science*. Vol. V, No. 17, 1954.

Friedmann, John. "Introduction" (to issue on regional planning). In *Journal of the American Institute of Planners*. Vol. 30, No. 2, 1964.

——— "Regional Planning as Field of Study." In *Regional Development and Planning*, eds. Friedmann, John and Alonso, William. Cambridge: The M.I.T. Press, 1964.

Frissell, Sidney S.; Lee, Robert G.; Stankey, George H.; and Zube, Ervin H. "A Framework for Estimating the Consequences of Alternative Carrying Capacity Levels in Yosemite Valley." In *Landscape Planning* 7, 1980.

Geddes, Patrick. 1915. *Cities in Evolution*. London: Ernest Benn Ltd.

Geiger, Rudolph. 1965. *The Climate Near the Ground*. Cambridge: Harvard University Press.

Geis, Aelred. "The New Town Bird Quadrille." In *Natural History*. Vol. XXXIII, No. 6, 1974.

Giedion, Sigfried. 1967. *Space, Time and Architecture*, 5th ed. Cambridge: Harvard University Press.

Gill, Don, and Bonnett, Penelope. 1973. *Nature in the Urban Landscape*. Baltimore: York Press.

Gilpin, M. S., and Diamond, J. M. "Subdivision of Nature Reserves and the Maintenance of Species Diversity." In *Nature* 285, 1980.

Gold, A. J. "Design with Nature: A Critique." In *Journal of the American Institute of Planners*. Vol. 40, No. 4, 1974.

Golley, F. B. "Energy Flux in Ecosystems." In *Ecosystem Structure and Function*, ed. Wiens, J. A. Corvallis: Oregon State University Press, 1971.

Gomez-Pompa, A.; Vazquez-Yanes, C.; and Guevara, S. "The Tropical Rain Forest: A Nonrenewable Resource." In *Science* 176, 1972.

Gottmann, Jean, 1961. *Megalopolis*. Cambridge: M.I.T. Press.

Haller, J. R. "Vegetation of Central America." Mimeograph. University of California, Santa Barbara, n.d.

Hartman, Frederick. "The Chicago Forestry Scheme." In *Natural History*. Vol. LXXXII, No. 9, 1973.

Hathaway, Melvin G. "The Ecology of City Squirrels." In *Natural History*. Vol. LXXXII, No. 9, 1973.

Higgs, A. J., and Usher, M. B. "Should Nature Reserves Be Large or Small?" In *Nature* 285, 1980.

Hills, Angus. "The Classification and Evaluation of Land for Multiple Uses." In *Forestry Chronicles*. Vol. 42, No. 2, 1966.

Holdridge, Leslie R. 1971. *Forest Environments in Tropical Life Zones*. New York: Pergamon Press.

Holdridge, Leslie R., and Tosi, Joseph A., Jr. *Report on the Ecological Adaptability of Selected Economic Plants for Small Farm Production in Six Regions of Costa Rica*. San José, Costa Rica: Tropical Science Center, n.d.

Krebs, Charles J. 1978. *Ecology: The Experimental Analysis of Distribution and Abundance*. New York: Harper & Row.

Küchler, A. W. In *Goode's World Atlas*, ed. Espenshade, E. B., Jr. Chicago: Rand-McNally, 1960.

——— 1964. *Potential Natural Vegetation of the Coterminous United States*. New York: American Geographical Society.

Kuehn, E. "Planning the City's Climate." In *Landscape*. Vol. 8, No. 3, 1959.

Landsberg, H. E. "The Climate of Towns." In *Man's Role in Changing the Face of the Earth*, ed. Thomas, William L. Chicago: University of Chicago Press, 1956.

Leopold, Aldo. "The Conservation Ethic." In *Journal of Forestry* 31, 1933.

——— 1936. *Game Management*. New York: Charles Scribner's Sons.

Lewis, Charles A. "Comment: Healing in the Urban Environment." In *American Planning Association Journal*, July, 1979.

Lewis, Phillip H., Jr. "Quality Corridors." In *Landscape Architecture*. Vol. 54, No. 1, 1964.

——— "The New Landscape Challenge." In *Agora*, Autumn, 1982.

Likens, Gene E., and Bormann, F. Herbert. "Nutrient Cycling in Ecosystems." In *Ecosystem Structure and Function*, ed. Weins, John A. Corvallis: Oregon State University Press, 1972.

Lösch, August. "The Nature of Economic Regions." In *Regional Development and Planning*, ed. Friedmann, John, and Alonso, William. Cambridge: The M.I.T. Press, 1964.

Lovelock, J. E. 1980. *Gaia: A New Look at Life on Earth*. New York: Oxford University Press.

Lovins, Amory B. 1977. *Soft Energy Paths: Toward a Durable Peace*. Cambridge: Ballinger Publishing Company.

Lowry, Ira S. "A Short Course in Model Design." In *Journal of the American Institute of Planners*. Vol. 21, No. 2, 1965.

Lyle, John, and von Wodtke, Mark. "An Information System for Environmental Planning." In *Journal of the American Institute of Planners*. Vol. 40, No. 6, 1974.

——— "Design Methods for Developing Environmentally Integrated Urban Systems." In *DMG-DRS Journal: Design Research and Methods*. Vol. 8, No. 3, 1974.

Lynch, Kevin. 1971. *Site Planning*, 2nd ed. Cambridge: M.I.T. Press.

MacArthur, R. H., and Wilson, E. O. 1967. *The Theory of Island Biogeography*. Princeton: Princeton University Press.

MacDougall, E. Bruce. "The Accuracy of Map Overlays." In *Landscape Planning*. Vol. 2, No. 1, 1977.

——— "Fuzzy Set Evaluation in Landscape Assessment." In *Regional Landscape Planning Proceedings*. Washington, D.C.: Annual Meeting of the American Society of Landscape Architects, 1981.

McHarg, Ian L. 1969. *Design with Nature*. New York: Natural History Press.

——— "Human Ecological Planning at Pennsylvania." In *Landscape Planning*. Vol. 8, No. 2, 1981.

McHarg, Ian L., and Sutton, Jonathan. "Ecological Plumbing for the Texas Coastal Plain." In *Landscape Architecture*. Vol. 65, No. 1, 1975.

MacKaye, Benton. 1962. *The New Exploration: A Philosophy of Regional Planning*. Urbana, Illinois: University of Illinois Press.

Margalef, Ramon. "On Certain Unifying Principles in Ecology." In *American Naturalist* 97, 1963.

——— 1968. *Perspectives in Ecological Theory*. Chicago: University of Chicago Press.

——— "Diversity, Stability and Maturity in Natural Ecosystems." In *Unifying Concepts in Ecology*, eds. van Dobben, W. H., and Lowe-McConnell, R. H. The Hague: Dr. W. Junk B.V. Publishers, 1975.

Marquis, Stewart. *Urban-Regional Ecosystems: Toward a Natural Science of Cities*. Unpublished manuscript. School of Natural Resources, University of Michigan, 1974.

Martin, Alexander C.; Zim, Herbert S.; and Nelson, Arnold L. 1951. *American Wildlife and Plants: A Guide to Wildlife Food Habits*. New York: McGraw-Hill Book Company, Inc.

Marušič, Ivan. "Landscape Planning Methods in the U.S.A.: An Outside View." In *Landscape Research*. Vol. 5, No. 1, 1979.

May, R. M. "Stability in Ecosystems: Some Comments." In *Unifying Concepts in Ecology*, eds. van Dobben, W. H. and Lowe-McConnell, R. H. The Hague: Dr. W. Junk B.V. Publishers, 1975.

——— 1973. *Stability and Complexity in Model Ecosystems*. Princeton: Princeton University Press.

Meadows, Donella H.; Meadows, Dennis L.; Randers, Jorgen; and Behrens, William W. 1972. *The Limits to Growth*. New York: Universe Books.

Meier, Richard L. 1974. *Planning for an Urban World: The Design of Resource Conserving Cities*. Cambridge: M.I.T. Press.

Meinig, D. W. "The Beholding Eye: Ten Versions of the Same Scene." In *Landscape Architecture*. Vol. 66, No. 1, 1976.

Meldan, R. "Besondere luftechnische Aufgaben der Industrie." *Staedtehygiene*. Vol. 3, No. 8, 1959.

Mesarovîc, Mihaljo, and Pestel, Edouard. 1974. *Mankind at the Turning Point*. New York: E. P. Dutton and Company.

Moore, Russell T.; Ellis, Scott L.; and Duba, David R. "Advantages of Natural Successional Processes on Western Reclaimed Lands." *Papers Presented at the Fifth Symposium on Surface Mining and Reclamation*. Louisville: National Coal Association, 1977.

Mumford, Lewis. "The Natural History of Urbanization." In *Man's Role in Changing the Face of the Earth*, ed. Thomas, William L. Chicago: University of Chicago Press, 1956.

Needham, Joseph. 1954. *Science and Civilization in China*. Vol. IV:3. Cambridge University Press.

Novikoff, Alex B. "The Concept of Integrative Levels and Biology." In *Science* 101, 1945.

Odum, Eugene P. "Strategy of Ecosystem Development." In *Science* 164, 1969.

——— 1971. *Fundamentals of Ecology*, 3rd ed. Philadelphia: W. B. Saunders Company.

——— "Diversity as a Function of Energy Flow." In *Unifying Concepts in Ecology*, eds. van Dobben, W. H., and Lowe-McConnell, R. H. The Hague: Dr. W. Junk B.V. Publishers, 1975.

Odum, Howard T. "Energy Quality and Carrying Capacity of the Earth." In *Tropical Ecology* 16, 1976.

——— "Net Energy Analysis of Alternatives for the United States." Testimony delivered before the U.S. House of Representatives Subcommittee on Energy and Power of the Committee on Interstate and Foreign Commerce, March 25 and 26, 1976.

Odum, Howard T., and Odum, Elizabeth C. 1975. *Energy Basis for Man and Nature*. New York: McGraw-Hill Book Company.

Orians, Gordon H. "Diversity, Stability and Maturity in Natural Ecosystems." In *Unifying Concepts in Ecology*, eds. van Dobben, W. H., and Lowe-McConnell, R. H. The Hague: Dr. W. Junk B.V. Publishers, 1975.

Ornstein, Robert E. 1972. *The Psychology of Consciousness*. New York: The Viking Press.

Pinkard, H. P. "Trees, Regulators of the Environment." In *Soil Conservation*, October, 1970.

Powell, John Wesley. *Report on the Arid Regions of the United States with a More Detailed Account of the Lands of Utah*. Edited by Stegner, Wallace. Cambridge: The Belknap Press of Harvard University Press, 1962.

Prigogine, Ilya. "Order through Fluctuation: Self-Organization and Social Systems." In *Evolution and Consciousness: Human Systems in Transition*, eds. Jantsch, Erich, and Waddington, Conrad H. Reading, Massachusetts: Addison-Wesley, 1976.

——— "From Being to Becoming." Austin, Texas: University of Texas Center for Statistical Mechanics and Thermodynamics, 1978.

——— "Biological Order, Structure, and Instabilities." In *Review of Biophysics*. Vol. 4, Nos. 2 and 3, 1978.

Reed, Sherwood; Bastian, Robert K.; and Jewell, William J. "Engineers Assess Aquaculture Systems for Wastewater Treatment." In *Civil Engineering*, July, 1981.

Regional Energy Consumption: Second Interim Report of a Joint Study. New York and Washington, D.C.: Regional Plan Association, Inc. and Resources for the Future, Inc., 1974.

Reichle, D. E.; O'Neill, R. V.; and Harris, W. F. "Principles of Energy and Material Exchange in Ecosystems." In *Unifying Concepts in Ecology*, eds. van Dobben, W. H., and Lowe-McConnell, R. H. The Hague: Dr. W. Junk B.V. Publishers, 1975.

Reilly, Michael. *Ecological Uses of Urban Trees*. Unpublished M.L.A. thesis, California State Polytechnic University, 1976.

Richards, P. W. 1952. *The Tropical Rain Forest*. London: Cambridge University Press.

Richardson, Harry Ward. 1979. *Regional Economics*. Urbana: University of Illinois Press.

Rigler, R. H. "The Concept of Energy Flow and Nutrient Flow between Trophic Levels." In *Unifying Concepts in Ecology*, eds. van Dobben, W. H., and Lowe-McConnell, R. H. The Hague: Dr. W. Junk B.V. Publishers, 1975.

Robinette, Gary. 1972. *Plants, People and Environmental Quality*. Washington, D.C.: U.S. Department of the Interior.

Robinson, A. C., et al. "An Analysis of the Market Potential of Water Hyacinth-Based Systems for Municipal Wastewater Treatment." Research Report No. BCL-OA-TFT-76-5. Columbus, Ohio: Battelle Columbus Laboratories, 1976.

Rodiek, Jon. "Wildlife Habitat Management and Landscape Architecture." In *Landscape Journal*. Vol. 1, No. 1, 1982.

Rodiek, Jon, and Wilen, Billy. "Landscape Inventory: A Key to Sound Environmental Planning." In *Landscape Research*. Vol. 5, No. 1, 1979–80.

Saurin, Jean-Pierre. "The Compatability of Conifer Afforestation with the Landscape of the Monts D'Arrei Region." In *Landscape Planning* 7, 1980.

Schlauch, Frederick C. "City Snakes, Suburban Salamanders." In *Natural History*. Vol. LXXXV, No. 5, 1976.

Schram, D., and Grist, C. "Building Site Selection to Conserve Energy by Optimizing Topoclimatic Benefits." *Proceedings for the National Conference on Technology for Energy Conservation*. Rockville, Maryland: Environmental Protection Agency, 1978.

Schweitzer, Albert. *Out of My Life and Work*. 1933. Reprint. New York: Holt, Rinehart and Winston, 1972.

Sharpe, Carl, and Williams, Donald L. "The Making of an Environmental Fit." In *Landscape Architecture*. Vol. 63,

No. 3, 1972.
Soemarwoto, Otto. "Rural Ecology and Development in Java." In *Unifying Concepts in Ecology*, eds. van Dobben, W. H., and Lowe-McConnell, R. H. The Hague: Dr. W. Junk B.V. Publishers, 1975.
Sperry, R. W. "The Great Cerebral Commissure." In *Scientific American,* January, 1964.
Steiner, Frederick; Brooks, Kenneth; and Struckmeyer, Kenneth. "Determining the Regional Context for Landscape Planning." In *Regional Landscape Planning: Proceedings of Three Educational Seminars*. Washington, D.C.: Annual Meeting of the American Society of Landscape Architects, 1981.
Steinitz, Carl. "Simulating Alternative Policies for Implementing the Massachusetts Scenic and Recreational Rivers Act: The North River Demonstration Project. In *Land-scape Planning* 6, 1979.
——— 1978. *Defensible Processes for Regional Landscape Design*. Washington, D.C.: Landscape Architecture Technical Information Series, American Society of Landscape Architects.
Steinitz, C., and Brown, H. J. "A Computer Modeling Approach to Managing Urban Expansion." *Geoprocessing* 1, 1981.
Steinitz, Carl; Brown, H. James; and Goodale, Peter. 1976. *Managing Suburban Growth: A Modeling Approach*. Cambridge: Landscape Architecture Research Office, Harvard University.
Steinitz, Carl; Murray, Timothy; Sinton, David; and Way, Douglas. 1969. *A Comparative Study of Resource Analysis Methods*. Cambridge Department of Landscape Architecture Research Office, Harvard University.
Stilgoe, John. "Suburbanites Forever: The American Dream Endures." In *Landscape Architecture*. Vol. 72, No. 3, 1982.
Stobaugh, Robert, and Yergin, Daniel. 1979. *Energy Future: Report of the Energy Project at the Harvard Business School*. New York: Random House.
Streeter, Robert G.; Moore, Russell T.; Skinner, Janet J.; Martin, Stephen G.; Terrel, Ted L.; Klimstra, Willard D.; Tate, James, Jr.; and Nolde, Michelle, J. "Energy Mining Impacts and Wildlife Management: Which Way to Turn." *Transactions of the 44th North American Wildlife and Natural Resources Conference*. Washington, D.C.: The Wildlife Management Institute, 1979.
Strong, Maurice F. "A Global Imperative for the Environment." In *Natural History*. Vol. 83, No. 3, 1974.
Stuart, Darwin G. *The Systems Approach in Urban Planning*. Planning Advisory Service Report No. 253. Chicago: American Society of Planning Officials, January, 1970.
Study of Man's Impact on Climate (SMIC). 1971. *Inadvertent Climate Modification*. Cambridge: M.I.T. Press.
Sullivan, Arthur L, and Shaffer, Mark L. "Biogeography of the Megazoo." In *Science* 189, 1975.
Tansley, A. G. "The Use and Abuse of Vegetational Concepts and Terms. In *Ecology* 16, 1935.
Teilhard de Chardin, Pierre. 1966. *Man's Place in Nature*. London: Collins Publishing Company.
Tosi, Joseph A. 1969. *Republica de Costa Rica: Mapa Ecológico*. San José, Costa Rica: Tropical Science Center.
——— "El Recurso Forestal como Base Pontencial para el Desarrollo Industrial de Costa Rica." In *La Nación*, March 20 and 21, 1971.
U.S. Department of the Interior. *Yosemite Draft General Management Plan*. Washington, D.C., 1978.
U.S. Department of the Interior. *Final Environmental Impact Report and Proposed Plan: California Desert Conservation Area*. Riverside, California, 1980.
van Dobben, W. H., and Lowe-McConnell, R. H. "Preface" to *Unifying Concepts in Ecology*. The Hague: Dr. W. Junk B.V. Publishers, 1975.
Van Dyne, George M. "Ecosystems, Systems Ecology and Systems Ecologists." In *Readings in Conservation Ecology*, ed. Cox, George W. New York: Appleton-Century-Crofts, 1969.
Van Til, Jon. "Spatial Form and Structure in a Possible Future." In *American Planning Association Journal*, July, 1979.
Wagstaff and Brady and Robert Odland Associates. *San Gorgonio Wind Resource Study—Draft Environmental Impact Report/Environmental Impact Statement*. Berkeley, California, 1982.
Wallace, McHarg, Roberts, and Todd. *Woodlands New Community Phase One: Land Planning and Design Principles*. Houston: Woodlands Development Corporation, 1972.
——— *Woodlands New Community: Guidelines for Site Planning*. Houston: Woodlands Development Corporation, 1973.
Watson, James D. 1968. *The Double Helix: A Personal Account of the Discovery of the Structure of DNA*. New York: New American Library.
Wheaton, William. "Metropolitan Allocation Planning." In *Regional Planning*, ed. Hufschmidt, M. New York: Praeger Publishers, 1967.
Whitehead, Alfred North. 1929. *The Aim of Education and Other Essays*. New York: The MacMillan Company.
Whittaker, R. H. "Gradient Analysis of Vegetation." In *Biological Review* 42, 1967.
——— "A Study of Summer Foliage Insect Communities in the Great Smoky Mountains." *Ecological Monographs* 22, 1966.
Williamson, Robert D. "Bird-and-People-Neighborhoods." In *Natural History*. Vol. LXXXII, No. 9, 1973.
Wittfogel, Karl A. "The Hydraulic Civilizations." In *Man's Role in Changing the Face of the Earth*, ed. Thomas, William L., Jr. Chicago: University of Chicago Press, 1956.
Wolman, Abel. "The Metabolism of Cities." In *Scientific American*. Vol. 213, No. 3, 1965.
Woodwell, George M. "Recycling Sewage Effluent through Plant Communities." In *American Scientist* 65, 1977.
——— "The Carbon Dioxide Question." In *Scientific American*. Vol. 238, No. 1, 1978.
Woodwell, G. M.; Whittaker, R. H.; Reiners, W. A.; Likens, G. E.; Delwicke, C. C.; and Botkin, D. B. "The Biota and the World Carbon Budget." In *Science* 199, 1978.

INDEX

Page numbers in **boldface** indicate case studies.

35501070776733
WITHDRAWN